Wetland Plants

Biology and Ecology

WETLAND PLANTS

BIOLOGY AND ECOLOGY

JULIE K. CRONK
M. SIOBHAN FENNESSY

Cover Photograph: A *Nymphaea odorata* (white water lily) flower surrounded by floating leaves of *Nuphar advena* (spatterdock). (Photo by Hugh Crowell.)

Library of Congress Cataloging-in-Publication Data

Cronk, J.K.
Wetland plants : biology and ecology / Julie K. Cronk and M. Siobhan Fennessy.
p. cm.
Includes bibliographical references (p.).
ISBN 1-56670-372-7 (alk. paper)
1. Wetland plants. 2. Wetlands. 3. Wetland ecology. I. Fennessy, M. Siobhan. II. Title.

QK938.M3 C76 2001
581.7′68—dc21 2001020390

Direct all inquiries to CRC Press LLC, 2000 N.W. Corporate Blvd., Boca Raton, Florida 33431.

Visit the CRC Press Web site at www.crcpress.com

Lewis Publishers is an imprint of CRC Press LLC

International Standard Book Number 1-56670-372-7
Library of Congress Card Number 2001020390
Printed in the United States of America 1 2 3 4 5 6 7 8 9 0
Printed on acid-free paper

Preface

The study of wetland plants has been of interest to botanists for many years, but the need to identify and understand these plants has expanded dramatically since the 1970s. At that time, ecologists began to make known the vital role that wetlands play in our landscapes. The image of wetlands has shifted from that of mosquito-ridden wastelands to natural areas of critical importance. Because the field of wetland ecology has expanded, so has the study of the plants that thrive there, and their role in ecosystem dynamics. Today, many professionals are expert in the identification of wetland plants and identification courses are regularly taught throughout the U.S. and elsewhere. Whether readers are working with wetlands in their professions, or novices to the field, we hope to convey an understanding of the habitat, life histories, and adaptations of these plants.

Wetland plants are interesting not only because they help us identify the boundaries of a wetland, but also because of their unique evolutionary strategies for coping with life in a saturated environment. Of approximately 250,000 described angiosperm species, only a small proportion has adapted to life in the water or saturated soils. The ways in which this evolution from land to water occurred are numerous and the group of plants we discuss here is far from uniform in this regard.

More than half of the wetlands of the U.S. have disappeared since the time of European settlement and many of the remaining areas are threatened by human alterations to the landscape. In Europe, virtually no wetlands are in their natural state. This rapid habitat loss has placed many wetland species on threatened and endangered species lists. And, as in other ecosystem types, invasive plants have displaced many native or more desirable species. In some ecosystems, invasives present almost as great a threat to wetland plants as outright destruction of the ecosystem. Gaining an understanding of wetland plants and their habitats is a critical first step in helping to combat these losses.

We refer to the plants covered here as *wetland plants, wetland macrophytes,* and *hydrophytes*. Our discussion includes vascular plants that grow in or on water or in saturated soils. These include submerged, emergent, floating, and floating-leaved species. The vast majority of vascular plants that grow in these conditions are angiosperms, and our discussion centers almost exclusively on them. We also discuss a few exceptions, such as *Taxodium distichum* (bald cypress) and *Larix laricina* (tamarack), both gymnosperms that inhabit wetlands. Some pteridophytes, or ferns and fern allies, are also adapted to wetland conditions and they are mentioned, though not extensively discussed. We include species of both freshwater and saline wetlands. Most of our discussion involves wetlands of the temperate zone; however, we have included mangrove forests, a subtropical and tropical wetland type. Species of algae are not discussed, but they are covered as a component of wetland primary productivity, and methods to measure phytoplankton and periphyton primary productivity are discussed. Bryophytes, or mosses, are discussed as the basis of peatland systems and as one of the driving forces in their substrate chemistry. However, species of bryophytes and their specific adaptations and reproduction are not covered. The plants adapted to flowing water environments and to marine habitats are not specifically discussed.

Plant names follow the U.S. Fish and Wildlife Service's National List of Plant Species that Occur in Wetlands (Reed 1997), and where plants outside the U.S. are named, we refer to the literature reference in which the plant is named. Family names follow Cook's 1996 book, *Aquatic Plant Book*. Cook sometimes provides equivalent or older names for families and we give these in parentheses following the family name. The names of orders are according to a recent re-classification of angiosperm families by the Angiosperm Phylogeny Group (1998). The names of species formerly all classified in the genus *Scirpus* (bulrush) have been undergoing a number of name changes. In a classification scheme proposed by Smith and Yatskievych (1996), the genus was divided into five genera (*Scirpus, Schoenoplectus, Bolboschoenus, Isolepsis,* and *Trichorphorum*). The recent literature is mixed regarding the adoption of the new names. For the species found in the U.S., we use the names as they appear in Reed 1997. For species found outside of the U.S., we use the name used by the authors of the papers we cite in each instance. In each chapter, the first time a species, genus, or family is mentioned, we give the scientific name first and follow it with the common name in parentheses. Subsequent mentions of the plant use only the scientific name, often with the genus abbreviated to the first letter (i.e., *Phragmites australis* becomes *P. australis* after the first time it is mentioned in any given paragraph or section). Some plants have no common name, or at least none that we were able to find in English, so for these, none is mentioned.

Wetland Plants: Biology and Ecology is a synthesis of current research on wetland plants and their communities. In our introductory section (Chapters 1 through 3), we present general information about the growth forms, evolution, distribution, and diminishing habitat of wetland plants. We also discuss wetland classifications and definitions and broad types of wetland ecosystems such as salt marshes, mangrove forests, riparian wetlands, and peatlands. To understand wetland plant evolution and life history strategies it is vital to understand the abiotic conditions that set the boundaries for their growth. A brief explanation of some important hydrological principles is provided in the first section of Chapter 3, with an emphasis on how wetland hydrology shapes the plant community. The second half of Chapter 3 covers other important factors for plant growth such as substrate type, salinity, and nutrient availability.

Part 2 is devoted to a discussion of the adaptations and reproduction of wetland macrophytes. In Chapter 4 we discuss the adaptations of wetland plants to anoxia, salinity, and other stressful conditions for growth. Chapter 5 covers wetland angiosperm reproduction, both sexual and asexual, as well as adaptations of pollen and pollination mechanisms, and methods of seed dispersal.

In Part 3, the function, dynamics, and potential disturbances of wetland plant communities are discussed. Chapter 6 provides background on the concept of primary productivity and the history and methods of its measurement. Primary productivity is of particular interest in wetland studies because some types of wetlands are among the most productive ecosystems in the world. We focus on methods in this chapter because the results depend so heavily on the method chosen. In Chapter 7 we discuss community dynamics. Specifically, we cover ecological succession, with a look at the classical idea that wetlands are a sere, or successional stage, between lake and terrestrial ecosystems; we look as well at material that refutes that idea. We also include competition in Chapter 7. Competition influences the diversity and composition of plant communities and many plant strategies have evolved to compete for both space and resources. In Chapter 8 we give examples of invasive plants and describe techniques used, with varying degrees of success, to control them. The ecological implications of invasive species are also discussed.

Applications of wetland plant study are discussed in the last two chapters (Part 4). We present research on the development of plant communities in newly restored or created wetlands, including the role of plants in wetlands constructed to improve water quality (Chapter 9). The interest in restoring degraded aquatic ecosystems is growing exponentially, and an understanding of wetland plant community dynamics is critical in planning successful restoration efforts. Indeed, it is often vegetation establishment that is used as a benchmark of success in restoration projects. Planting and seeding techniques, the use of seed banks, including the use of salvaged soils, and the design aspects of restoration planning are covered. The uses of wetland plants as indicators of ecological integrity and of wetland boundaries (delineation) are covered in Chapter 10. The use of wetland plants as biological indices of ecosystem integrity is currently under study and we present methods for choosing and testing plant indicators. We also discuss the history of wetland delineation, the ecological principles behind it, and its current status.

Wetland Plants: Biology and Ecology is intended for wetland professionals, academicians, and students. Professionals whose plant identification skills may be well honed from delineation experience will be interested in a comprehensive reference on the ecology of aquatic plants. The book may also serve as a text for courses on wetland plants, aquatic botany, or wetland ecology. This book will be best for upper-level undergraduates or graduate students. A textbook for wetland plant courses has not been available in the past. We have found that without a textbook, students are at a disadvantage to understand and integrate course material. For this reason, we have tried to gather the information necessary for such a course under one title. To use this book, a basic knowledge of botany and ecology is helpful, but not essential, as we try to provide enough background for those who are learning on the job or who are catching up on background material as they learn new subject matter.

Many of our colleagues provided helpful suggestions, information, and critical comments on portions of the book. Brian Reeder reviewed the entire manuscript and provided useful constructive comments, suggestions, and references. We very much appreciate the time, enthusiasm, and energy he devoted to this project; even more, we are grateful for his generosity and friendship. We would also like to thank Bob Lichvar, James Luken, John Mack, Irving Mendelssohn, Bob Nairn, Diane Sklensky, and Courtenay Willis who each took the time to carefully review chapters. Brad Walters provided constructive comments on one of our case studies as well as a number of helpful articles and photographs. Andy Baldwin, Ernie Clarke, Joe Ely, Mark Gernes, Stan Smith, and Gerald van der Velde sent figures, photographs and/or useful articles and information. Donald Hey kindly allowed us to use a photograph from Wetlands Research, Inc. The biology department at Kenyon College provided logistical support for which we are thankful. John Schimmel, the director of the Ebersole Environmental Center, was generous in allowing J. Cronk freedom and time to work on this project. Portions of the chapter on primary productivity were originally conceived as a review article and we appreciate the comments of two anonymous reviewers of that manuscript. Two anonymous reviewers provided helpful comments on the proposal for this book, and we used several of their ideas. We are also grateful to Randi Gonzalez, the late Arline Massey, Bob Caltagirone, and Jane Kinney (formerly with CRC Press) at Lewis Publishers/CRC Press.

Our students inspired us to write in the first place. Their expectations for excellence are the impetus for our search for answers. We would particularly like to mention the contributions, ideas, and inspiration provided by Jessen Book (who also made excellent editorial comments), Christina Bush, Clement Coulombe, Eric Crooks, Brenda Cruz, Julie Latchum, Amanda Nahlik, Laura Marx, and Abby Rokosch.

Our friends and families have been supportive and helpful throughout the years it took to write this book. Hugh Crowell was instrumental in the completion of this book; he carefully edited every chapter, table, and figure. He provided critical comments, found new references, and suggested many ways to improve the book. His help and support and his knowledge of wetland science and botany have been crucial every step of the way. Hugh took the great majority of the original photographs for this book, sacrificing three years of Saturdays and vacations to finding plants and taking their pictures. Hugh solved the many computer-related problems that arose along the way, as well. We thank Ted Rice for his moral support, and for creating space in which S. Fennessy could write. His boundless belief in our abilities inspired us. Ted also contributed many of his exceptional photos to this volume. We thank Kay Irick Moffett for photographing *Tamarix ramosissima* and for her steadfast support of this project. Carolyn Crowell's knowledge of the ecology of Cape Cod's salt marshes enhanced our own and led to several photographs used in the book. We are grateful to Barb Zalokar, who applied her exceptional skill and talent to several original figures for the book. Dean Greenberg graciously allowed us to use one of his photographs. We especially thank our children, Seth Crowell and Nora and Thomas Rice, for their patience, help, and wonderful ideas.

Our love of wetland ecology was originally inspired by William J. Mitsch. For all the advice, enthusiasm, and encouragement that he has given us over the years, we are grateful. We dedicate this book to him in recognition of all that he has given us.

Julie K. Cronk
M. Siobhan Fennessy

Authors

Julie K. Cronk, who is currently a private consultant in wetland ecology and restoration, earned a Ph.D. in environmental biology from The Ohio State University in 1992. Her dissertation research was on water quality and algal primary productivity in four constructed riparian emergent marshes at the Des Plaines River Wetlands Demonstration Project outside Chicago, Illinois. She worked as an assistant professor in the Department of Biological Resources Engineering at the University of Maryland from 1993 to 1995. Her primary research interests have been wetland plant primary productivity, the development of plant communities in new wetlands, and the improvement of water quality in constructed wetlands to treat domestic and animal wastewater. She is author or co-author of several peer-reviewed journal articles on wetland-related topics and she has given presentations at conferences for the Society of Wetland Scientists, INTECOL, and the American Society of Agricultural Engineers. Dr. Cronk has taught wetland ecology, aquatic plants, plant biology, and water quality courses, as well as seminars on constructed wetlands at the University of Maryland, Grand Valley State University in Allendale, Michigan, and at The Ohio State University. She is a member of the Society of Wetland Scientists.

M. Siobhan Fennessy is assistant professor of biology at Kenyon College where she teaches, supervises students, and conducts research on wetland ecosystems and their plant communities. She received a Ph.D. in environmental biology from The Ohio State University in 1991. Her dissertation research focused on the development of wetland plant communities in restored wetlands, and the impact of different hydrologic regimes on plant species establishment and primary productivity. Dr. Fennessy previously served on the faculty of the Geography Department of University College London and held a joint appointment at the Station Biologique de la Tour du Valat (located in southern France) where she conducted research on Mediterranean wetlands. She subsequently worked at the Ohio Environmental Protection Agency where she developed water quality standards for wetlands and began a wetland assessment program. She has published numerous peer-reviewed and technical papers on the ecology of wetland plant communities, wetland biogeochemistry, and the use of plants as biological indicators of wetland ecosystem integrity. She is a member of the U.S. EPA's Biological Assessment of Wetlands Workgroup, a technical committee working to develop biological assessment techniques. Dr. Fennessy is also a member of the Society of Wetland Scientists, the Society for Ecological Restoration, and the Ecological Society of America.

Contents

Part I Introduction

Chapter 1 Introduction to Wetland Plants
I. Wetlands and Wetland Plants 4
II. What Is a Wetland Plant? 5
III. Types of Wetland Plants 7
A. Emergent Plants 7
B. Submerged Plants 12
C. Floating-Leaved Plants 13
D. Floating Plants 14
IV. Wetland Plant Distribution 16
V. The Evolution of Wetland Plants 17
A. Changes in Angiosperm Classification and Phylogeny 17
B. Evolutionary Processes in Wetland Plants 20
VI. Threats to Wetland Plant Species 20
A. Hydrologic Alterations 21
B. Exotic Species 21
C. Impacts of Global Change 22
D. Threatened and Endangered Species 23
Summary 27

Chapter 2 Wetland Plant Communities
I. Wetland Plant Habitats 29
II. Wetland Definitions and Functions 29
A. Ecological Definition 30
B. Legal Definitions 30
1. U.S. Army Corps of Engineers' Definition 30
2. U.S. Fish and Wildlife Classification of Wetlands 31
3. International Definition 31
C. Functions of Wetlands 32
1. Hydrology 32
a. Groundwater Supply 33
b. Flood Control 33
c. Erosion and Shoreline Damage Reduction 33
2. Biogeochemistry 33
3. Habitat 34
a. Wildlife and Fish Habitat 34

b. Plant Habitat . . . 34
III. Broad Types of Wetland Plant Communities . . . 34
A. Marshes . . . 36
1. Coastal Marshes . . . 36
a. Salt Marshes . . . 36
b. Tidal Freshwater Marshes . . . 39
2. Inland Marshes . . . 39
a. Lacustrine Marshes . . . 41
b. Riverine Marshes . . . 42
c. Depressional Marshes . . . 42
B. Forested Wetlands . . . 44
1. Coastal Forested Wetlands: Mangrove Swamps . . . 44
2. Inland Forested Wetlands . . . 48
a. Southern Bottomland Hardwood . . . 48
b. Northeastern Floodplain . . . 49
c. Western Riparian Zones . . . 50
d. Cypress Swamps . . . 51
C. Peatlands . . . 52
Summary . . . 59

Chapter 3 The Physical Environment of Wetland Plants

I. An Introduction to the Wetland Environment . . . 61
II. The Hydrology of Wetlands . . . 61
A. Hydroperiod and the Hydrologic Budget . . . 62
1. Transpiration and Evaporation . . . 64
2. Measuring Transpiration and Evaporation . . . 65
B. The Effects of Hydrology on Wetland Plant Communities . . . 67
1. Hydrology and Primary Productivity . . . 67
2. Hydrologic Controls on Wetland Plant Distribution . . . 69
3. The Effects of Water Level Fluctuation on Wetland Plant Diversity . . . 70
4. Riparian Wetland Vegetation and Stream Flow . . . 72
C. Hydrological and Mineral Interactions and Their Effect on Species Distribution . . . 72
III. Growth Conditions in Wetlands . . . 74
A. Anaerobic Sediments . . . 74
1. Reduced Forms of Elements . . . 75
a. Nitrogen . . . 75
b. Manganese . . . 76
c. Iron . . . 77
d. Sulfur . . . 77
e. Carbon . . . 78
2. Nutrient Availability under Reduced Conditions . . . 78
3. The Presence of Toxins under Reduced Conditions . . . 79
B. Substrate Conditions in Saltwater Wetlands . . . 79
C. Substrate Conditions in Nutrient-Poor Peatlands . . . 80
D. Growth Conditions for Submerged Plants . . . 81
1. Light Availability . . . 81

2. Carbon Dioxide Availability 83
Summary..83

Part II Wetland Plants: Adaptations and Reproduction

Chapter 4 Adaptations to Growth Conditions in Wetlands

I. Introduction 87
A. Aerobic Respiration and Anaerobic Metabolism 87
B. Upland Plant Responses to Flooding 88
II. Adaptations to Hypoxia and Anoxia 88
A. Structural Adaptations...88
1. Aerenchyma 88
a. Aerenchyma Formation 89
b. Aerenchyma Function 91
2. Root Adaptations 91
a. Adventitious Roots 91
b. Shallow Rooting 93
c. Pneumatophores 93
d. Prop Roots and Drop Roots...95
3. Stem Adaptations 95
a. Rapid Underwater Shoot Extension....................................95
b. Hypertrophy 96
c. Stem Buoyancy 96
4. Gas Transport Mechanisms in Wetland Plants 97
a. Passive Molecular Diffusion 97
b. Pressurized Ventilation 97
c. Underwater Gas Exchange 101
d. Venturi-Induced Convection..101
5. Radial Oxygen Loss 102
6. Avoidance of Anoxia in Time and Space 104
7. Development of Carbohydrate Storage Structures 104
B. Metabolic Processes...104
1. Anaerobic Metabolism and the Pasteur Effect 106
2. Hypotheses Concerning Metabolic Responses to Anaerobiosis 106
a. McManmon and Crawford's Hypotheses.................................106
b. Davies' Hypothesis...108
3. Other Metabolic Responses to Anoxia 109
III. Adaptations in Saltwater Wetlands 110
A. Adaptations to High Salt Concentrations...................................110
1. Water Acquisition 110
2. Salt Avoidance 111
a. Exclusion...111
b. Secretion 111
c. Shedding..113

d. Succulence 113
B. Adaptations to High Sulfide Levels 113
IV. Adaptations to Limited Nutrients 114
A. Mychorrhizal Associations 114
B. Nitrogen Fixation 116
C. Carnivory 117
1. Habitat and Range of Carnivorous Plants 117
2. Types of Traps 118
a. Pitfall Trap 118
b. Lobster Pot Trap 119
c. Passive Adhesive Trap 120
d. Active Adhesive Trap 122
e. Bladder Trap 123
f. Snap-Trap 124
3. Benefits and Costs of Carnivory 126
D. Nutrient Translocation 126
E. Evergreen Leaves 127
V. Adaptations to Submergence 127
A. Submerged Plant Adaptations to Limited Light 127
B. Submerged Plant Adaptations to Limited Carbon Dioxide 129
1. Use of Bicarbonate 129
2. Aquatic Acid Metabolism 130
3. Lacunal Transport 131
4. Sediment-Derived CO_2 131
C. Adaptations to Fluctuating Water Levels 131
VI. Adaptations to Herbivory 134
A. Chemical Defenses 135
B. Structural Defenses 135
VII. Adaptations to Water Shortages 136
Summary 138
Case Studies 139
4.A. Factors Controlling the Growth Form of *Spartina alterniflora* 139
4.B. Carnivory in *Sarracenia purpurea* (Northern Pitcher Plant) 142

Chapter 5 Reproduction of Wetland Angiosperms

I. Introduction 147
A. A Brief Review of Floral Structures Involved in Reproduction 147
B. Challenges to Sexual Reproduction in Wetland Habitats 148
II. Sexual Reproduction of Wetland Angiosperms 150
A. Pollination Mechanisms 150
1. Insect Pollination 150
2. Wind Pollination 152
3. Water Pollination 154
a. Planes of Water Pollination 155
b. Hydrophilous Pollen Adaptations 162
c. Hydrophilous Stigma Adaptations 163
d. The Evolution of Hydrophily 164
4. Self-Pollination 166
B. Fruits and Seeds 167

1. Types of Fruits Produced by Wetland Plants 167
2. Seed and Fruit Dispersal 171
3. Seed Dormancy and Germination 173
C. Seedling Adaptations 175
1. Seedling Dispersal and Establishment 175
2. Vivipary 176
III. Asexual Reproduction in Wetland Angiosperms 177
A. Structures and Mechanisms of Cloning..................................178
1. Shoot Fragments 178
2. Modified Buds 180
a. Turions 180
b. Pseudoviviparous Buds 182
c. Gemmiparous Buds 183
3. Modified Stems 183
a. Layers 183
b. Runners 184
c. Stolons 184
d. Rhizomes 184
e. Stem Tubers 184
4. Modified Shoot Bases 185
a. Bulbs 185
b. Corms 185
5. Modified Roots 185
a. Creeping Roots 185
b. Tap Roots 185
c. Root Tubers...185
B. Occurrence and Success of Cloning among Wetland Plants................186
Summary...188

Part III Wetland Plant Communities: Function, Dynamics, Disturbance

Chapter 6 The Primary Productivity of Wetland Plants

I. Introduction 191
A. Definition of Terms...191
1. Standing Crop 191
2. Biomass 192
3. Peak Biomass 193
4. Primary Production 193
5. Respiration 194
6. Primary Productivity 195
a. Gross Primary Productivity 195
b. Net Primary Productivity 195
7. Turnover 195
8. P/B Ratio 196
B. Reasons for Measuring Wetland Primary Productivity...................196

1. To Quantify an Ecosystem Function .196
2. To Make Comparisons within a Wetland .196
3. To Make Comparisons among Wetlands .197
4. To Determine Forcing Functions and Limiting Factors of Primary Productivity .197
II. Methods for the Measurement of Primary Productivity in Wetlands .197
A. Phytoplankton 199
1. Dissolved Oxygen Concentration .199
a. Diurnal Dissolved Oxygen Method 199
b. Light Bottle/Dark Bottle Dissolved Oxygen Method 200
2. Carbon Assimilation: The ^{14}C Method .201
B. Periphyton 202
C. Submerged Macrophytes 204
1. Biomass .204
2. Oxygen Production .204
3. Carbon Assimilation .205
D. Emergent Macrophytes 205
1. Aboveground Biomass of Emergent Plants .205
a. The Peak Biomass Method 210
b. The Milner and Hughes Method 211
c. The Valiela et al. Method 212
d. The Smalley Method 213
e. The Wiegert and Evans Method 213
f. The Lomnicki et al. Method 215
g. The Allen Curve Method 216
h. The Summed Shoot Maximum Method 219
2. Belowground Biomass of Emergent Wetland Plants .219
a. Harvest Method 220
b. Decomposition Method 220
E. Floating and Floating-Leaved Plants 220
F. Trees 221
1. Measures of Dimension Analysis .221
a. Diameter at Breast Height 221
b. Height 222
2. Parameters Based on Dimension Analysis .222
a. Basal Area 222
b. Basal Area Increment 223
3. Calculations of NPP of Trees .223
a. Stem Production 223
b. Leaf Production 224
c. Branch Production 224
d. Root Production 224
4. Community Primary Productivity of Forested Wetlands .225
G. Shrubs 225
H. Moss 226
Summary .227
Case Studies .228
6.A. Salt Marsh Productivity: The Effect of Hydrological Alterations in Three Sites in San Diego County, California 228

6.B. Mangrove Productivity: Laguna de Terminos, Mexico. 230
6.C. Peatland Productivity: Forested Bogs of Northern Minnesota 232

Chapter 7 Community Dynamics in Wetlands

I. An Introduction to Community Dynamics 237
II. Ecological Succession 237
A. Holistic and Individualistic Approaches to Ecological Succession 238
B. The Replacement of Species 239
C. Developing and Mature Ecosystems 240
III. Ecological Succession in Wetlands 241
A. Models of Succession in Wetlands 241
1. Hydrarch Succession 241
2. Succession in Coastal Wetlands 246
3. The Environmental Sieve Model 248
B. The Role of Seed Banks in Wetland Succession. 250
1. The Relationship of the Seed Bank to the Existing Plant Community .250
2. Factors Affecting Recruitment from the Seed Bank 253
IV. Competition and Community Dynamics 253
A. Intraspecific Competition 254
B. Interspecific Competition. 255
1. Competition and Physiological Adaptations 256
2. Competition and Life History Characteristics 257
3. Resource Availability and Competitive Outcome 261
4. Light in Submerged Communities 262
5. Light in Emergent Communities. 263
6. Competition and Salt Marsh Communities 263
C. Allelopathy. 265
V. The Role of Disturbance in Community Dynamics 266
A. Hydrologic Disturbances. 266
B. Severe Weather. 269
1. Floods 269
2. Hurricanes 270
C. Fire 270
D. Biotic Disturbance. 271
E. Human-Induced Disturbance 272
Summary 273
Case Studies 275
7.A. Successional Processes in Deltaic Lobes of the Mississippi River 275
7.B. Eutrophication of the Florida Everglades: Changing the Balance of Competition. 276

Chapter 8 Invasive Plants in Wetlands

I. Characterization of Invasive Plants 279
II. The Extent of Exotic Invasions in Wetland Communities .282
III. Implications of Invasive Plant Infestations in Wetlands .284
A. Changes in Community Structure 284
B. Changes in Ecosystem Functions 286
C. Effects on Human Endeavors 287
IV. The Control of Invasive Plants in Wetlands 288

A. Habitat Alterations 288
1. Shading the Water's Surface 288
2. Shading the Sediment Surface 289
3. Dredging Sediments 289
4. Altering Hydrology 289
B. Mechanical Controls 290
C. Chemical Controls 294
D. Biological Controls 296
1. Insects 297
2. Fish 298
3. Pathogens 298
4. Fungi 299
5. Other Organisms 299
V. Case Studies of Invasive Plants in Wetland Communities 299
A. *Myriophyllum spicatum* (Eurasian Watermilfoil) 299
1. Biology 299
2. Origin and Extent 300
3. Effects in New Range 301
4. Control 301
5. The Natural Decline of Some *Myriophyllum spicatum* Populations . . . 302
B. *Hydrilla verticillata* (Hydrilla) 303
1. Biology 303
2. Origin and Extent 304
3. Effects in New Range 305
4. Control 305
C. *Eichhornia crassipes* (Water Hyacinth) 306
1. Biology 306
2. Origin and Extent 307
3. Effects in New Range 308
4. Control 309
D. *Lythrum salicaria* (Purple Loosestrife) 310
1. Biology 310
2. Origin and Extent 310
3. Effects in New Range 312
4. Control 313
E. *Phragmites australis* (Common Reed) 313
1. *Phragmites australis* as an Invasive Species in North America 313
a. Biology 315
b. Origin and Extent 315
c. Effects on the Habitat 316
d. Control 317
2. *Phragmites australis* as a Declining Species in Europe 317
a. Extent of the Problem 317
b. Causes of the Decline 319
c. Solutions to the *Phragmites australis* Decline 319
Summary 321

Part IV Applications of Wetland Plant Studies

Chapter 9 Wetland Plants in Restored and Constructed Wetlands

I. Wetland Restoration and Creation 326
A. The Development of Plant Communities in Restored and Created Wetlands 327
1. Environmental Conditions 327
2. Self-Design and Designer Approaches 329
3. Seed Banks in Restored Wetlands 331
B. Planting Recommendations for Restoration and Creation Projects 332
II. Treatment Wetlands 333
A. Removal of Wastewater Contaminants 336
1. Nitrogen Removal 336
2. Phosphorus Retention 337
a. Biotic Uptake of Phosphorus 337
b. Sorption onto Soil Particles 337
c. Accretion of Wetland Soils 338
3. Pathogen Removal 338
4. Metal Removal 339
a. Plant Uptake of Metals 340
b. Phytoremediation 340
B. The Role of Vascular Plants in High-Nutrient Load Treatment Wetlands 341
1. Vegetation as a Growth Surface and Carbon Source for Microbes 341
2. Physical Effects of Vegetation 343
3. Nutrient Uptake 343
a. Tissue Nutrient Content of Wetland Plants 346
b. Factors Affecting Nutrient Uptake 347
c. The Accretion of Organic Sediments 347
4. Vegetation as a Source of Rhizospheric Oxygen 348
5. Wildlife Habitat and Public Recreation 349
C. Species Commonly Used in Treatment Wetlands 350
D. The Establishment and Management of Plants in Wastewater Treatment Wetlands 354
Summary 355
Case Studies 356
9.A. Integrating Wetland Restoration with Human Uses of Wetland Resources . 356
9.B. Restoring the Habitat of an Endangered Bird in Southern California 359
9.C. Vegetation Patterns in Restored Prairie Potholes 360

Chapter 10 Wetland Plants as Biological Indicators

I. Introduction 363
II. Wetland Plants as Indicators of Wetland Boundaries 363
A. Hydrophytic Vegetation as a Basis for Delineation 368
B. Wetland Boundaries and Wetland Functions 369
C. The Use of Remotely Sensed Data in Wetland Identification and Classification 370
III. Wetland Plants as Indicators of Ecological Integrity 371

A. An Operational Definition of Ecological Integrity . . . 372
B. Wetland Plant Community Composition as a Basis for Indicator Development . . . 374
C. General Framework for Wetland Biological Indicator Development . . . 375
D. Vegetation-Based Indicators . . . 377
E. The Floristic Quality Assessment Index for Wetland Assessment . . . 378
F. Using Biological Indicators to Assess Risk . . . 382
Summary . . . 383
Case Study . . . 384
10.A. The Development of a Vegetation IBI . . . 384

References . . . 389

Index . . . 439

Part I

Introduction

1

Introduction to Wetland Plants

Wetland plants are found throughout the world, in swamps and marshes, in peatlands, billabongs, and sloughs, at the margins of lakes, streams, and rivers, in bays and estuaries, and along protected oceanic shorelines. In short, they are found wherever there are wetlands and they are often the most conspicuous component of the ecosystem. Emergent taxa such as *Carex* (sedge), *Juncus* (rush), *Typha* (cattail), and *Polygonum* (smartweed) dominate the freshwater marshes of North America; *Phragmites australis* (common reed) provides the name for the reedswamps of Europe; *Spartina* species (cordgrass) dominate many temperate coastal salt marshes; and *Taxodium distichum* (bald cypress) is found in the deepwater swamps of the southeastern U.S. An interest in wetland plants, their ecology and distribution, often begins with an appreciation of their appearance. From a biological standpoint, wetland plants have multiple roles in the functioning of wetlands. They, like all photosynthetic organisms, are crucial in fixing the energy that powers all other components of the system. They supply oxygen to other biota and contribute to the physical habitat. Although wetland plants are defined by their ability to inhabit wet places, they represent a diverse assemblage of species with different adaptations, ecological tolerances, and life history strategies that enable their survival in saturated or flooded soils. These differences have implications for their conservation, management, and restoration.

Our understanding of the ecology of wetland plants has increased dramatically over the past several decades. Much of this understanding has been fueled by the surge of interest in wetland ecosystems more generally. Research has documented the high levels of biological diversity that wetlands support as well as the unique ecological processes, or functions, that occur there. As information on wetlands has increased, so too has the literature on wetland plants: field guides and manuals have been completed for many geographical areas, numerous magazines and scholarly journals are devoted solely to their study, and a growing horticultural and aquarium trade is based on their cultivation and sale. The use of wetland plants in the delineation of wetlands in the U.S., as well as the relatively new field of wetland restoration, has created a demand for people knowledgeable in their taxonomy and ecology. In addition, concern about the invasive potential of some species, such as *Eichhornia crassipes* (water hyacinth), *Hydrilla verticillata* (hydrilla), and *Lythrum salicaria* (purple loosestrife), has driven research and the development of management techniques designed to reduce their abundance (Barrett et al. 1993; see Chapter 8, Invasive Plants in Wetlands). Despite the importance of wetland plants in a number of research and management fields, very few texts on wetland plant ecology have been written. This volume provides a comprehensive discussion of the ecology of wetland plants at levels of biological organization ranging from the individual to the role of wetland plants in ecosystem function.

I. Wetlands and Wetland Plants

One key to understanding the unique characteristics of wetland plants is to understand the contribution they make to wetland ecosystems. They are vitally important for many reasons (Wiegleb 1988):

- Wetland plants are at the base of the food chain and, as such, are a major conduit for energy flow in the system. Through the photosynthetic process, wetland plants link the inorganic environment with the biotic one. The primary productivity of wetland plant communities varies, but some herbaceous wetlands have extremely high levels of productivity, rivaling those of tropical rain forests. And unlike many terrestrial ecosystems, much of the organic matter produced is not used directly by herbivores but instead is transferred to the detrital food chain.
- They provide critical habitat structure for other taxonomic groups, such as epiphytic bacteria, periphyton, macroinvertebrates, and fish (Carpenter and Lodge 1986; Wiegleb 1988; Cronk and Mitsch 1994b). The composition of the plant community has implications for diversity in these other taxonomic groups.
- They strongly influence water chemistry, acting as both nutrient sinks through uptake, and as nutrient pumps, moving compounds from the sediment to the water column. Their ability to improve water quality through the uptake of nutrients, metals, and other contaminants is well documented (Gersberg et al. 1986; Reddy et al. 1989; Peverly et al. 1995; Rai et al. 1995; Tanner et al. 1995a, b). Submerged plants also release oxygen to the water that is then available for respiration by other organisms.
- They influence the hydrology and sediment regime of wetlands through, for example, sediment and shoreline stabilization, or by modifying currents and helping to desynchronize flood peaks. Vegetation can control water conditions in many ways including peat accumulation, water shading (which affects water temperatures), and transpiration (Gosselink and Turner 1978). For instance, bog plants can build peat to the point that surface water no longer flows into the wetland. Some wetland tree species, including *Melaleuca quinquenervia,* which has invaded the Everglades, transpire at very high rates and are capable of drawing down the groundwater table.

Wetland plants are also among the tools used by wetland managers and researchers in the conservation and management of wetland areas, for example:

- They are routinely used to help identify or delineate jurisdictional boundaries of wetlands in the U.S. and elsewhere (U.S. Army Corps of Engineers 1987).
- Increasingly, the composition of the plant community and the predictable changes in community structure that result from anthropogenic disturbance are being investigated for their ability to act as biological indicators of the "health" or ecological integrity of the wetland (Adamus 1996; Karr and Chu 1997; Fennessy et al. 1998a; Carlisle et al. 1999; Gernes and Helgen 1999; Mack et al. 2000; see Chapter 10, Wetland Plants as Biological Indicators). This kind of information has many potential applications including monitoring wetland condition over time or setting goals for wetland restoration or mitigation projects.
- Wetland plants are often used to help organize environmental inventories and research programs, and to set goals for management programs or restoration projects (Cowardin et al. 1979; Britton and Crivelli 1993; Brinson 1993a; Reed 1997).

Thus, wetland plants have major effects in terms of the physical (temperature, light penetration, soil characteristics) and chemical environment of wetlands (dissolved oxygen, nutrient availability), and provide the basis of support for nearly all wetland biota. They are drivers of ecosystem productivity and biogeochemical cycles, in part because they occupy a critical interface between the sediments and the overlying water column (Carpenter and Lodge 1986). Although some of the adaptations possessed by wetland plant species are also found in related terrestrial species, many attributes are unique or, if shared, have reached a high degree of specialization.

II. What Is a Wetland Plant?

Most of the terminology used to describe wetland plants is based on the hydrological regime that a species requires. In general, there exists a continuum of tolerance among all vascular plant species ranging from those adapted to extremely dry conditions (xeric terrestrial species) to those species that complete their entire life cycle (from seed to seed) underwater. The latter never come into direct contact with the atmosphere. Along this continuum there are no discrete categories in terms of moisture requirements, and although it is not possible to make a division where terrestrial plants end and wetland species begin, many operational definitions exist. Wetland plants, which we consider to be synonymous with *wetland hydrophytes*, are commonly defined as plants "growing in water or on a substrate that is at least periodically deficient in oxygen as a result of excessive water content" (Cowardin et al. 1979). This term includes both herbaceous and woody species (Table 1.1).

The definition of the term hydrophyte has evolved since its inception in the late 19th century. Originally used by Europeans in the late 1800s, it was used to denote plants that grew in water, or with their perennating organs submerged in water (Sculthorpe 1967; Tiner 1991; U.S. National Research Council 1995). Warming (1909, as reported in Tiner 1999) is credited as the first to arrange plant communities according to their hydrological preferences. Aquatic plants were defined as submerged species or those with floating leaves, while marsh plants were categorized as terrestrial plants. He further organized vegetation into various "oecological classes" based on soil conditions. Very wet soils supported two classes of plants, the hydrophytes (those in water) and the helophytes (those in marshes, i.e., emergent plants). Penfound (1952) developed a classification scheme recognizing two groups, the terrestrial plants and the hydrophytes, the latter of which included both submerged and emergent species (U.S. National Research Council 1995). Under these definitions, terrestrial species cannot tolerate flooding or soil saturation during the growing season. Aquatic species require flooding and cannot tolerate dewatering, while wetland species tolerate both (U.S. National Research Council 1995). Sculthorpe (1967) also adopted this broad definition of hydrophyte.

Many authors do not make a distinction between *wetland plants* and *aquatic plants*. For example, Barrett and others (1993) use the term aquatic plant in its broadest sense to include all plants that occur in permanently or seasonally wet environments. However, other authors such as Cook (1996) define (vascular) aquatic plants as those Pteridophytes (ferns and fern allies) and Spermatophytes (seed-bearing plants) whose photosynthetically active parts are permanently or semi-permanently submerged in water or float on the surface. Other authors make a similar distinction with regard to species they consider to be *true aquatics*, a term sometimes used to denote species that complete their life cycle with all vegetative parts submerged or supported by the water (Best 1988). Examples of families with submerged and floating-leaved species that fall in this category include the Nymphaeaceae (water lilies), Potamogetonaceae (pondweeds),

TABLE 1.1

A History of the Definition of Wetland Plants[a]

"Any plant growing in a soil that is at least periodically deficient in oxygen as a result of excessive water content."

— Daubenmire 1968

"Any plant growing in water or on a substrate that is at least periodically deficient in oxygen as a result of excessive water content."

— Cowardin et al. 1979

"Any macrophyte that grows in water or on a substrate that is at least periodically deficient in oxygen as a result of excessive water content; plants typically found in wet habitats."

— U.S Army Corps of Engineers 1987

"Large plants (macrophytes) … that grow in permanent water or on a substrate that is at least periodically deficient of oxygen as a result of excessive water content. This term includes both aquatic plants and wetland plants."

— Sipple 1988

"… an individual plant adapted for life in water or periodically flooded and/or saturated soils … (which) may represent the entire population of a species or only a subset of individuals so adapted … "

— Tiner 1988

"Any macrophyte that grows in water or on a substrate that is at least periodically deficient in oxygen as a result of excessive water content; plants typically found in wetlands and other aquatic habitats."

— Federal Interagency Committee for Wetland Delineation 1989

"… plants that live in conditions of excess wetness … macrophytic plant life growing in water or on submerged substrates, or in soil or on a substrate that is at least periodically anaerobic … "

— Proposed Revisions 1991

[a] Based here on the term 'wetland hydrophyte.'

From Tiner 1991; U.S. National Research Council 1995.

Lentibulariaceae (bladderworts), and Najadaceae (naiads). Terms other than hydrophyte that have been used to describe wetland plants include: *limnophyte* (freshwater plant), *aquatic macrophyte* (plant visible to the naked eye), *amphiphyte* (species capable of growing on land or in water), *helophyte* (emergent plant), and *amphibious species*.

For the purposes of this book, we define wetland plants as those species that are normally found growing in wetlands, i.e., in or on the water, or where soils are flooded or saturated long enough for anaerobic conditions to develop in the root zone, and that have evolved some specialized adaptations to an anaerobic environment. We restrict our treatment to vascular plants, often called macrophytes. Wetland plants may be floating, floating-leaved, submerged, or emergent (Sculthorpe 1967), and may complete their life cycle in still or flowing water or on inundated or non-inundated hydric soils. The vast majority of species that grow in these conditions are angiosperms, although there are exceptions such as *Taxodium distichum* (bald cypress) and *Larix laricina* (tamarack), both gymnosperms. Both freshwater and saltwater species are included here, and their distribution ranges from cold-temperate to tropical latitudes. Classes of species that include relatively fewer numbers of wetland species such as the Pteridophytes (ferns, such as *Osmunda regalis*, royal fern, or *Azolla* species) and Bryophytes (mosses, such as *Sphagnum* species) are included where relevant, but are discussed in less detail than the angiosperms.

There are approximately 250,000 described angiosperm species (including terrestrial species), and an estimated 50,000 to 250,000 species that have not yet been described (Savage 1995). Variations in estimates for the taxonomic richness of wetland plant species reflect the range of definitions used to identify them. Of the known species of angiosperms, between 2 and 3% are considered to be true aquatics, placing their total number between 4,700 and 7,500 species (Cook 1996; Philbrick and Les 1996); however, these authors do not include woody or many emergent species. Reed (1997), who does include woody species and the range of species that we cover in this text, places the estimate of wetland plants found in the U.S. alone at nearly 7,500. A recent U.S. Environmental Protection Agency report estimates nearly 7,000 North American wetland plant species (calculated from Adamus, in review).

III. Types of Wetland Plants

Wetland vascular plants are generally categorized based on their growth form. This scheme is independent of phylogenetic relationships; it is based solely on the way in which the plants grow in physical relationship to the water and soil. Many different classification schemes have been developed, based on variations in plant form, the means by which they grow and reproduce, or adaptations for surviving inundated or saturated conditions (Hutchinson 1975; Cook 1996). We follow Sculthorpe (1967) in adopting the simplest scheme with the least amount of terminology. The categories used to group wetland plants include *emergent, submerged, floating-leaved,* and *floating*. The general characteristics of each group are described below.

A. Emergent Plants

Emergent plants are rooted in the soil with basal portions that typically grow beneath the surface of the water, but whose leaves, stems (photosynthetic parts), and reproductive organs are aerial. Most of the plants in this group are herbaceous, but we also include woody wetland species here. Where saturated soils are present rather than standing water, all the aboveground portions of the plant are aerial. Among all the types of wetland plants, emergents are perhaps the most similar to terrestrial species, relying on aerial (above the water) reproduction and on the soil as their exclusive source of nutrients. Emergent herbaceous plants often inhabit shallow waters in marshes, along lakeshores or stream banks, and because of their ability to intercept sunlight before it reaches the water's surface, they often dominate, outcompeting floating-leaved and submerged plants in these habitats.

Perhaps the most common emergent species are found in the large families of monocotyledons that tend to dominate both freshwater and saltwater marshes, i.e., the Poaceae (grasses), Cyperaceae (sedges, e.g., *Carex, Cyperus*), Juncaceae (rushes), and the Typhaceae (cattail). Other families with frequently encountered emergent species are the Alismataceae (water plantain), Araceae (arum), Asteraceae (aster), Lamiaceae (mint, e.g., *Lycopus, Mentha*), Polygonaceae (smartweed), and Sparganiaceae (bur reed; Figure 1.1a–d).

Woody wetland species include the tree and shrub species found in riparian wetlands, forested bottomlands, swamp forests, and peatlands (Figure 1.2a–d). Typical bottomland and swamp forest tree species in the U.S. include *Taxodium distichum* (bald cypress), *Nyssa*

FIGURE 1.1a
An emergent species of freshwater wetlands, *Phalaris arundinacea* (reed canary grass) is shown here in flower with a yellow-headed blackbird (*Xanthocephalus xanthocephalus*) perched on its stems. (Photo by T. Rice.)

FIGURE 1.1b
Scirpus cyperinus (wooly bulrush) grows in freshwater wetlands. (Photo by H. Crowell.)

FIGURE 1.1c
Aster novae-angliae (New England aster) is an emergent of freshwater wetlands. (Photo by M.S. Fennessy.)

FIGURE 1.1d
Limonium carolinianum (sea lavender) is a common emergent of the high marsh areas of many U.S. east coast salt marshes. (Photo by H. Crowell.)

aquatica (water tupelo), *Acer rubrum* (red maple), and members of the *Fraxinus, Quercus, Salix,* and *Populus* genera. Common families containing wetland shrubs include the Rosaceae (rose), Cornaceae (dogwood), Rubiaceae (madder, e.g., *Cephalanthus*), Betulaceae (alder, e.g., *Alnus*), Caprifoliaceae (honeysuckle, e.g., *Viburnum*), and particularly in bogs, the Ericaceae (heath, e.g., *Vaccinium, Chamaedaphne*).

Temperate coastal zones are fringed by salt marshes that are regularly flooded with saline or brackish water. The dual stresses of flooding and salt limit the number of plants

FIGURE 1.2a
Cephalanthus occidentalis (buttonbush) is a shrub of peatlands. (Photo by H. Crowell.)

FIGURE 1.2b
The foliage of *Rhizophora mangle* (red mangrove), which grows on the seaward edge of mangrove forests of the western hemisphere. (Photo by H. Crowell.)

that can survive there. Those that can survive include *Spartina alterniflora* (cordgrass) and *Juncus roemerianus* (black needlerush). These species grow successfully in salt marshes, in large part because they have little competition with other plants (Bertness 1991b).

Tropical and subtropical coastal areas are dominated not by the salt marsh grasses found at higher latitudes, but by coastal forests of halophytic mangroves. Like their temperate counterparts, mangroves are often the only group that can tolerate the combination of high salinity levels and flooding. The name *mangrove* actually refers to an ecological grouping of plants belonging to up to 16 families with a high degree of similarity in physiological characteristics and structural adaptations. Historically, between 60 and 75% of the earth's tropical coastlines were once lined with mangrove forests. The term mangrove encompasses an estimated 50 to 79 species of trees, shrubs, palms, and ferns in 9 to 33 different genera. The wide variation in numbers reflects the inexact definition of the term. The

FIGURE 1.2c
The extensive aerial root system of *R. mangle*. (Photo by H. Crowell.)

FIGURE 1.2d
Taxodium distichum (bald cypress) is the dominant species of many southeastern U.S. riparian and depressional forested wetlands. (Photo by H. Crowell.)

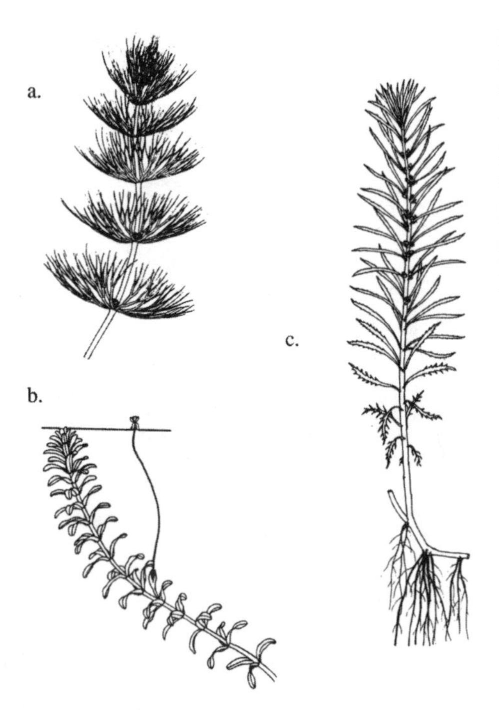

FIGURE 1.3
(a) *Ceratophyllum demersum* (hornwort) is a submerged rootless species (its leaves are about 1 cm in length). (b) *Elodea canadensis* (water weed) is a rooted submerged plant that grows in fresh waters in many areas of the world (leaves are 1 to 2 cm long). (c) *Myriophyllum oliganthum* (water milfoil) is a freshwater submerged plant (leaves are 1 to 2 cm long). (From Cook, C.D.K. 1996. *Aquatic Plant Book.* The Hague. SPB Academic Publishing/Backhuys Publishers. Reprinted with permission.)

families containing strict or "true" mangroves, which occur only in intertidal mangrove forests and do not extend into upland communities, include the Avicenniaceae, Combretaceae, Palmae, Rhizophoraceae, and Sonneratiaceae (Lugo and Snedaker 1974; Tomlinson 1986).

B. Submerged Plants

With the possible exception of flowering, submerged plants typically spend their entire life cycle beneath the surface of the water and are distributed in coastal, estuarine, and freshwater habitats (Figure 1.3). Nearly all are rooted in the substrate, although there are several rootless species that float free in the water column, including *Ceratophyllum demersum* (hornwort). In submerged species, all photosynthetic tissues are normally underwater (Cook 1996). Stems and leaves of submerged species tend to be soft (lacking lignin) with leaves that are either elongated and ribbon-like, or highly divided, making them flexible enough to withstand water movement without damage. Generally, the terminal portion of the plant does not reach the water's surface although it may lie in a horizontal position just beneath it (e.g., *Vallisneria americana*, water celery). In most species flowers are aerial (borne above the water) and pollination occurs via wind or insects (e.g., *Utricularia* and *Myriophyllum*). However, for approximately 125 to 150 species in this group, pollen transport occurs on or below the water's surface (see Chapter 5, Section II.A.3, Water Pollination).

Submerged plants take up dissolved oxygen and carbon dioxide from the water column, and many are able to use dissolved bicarbonate (HCO_3^-) in photosynthesis as well. Rooted submerged species acquire the majority of their nutrients from the sediments, although some nutrients, particularly micronutrients, may be absorbed from the water column (Barko and Smart 1980, 1981b). Rootless species are dependent on the water column as their sole nutrient source.

Examples of families in which all or nearly all of the species are submerged include the Callitrichaceae (water starwort), Ceratophyllaceae (hornwort), Haloragaceae (water milfoil), Potamogetonaceae (pondweeds), and Lentibulariaceae (bladderworts). The largest family, with 17 genera and about 75 known species, all of which are submerged, is the Hydrocharitaceae (frogbit).

C. Floating-Leaved Plants

The leaves of floating-leaved species (also known as floating attached) float on the water's surface while their roots are anchored in the substrate (Figure 1.4a and b). Petioles (as in the case of the Nymphaeaceae, water lily) or a combination of petioles and stems (as in some pondweeds, Potamogetonaceae) connect the leaves to the bottom. Most floating-leaved species have circular, oval, or cordate leaves with entire margins that reduce tearing, and a tough leathery texture that helps prevent both herbivory and wetting (Guntenspergen et al. 1989). The *stomata*, through which gas exchange occurs, are located on the aerial side of the leaf.

The long flexible petioles of the waterlilies allow the leaves to spread out into open areas of water, forming a cover over the water's surface that can reduce evaporative losses. Floating-leaved species shade the water column below and are often able to outcompete submerged species for light, particularly when turbidity levels are high and light penetration is reduced (Haslam 1978). Inflorescences either float, as in the Nymphaeaceae (water lily), or are borne on the water's surface on emergent *peduncles* (flower stalks), as seen in the Nelumbonaceae (water lotus).

Some species, for example, *Ranunculus flabellaris*, have underwater leaves in addition to floating leaves. Generally these leaves differ in form, with underwater leaves being finely divided while floating leaves are entire — a condition known as *heterophylly* (Sculthorpe 1967; see Chapter 4, Section V.C, Adaptations to Fluctuating Water Levels). Some floating-leaved plants also produce emergent leaves, including *Nymphaea alba*, *Nymphoides peltata*, and some species of *Nuphar* and *Potamogeton*. Floating-leaved plants

FIGURE 1.4a
The leaves and flowers of *Nymphaea odorata* (white water lily) float on the surface of freshwater wetlands. (Photo by T. Rice.)

FIGURE 1.4b
Nelumbo lutea (American water lotus) has both floating and emergent leaves. The flower is emergent on an erect petiole. (Photo by J. Ely.)

that produce emergent leaves are able to persist when the water level decreases. The aerial leaves are capable of surviving for some time out of the water (Sculthorpe 1967). In another interesting variation on leaf form, emergent plants sometimes produce floating leaves during juvenile stages (e.g., some *Sagittaria*). The formation of floating leaves can also be triggered by an increase in water level in normally emergent plants such as *Ranunculus sceleratus* and *Sparganium eurycarpum* (Kaul 1976; Maberly and Spence 1989).

D. Floating Plants

The leaves and stems of floating plants (also known as floating unattached) float on the water's surface. If roots are present, they hang free in the water and are not anchored in the sediments (Figure 1.5a–c). Floating plants move on the water's surface with winds and water currents. A widespread family of free-floating plants is the Lemnaceae, which includes the genera *Lemna* (duckweed), *Spirodela* (greater duckweed), and *Wolffiella* and *Wolffia* (water meal). The Lemnaceae contain some of the smallest angiosperms; some are so tiny that they are supported by the surface tension of the water alone. *Wolffia* is the smallest known angiosperm, having a subspherical shape and lacking roots.

Also included in the floating plants are larger species, such as *Eichhornia crassipes* (water hyacinth) and *Pistia stratiotes* (water lettuce), some of which have become the most troublesome invasive species in tropical and subtropical wetlands. *E. crassipes* has an inflated petiole that serves as a float, while *P. stratiotes* has broad, flat, water-resistant leaves. Both have extensive branching roots that hang down into the water column. Besides the roots' role in absorbing nutrients, they also serve as a weight that helps stabilize the plant on the water. Floating wetland plants commonly exhibit extensive vegetative growth. For example, *E. crassipes* and *P. stratiotes* both form daughter rosettes at the end of long stolons that easily separate from the parent plant.

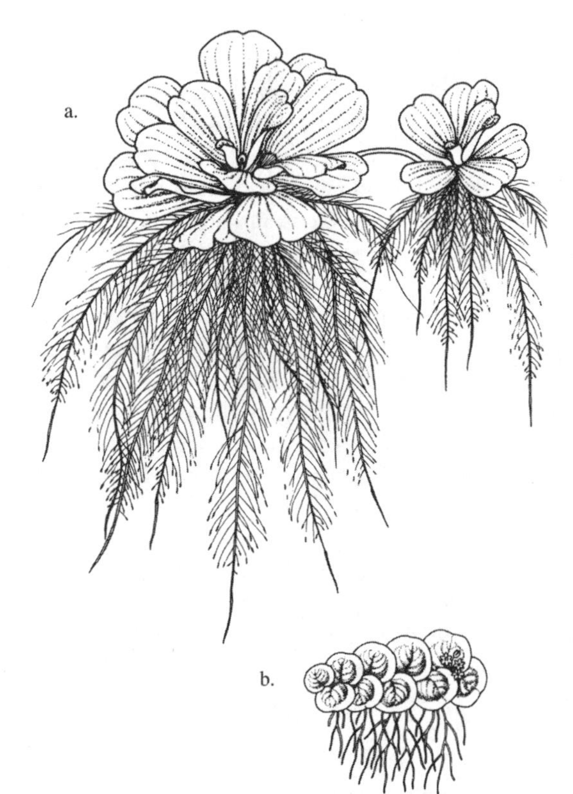

FIGURE 1.5
(a) *Pistia stratiotes* (water lettuce), a free-floating species of warm fresh waters with extensive fibrous roots (the diameter of the rosettes are up to 15 cm). (b) *Phyllanthus fluitans* is a free-floating South America plant with leaves 1 to 2 cm in diameter. (From Cook, C.D.K. 1996. *Aquatic Plant Book.* The Hague. SPB Academic Publishing/Backhuys Publishers. Reprinted with permission.)

FIGURE 1.5c
Floating species of the Lemnaceae including *Lemna minor, Spirodela polyrrhiza,* and *Wolffia* sp. (the largest leaves are about 1 cm in diameter). (Photo by H. Crowell.)

IV. Wetland Plant Distribution

The distribution of wetland plants depends on the distribution of wetland ecosystems themselves. The primary environmental factors that explain the distribution and types of wetlands on a global scale include climate, topography, and geology, and in coastal areas, tides. Wetlands occur in many geomorphological settings including river deltas, coastal lagoons and intertidal zones, river floodplains and headwaters, inland lakes, and inland depressions and flats (Brinson 1993a; Britton and Crivelli 1993; Mitsch and Gosselink 2000). On a global scale wetlands are ubiquitous, found on every continent except Antarctica, and in every climate. More than half of the world's total wetland area is found in tropical and subtropical regions, while a large proportion of the rest is boreal peatland (Mitsch and Gosselink 2000).

Some wetland species have extensive geographical distributions that range over several continents, leading them to be classified as *cosmopolitan*. Sculthorpe (1967) estimated that approximately 60% of aquatic species have ranges that span more than one continent. The most widely dispersed species tend to be monocots. For example, *Phragmites australis* has been called the most widely distributed angiosperm; its range extends as far north as 70°N. It is common in temperate latitudes and, although less common, is also found in tropical regions. *Lemna minor* is an example of a floating species that is cosmopolitan, absent from only a few areas in the tropics and polar regions (Sculthorpe 1967). Examples of cosmopolitan (or nearly so) submerged species include *Ceratophyllum demersum*, *Potamogeton crispus* (curly pondweed), and *P. pectinatus* (sago pondweed). Their widespread distribution indicates a well-developed facility for long-distance dispersal of seeds and vegetative parts over inhospitable territory such as land and sea. Mechanisms of dispersal include wind and water transport, movement by migratory birds, and, increasingly, transport by humans.

While most wetland plant species are not cosmopolitan, many still cover a wide latitudinal gradient relative to land plants. Their larger ranges are attributed to the moderating effect of water on environmental conditions. The distribution of many species tends to follow predictable patterns, with geographic ranges focused across large regions such as Eurasia–North Africa, continental Africa, or the tropical and subtropical latitudes of the Americas. There is also an interesting distribution pattern in which species inhabit the temperate latitudes of both North and South America. In this case, the same species, such as *Sagittaria montevidensis*, occurs in both northern and southern locations, but not necessarily in-between (Sculthorpe 1967). Migratory waterfowl, which aid in seed dispersal, are thought to contribute to this pattern.

In contrast, there are also endemic wetland species that are, by definition, confined to small geographical areas. Endemic species are those that are known to exist only in restricted areas; their limited distribution is often the result of barriers to dispersal or restriction to specific soil or climatic conditions. The geographic distribution of wetland plants with smaller ranges is in part a function of the patchy nature of the distribution of some wetland types. In geographically isolated wetlands such as the mountain bogs of Venezuela (Slack 1979) and the vernal pools of California (Baskin 1994), there is a high incidence of endemics. Tropical South America is particularly rich in endemic wetland species, as are the tropics and subtropics of Africa and Asia (Sculthorpe 1967). Some genera, such as *Sagittaria* and *Echinodorus*, display a high rate of endemism. For example, *S. sanfordii* has been found only in the Great Valley of California.

V. The Evolution of Wetland Plants

Unlike terrestrial plants, the evolution of wetland plants has received relatively little attention. Much of the study of wetland plants has centered on their systematics and ecology, while much less work has been done to understand their phylogenetic relationships or evolution. Consequently we know much less about their population genetics or the evolutionary implications of their life history characteristics (such as the predominance of vegetative reproduction or the high frequency of selfing in some species) in comparison with terrestrial species (Barrett et al. 1993).

While much remains to be determined about their evolutionary relationships, one thing is clear — wetland plants are derived from terrestrial ancestors. The evolutionary pathway of wetland plants begins and ends in the water. Initially, terrestrial vascular plants, derived from green algae, made the transition to land from nearshore estuarine or freshwater environments. This transition required the evolution of structures to obtain and transport water (e.g., roots, vascular tissue), minimize water loss (stomata, cuticle), and provide structural support (cellulose, lignin). These evolutionary innovations were probably derived from a class of green algae, the descendants of which are now included in the Charophyceae, beginning in the Ordovician period (510 million years ago). As plants radiated onto land and angiosperms evolved, the adaptive radiation continued and eventually plants moved back into aquatic habitats. Both fresh and salt waters were invaded. Fossil evidence suggests that there were a few primitive species of angiosperms developing distinctively aquatic habits by the upper Cretaceous (115 million years ago; Ingrouille 1992). It is interesting to note that the evolution of angiosperms into water occurred more than once. Terrestrial species have reportedly invaded aquatic habitats in an estimated 50 to 100 separate events (Cook 1996), illustrating that although they share similar habitats, wetland plant species have arrived there by very different evolutionary pathways (Philbrick and Les 1996).

The colonization of aquatic habitats by angiosperms presented numerous physiological challenges, in part because environmental conditions in saturated or flooded environments can be extremely harsh to plant growth and reproduction. Adaptations include the development of *aerenchyma* (tissue with large intercellular air spaces) and the diffusion of oxygen from the roots to the sediments (*radial oxygen loss*) that can detoxify potential phytotoxins that accumulate in reduced soils (see Chapter 4, Section II, Adaptations to Hypoxia and Anoxia).

One line of evidence that supports the theory that wetland plants evolved from terrestrial species is the fact that most wetland plant groups have retained characteristics typical of terrestrial plants. This includes traits such as flowers that are borne above the water's surface, pollination that depends on wind or insects, and, particularly in the case of emergent species, well-developed structural tissues (Moss 1988; Guntenspergen et al. 1989). By contrast, many floating-leaved and submerged taxa have lost terrestrial features such as well-developed secondary leaf thickening, elaborate leaf structures, or the function of some structures such as stomata.

A. Changes in Angiosperm Classification and Phylogeny

Until recently, angiosperms were divided into two classes, the monocotyledons and dicotyledons. New genetic evidence has caused this classification to be revised to include a third and separate group, the *magnoliids,* a group of angiosperms possessing the most primitive angiosperm features. Traditionally the magnoliids were classified with the dicots, in spite of the fact that they have many features uncharacteristic of dicots such as pollen with a single aperture. Two groups of magnoliids have been identified: those that are woody and a diverse assemblage of plants called the paleoherbs. The paleoherbs are now considered to be the ancestors of the monocots (which arose sometime before the end of the Cretaceous period, more than 120 million years ago) and the *eudicotyledons* (i.e., dicots minus the magnoliids). Members of the Nymphaeaceae (water lily family) are considered to be living paleoherbs, and as such are angiosperms with primitive features (Raven et al. 1999).

Many primitive monocot families are aquatic, suggesting their early adaptive radiation into wet places. In fact, it was once thought that all monocotyledons originated as aquatic plants, although this hypothesis has not been borne out (Les and Schneider 1995). Wetland plants are substantially more frequent among the monocots as compared with the eudicots, however. Les and Schneider (1995) estimate that while only 14% of eudicot families contain aquatic plants (defined broadly to include submerged, floating-leaved, floating, and emergent), 52% of monocot families do. Ultimately, in any plant family containing terrestrial and wetland species, the wetland plants are probably of more recent origin.

The use of DNA to investigate evolutionary relationships promises to reveal unexpected relationships. One such surprise has been shown for the submerged plant, *Ceratophyllum*. This genus is classified in a family all its own (Ceratophyllaceae), and is somewhat notorious among taxonomists for the difficulty it has presented in distinguishing its evolutionary relationships. *Ceratophyllum* is considered to have many specialized characteristics including no roots, highly reduced leaves, and underwater reproduction (including underwater pollination). At the same time it has many primitive features including a lack of vessels (xylem) and no petals or sepals. Recently these traits have come to be viewed as highly specialized adaptations to a long-standing aquatic habit, derived over a long evolutionary time from an ancestor that appears to predate many angiosperms. It is now thought that *Ceratophyllum* became aquatic long ago, before the majority of wetland plant species. The new molecular evidence has resulted in a revised cladogram that puts *Ceratophyllum* at the base of the angiosperms (Figure 1.6), suggesting that its current

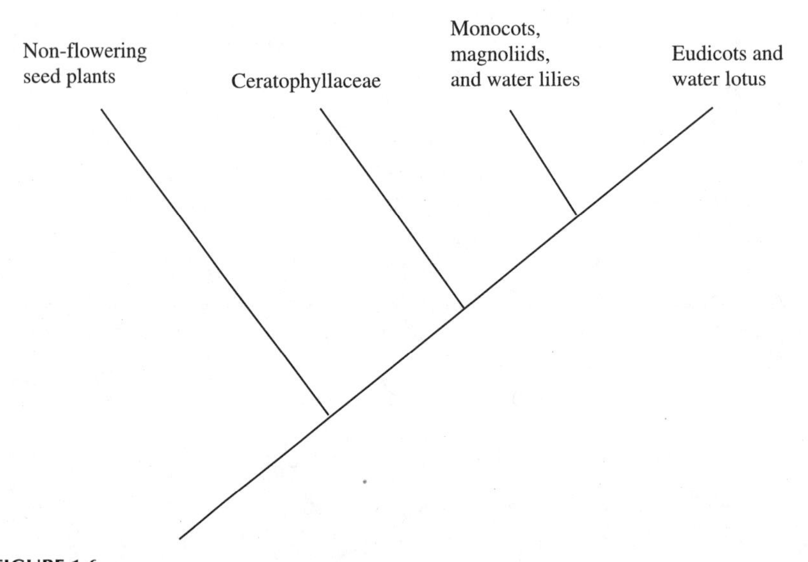

FIGURE 1.6
Cladogram showing the phylogenetic relationship between the Ceratophyllaceae and other plant groups. (From Raven et al. 1999. *Biology of Plants.* New York. W.H. Freeman and Company. Redrawn with permission.)

TABLE 1.2

Summary of Some Distinctive Ecological Features Possessed by Wetland Plants with Their Potential Evolutionary Consequences, and Examples of Genera Possessing These Characteristics

Characteristic	Groups That Tend to Exhibit This Trait	Potential Evolutionary Consequences	Examples of Genera Possessing This Trait
High phenotypic plasticity	Most groups	Acts as a buffer against variable habitats, thus reducing selection pressure	*Spartina, Alisma*
High rates of vegetative/clonal reproduction	Perennial species	Can lead to genetically homogeneous populations	*Typha, Myriophyllum, Salicornia*
Low incidence of sexual reproducton	Perennial species	Limits genetic recombination and can reduce genetic diversity	*Spartina, Puccinellia*
Accelerated reproduction	Annual species in variable habitats	Allows dispersal into unpredictable environments	*Alisma, Cyperus, Polygonum*
Water pollination	Submerged species	Limits gene flow to the wetland where population is located	*Ceratophyllum, Vallisneria*
Water-dispersed propagules	Many groups	Facilitates dispersal	*Rhizophora, Xylocarpus, Nelumbo*

Adapted from Barrett et al. 1993, with some examples from Sculthorpe 1967, Vince and Snow 1984, and Jackson et al. 1986.

features are a combination of both ancient and more recently derived traits (Les et al. 1991). Future molecular studies will undoubtedly contribute more to our understanding of wetland plant evolution.

B. Evolutionary Processes in Wetland Plants

The unique adaptations that have evolved in many wetland plant species present some potential evolutionary consequences that many terrestrial species do not face. Table 1.2 summarizes these characteristics as well as their implications. These traits represent life-history strategies that are adapted to the physical and chemical conditions of wetlands including anaerobic soils and fluctuating water levels. Some of the most common adaptations are the timing of seed production to coincide with suitable conditions for germination, and the avoidance of sexual reproduction for long periods during which vegetative reproduction dominates.

One general evolutionary response to life in variable environments is a relatively high degree of phenotypic plasticity. This allows a plant to respond rapidly to changing environmental conditions by, for example, altering growth rates or the timing of flowering. The phenotype of an individual can vary without an associated genetic change. This is a protective mechanism, allowing species to cope with short-term vagaries of the environment. Phenotypic plasticity provides an evolutionary buffer of sorts, insulating the species from selection pressures that may, in the long run, be maladaptive (Barrett et al. 1993). For example, the rate of shoot elongation in many grass and grass-like species such as *Phalaris arundinacea* (reed canary grass) increases as water depth increases, enabling the upper parts of the plant to remain above the water's surface (see Chapter 4, Section II.A.3.a, Rapid Underwater Shoot Extension). Early (or precocious) reproduction may evolve when short-lived plants occur in ephemeral habitats. In this case, early dry periods exert strong selection pressures on the timing of growth and reproduction such that flowering occurs more rapidly than normal. Wetland plants face other evolutionary challenges such as the difficulty of moving pollen and propagules when water is present. In some species underwater pollination and water-dispersed propagules have evolved, but the implications of these characteristics for gene flow within plant populations are not well understood (Barrett et al. 1993).

VI. Threats to Wetland Plant Species

It is estimated that 6.4% of the world's land area, or nearly 9 million km^2, is wetland (Maltby 1986; Mitsch et al. 1994; Mitsch and Gosselink 2000). Approximately 5% of the land in the contiguous U.S. is wetland, making wetland plant communities relatively rare from a landscape perspective. Wetland plants are threatened by the same forces that threaten wetland ecosystems generally, including human activities such as wetland draining or filling, hydrologic alterations, chronic degradation due to nonpoint source pollution, and the invasion of exotic species. There is also increasing concern about the effects of global change (including climatic changes and possible sea level rise) on plant populations.

Wetland losses around the world have been extensive and examples abound of dramatic declines in area, both on a geographic basis and in terms of wetland type. In Canada, for example, wetland area has been reduced by 15%, with losses of nearly 70% reported in highly populated areas such as southwestern Ontario or in the Pacific estuary marshes (Lovett-Doust and Lovett-Doust 1995). Riparian wetland forests have essentially disappeared in southern Europe, and it is estimated that nearly 75% of U.S. riparian forests have been cleared or altered in some way (Britton and Crivelli 1993; Kentula 1997). In the U.S., where much time and energy have gone into documenting wetland losses, more than 50%

of the original wetland area has reportedly been lost in the lower 48 states. In many inland agricultural areas, losses of 80 to 90% are not uncommon (Dahl 1990). Conversion to agriculture is cited as the most significant agent of land use change, accounting for 80% of wetland conversions in the U.S. between 1955 and 1975 (U.S. National Research Council 1992; Mathias and Moyle 1992). The declines in wetland area have led to decreases in wetland plant species diversity, and, as a result, wetlands are home to a disproportionately large number of rare plant species. It is estimated that nearly one third of threatened and endangered plant species in the U.S. depend on wetlands for their survival (Niering 1988; Murdock 1994).

A. Hydrologic Alterations

Hydrologic changes that result from human activities, such as agriculture or flood control, often lead to a decrease in wetland area or a change in the hydrologic regime of the area that remains (Mathias and Moyle 1992). For example, water diversions (e.g., dams, groundwater pumping, or irrigation projects) can significantly alter the *hydroperiod* (water level over time) of associated wetlands and change the distribution of wetland species. In arid areas, wetland ecosystems and the species that inhabit them are often in direct competition with humans and human activities that consume water. In such areas, many wetland species are endangered. Groundwater depletion is a threat to many riparian wetlands. The availability of shallow groundwater has been shown to structure riparian plant communities, and as the distance to the groundwater increases, the abundance of herbaceous wetland plants declines dramatically (Stromberg and Patten 1996).

Stream channelization projects have altered or eliminated stream-side wetlands and their plant communities. Many miles of streams and rivers have been channelized in order to expedite the drainage of water from uplands, control flooding, move water away from agricultural fields or urban areas, and reduce meandering. Channelization projects have been shown to alter riparian zone geomorphology and wetland hydroperiods, leading to significant changes in plant community structure and decreased plant species diversity (Carpenter et al. 1992).

B. Exotic Species

Non-indigenous, or *exotic*, species are considered a major threat to the biological diversity of many types of ecosystems, including wetlands. The impact of exotic species can be severe and includes the alteration of nutrient cycles, development of monocultures, and the extirpation or extinction of native species, resulting in severe losses of native biodiversity (D'Antonio and Vitousek 1992; Gordon 1998; Wilcove et al. 1998). For example, Gordon (1998) estimates that from 32 to 39% of the most aggressive invasive species in various Florida ecosystems have significant impacts on biogeochemical processes in the ecosystems they dominate. E. O. Wilson (1992) has called the spread of exotic species one of the four "mindless horsemen of the environmental apocalypse" (the others being overexploitation, habitat destruction, and the spread of diseases carried by exotic species). As the capacity and speed of human travel have increased, the rate of biological invasions has also increased. Humans have introduced species in several ways: some have been introduced purposefully (e.g., as ornamental plants), some have escaped into natural areas after import (for instance *Eichhornia crassipes*), while others have entered new areas accidentally (e.g., by "hitchhiking" in ballast water, on travelers, or in packages or trade goods; Ruesink et al. 1995; see Chapter 8, Invasive Plants in Wetlands). These agents of introduction have led to the accumulation of many new species in areas where they did not originate.

Where invasive exotic species have become established, their tendency to expand rapidly has caused a reduction in the abundance and species richness of native macrophytes and associated invertebrate communities. The dense growth of some exotic species can also limit recreational fishing, boating, and swimming. For instance, *Myriophyllum spicatum* (Eurasian watermilfoil) is a widespread nuisance submerged species in North America (Sheldon and Creed 1995). This species has life history characteristics that are common to invasive species: a high photosynthetic efficiency, rapid nutrient uptake from the sediments or water, and rapid vegetative reproduction. It has been estimated that a single plant can produce 250 million ramets by repeated fragmentation. Once it is established in an area, it can spread rapidly on boat propellers, center boards, and trailers. The invasion of *Lythrum salicaria* (purple loosestrife) has been shown to modify wetland ecosystem processes, in part due to its ability to shade out other species and to rapidly take up available nutrients (Thompson et al. 1987). The invasion of exotic species tends to increase as ecosystems become degraded. As nutrient enrichment or hydrological modifications occur, the likelihood of invasion and successful establishment of exotic plants increases.

C. Impacts of Global Change

Human activities are having a profound negative impact on land use patterns, atmospheric chemistry, and increasingly, climate (Vitousek 1994). Increasing levels of a number of atmospheric gases, including CO_2, methane, and nitrous oxide, are predicted to change world climates. The predicted increase in mean annual temperatures is expected to impact wetland plants. Warming may result in both the expansion of some wetland areas and the retraction of others, depending on the type and location of the wetland. Predicted changes to the hydrological cycle will drive many changes in wetland plant communities. For example, groundwater is an important source of water to many wetlands, and changes in the balance between precipitation and evapotranspiration will lead to groundwater level changes. In regions that become drier, water will not be as plentiful for groundwater recharge, and as water tables decline, wetland desiccation will result. Peatlands may be particularly vulnerable, especially those associated with permafrost in high latitudes. The diversity of wetland plant communities will change as temperatures warm and growing seasons lengthen.

Sea level rise is a threat to many wetlands in coastal areas (Baldwin and Mendelssohn 1998). In the U.S. alone, there are about 8,000 km^2 of dry land within 50 cm of high tide, 80% of which is currently undeveloped. A 50-cm rise in sea level would eliminate between 17 and 43% of U.S. coastal wetlands, with half of this loss occurring in Louisiana (Titus and Narayanan 1995). Projections made by the U.S. EPA anticipate that we will experience a 15- to 34-cm sea level rise during the next century, with up to 65-cm rise possible (Titus and Narayanan 1995). This would inundate coastal wetlands, the majority of which are less than a meter above sea level. Titus and others (1991) estimate that an area approximately the size of Massachusetts will need to be abandoned if coastal wetlands are to be allowed to migrate with rising water levels, thus preserving the species present in these systems. Whether or not these wetlands are allowed to migrate inland as water levels rise, total wetland area is predicted to diminish because the slope above the wetlands is generally steeper than the slope where the wetlands are currently located. Therefore, sea level rise is predicted to cause a net loss of wetland area and a loss of wetland plant diversity.

Vegetation has also been shown to affect the rate of greenhouse gas emissions from wetlands. For example, it has been demonstrated that in some wetlands emergent plants are responsible for between 50 and 90% of total methane efflux (Cicerone and Shetter 1981; Grosse et al. 1996). The rates of methane emission are greatest from those species that have pressurized convective ventilation (see Chapter 4, Section II, Adaptations to Hypoxia and Anoxia). Many species have been shown to actively transport gasses from the root zone, including *Nuphar advena* (yellow water lily; Dacey 1981), *Typha latifolia* (Sebacher et al. 1985), *Eleocharis* sp. (spike rush; Sorrell and Boon 1994), and *Phragmites australis* (Brix 1993).

D. Threatened and Endangered Species

The extensive loss of wetland area, combined with the degradation of many of those that remain, has contributed to the number of wetland plant species that are rare, threatened, or endangered. Table 1.3 is a compilation of threatened and endangered wetland plant species in the U.S. as listed under the Endangered Species Act. The species shown here are those with an indicator status of obligate (OBL), facultative wetland (FACW), or facultative (FAC) according to Reed (1997; see Chapter 10, Wetland Plants as Biological Indicators), and are also on the U.S. list of endangered (a plant species in danger of extinction within the foreseeable future throughout all or a significant portion of its range) and threatened species (plants likely to become endangered within the foreseeable future).

A species may be described as rare for many reasons including the extent of its geographic range, habitat specificity, and local population size. The fact that some wetland plant species have very specific habitat requirements has implications for their abundance: they are restricted to those areas that meet their specific needs, and they tend to be vulnerable to disturbance (A.F. Davis 1993; Lentz and Dunson 1999). There is also a positive correlation between unique types of wetlands and the presence of rare and endangered plants.

Generally, rare species tend to have highly specific requirements, persisting only under a narrow set of wetland conditions. For example, *Howellia aquatilis*, an annual plant that inhabits ephemeral wetlands in the Pacific Northwest, is considered extirpated or endangered throughout its range. It has very narrow requirements (for both germination and growth) in terms of water depth, the concentration of dissolved substances in the water, and soil texture (Lesica 1992). In addition, this species does not form a persistent seed bank, making it prone to large fluctuations in population size as a result of environmental fluctuations. These characteristics predispose *H. aquatilis* to rarity. *Scirpus ancistrochaetus* (northeastern bulrush) is another example of a species with very specific requirements, including fluctuating water levels (A.F. Davis 1993), high soil organic matter content (with a mean of 50.8% at sites supporting this species; Lentz and Dunson 1999), relatively high soil-exchangeable sodium, and low water pH. Knowledge of the basic autecology of a threatened or endangered species is vital for the successful conservation or reintroduction of rare species.

Certain species may also become threatened or endangered because they depend on a unique wetland type for their habitat. For instance, in California, where 93 to 97% of vernal pool wetlands have been destroyed, many endemic species such as the mint *Pogogyne abramsii*, and grasses in the genera *Neostapfia, Tuctoria,* and *Orcuttia* are now considered rare or endangered (Griggs and Jain 1983; Keeley 1988; Baskin 1994). *Helenium virginicum*, listed in the U.S. as a candidate for endangered or threatened status, is a narrow endemic, limited to 25 sinkholes in west-central Virginia (Messmore and Knox 1997).

TABLE 1.3

Endangered and Threatened Wetland Plant Species of the U. S., Their Status (E = endangered and T = threatened), and Region

Species	Status	Region(s)
Acaena exigua	E	Hawaii[1]
Acanthomintha ilicifolia	T	California
Aeschynomene virginica	T	Northeast,[2] Southeast[3]
Alsinidendron lychnoides	E	Hawaii
Alsinidendron viscosum	E	Hawaii
Amaranthus pumilus	T	Northeast, Southeast
Amphianthus pusillus	T	Southeast
Arenaria paludicola	E	Northwest,[4] California
Argyroxiphium kauense	E	Hawaii
Astragalus lentiginosus var. *piscinensis*	T	California
A. phoenix	T	Inter-Mountain[5]
Blennosperma bakeri	E	California
Brodiaea filifolia	T	California
B. pallida	T	California
Callicarpa ampla	E	Caribbean[6]
Calyptronoma rivalis	T	Caribbean
Campanula robinsiae	E	Southeast
Cardamine micranthera	E	Southwest[7]
Carex albida	E	California
C. specuicola	T	Southwest
Castilleja campestris ssp. *succulenta*	T	California
Centaurium namophilum	T	Inter-Mountain, California
Chamaesyce hooveri	T	California
Cirsium vinaceum	T	Southwest
Clermontia drepanomorpha	E	Hawaii
C. samuelii	E	Hawaii
Conradina verticillata	T	Northeast, Southeast
Cordia bellonis	E	Caribbean
Cordylanthus palmatus	E	California
C. mollis ssp. *mollis*	E	California
Cornutia obovata	E	Caribbean
Crescentia portoricensis	E	Caribbean
Cyperus trachysanthos	E	Hawaii
Cyrtandra viridiflora	E	Hawaii
Deeringothamnus pulchellus	E	Southeast
Dubautia pauciflorula	E	Hawaii
Eryngium constancei	E	California
Eugenia haematocarpa	E	Caribbean
Eutrema penlandii	T	Inter-Mountain
Exocarpos luteolus	E	Hawaii
Geranium multiflorum	E	Hawaii
Gesneria pauciflora	T	Caribbean
Grindelia fraxino-pratensis	T	Inter-Mountain, California
Halophila johnsonii	T	Southeast
Harperocallis flava	T	Hawaii

Helianthus paradoxus	T	South Plains[8]
Helonias bullata	T	Northeast, Southeast
Howellia aquatilis	T	Northwest
Hymenoxys texana	E	South Plains
Ilex cookii	E	Caribbean
I. sintenisii	E	Caribbean
Iris lacustris	T	North Central[9]
Ischaemum byrone	E	Hawaii
Isotria medeoloides	T	Northeast, Southeast, North Central
Justicia cooleyi	E	Southeast
Lasthenia burkei	E	California
L. conjugens	E	California
Lembertia congdonii	E	California
Lesquerella pallida	E	South Plains
Lilium occidental	E	California
Limnanthes vinculans	E	California
Lindera melissifolia	E	Southeast, North Central
Lobelia oahuensis	E	Hawaii
Lomatium bradshawii	E	Northwest
Lysimachia asperulaefolia	E	Southeast
L. filifolia	E	Hawaii
Macbridea alba	T	Southeast
Marshallia mohrii	T	Southeast
Melicope lydgatei	E	Hawaii
Mentzelia leucophylla	T	Inter-Mountain
Monardella linoides ssp. *viminea*	E	California
Myrsine juddii	E	Hawaii
Navarretia fossalis	T	California
Neostapfia colusana	T	California
Nitrophila mohavensis	E	Inter-Mountain, California
Orcuttia californica	E	California
O. inaequalis	T	California
O. pilosa	E	California
O. tenuis	T	California
O. viscida	E	California
Oxypolis canbyi	E	Northeast, Southeast
Pedicularis furbishiae	E	Northeast
Pinguicula ionantha	T	Southeast
Plagiobothrys strictus	E	California
Platanthera holochila	E	Hawaii
P. leucophaea	T	Northeast, Southeast, North Central, Central Plains,[10] South Plains
Pleodendron macranthum	E	Caribbean
Poa mannii	E	Hawaii
P. napensis	E	California
P. sandvicensis	E	Hawaii
P. siphonoglossa	E	Hawaii
Pogogyne abramsii	E	California

– continued

TABLE 1.3 continued

P. nudiuscula	E	California
Potamogeton clystocarpus	E	South Plains
Potentilla hickmanii	E	California
Pritchardia affinis	E	Hawaii
P. kaalae	E	Hawaii
P. munroi	E	Hawaii
P. viscosa	E	Hawaii
Ptilimnium nodosum (=fluviatile)	E	Northeast, Southeast
Rhododendron chapmanii	E	Southeast
Rhynchospora knieskerni	T	Northeast
Ribes echinellum	T	Southeast
Rorippa gambellii	E	California
Sagittaria fasciculata	E	Southeast
Sanicula purpurea	E	Hawaii
Sarracenia oreophila	E	Southeast
Scirpus ancistrochaetus	E	Northeast
Scutellaria floridana	T	Southeast
Sidalcea nelsoniana	T	Northwest
Solidago houghtonii	T	Northeast, North Central
Spiraea virginiana	T	Northeast, Southeast
Spiranthes diluvialis	T	Inter-Mountain
Stahlia monosperma	T	Caribbean
Styrax portoricensis	E	Caribbean
Suaeda californica	E	California
Taraxacum californicum	E	California
Ternstroemia luquillensis	E	Caribbean
T. subsessilis	E	Caribbean
Tetraplasandra gymnocarpa	E	Hawaii
Thalictrum cooleyi	E	Southeast
Thelypodium stenopetalum	E	California
Trematolobelia singularis	E	Hawaii
Trifolium amoenum	E	California
Tuctoria greenei	E	Calfornia
T. mucronata	E	California
Verbena californica	T	California
Viola helenae	E	Hawaii
V. oahuensis	E	Hawaii
Zizania texana	E	South Plains

Note: These species are on both the November 1999 U.S. Fish and Wildlife Service list of endangered and threatened species (U.S. FWS 1999) and the U.S. FWS list of Wetland Plants (Reed 1997). Subspecies and varieties were omitted if they did not match exactly. Plants were included if they were obligate wetland plants (occur in wetlands >99% of the time) to facultative upland plants (found in wetlands 1 to 33% of the time).

[1] Hawaii: Hawaiian Islands, American Samoa, Federated States of Micronesia, Guam, Marshall Islands, Northern Mariana Islands, Palau, U.S. Minor Outlying Islands

[2] Northeast: CT, DE, KY, MA, MD, ME, NH, NJ, NY, OH, PA, RI, VA, VT, WV

[3] Southeast: AL, AR, FL, GA, LA, MS, NC, SC, TN

[4] Northwest: ID, MT (western), OR, WA, WY (western)

[5] Inter-Mountain: CO (western), NV, UT

[6] Caribbean: Puerto Rico, U.S. Virgin Islands

[7] Southwest: AZ, NM

[8] South Plains: OK, TX

[9] North Central: IA, IL, IN, MI, MO, MN, WI

[10] Central Plains: CO (eastern), KS, NE

The destruction of habitat that has occurred in once common types of wetlands has also led to the decline of some species (A.F. Davis 1993). In Pennsylvania, more than 50% of the 579 species of special concern are considered to be wetland species. In a large wetland complex in Put-in-Bay Harbor, Lake Erie, more than 50% of the flowering plant species disappeared between 1898 and 1970 (Stuckey 1971). This area had been considered one of the most diverse communities in the Great Lakes region. Of the species that remain, several are rare and in danger of being eliminated. Most of those still present are able to tolerate warm, turbid, poorly oxygenated water. The dune slacks of the European coast provide another example. The dunes are low-lying areas within the coastal dune system where the water table is near the surface but seasonal fluctuations are extreme. The plant communities found here are similar to those found in lowland marshes or fens and include *Littorella uniflora, Schoenus nigricans,* and many *Carex* species. The dune slacks are highly ranked on the international conservation agenda because they are habitat for many endangered and endemic species. The expansion of tourism, afforestation (forest regrowth), and increased drinking water use, all of which have contributed to lower water tables, threaten the dune slacks. Changes in groundwater levels are in large part responsible for the rapid loss of dune slack plant diversity (Grootjans et al. 1998).

Summary

Wetland plants are defined as those species normally found growing in wetlands, either in or on the water, or where soils are flooded or saturated long enough for anaerobic conditions to develop in the root zone. Wetland plants include floating, floating-leaved, submerged, and emergent species. We include woody wetland plants (trees and shrubs) with the emergents. Of the known angiosperms, only a small proportion is adapted to the wetland environment.

Many wetland plants are widely distributed, particularly many of the monocots. However, some are endemic to small areas or certain wetland types. Wetland plants evolved from terrestrial plants, developing a number of adaptations to the aquatic environment including the formation of aerenchyma and the timing of reproduction. Wetland plants are threatened by the same forces that threaten wetland ecosystems, including agriculture, hydrologic alterations, pollution, development, and the invasion of exotic species. Global climate change may result in a severe decrease in wetland area. A disproportionate share of wetland species is either threatened or endangered as compared to terrestrial species, particularly those with narrow ecological tolerances.

2

Wetland Plant Communities

I. Wetland Plant Habitats

Wetland plants grow in a variety of climates, from the tropics to polar regions — wherever the water table is high enough, or the standing water is shallow enough, to support them. Each species is adapted to a range of water depths and many do not survive outside of that range for extended periods. For example, *Hydrilla verticillata* (hydrilla) thrives when fully submerged; *Typha angustifolia* (narrow-leaved cattail) can grow in water over 1 m in depth, but its leaves are emergent; and others, like *Larix laricina*, the tamarack tree of northern peatlands, are fully emergent and normally do not grow where water covers the soil surface. All rooted wetland plants are adapted to at least periodically saturated substrates where soil oxygen levels are low to non-existent.

The terms for different types of wetlands help to pinpoint the differences between wetland communities and can be defined, at least in part, by the type of vegetation that grows there. For example, *swamp* denotes a wet area where trees or shrubs dominate the canopy, such as a cypress swamp, while a *marsh* is dominated by herbaceous species, such as a cattail marsh. Names given to some wetland types denote either the source or the chemistry of the water, such as *riparian wetland,* or *salt marsh.*

Wetlands are recognized as vital ecosystems that support a wide array of unique plants especially adapted to wet conditions. Wetland plants, in turn, support high densities of fish, invertebrates, amphibians, reptiles, mammals, and birds. Wetland conditions such as shallow water, high plant productivity, and anaerobic substrates provide a suitable environment for important physical, biological, and chemical processes. Because of these processes, wetlands play a vital role in global nutrient and element cycles. Wetlands also provide key hydrologic benefits: flood attenuation, shoreline stabilization, erosion control, groundwater recharge and discharge, and water purification (Mitsch and Gosselink 2000). In addition, they provide economic benefits by supporting fisheries, agriculture, timber, recreation, tourism, transport, water supply, and energy resources such as peat (T.J. Davis 1993).

II. Wetland Definitions and Functions

The term *wetland* envelops a wide variety of habitats, from mangroves along tropical shorelines to peatlands that lie just south of the Arctic. The following definitions help identify commonalties among these vastly different ecosystems.

A. Ecological Definition

The determining factor in the wetland environment is water. To a great extent, hydrology determines soil chemistry, topography, and vegetation. All wetlands have water inputs that exceed losses, at least seasonally. It is difficult to say exactly how much water an area must have at any given time in order to be a wetland. Indeed, Cowardin and others (1979) state that a "single, correct, indisputable, and ecologically sound" definition of wetlands does not exist, mostly because the line between wet and dry environments is not easily drawn. Moisture levels vary along a continuum that shifts in time and space. Wetlands may have standing water throughout the year, or only during a portion of the year. Those influenced by tide may have water at each high tide, or only at each spring tide.

In all wetlands, the substrate is saturated enough of the time that plants not adapted to saturated conditions cannot survive. Saturated conditions lead to low oxygen (*hypoxia*) or a lack of oxygen (*anaerobiosis* or *anoxia*) in the soil pore spaces. Scarcity of oxygen brings about reducing conditions, in which reduced forms of elements (e.g., nitrogen, manganese, iron, sulfur, and carbon) are present (Gambrell and Patrick 1978). Such substrates are termed *hydric* soils. Wetland plants have adaptations to waterlogging and hydric soils that allow them to persist. Wetlands, then, are ecosystems in which there is sufficient water to sustain both hydric soils and the plants that are adapted to them.

B. Legal Definitions

Legal or formal definitions of wetlands have been adopted in a number of countries. In the U.S., a legal definition of wetlands is needed because wetlands are protected areas, regulated by government agencies. Wetland definitions help classify areas so that the appropriate protections or uses can be determined. Many nations have wetland definitions, and each country's definition tends to focus on the characteristics of that country's wetlands (Scott and Jones 1995). The international definition adopted by the Ramsar Convention of 1971 (Matthews 1993) is often the basis for the definition used by individual countries.

1. United States Army Corps of Engineers' Definition

In the U.S., wetlands are legally defined by government agencies actively involved in wetland identification, protection, and the issuance of permits to people who seek to alter wetlands. The U.S. Army Corps of Engineers and the U.S. Environmental Protection Agency define wetlands as:

> ... those areas that are inundated or saturated by surface or ground water at a frequency and duration sufficient to support, and that under normal circumstances do support, a prevalence of vegetation typically adapted for life in saturated soil conditions. Wetlands generally include swamps, marshes, bogs, and similar areas (Federal Interagency Committee for Wetland Delineation 1989).

This definition is used for the delineation of wetlands throughout the U.S. Disputes concerning wetland boundaries often arise because wetlands do not have distinct edges.

Three components of the wetland ecosystem are taken into consideration by the U.S. definition: hydrology, soil, and vegetation (see Chapter 10, Wetland Plants as Biological Indicators). Specific indicators of all three must be present during some part of the growing season for an area to be a wetland, unless the site has been significantly altered. Indicators of wetland hydrology include the presence of standing or flowing water or tides, but water may also be below the soil surface in a wetland. Secondary indicators of water level may also be used to establish wetland hydrology, such as water marks, drift lines, debris lodged in trees or elsewhere, and layers of sediment that form a crust on the soil surface. Hydric soils develop under low oxygen conditions that bring about diagnostic soil colors, textures, or odors. Other soil indicators include partially decomposed plant material in the soil profile (as in peatlands) or decomposing plant litter at the surface of the soil profile. The dominance of wetland vegetation (and the absence or rarity of upland vegetation) indicates a wetland.

2. U.S. Fish and Wildlife Classification of Wetlands

For the purpose of wetland and deepwater habitat classification, the U.S. Fish and Wildlife Service (Cowardin et al. 1979) defined wetlands as:

> ... lands transitional between terrestrial and aquatic systems where the water table is usually at or near the surface or the land is covered by shallow water. For purposes of this classification, wetlands must have one or more of the following three attributes: (1) at least periodically, the land supports predominantly hydrophytes; (2) the substrate is predominantly undrained hydric soil; and (3) the substrate is nonsoil and is saturated with water or covered by shallow water at some time during the growing season of each year.

This definition is the basis for a detailed classification of wetlands (the Cowardin system, 1979) that was a first step in compiling an inventory of all U.S. wetlands (the National Wetlands Inventory).

3. International Definition

In 1971, an international convention on wetlands was held in Ramsar, Iran by the International Union for the Conservation of Nature and Natural Resources (IUCN). An international treaty on wetlands, the Convention on Wetlands of International Importance Especially as Waterfowl Habitat, also known as the Ramsar Convention, was signed there. It "provided the framework for international cooperation for the conservation and wise use of wetlands and their resources" (Matthews 1993). Under the Ramsar Convention wetlands are defined as:

> ... areas of marsh, fen, peatland or water, whether natural or artificial, permanent or temporary, with water that is static or flowing, fresh, brackish or salt, including areas of marine water the depth of which at low tide does not exceed six meters.

In addition, wetlands "may incorporate riparian and coastal zones adjacent to the wetlands, and islands or bodies of marine water deeper than six meters at low tide lying within the wetlands."

The Ramsar Convention definition of wetlands is broader than the U.S. Army Corps of Engineers' definition as it includes coral reefs and other deeper water habitats. The inclusion of more habitat types in the definition allows the convention to protect a greater area. All signatory nations agree to designate at least one site for inclusion on the Ramsar List. Inclusion confers international recognition on a site and obliges the government to maintain and protect the wetland. As of February 2000, there were 118 contracting parties with 1,016 sites on the Ramsar List for a total area of over 72.8 million ha (Ramsar Convention Bureau 2000).

The Ramsar Convention emphasizes the "wise use" and "sustainable development" of wetlands rather than conservation. They define wise use as the "sustainable utilization [of wetlands] for the benefit of mankind in a way compatible with the maintenance of the natural properties of the ecosystem." Sustainable utilization of a wetland is defined as "human use of a wetland so that it may yield the greatest continuous benefit to present generations while maintaining its potential to meet the needs and aspirations of future generations" (T.J. Davis 1993). In order to use a wetland wisely, a thorough understanding of its functions within the landscape is essential.

C. Functions of Wetlands

Whether wetlands are bordered by upland forest, desert, tundra, agricultural land, urban areas, or ocean, they often perform similar roles, or *functions*, within the broader landscape. All wetland functions are related to the presence, quantity, quality, and movement of water in wetlands (Carter et al. 1979). Functions are linked to the self-maintenance of the wetland and its relationship to its surroundings (Mitsch and Gosselink 2000). The functions of wetlands can be categorized into three main categories: hydrology, biogeochemistry, and habitat (Walbridge 1993). Wetland functions do not necessarily affect humans directly. Another term, *values*, refers to the benefits society derives from wetlands. Wetland values are closely tied to functions (Table 2.1).

1. Hydrology

Hydrologic functions of wetlands include the recharge and discharge of ground water supplies, floodwater conveyance and storage, and shoreline and erosion protection.

TABLE 2.1

Functions and Values Commonly Attributed to Wetlands

Function	Societal Value
Hydrology	Flood mitigation Groundwater recharge Shoreline protection
Biogeochemistry	
Sediment deposition Phosphorus sorption Nitrification Denitrification Sulfate reduction Nutrient uptake Sorption of metals	Improved water quality
Carbon storage Methane production	Global climate mitigation
Plant and animal habitat	Timber production Agricultural crops (rice, cranberries, etc.) Animal pelts (furs and skins) Commercial fish/shellfish production Recreational hunting and fishing

Adapted from Walbridge 1993.

a. Groundwater Supply
Groundwater may move into a wetland via springs or seeps (groundwater discharge) and water from the wetland may seep into the groundwater (groundwater recharge). Groundwater can be recharged from depressional wetlands if the water level in the wetland is above the water table of the surrounding soil. Recharge is important for replenishing aquifers for water supply. At some sites, both recharge and discharge occur. For example, in Florida cypress ponds, the water level is continuous with the water table of the surrounding landscape. When the water table rises due to rainfall, groundwater moves into the cypress pond. In dry periods, the water movement is reversed, and the aquifer is recharged (Ewel 1990a).

b. Flood Control
Wetlands can temporarily store excess water and release it slowly over time, thus buffering the impact of floods. Intact and undeveloped riparian wetlands can prevent damaging floods along rivers (Sather and Smith 1984). Depressional wetlands such as cypress ponds or prairie potholes have the capacity to receive and store at least twice as much water as a site filled with soil (Ewel 1990a). Some wetlands are not able to store excess water. If wetlands are impounded in order to store more floodwater than they normally would, significant changes in the plant community can result (Thibodeau and Nickerson 1985).

c. Erosion and Shoreline Damage Reduction
Wetlands along rivers, lakes, and seafronts can protect the shoreline by absorbing the energy of waves and currents. Wetlands along shorelines are dynamic systems, generally reaching equilibrium between accretion and erosion of substrate. Structures used for shoreline protection, such as bulkheads or jetties, can destroy the shoreline habitat by interrupting this equilibrium. These structures can also channel sediment into navigable waterways, where the cost of dredging is added to the cost of shoreline protection (Adamus and Stockwell 1983). Mangroves along tropical shorelines provide a good example of the erosion protection that wetlands can provide. Their extensive roots help stabilize sediments and prevent wave damage to inland areas (Odum and McIvor 1990). In China, wetlands have been created for shoreline reclamation and stabilization using vast plantings of *Spartina alterniflora* (cordgrass [Chung 1993]).

2. Biogeochemistry
A number of important biogeochemical processes are favored in wetlands due to shallow water (which maximizes the sediment-to-water interface), high primary productivity, the presence of both aerobic and anaerobic sediments, and the accumulation of litter (Mitsch and Gosselink 2000). These conditions often lead to a natural cleansing of the water that flows into wetlands. Incoming suspended solids settle from the water column due to the reduced water velocity found in wetlands (Johnston et al. 1984; Fennessy et al. 1994b). Materials associated with solids, such as phosphorus, are also removed from the water column in wetlands (Johnston 1991; Mitsch et al. 1995). Nitrogen is transformed through microbial processes (e.g., nitrification followed by denitrification; Faulkner and Richardson 1989) which require the presence of both aerobic and anaerobic substrates. Plant uptake and plant tissue accumulation can also remove nitrogen and phosphorus from the water; however, this process can be reversed when plants die back after the growing season (Howarth and Fisher 1976; Richardson 1985; Peverly 1985). Wetlands also play a role in the global cycling of sulfur and carbon as their anaerobic forms are produced under wetland conditions (see Chapter 3, Section III.A.1, Reduced Forms of Elements).

The capacity of wetlands to purify water is one of the most important societal values wetlands provide. Water quality improvements within wetlands are well documented (Engler and Patrick 1974; Odum et al. 1977; Mitsch et al. 1979; Dierberg and Brezonik 1983; Nichols 1983; Kadlec 1987; Knight et al. 1987; Brodrick et al. 1988; Mitsch 1992; Mitsch et al. 1995; Cronk 1996; Fennessy and Cronk 1997) and both natural and constructed wetlands are used worldwide to treat wastewater from industrial, agricultural, and domestic sources (Kadlec and Knight 1996; see Chapter 9, Section II, Treatment Wetlands).

3. *Habitat*

a. Wildlife and Fish Habitat

Because many wetlands are highly productive ecosystems, they support a large number of fish and wildlife species. Some animals, such as many fish, reptiles, and amphibians, depend exclusively on wetland habitats. Others utilize wetlands for only short periods of their life cycles (breeding, resting grounds) and some use wetlands as a source of food and water. Wetlands provide a habitat for many endangered and threatened animal species such as whooping cranes (*Grus americana*; U.S. Fish and Wildlife Service 1980), wood storks (*Mycteria americana*), crocodiles (*Crocodylus acutus*), snail kites (*Rostrhamus sociabilis*; U.S. National Park Service 1997), and Florida panthers (*Puma concolor coryi*; Maehr 1997). Hunters use wetland areas extensively for both waterfowl and deer, and their activities provide an economic value to the wildlife function of wetlands. Many animals such as muskrats, beavers, mink, and alligator are harvested for the fur and leather industries, worth millions of dollars annually. Both commercial and sports fisheries depend on the fish and shellfish of wetlands.

b. Plant Habitat

Wetland plant communities are among the most highly productive ecosystems in the world (Mitsch and Gosselink 2000). The production of biomass and the export of organic carbon to downstream areas make wetlands an integral part of a landscape's food web. The high usage of wetlands by wildlife attests to wetland plants' importance and diversity. Wetland plant products such as timber from bottomland swamps, peat from bogs, and many plant food products such as *Oryza sativa* (rice), *Trapa bispinosa* (water chestnut), and various species of *Vaccinium* (blueberries and cranberries) are harvested throughout the world. In many areas, farm animals graze wetland plants.

Wetland plant habitat is threatened by changes in wetland hydrology, eutrophication, the invasion of exotic plants, and other human-induced disturbances such as agriculture and development (Wisheu and Keddy 1994). Although many wetland plants are listed by the U.S. Fish and Wildlife Service as rare or endangered, wetland management plans rarely mention the conservation of rare species (Lovett-Doust and Lovett-Doust 1995; see Chapter 1, Table 1.3).

III. Broad Types of Wetland Plant Communities

One of the challenges wetland ecologists face is classifying wetlands so that plant communities, soil types, and hydrologic influences can be described, managed, mapped, or quantified. The variety of wetland types is enormous, and all wetland classifications must impose subjective boundaries on types. The sources and amounts of water vary over a wide range even within the same type of wetland. In addition, wetlands are found along successional gradients, further complicating their classification. Nonetheless, classification of wetlands is useful in order to describe their characteristics and manage them effectively (Cowardin et al. 1979).

Several wetland classification schemes have been used, some for specific regions, countries, or states, and some for certain types of wetlands, such as peatlands (Shaw and Fredine 1956; Taylor 1959; Bellamy 1968; Stewart and Kantrud 1971; Golet and Larson 1974; Cowardin et al. 1979; Beadle 1981; Zoltai 1983). Internationally, a number of nations have classified and inventoried their wetlands, including Canada, Greece, Indonesia, and South Africa. Some of these countries have used the Ramsar definition as a starting point and adapted it to local conditions. For example, Canada's classification system has five wetland classes and 70 wetland forms, half of which are types of northern peatlands. Indonesia has classified wetlands into six mangrove forest types and eight freshwater forested wetland types (Scott and Jones 1995).

In the U.S., the first well-known official wetland classification was published by the U.S. Fish and Wildlife Service in 1956 (Shaw and Fredine 1956). In this publication, known as Circular 39, wetlands were categorized into four broad types: *inland fresh areas, inland saline areas, coastal fresh areas,* and *coastal saline areas.* Each of these was further divided for a total of 20 wetland types. This classification scheme was influential in the beginning of federal wetland protection. Other classifications were statewide and were based on regional wetland characteristics.

In order to better define and inventory the wetlands of the U.S., the U.S. Fish and Wildlife Service developed a classification of wetlands and deepwater habitats based on the geologic and hydrologic origins of wetlands (Cowardin et al. 1979). This classification is beneficial because it eliminates the reliance on regional terms that may be meaningless in other parts of the country. In the Cowardin classification scheme, the major *systems* of wetland and deepwater habitat types are *marine, estuarine, lacustrine, palustrine,* and *riverine.* Systems are wetlands that share similar hydrologic, geomorphologic, chemical, or biological factors. The Cowardin system includes deepwater habitats (e.g., coral reefs), and those where plants do not grow, such as coastal sand flats or rocky shores.

A more recently developed classification scheme, called the hydrogeomorphic (HGM) setting of a wetland, is based on three parameters: the wetland's geomorphic setting within the landscape (i.e., riverine, depressional, lacustrine fringe), its water source, and the internal movement of water within the wetland, known as its hydrodynamics. As a classification system, the HGM approach emphasizes the topographic setting and the hydrology of the wetland that in turn affect its functions (Brinson 1993a). In this scheme, the presence of vegetation is seen as a result of the long-term interaction of climate and landscape position that also control wetland hydrology.

Alternatively, an approach based on the hydrogeologic setting (HGS) refers to the factors, both regional and local, that drive wetland hydrology and chemistry. It places an emphasis on the surface and subsurface features of the landscape that cause water flow into wetlands, thus determining the quantity and quality of water that a wetland receives (Bedford 1999). Winter (1992) defined the HGS in terms of surface relief and slope, soil thickness and permeability, and the stratigraphy, composition, and hydraulic conductivity of the underlying geologic materials. He used these parameters to classify sites into one of 24 "type settings" based on unique combinations of physiography and climate. This framework has a landscape basis and has been proposed for use in classifying wetlands for research into their diversity and ecological functions.

For the purposes of this book, we describe broad types of systems where wetland plants grow. We have categorized wetlands into three major wetland plant communities: (1) *marshes,* where herbaceous species dominate; (2) *forested wetlands,* where trees or shrubs dominate; and (3) *peatlands*, where the decomposition of plant matter is slow enough to allow peat to accumulate. Within these three categories, we further divide our description of plant communities based on hydrology, salinity, and pH.

A. Marshes

Marshes are dominated by herbaceous species which can include emergent, floating-leaved, floating, and submerged species. The term *marsh* covers a broad range of habitat types, and marshes can be found around the world in both inland and coastal areas. Further classification is based on hydrology and specific herbaceous type. Many names for marshes exist due to the numerous possible local plant associations in marshes. For example, in the state of Florida, a marsh can be classified as a water lily marsh, a cattail marsh, a flag marsh, or a sawgrass marsh (after the dominant plant), or a submersed marsh or wet prairie (after the community type; Kushlan 1990). Coastal marshes and inland marshes are discussed in more detail below.

1. Coastal Marshes

a. Salt Marshes

Salt marshes occur in coastal areas and are usually protected from direct wave action by barrier islands, or because they are located within bays or estuaries, or along tidal rivers (Figure 2.1). However, some are in direct contact with ocean waves on low-energy coastlines such as the Gulf of Mexico coast in west Florida and parts of Louisiana, the north Norfolk coast of Britain, and the coast of the Netherlands (Pomeroy and Wiegert 1981). Most salt marshes are found north and south of the tropics. In the tropics, mangroves are able to outcompete marsh plants (Kangas and Lugo 1990), although salt marshes do persist inland from mangroves in tropical (northern) Australia (Finlayson and Von Oertzen 1993) and alongside mangroves in some coastal areas of Mexico (Olmsted 1993). Salt marshes occur as far north as the subarctic and are particularly extensive around the Hudson and James Bays of Canada (approximately 300,000 km^2; Glooschenko et al. 1993).

FIGURE 2.1
Salt marsh in Cape Cod, Massachusetts with *Spartina patens* (salt marsh hay) in the foreground and *S. alterniflora* (cordgrass) near the tidal creek. (Photo by H. Crowell.)

The plant communities of salt marshes are subjected to daily and seasonal water level fluctuations due to tides, and to variations in freshwater inputs from overland runoff. In addition, plants are adapted to low soil oxygen levels that can lead to high levels of sulfide (Valiela and Teal 1974). Some salt marsh plants are able to withstand salt concentrations in the soil pore water that are sometimes higher than that of seawater (i.e., 35 ppt) due to the deposition of salt and evaporation (Wijte and Gallagher 1996a).

In North America, some of the major remaining areas of salt marshes are on the Atlantic coast and along the Gulf of Mexico. Along the northern Atlantic shore, the coasts of Labrador, Newfoundland, and Nova Scotia harbor salt marshes in river deltas and where the wave energy is low (0 to 2 m amplitude; Roberts and Robertson 1986). South of this region, salt marshes have been divided into three major types (Chapman 1974; Mitsch et al. 1994):

1. The Bay of Fundy marshes in Canada: These marshes are influenced hydrologically by rivers and a high tidal range (up to 11 m; Gordon and Cranford 1994) that erodes the surrounding rocks. The substrate is predominantly red silt.
2. New England marshes (from Maine to New Jersey): These marshes were formed on marine sediments and marsh peat without as much upland erosion as in the Bay of Fundy marshes.
3. Coastal Plain marshes: These marshes extend from New Jersey south along the Atlantic and along the Gulf of Mexico coast to Texas. The tidal range is smaller and the inflow of silt from the coastal plain is high. Included among these are the Mississippi River delta wetlands, which are the largest salt marshes in the U.S.

All three of these salt marsh types are dominated by *Spartina alterniflora* (Figure 2.2). *S. alterniflora* is a perennial grass that usually occurs along the seaward edge of salt marshes (Metcalfe et al. 1986) and can grow in water salinities as high as 60 ppt (Wijte and

FIGURE 2.2
Spartina alterniflora (cordgrass), the dominant plant of many U.S. east coast and Gulf of Mexico salt marshes. (Photo by H. Crowell.)

Gallagher 1996a). Two forms of *S. alterniflora* often coexist within the same marsh: the tall and short forms. The tall form (1 to 3 m) grows along the banks of tidal creeks, in the lowest part of the marsh, closest to the sea. The short form (10 to 80 cm) grows inland from there (Valiela et al. 1978; Anderson and Treshow 1980; Niering and Warren 1980). More stressful conditions in the inland area of the low marsh, such as nitrogen limitation (Valiela and Teal 1974; Gallagher 1975), high salinity (Anderson and Treshow 1980), and low soil oxygen levels (Howes et al. 1981), may cause the height difference (see Chapter 4, Case Study 4.A, Factors Controlling the Growth Form of *Spartina alterniflora*).

Salt marshes provide a striking example of plant species zonation in response to environmental variation, with different species occurring at different marsh elevations. Each species' habitat can be explained by its tolerance to salinity levels, tidal regime, soil oxygen availability, sulfur levels, or other factors (Partridge and Wilson 1987). In many eastern U.S. and Gulf coast salt marshes, a zone of *Spartina patens* (salt marsh hay) is located inland from the zone of both forms of *S. alterniflora* (Bertness and Ellison 1987; Gordon and Cranford 1994). *S. patens* may dominate in the better drained and less saline areas of salt marshes because it outcompetes *S. alterniflora* in those sites (Bertness and Ellison 1987; Bertness 1991a, b). Although east coast salt marshes of the U.S. appear to be monospecific within each of these zones, other salt marsh species are present in smaller numbers, such as *Juncus gerardii* (rush), *Distichlis spicata* (spike grass), and *Salicornia europaea* (glasswort; Bertness and Ellison 1987).

On the Pacific coast of the U.S. and Canada, salt marshes are less extensive than in the east, mostly because the geophysical conditions are not suitable for salt marsh formation. Crustal rise has resulted in shoreline emergence and a coastline with cliffs and few wide flat river deltas and estuaries. The majority of Pacific coast salt marshes that did exist have been filled for development (over 90% in some areas; Dahl and Johnson 1991; Chambers et al. 1994). Salt marshes still exist in estuaries or protected bays like Tijuana Estuary near San Diego (Zedler 1977), in northern San Francisco Bay (Mahall and Park 1976), Tomales Bay north of San Francisco (Chambers et al. 1994), Nehalem Bay in northern Oregon (Eilers 1979), Puget Sound in Washington (Burg et al. 1980), at the head of fjords and on the Queen Charlotte Islands in British Columbia (Glooschenko et al. 1993), and in Cook Inlet near Anchorage, Alaska (Vince and Snow 1984).

The plant communities of western salt marshes tend to be more diverse than Atlantic coast and Gulf of Mexico marshes. Like Atlantic salt marshes, many west coast salt marshes are dominated by grasses. For example, *Spartina foliosa* dominates some southern California marshes (Zedler 1977) as well as marshes near San Francisco (Mahall and Park 1976). Other northern California marshes are dominated by *Distichlis spicata* (Chambers et al. 1994), while *Salicornia virginica* (glasswort) is a dominant species in marshes of both northern and southern California (Callaway et al. 1990; Zedler 1993; Chambers et al. 1994). In Oregon, Washington, and British Columbia, the sedge, *Carex lyngbyei*, dominates salt marshes (Eilers 1979; Burg et al. 1980; Glooschenko et al. 1993). Alaskan salt marshes are dominated by the grass, *Puccinellia phryganodes*, and by various species of *Carex* (Jefferies 1977; Vince and Snow 1984). Diversity tends to be highest in better drained and less saline locations (MacDonald and Barbour 1974; Vince and Snow 1984; Chambers et al. 1994).

In western and northern Europe, salt marshes are found along the Atlantic coasts of Spain, Portugal, France, and Ireland, and along the North Sea and the Baltic Sea. In southern Europe, salt marshes are located within the watershed of the Mediterranean Sea and in the Rhone River delta (the Camargue; Chapman 1974). Mediterranean salt marshes also fringe northern Africa along the Tunisian, Moroccan, and Algerian coasts (Britton and Crivelli 1993). The seaward portions of European salt marshes are often tidal mudflats, with sparse vegetation. The equivalent area in eastern U.S. salt marshes is heavily vegetated and dominated by *Spartina alterniflora.* The difference is due to higher tidal fluctuations in many European salt marshes (up to 15 m). While eastern North American salt marshes are flooded twice daily, many European marshes are only partially flooded, with their highest areas flooded only during spring tides. The lowest areas of the marsh tend to be dominated by *Spartina maritima* in Portugal, *Salicornia europaea* and *Spartina anglica* in France and the United Kingdom, and *Salicornia dolichostachya* in the Netherlands (Lefeuvre and Dame 1994).

b. Tidal Freshwater Marshes

Tidal freshwater wetlands are influenced by the daily flux of tides, yet they have a salinity of less than 0.5 ppt. They are usually located in upstream reaches of rivers that drain into estuaries or oceans. Their position within the landscape places them at the interface between the upstream sources of fresh water and the downstream sources of tides. They occur worldwide, wherever these conditions are met. Tidal freshwater wetlands cover an estimated 632,000 ha in the U.S., with the majority along the Gulf of Mexico (468,000 ha), primarily in Louisiana (Mitsch and Gosselink 2000). Along the Atlantic coast, there are about 164,000 ha with over half (89,000 ha) in New Jersey, and most of the remaining in the Chesapeake Bay watershed (Odum et al. 1984).

The organisms that inhabit tidal freshwater wetlands originate in upstream freshwater or in downstream brackish areas. Because of the heterogeneity of habitat conditions, tidal freshwater wetlands harbor diverse communities of plants and animals. Since salinity and sulfur stresses are not as profound, macrophyte diversity is higher in tidal freshwater systems than in salt marshes. Tidal freshwater communities tend to have several plant forms such as shrubs, floating plants, grasses, and forbs, rather than the monotypic stands of grasses typical of salt marshes (Whigham et al. 1978; Simpson et al. 1983a; Odum et al. 1984). Many of the plants of tidal freshwater marshes are also found in inland marshes.

Tidal freshwater marshes show distinct vegetation patterns according to moisture levels (Odum et al. 1984; Leck and Simpson 1994). For example, along the Delaware River in New Jersey, *Acnida cannabina* (salt marsh water hemp) and *Ambrosia trifida* (great ragweed) grow along banks and levees. *Polygonum punctatum* (water smartweed) and *Bidens laevis* (larger bur marigold) are common along stream channels. *B. laevis* also grows on the high marsh with *Impatiens capensis* (spotted touch-me-not), *Peltandra virginica* (arrow arum), *Phalaris arundinacea* (reed canary grass), *Sium suave* (water parsnip), and the parasitic vine, *Cuscuta gronovii* (common dodder). *Nuphar advena* (spatterdock; Figure 2.3) and *Acorus calamus* (sweetflag) grow in the tidal channel and adjacent banks. *Pilea pumila* (clearweed) grows in elevated sites, *Sagittaria latifolia* (arrowhead; Figure 2.4) is scattered in all areas except the stream channel, and the vine, *Polygonum arifolium* (halberd-leaved tearthumb), occurs along the entire moisture gradient (Leck and Simpson 1994). In other Chesapeake Bay area tidal freshwater marshes, tall emergents such as *Zizania aquatica* (wild rice) and various species of *Typha* (cattail) also grow, often in dense stands (Odum et al. 1984).

2. Inland Marshes

Inland freshwater marshes are a diverse group of wetlands that, in the U.S., range in size from quite small (<1 ha) to the size of the Everglades (currently 607,000 ha). They are found worldwide wherever hydrologic and geologic conditions allow for their formation. Many different kinds of freshwater marshes have been defined, and they are often named for the dominant vegetation. In this book we divide marshes into three broad categories based on their landscape position: lacustrine marshes, riverine marshes, and depressional marshes. Inland marshes that accumulate peat, commonly called fens and bogs, are discussed in Section III.C, Peatlands.

Several thousand plant species are adapted to freshwater marshes (Cook 1996; Reed 1997) and whether the marshes are lacustrine, riverine, or depressional, the plants are often the same. The plant communities of freshwater marshes tend to be diverse and highly productive (Herdendorf 1987; Keeley 1988; Kantrud et al. 1989; Galatowitsch and van der Valk

FIGURE 2.3
Nuphar advena (spatterdock) of the Nymphaeaceae (water lily family) has thick, spongy roots and produces both emergent and floating leaves. Its yellow blossoms float or are held above the water's surface on rigid stalks. *N. advena* grows in both tidal and inland freshwater marshes in the eastern and midwestern U.S. (Photo by H. Crowell.)

FIGURE 2.4
Sagittaria latifolia (arrowhead) of the *Alismataceae* (water plantain family) develops floating leaves when it is immature or when the water level rises (as seen here). It is often seen with fully emergent leaves. *S. latifolia* grows in both tidal and depressional freshwater wetlands. (Photo by T. Rice.)

1994). The structure of plant communities varies with climate, substrate type, flooding regime, water depth, and nutrient availability. Some of the most common emergent plants of freshwater marshes are monocots, many in three major families: Poaceae (grass), Cyperaceae (sedge), and Juncaceae (rush). Other common emergent families are Typhaceae (cattail), Sparganiaceae (bur reed), Alismataceae (water plantain), Butomaceae (flowering rush), Araceae (arum), Pontederiaceae (pickerelweed), Iridaceae (iris), Polygonaceae (smartweed), Lythraceae (loosestrife), Apiaceae (=Umbelliferae; parsley), and Lamiaceae (=Labiatae; mint). Commonly encountered submerged species are in the Najadaceae (naiads), Potamogetonaceae (pondweed), Zannichelliaceae (horned pondweed), Hydrocharitaceae (frogbit), Ceratophyllaceae (hornwort), Ranunculaceae (buttercup), and Haloragaceae (water milfoil). Floating species include those in the Lemnaceae (duckweed). Common floating-leaved plants are in the Nymphaeaceae family (water lily; Figure 2.5). Often, a zone of shrubs surrounds depressional wetlands. Frequently found shrub genera include *Salix* (willow), *Spiraea* (meadow sweet), *Rosa* (rose), *Cephalanthus* (buttonbush), *Alnus* (alder), and *Cornus* (dogwood).

FIGURE 2.5
Nymphaea odorata (fragrant water lily) of the Nymphaeaceae (water lily family) is a floating-leaved plant that grows in freshwater marshes and lake edges throughout the eastern U.S. and southeastern Canada. (Photo by H. Crowell.)

a. Lacustrine Marshes

As defined by Cowardin and others (1979) the lacustrine system is divided into the limnetic, or deep water habitat, and the littoral, or edge habitat. Lacustrine wetlands include littoral aquatic beds, dominated by submerged and floating-leaved species, and emergent marshes slightly upland (Figure 2.6). A lake's shape dictates whether lacustrine wetlands can exist along its fringes. A steep-sided V- or U-shaped lake, generally formed by tectonic forces, has less water in contact with sediments, a more abrupt drop from edge to deep water, and generally supports little, if any, littoral vegetation. Lacustrine wetlands are more likely to occur along shallow lake basins, often formed by glaciation (Wetzel 1983a). The size and depth of littoral wetlands shift with changes in water level due to precipitation or changes in drainage or runoff.

Lacustrine wetlands are located around the world, along the edges of both small and large lakes. Most of the world's lakes are small with a high ratio of lacustrine marsh area to open water (Wetzel and Hough 1973). Along large lakes with tides, such as the Laurentian Great Lakes of the U.S. and Canada, wetlands occur in coastal lagoons behind barrier beaches, in tributary mouths, and as managed marshes protected by dikes (Herdendorf 1987; Glooschenko et al. 1993). Lacustrine wetlands along the Great Lakes tend to fall into one of three categories, according to depth: *wet meadow, marsh,* or *aquatic.* The plant communities of these three categories are dominated, respectively, by sedges and grasses, emergents such as *Typha,* and submerged species (Glooschenko et al. 1993).

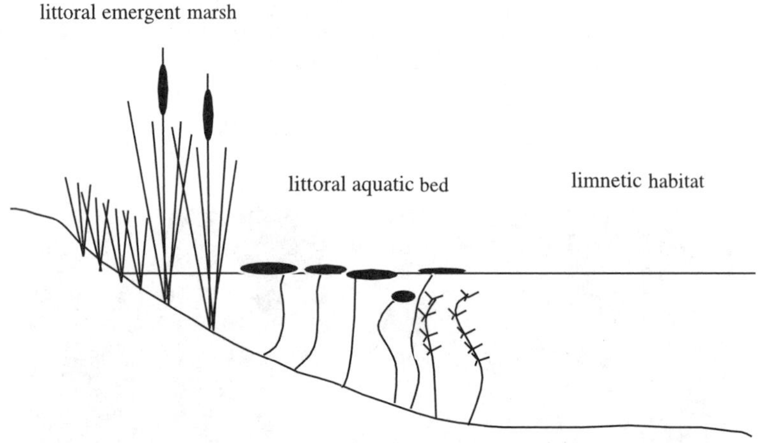

FIGURE 2.6
Lacustrine marsh with limnetic zone dominated by phytoplankton, a littoral aquatic bed dominated by submerged and floating-leaved plants, and a littoral emergent marsh.

b. Riverine Marshes

Riverine marshes form along rivers and streams, in the lowlands behind river levees, or in old oxbows of rivers (Figure 2.7). While many riverine wetlands are forested, herbaceous wetlands can be found at the edges of forests, or in newly opened areas such as beaver-formed wetlands (van der Valk and Bliss 1971; Johnston and Naiman 1990; Johnston 1994). The most extensive riverine marshes in the U.S. are found in the Mississippi River floodplain. Marshes that fringe streams or are flooded by river water are subjected to flowing water which carries a higher mass input of sediments and nutrients and allows increased export of waste products. Such wetlands often have higher plant productivity than still-water wetlands (Brinson et al. 1981; Lugo et al. 1988).

c. Depressional Marshes

Depressional wetlands are lowlands, or basins, that are either hydrologically connected to other wetlands or bodies of water, or hydrologically isolated. Depressions include former lake basins, shallow peat-filled valleys between existing lakes, and glacially formed basins (Kushlan 1990; Galatowitsch and van der Valk 1994). Depressional wetlands are found worldwide, at all latitudes, and may be forested wetlands, marshes, or peatlands (forested depressional wetlands and peatlands are discussed below). Examples of depressional marshes within the U.S. and Canada include prairie potholes, playas, and vernal pools.

FIGURE 2.7
Riverine marsh along the upper Mississippi River, Wisconsin. (Photo by H. Crowell.)

One extensive region of depressional wetlands is found in the prairie pothole region of Iowa, Minnesota, North and South Dakota in the U.S., and Alberta, Saskatchewan, and Manitoba in Canada (see Chapter 9, Case Study 9.C, Figure 9.C.1). This area encompasses over 70 million ha of land (Kantrud et al. 1989) with millions of prairie potholes scattered throughout (Shay and Shay 1986). Depressions within the prairie landscape were formed during the last glaciation. As the glaciers retreated, basins were left behind where caves and tunnels in the ice sheet had been. The basins range in size from several meters in diameter to lakes of several hundred hectares (Glooschenko et al. 1993). In southern Minnesota and northern Iowa the basins are shallow depressions linked by drainage ways. In other parts of the prairie pothole region, there are fewer surface links between the potholes (Galatowitsch and van der Valk 1994).

Five different pothole habitat types have been identified as a function of water depth: *wet prairie, sedge meadow, shallow marsh, deep marsh,* and *permanent open water.* Some potholes are only shallow depressions and may support only wet prairie and sedge meadow, while others are permanent ponds and lakes. Because the physical aspects of the habitat are so diverse, potholes support a wide array of wetland vegetation. In the southern pothole region of Iowa and southern Minnesota, nearly 350 plant species can be found and up to one third of those may be found within a single basin (Galatowitsch and van der Valk 1994).

Playas are ephemeral freshwater ponds that vary in area from a few square meters to hundreds of hectares and in depth from a few centimeters to a few meters (MacKay et al. 1992). Over 25,000 playas are scattered throughout an 8.2 million-ha region on the high plains of New Mexico and northern Texas. Some playas may have been formed by prairie wind erosion 10,000 to 15,000 years ago (Bolen et al. 1989). This theory is supported by the presence of large lee-side dunes adjacent to some playa basins. The dunes' volume approximates the volume of material removed from the playas. Not all playas have dunes, and some may have been formed where geologic joints provided paths of weakness for surface drainage and water accumulation (Zartman and Fish 1992).

Playas support a high diversity of wetland plants. Many are the same wetland emergents found in prairie potholes, such as species of *Typha, Scirpus* (bulrush), and *Polygonum* (smartweed). Often, *Potamogeton* species (pondweed) dominate the open water areas while *Echinochloa crus-galli* (barnyard grass) and *Leptochloa filiformis* (red sprangletop) grow in drier zones, called wet meadows. During dry years the vegetation is more characteristic of the surrounding grasslands. Many playas are cultivated or modified for other agricultural purposes, such as catchments for feedyard runoff, disposal areas, irrigation pits, and sewage treatment facilities (Bolen et al. 1989; Pezzolesi et al. 1998).

California's vernal pools are ephemeral wetlands, flooded during winter and spring and dry throughout the summer. The Mediterranean-type climate is unpredictable and in some years the pools do not fill. Vernal pools are what remains of the inland sea that once covered the Central Valley of California. As California's coastal mountains began to rise about 10,000 years ago, the inland sea receded and became several freshwater lakes. These, in turn, were fragmented into vernal pools that are generally less than one half hectare in size. Only 3 to 7% of this habitat remains, so many of the endemic plant species of vernal pools are threatened or endangered (see Chapter 1, Table 1.3). Some genera such as *Pogogyne* and *Orcuttia* are found almost exclusively in vernal pools (Baskin 1994).

B. Forested Wetlands

Forested wetlands are dominated by woody vegetation of all sizes. Forested wetlands can include trees over 50 m tall, dwarf trees 1 m in height in areas of environmental stress, or shrubs (Lugo 1990). They are found from boreal regions to the tropics, along coastlines, and in alpine regions, and they cover an estimated 250 to 330 million ha worldwide (Lugo et al. 1988; Lugo 1990). The U.S. (including Alaska) has more than 26 million ha of forested wetlands (Dahl and Johnson 1991; Hall et al. 1994). Like marshes, forested wetlands can be further categorized by hydrology and dominant plant species. We have divided this section into two major types: coastal and inland forested wetlands.

1. Coastal Forested Wetlands: Mangrove Swamps

Mangrove swamps are the only forested wetlands found along coastlines. The term *mangrove* refers to both the forest type and the trees that grow there. Mangrove is the common name of some 50 to 79 plant species, the most common of which are in the genera *Avicennia, Bruguiera, Ceriops, Laguncularia, Lumnitzera, Kandelia, Rhizophora,* and *Sonneratia.* Different numbers of mangrove species are sometimes reported because the term *mangrove* is variably interpreted. All mangrove species share an ability to survive in shallow and fluctuating salt water. Their primary physiological adaptations to the stress of salinity include the ability to exclude or excrete salt from their tissues and sprout aerial roots that aid in gas exchange (Tomlinson 1986). The highest diversity of mangrove species is found in Asia along Indian and Pacific Ocean coastlines. In the western hemisphere, there are eight species, and only three of these are common: *Rhizophora mangle* (red mangrove; Figure 2.8), *Avicennia germinans* (black mangrove; Figure 2.9), and *Laguncularia racemosa* (white mangrove; Lugo and Snedaker 1974; Tomlinson 1986; Rutzler and Feller 1996). Other than mangrove trees, few vascular species grow in western hemisphere mangrove swamps. The lack of understory may be due to the high number of natural stresses in the mangrove environment: salinity, high sulfide concentrations in the soil, root zone anoxia, low light due to shading, and crab predation of seeds and propagules (Janzen 1985; Corlett 1986; Lugo 1986; Snedaker and Lahmann 1988; Alongi 1998).

Mangroves are restricted to areas where the mean water temperature is above 23 to 24°C (Tomlinson 1986; Rutzler and Feller, 1996; Figure 2.10) and they cover 17 to 24 million ha worldwide (Field 1995; Ramsar Convention Bureau 2000). Mangroves occupy the same niche in the tropics that salt marshes occupy to the north and south. Outside of the tropics, frost limits their establishment and causes periodic diebacks (Kangas and Lugo 1990).

FIGURE 2.8
Rhizophora mangle (red mangrove) growing at water's edge with prop roots extending into the sediments at the Everglades National Park, Florida. (Photo by H. Crowell.)

FIGURE 2.9
Avicennia germinans (black mangrove) with aerial roots called pneumatophores growing above the water's surface. Pneumatophores aid in root gas exchange (Everglades National Park, Florida). (Photo by H. Crowell.)

Mangrove forests are among the most threatened of wetland types as they are harvested for wood products, used as waste and garbage dumps, and cleared for tourism and other types of development. In many countries few, if any, protective measures for mangroves are in place (Farnsworth and Ellison 1997; Li and Lee 1997; Walters 1997).

In many mangrove wetlands, hurricanes play a major role in the size and structure of the community (Pool et al. 1977). The mangrove forests of Puerto Rico, the Bahamas, Cuba,

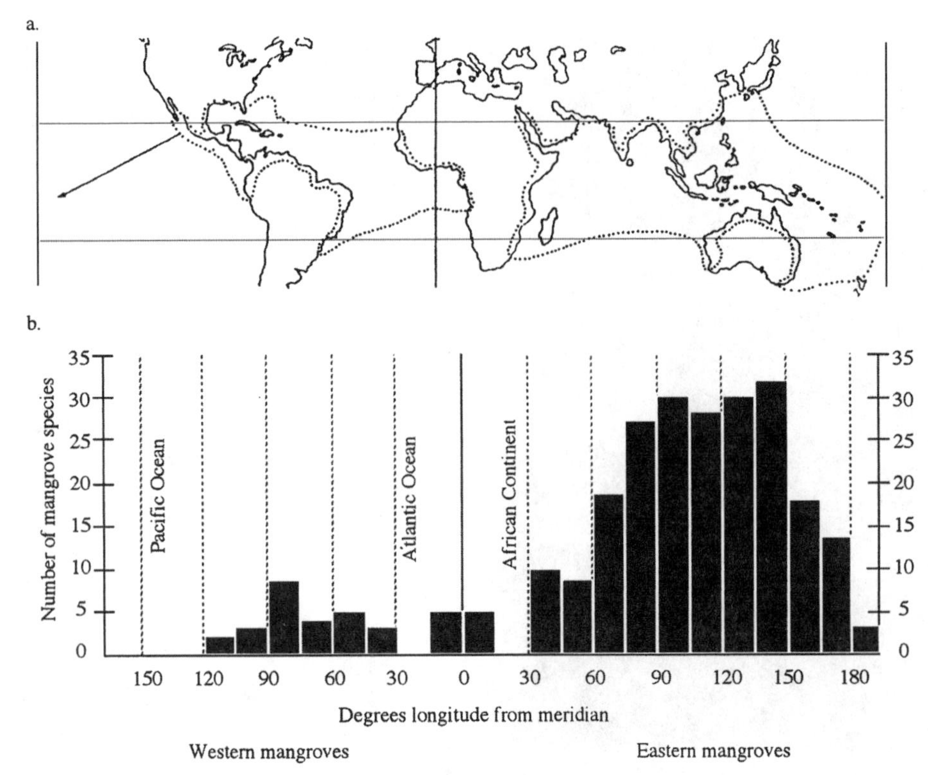

FIGURE 2.10
(a) The worldwide distribution of mangroves with the approximate northern and southern limits for all species. The eastern and western groups do not overlap except for a possible extension (arrow) in the western Pacific (*Rhizophora samoensis*). (b) Histogram of the approximate species number per 15° of longitude, showing the species richness of the eastern group. (From Tomlinson, P.B. 1986. *The Botany of Mangroves*. London. Cambridge University Press. Map reprinted and histogram redrawn with permission.)

and South Florida are frequently hit by destructive hurricanes. In other regions such as parts of Central America and northern South America, mangroves tend to be much larger trees, probably due to the absence of hurricanes (Odum and McIvor 1990; Lugo 1997). The most extensive mangrove swamps with the largest trees occur where there is significant tidal amplitude, high rainfall, heavy terrestrial runoff, and no (or rare) hurricanes, such as on the west coast of Panama (Odum and McIvor 1990), the east coast of Costa Rica (Pool et al. 1977), the River Niger of Western Africa (Denny 1993), the coastline of French Guiana (Fromard et al. 1998), or the western Yucatan Peninsula (Figure 2.11). In the absence of ecological constraints such as high salinity or low annual rainfall, mangrove forests near the equator tend to develop significantly greater biomass than in the northern or southern extent of their range (Fromard et al. 1998).

The mangrove swamps of the U.S. are in Florida, Texas, Hawaii, the U.S. Virgin Islands, and Puerto Rico (Lugo 1997). In Florida and Central America, the three common mangrove species occupy distinct zones (Figure 2.12). In the area closest to the sea is *Rhizophora mangle*. Just inland from the zone of *R. mangle* is *Avicennia germinans*, and immediately to the interior is *Laguncularia racemosa*. *Conocarpus erectus* (buttonwood or grey mangrove) is sometimes found between the mangroves and upland ecosystems, or among the *L. racemosa* trees. Each species occupies a different zone because of different adaptations. *R. mangle* grows a web of stilt-like prop roots that support the tree in standing water and protect

FIGURE 2.11
Rhizophora mangle (red mangrove) growing on the west side of the Yucatan Peninsula, reaching greater than average height due in part to a lack of hurricanes (Celustun, Mexico). (Photo by H. Crowell.)

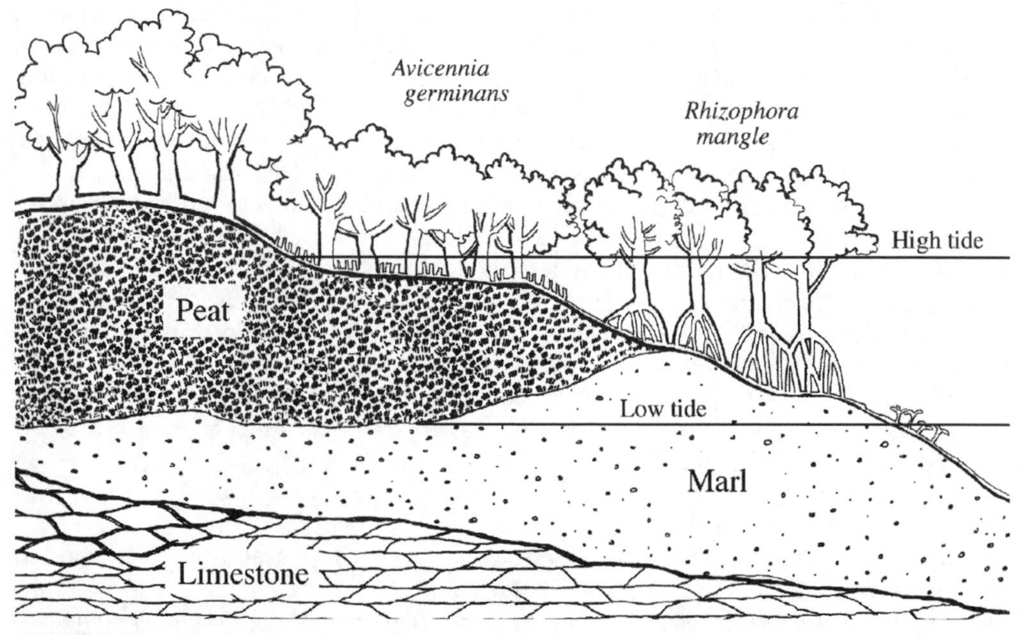

FIGURE 2.12
The zonation pattern of the common mangrove species of South Florida. *Rhizophora mangle* (red mangrove) grows closest to the sea. *Avicennia germinans* (black mangrove) grows inland from *R. mangle*, and *Laguncularia racemosa* (white mangrove) as well as *Conocarpus erectus* (buttonwood or grey mangrove; sometimes not included in the zonation pattern) inhabit the area inland from the zone of *A. germinans.* (Drawing by B. Zalokar.)

it from tides. *A. germinans* sends up hundreds of pencil-shaped roots called pneumatophores that aid in gas exchange (Tomlinson 1986). *L. racemosa* outcompetes the first two in less saline and better drained locations. On the western side of the continent, in Baja California, Mexico, dwarf mangrove forests are found in highly saline environments, but with the same species and zonation pattern as in Florida (Toledo et al. 1995).

Zonation may vary in some mangrove swamps due to local differences in substrate type, salinity, wave energy, and the effects of rising sea level (Odum and McIvor 1990). For example, *A. germinans* has the greatest tolerance for salt of the three dominant western hemisphere mangrove species. It tends to dominate in mangrove swamps where salt accumulates due to infrequent tidal flushing and little freshwater input (Tomlinson 1986; Brown 1990; Rutzler and Feller 1996). In the eastern hemisphere, more mangrove species co-exist, and therefore a greater variety of associations are formed. For example, in Australia there are 11 common mangrove species that are found in 30 combinations (Finlayson and Von Oertzen 1993).

2. Inland Forested Wetlands

Inland forested wetlands are referred to as either *basin* wetlands or *riverine* wetlands, according to their location in the landscape and their sources of water (Lugo et al. 1988). Riverine forests are heavily influenced by flooding and sediment transport in the adjacent river or stream. Due to these energy influxes, freshwater riverine forests tend to have higher productivity and greater species richness than basin forested wetlands. Forested wetlands along a single river can change in structure and productivity as the river changes from headwaters to mouth. Riverine wetlands are highly dynamic ecosystems, since a single large flood can change the floodplain topography sufficiently to alter or eliminate the forest habitat (Brinson 1990).

The terms used here, *floodplain, bottomland, riparian,* and *riverine,* all refer to land along a river or stream. *Bottomland* and *floodplain* both refer to the low areas adjacent to a river or stream. The term *bottomland* is most often used to refer to forested wetlands in the southern U.S. *Riparian* areas include the lowlands as well as the levees adjacent to rivers and this term is used for any such area worldwide. The literature dealing with western rivers usually uses the term *riparian.* The term *riverine* can refer to the U.S. FWS classification system of wetlands (Cowardin et al. 1979), which restricts *riverine* wetlands to those found in the stream channel and the immediate bank or levee. Here, *riverine* should be interpreted in the sense given by Lugo and others (1988) to describe wetlands that are influenced by rivers or streams (i.e., floodplain and riparian areas as well as wetlands within the stream channel).

With an emphasis on North America, we discuss three major riverine forested wetland types here: bottomland hardwoods of the southern U.S., northeastern floodplains, and western riparian zones. Cypress wetlands occur in both riverine and basin habitats, and both are described below. Some inland basin forested wetlands are peatlands, which are discussed in the following section.

a. Southern Bottomland Hardwood

Many wide floodplains of the southern U.S. have a variable topography characterized by depressions and ridges behind the levees that border the stream channel. The topography is determined by past meanderings of the current stream across the floodplain. The first area behind the levee, sometimes called the backswamp or backwater, is prone to frequent and longer periods of flooding than those behind subsequent ridges. Low elevation wet sites lie upland from the backswamp and may only be flooded during the winter or spring. Behind these low sites, slightly higher elevation forests tend to be only periodically flooded and therefore better drained (Hodges 1997).

Southern bottomland hardwood forests are found in both major and minor watersheds from the Atlantic coast to eastern Texas and Oklahoma and as far north as the Ohio and Wabash Rivers. Some of the largest floodplains in which bottomland forests grow are along the lower Mississippi and its tributaries, the Arkansas, Red, Ouachita, Yazoo, and St. Francis Rivers. Other rivers that flow toward the Gulf of Mexico as well as those that drain the southern and middle Atlantic coast also support bottomland hardwood forested wetlands (Brinson 1990).

The structure of bottomland forest communities is determined, to a great extent, by the hydroperiod. The first depression behind the levee may support *Taxodium distichum* (bald cypress) and *Nyssa aquatica* (water tupelo) forests, both adapted to prolonged flooding and frequently found in areas flooded 10 to 12 months per year (Conner and Day 1976; Visser and Sasser 1995; Kellison and Young 1997). Depressions farther inland and riverine forests with less frequent and shorter inundation are dominated by *Quercus* (oak), *Acer* (maple), and *Salix* (willow) species, as well as *Platanus occidentalis* (sycamore), *Liquidambar styraciflua* (sweetgum), *Nyssa sylvatica* (black tupelo or swamp black gum), and *Liriodendron tulipifera* (yellow poplar; Conner and Day 1976; Brinson 1990; Kellison and Young 1997; Hodges 1997). A number of shrubs coexist with the trees in bottomland hardwood swamps, such as *Cornus* (dogwood) and *Ilex* (holly) species as well as *Sambucus canadensis* (elderberry) and *Cephalanthus occidentalis* (buttonbush; Conner and Day 1976; Conner et al. 1981). Species composition and tree size reflect the fact that bottomland forests have been and continue to be extensively harvested (Brinson 1990; Kellison and Young 1997).

b. Northeastern Floodplain

The northeastern floodplain forests lie north of the range of *Taxodium distichum* (bald cypress) in Maryland and Virginia and south of the peatlands of northern Michigan, Wisconsin, upper New England, and Canada. They extend east from the upper Mississippi River valley to the Atlantic (Figure 2.13). Some of the largest watersheds are the Ohio, Susquehanna, Shenandoah, Delaware, and upper Mississippi Rivers (Brinson 1990). With the exception of streams in the Appalachian mountains, northeastern floodplains tend to

FIGURE 2.13
A northeastern floodplain forest in a backwater of the Kalamazoo River in western Michigan. (Photo by H. Crowell.)

be wide, with several ridges and low lying areas, and they are subject to periodic flooding and sediment deposition. The natural surroundings of northeastern floodplain forests are upland deciduous forests and they share many of the same species. Like riparian areas elsewhere, the northeastern forests have been heavily impacted by development, forestry, channelization of streams, and draining for agriculture. Almost all of the states of this region have lost over half of their original wetland area, with forested wetlands the most heavily impacted (Dahl and Johnson 1991).

Maples, particularly *Acer rubrum* (red maple), along with *Ulmus americana* (American elm), *Fagus grandifolia* (American beech), and *Platanus occidentalis,* are a few of the trees that are frequently found in eastern floodplain forests (Brinson 1990). In the western portion of this region, along the Wabash River in Indiana, the genera found in the lowest elevations are *Populus* (cottonwood) and *Salix* (willow). On slightly higher ground, *Acer rubrum, Aesculus glabra* (Ohio buckeye), *Cercis canadensis* (Eastern redbud), *F. grandifolia,* and *U. americana* are the dominant species (Lindsey et al. 1961). In Wisconsin, *Acer saccharinum* (silver maple) and *Fraxinus pennsylvanica* (green ash) dominate floodplain forests (Dunn and Stearns 1987). In Ohio, many forested floodplains are dominated by *Populus deltoides* (cottonwood), *A. saccharinum, A. negundo* (box elder), *P. occidentalis,* and *Salix* species. In the unglaciated southeastern portion of Ohio, *Betula nigra* (river birch), *A. negundo, A. rubrum,* and *A. saccharinum* form the most common floodplain association (Anderson 1982).

c. Western Riparian Zones

In the arid regions of the western U.S., riverine areas are often the only areas moist enough to provide suitable habitat for tree and shrub species. Because western riparian zones are oases of water, food, and cover, they tend to have a high concentration of wildlife, particularly birds (Knopf and Samson 1994). The habitat along a single river may shift from wet to mesic to xeric, and the plant communities change accordingly. Western riparian wetlands tend to be sparsely vegetated and have far fewer species than eastern forested wetlands. For example, forested wetlands near the South Platte River in Colorado have only two tree species, while a wetland forest at the same latitude in New Jersey has 24 (Brinson 1990).

Western riparian zones are located in the southwestern U.S. and the plains grasslands of the central states. Major rivers in the southwest that support riparian zones are the San Joaquin, Sacramento, Salt-Gila, and Rio Grande-Pecos. In the Plains States in central to western U.S., some of the major watersheds are the Missouri, Platte, upper Arkansas, and Canadian (Brinson 1990). The riparian zones of these rivers tend to be narrower than in the east because the streams have steeper gradients, which leads to more severe flooding of surrounding areas (Mitsch and Gosselink 2000). Less soil development, less deposition of silt, and less stream meandering have occurred in the west than in the east, so the floodplains tend to have few if any ridges and backswamps. Only about 1% of the west is riparian habitat and most riparian areas in the western U.S. (in southern California over 95%) have been destroyed by development, agriculture, grazing, logging, or manipulation of water resources, such as damming (Faber et al. 1989; Brinson 1990; Knopf and Samson 1994). Western riparian areas may be "the most modified land type in the western U.S." (Wilen and Tiner 1993).

Typical trees in central U.S. arid riparian zones are species of *Populus, Salix,* and *Fraxinus* (ash), as well as *Platanus wrightii* (sycamore) and *Acer negundo* (Keammerer et al. 1975; Faber et al. 1989). In the southwest, a number of riparian areas are dominated by *Prosopis* species (mesquite) and *Populus,* as well as the invasive *Elaeagnus angustifolia* (Russian olive) and *Tamarix* (saltcedar). The most frequently encountered species of saltcedar, *Tamarix*

FIGURE 2.14
Tamarix ramosissima (saltcedar), an invasive species of western riparian zones, growing along the Colorado River in Arizona. (Photo by K. Irick Moffett.)

ramosissima and *T. chinensis*, have displaced native riparian tree and shrub species throughout the western U.S. (Figure 2.14; Brinson 1990; Busch and Smith 1995).

d. Cypress Swamps

Forested wetlands dominated by *Taxodium* species (cypress) are found throughout the southeastern U.S. There are two types of cypress trees, commonly called bald cypress and pond cypress. There is some disagreement about whether or not these are separate species, called *Taxodium distichum* and *Taxodium ascendens*, respectively, or two varieties: *Taxodium distichum* var. *distichum* and *Taxodium distichum* var. *nutans* (Ewel 1990a). We refer to them as *T. distichum* and *T. ascendens*, after Reed (1997). The two trees are generally found in different habitats: bald cypress is found in flowing water swamps and pond cypress grows in stillwater swamps. However, bald and pond cypress do grow together in some areas. Bald cypress grows where water flows through a deeper area, as along a channel, and pond cypress may grow in the same depression, but in shallower areas, away from the main water flow (Ewel 1990b).

Cypress wetlands are found in Florida and in the Atlantic coastal zone as far north as Delaware. They also grow in the Gulf of Mexico coastal zone into Louisiana and Texas and north as far as southern Illinois (Ewel 1990a). The largest remaining cypress forest is in Big Cypress National Preserve, north of the Everglades in Florida. About one third of the 620,000-ha preserve is dominated by cypress (U.S. National Park Service 1997). Many cypress wetlands were harvested by early settlers and harvesting continued through the early 1900s. Cypress wood was considered desirable for building, boats, bridges, and docks because of its resistance to decay. It can be a very long-lived species and reports of harvests at the beginning of the 1900s indicated that many trees were 400 to 600 years old, with some as old as 915 years (Mattoon 1915, as reported by Visser and Sasser 1995). Today one can occasionally find a 1000-year-old cypress tree in protected areas (Figure 2.15).

Many of the backswamps closest to southeastern rivers are habitats for bald cypress, often in association with *Nyssa aquatica* (water tupelo). These riparian cypress stands are considered within the category of southern bottomland hardwoods (Brinson 1990; Ewel 1990a; Visser and Sasser 1995). Monospecific stands of bald cypress can be found in some

FIGURE 2.15
Taxodium distichum (bald cypress) estimated to be 1000 years old in the Francis Beidler Forest, South Carolina. (Photo by J. Cronk.)

southern floodplains and along the edges of lakes (Ewel 1990b), or in wetlands known as *cypress strands,* which are linear swamps with slowly flowing water (Ewel and Wickenheiser 1988). The water level of all cypress swamps fluctuates throughout the year. In Florida, for example, most cypress swamps are inundated from 5 to 9 months each year. Drawndown periods are vital to cypress swamps because cypress seeds will not germinate under standing water. Inundation also reduces competition from plants that grow more quickly under drier conditions (Mitsch and Ewel 1979; Ewel 1990a).

Pond cypress is often found in swamps with little or no water throughflow; these are called *cypress ponds, domes,* or *heads*. The term *dome* refers to the shape of these forests seen from a distance: the largest trees are in the center and the smaller ones are around the edges of the pond (Brown 1990). Cypress ponds often have standing water for more than 6 months per year and there is generally no streamflow into or out of them. They vary in size from about 1 to 10 ha (Ewel and Wickenheiser 1988). Pond cypress can also grow in poorly drained savannas where their growth is stunted. In this environment they are referred to as dwarf cypress (Ewel 1990a).

C. Peatlands

Peatlands are wetlands in which dead plant matter (peat) accumulates due to slow decomposition. While many wetlands accumulate organic litter, peatlands can be distinguished by a deep accumulation (generally >30 cm; Glaser 1987) and by the fact that they are found in areas with short growing seasons. Peat can be any decaying vegetative matter, but very often it is dominated by moss, usually one of the *Sphagnum* species (of which there are at least 185 worldwide; Figure 2.16; Crum 1992). Peat forms a variety of domed or raised

FIGURE 2.16
An example of a *Sphagnum* moss, *S. papillosum*. This moss usually grows in wet acidic conditions with some groundwater input (Crum 1992) and was found in exactly such conditions at Miner Lake in southwestern Michigan. (Photo by H. Crowell.)

shapes as it accumulates. Peat decomposes slowly because of cold temperatures and low levels of the oxygen needed by decomposers.

Peatlands have been classified into two main types based upon the source of water. *Fens* are fed by groundwater that carries minerals from the surrounding soil, and are sometimes called *minerotrophic* after their water source. The calcium concentration and the pH of fens tend to be relatively high. *Bogs* receive mostly rainwater (*ombrotrophic*) and tend to be much poorer in nutrients and minerals and have a lower pH (Bellamy 1968; Moore and Bellamy 1974; Wassen et al. 1990). Within the categories of bogs and fens, there is a continuum of water and substrate chemistry brought about by the different sources of water. Peatlands have been categorized according to the pH of the interstitial water (Figure 2.17). The more acidic (pH < 5) have been called *Sphagnum bogs, extreme poor fens,* or simply *bogs.* At higher pH values where calcium carbonate inputs buffer the acidic water, the vegetation is often dominated by sedges, and the term *fen* is used. More detailed categories of fens according to pH are intermediate fen (pH 5.2 to 6.4), transitional rich fen (pH 5.8 to

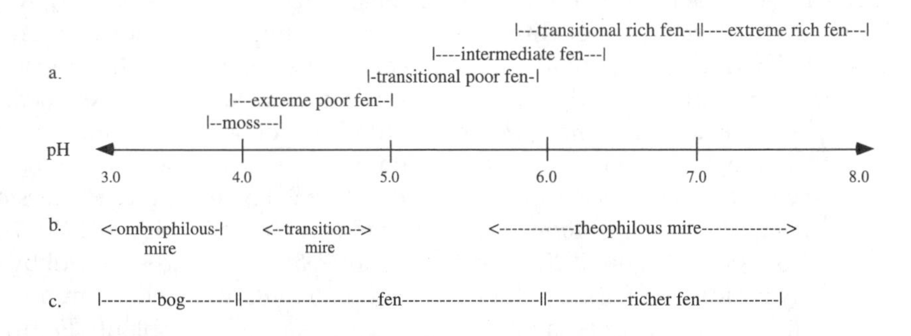

FIGURE 2.17
Three classifications of peatlands according to pH value (names and their pH ranges are from [a] Sjors 1950, [b] Bellamy 1968, and [c] Crum 1992).

FIGURE 2.18
Parnassia glauca (grass of Parnassus) is usually found in mineral-rich fens (Ebersole Center, Jackson Lake in southwestern Michigan). (Photo by H. Crowell.)

7.0) and extreme rich fen (pH 7.0 to 8.0; Sjors 1950). In many studies of peatlands, only one or two of the defining parameters (pH/alkalinity, hydrology, nutrient availability, plant community structure) are measured (Bridgham et al. 1996). The other parameters are usually inferred from plants present, or other indicators. In the cases where all components are measured, they are often contradictory and suggest that the system involved is both a traditional fen and a traditional bog. For this reason, Bridgham and others (1996) suggest that only the term *peatland* be used and that ambiguous terms such as rich or poor fen be dropped from usage.

Peatlands are found around the world wherever cold climates and high humidity coincide, mostly in the northern hemisphere (Mitsch and Gosselink 2000). In northern areas, such as in Canada, Russia, and Scandinavia, peatlands may be vast, extending for thousands of hectares. Canada has approximately one third of the world's peatlands, or 112 million ha (88% of Canada's wetlands are peatlands; Glooschenko et al. 1993; Rubec 1994). Over 200 million ha of peatlands are found in the eastern hemisphere, in northern Russia, Eastern Europe, Scandinavia, the United Kingdom, and Ireland (Mitsch et al. 1994). In the U.S. most peatlands are found in the northern states, especially Minnesota, Wisconsin, Michigan (Glaser 1987), and Maine (Damman and French 1987), but they can be found as far south as the Appalachian Mountains of Virginia, Maryland, and West Virginia, where high altitude excludes the warmer climate of the surrounding land. Many peatlands in the U.S. are isolated and relatively small, unlike the extensive peatlands to the north.

The plant habitat in peatlands ranges from calcareous standing water to the acidic interstitial spaces created by a *Sphagnum* mat. Hundreds of species are adapted to peatlands and most grow in a range of conditions. Some, however, can be indicative of certain conditions since they are rarely found outside of them. For example, *Parnassia glauca* (grass of *Parnassus;* Figure 2.18), *Tofieldia glutinosa* (sticky tofieldia, Figure 2.19), *Triglochin maritimum* (arrow grass; Figure 2.20), and the shrub, *Potentilla fruticosa* (shrubby cinquefoil), are usually indicative of areas of high calcium content (Crum 1992). A number of carnivorous plants in several genera including *Sarracenia* (pitcher plant; Figure 4.B.1), *Utricularia* (bladderwort; Figure 4.23), *Drosera* (sundew; Figure 4.22), and *Pinguicula* (butterwort), are usually found in areas of low pH. These generalizations should be taken with

FIGURE 2.19
Tofieldia glutinosa (sticky tofieldia) grows in mineral-rich fens (Ebersole Center, Jackson Lake in southwestern Michigan). (Photo by H. Crowell.)

FIGURE 2.20
Triglochin maritimum (arrow grass) is often found in mineral-rich peatlands, or fens (Ebersole Center, Jackson Lake in southwestern Michigan). (Photo by H. Crowell.)

caution, since plants are not always good indicators of peatland type (Glaser 1992; Bridgham et al. 1996). For example, one third of the species that grow in the bogs of the southern maritime region of Canada are restricted to fens or mineral uplands inland. The plants' distribution may be due more to a physiological adaptation to the maritime climate than to the calcium and pH conditions normally thought to influence peatland plant distribution (Glaser 1992).

A range of water availability, calcium content, and pH can occur within the same basin, particularly in areas where decaying vegetation is accumulating and raising the substrate above the influence of incoming groundwater. In many bogs (Figure 2.21), open water, where *Nymphaea odorata* (white water lily) and submerged species such as *Utricularia macrorhiza* (bladderwort) grow, is rimmed by a *Carex* (sedge) or *Scirpus* (bulrush) mat that gives way to a shrubbed area of less moisture with, typically, shrubs of the Ericaceae (heath) family. These shrubs include *Vaccinium corymbosum* (highbush blueberry; Figure 2.22), which often grows upslope from the water, as well as several other *Vaccinium* species, such as *V. macrocarpon* (cranberry) which grows in or near the open water. Other common bog shrubs also of the Ericaceae are *Chamaedaphne calyculata* (leatherleaf; Figure 2.23) and *Andromeda glaucophylla* (bog rosemary; Figure 2.24). In many bogs as well as fens, shrubs are dominant features of the plant community; in others they occur in narrow zones around the edges of open water or a zone of herbaceous plants. Upland from the shrubs there are often trees. Two common peatland species are *Larix laricina* (tamarack; Figure 2.25) and *Picea mariana* (black spruce). As the water level rises and more of the bog fills, the area can become increasingly forested. Such forests are *basin* forests (*sensu* Lugo et al. 1988), which tend to have low species diversity and decreasing stand height with increasing latitude (Brown 1990).

FIGURE 2.21
A bog habitat in western Michigan with open water to the far right of the photo where *Nymphaea odorata* (white water lily) grows, with a zone of *Scirpus acutus* (hardstem bulrush) at the edge of the open water. Slightly upland from the bulrush zone is a zone of shrubs followed by a tree zone dominated by *Larix laricina* (tamarack). (Photo by H. Crowell.)

FIGURE 2.22
Vaccinium corymbosum (highbush blueberry) of the Ericaceae (heath family) frequently grows around the fringe of peatlands (Miner Lake in southwestern Michigan). (Photo by H. Crowell.)

FIGURE 2.23
Chamaedaphne calyculata (leatherleaf) of the Ericaceae is often a dominant shrub in bogs of the Great Lakes Region (Miner Lake in southwestern Michigan). (Photo by H. Crowell.)

FIGURE 2.24
Andromeda glaucophylla (bog rosemary) usually grows in the wetter portions of floating mats, closer to the water than other Ericaceae (Miner Lake in southwestern Michigan). (Photo by H. Crowell.)

FIGURE 2.25
Larix laricina (tamarack) is a deciduous gymnosperm whose needles turn yellow and drop each fall and grow back in dense clusters in the spring. *L. laricina* often grows on hummocks between the floating peat mat and more solid ground (Ebersole Center, Jackson Lake in southwestern Michigan). (Photo by H. Crowell.)

Summary

Wetlands are defined using three major components: hydrology, soils, and plants. The hydrology of wetlands varies daily in coastal wetlands, and seasonally in others. There must be sufficient water to form hydric soils, which are low in oxygen. Wetland plants are adapted to hydric soils or they are unrooted and live in the water column. The functions of wetlands can be categorized under the headings hydrology, biogeochemistry, and habitat.

We categorize wetlands into marshes, which are dominated by herbaceous vegetation, forested wetlands, and northern peatlands. Marshes are further divided into coastal marshes, tidal freshwater marshes, and inland marshes. Coastal marshes, which are subject to tidal influences and salt water, are often dominated by grasses or other monocots. Tidal freshwater marshes are influenced by tides, but they also have a high influx of fresh water, and their plant communities are more diverse than those of coastal marshes. Inland marshes may be found at the edge of lakes (lacustrine) or rivers (riverine) or they may be depressional wetlands such as prairie potholes, vernal pools, and playas. They tend to have very productive and diverse plant communities.

Forested wetlands are found along rivers (in the U.S., these are called southern bottomland hardwoods, northeastern floodplains, and western riparian zones) as well as in depressional areas (often cypress swamps or forested peatlands). Mangrove forests are found along tropical coasts. The diversity of mangrove species is greater in the eastern hemisphere. In the U.S., mangrove forests are dominated by three species, *Rhizophora mangle, Avicennia germinans,* and *Lagunularia racemosa,* that are usually found in distinct zones.

Peatlands occur where organic matter accumulates and forms a peat substrate. The growing season is generally short. Peatlands have been categorized into fens and bogs. Fens tend to have groundwater inputs and high calcium concentrations while bogs are often dominated by *Sphagnum* moss and have a low pH (<5). Both woody and herbaceous plants are found in peatlands.

3

The Physical Environment of Wetland Plants

I. An Introduction to the Wetland Environment

Water is one of the primary factors that organizes the landscape, doing so through processes such as transport, erosion, leaching, solution, and evapotranspiration (Brown 1985). The hydrologic regime of a wetland is one of the key variables that determine the composition, distribution, and diversity of wetland plants. Hydrologic conditions affect species composition, successional trends, primary productivity, and organic matter accumulation (Gosselink and Turner 1978; Brinson et al. 1981; Howard-Williams 1985; van der Valk 1987). Factors related to the hydrologic regime that affect wetland plant communities include water depth (Spence 1982; Grace and Wetzel 1982, 1998), water chemistry (Ewel 1984; Pip 1984; Rey Benayas et al. 1990; Rey Benayas and Scheiner 1993), and flow rates (Westlake 1967; Lugo et al. 1988; Nilsson 1987; Carr et al. 1997). Hydrology also influences the plant community composition and primary productivity by influencing the availability of nutrients (Neill 1990), soil characteristics (Barko and Smart 1978, 1983), and the deposition of sediments (Barko and Smart 1979). The hydrologic regime can be thought of as a master variable with respect to all these factors since it not only determines the hydroperiod, but it is also instrumental in carrying nutrients and sediment (and so modifying soil type) into a wetland.

In this chapter, we focus on the ways in which hydrology controls plant community structure. We describe the hydrologic budget, with an emphasis on transpiration and its measurement. We also discuss how hydrologic forces affect species distribution, community composition, and primary productivity.

Following the section on hydrology, we discuss the characteristics of saturated soils that render them inhospitable to plants. Low oxygen levels stress plant roots, which require oxygen to maintain cellular respiration. High concentrations of toxic forms of metals accumulate, and nutrients may become less bioavailable. In saltwater ecosystems, these obstacles to growth and establishment are compounded by osmotic stresses. We discuss special conditions in nutrient-poor peatlands and the influences on substrate pH and nutrient availability. Finally, in the underwater environment, light and carbon dioxide may become limiting factors for submerged plants.

II. The Hydrology of Wetlands

Wetlands exist in geologic settings that favor the accumulation of water (Winter 1992). A wetland's hydrology is a major influence on vegetation composition, which in turn determines the value of the wetland to other organisms. Differences in wetland type, soil type, and vegetation composition are the result of the geology of an area, its topography and climate (Bedford 1996, 1999). Ultimately, the hydrologic budget and local geology determine the quantity and chemistry of water in a wetland. The distribution, abundance, and type of plants in a wetland are related to the timing and duration of flooding, the timing and duration of soil saturation, and soil characteristics.

A. Hydroperiod and the Hydrologic Budget

An understanding of hydrology provides a basis for understanding the ecology of wetland plants, particularly their association with flooded or saturated conditions. Plant establishment is influenced by a number of hydrologic processes including inflow rates, water depth, internal flow rates and patterns, the timing and duration of flooding, and groundwater exchanges. Changes in water level over time are referred to as the hydroperiod (Mitsch and Gosselink 2000). The hydroperiod is a result of the hydrologic budget, or the balance of a wetland's water inflows and outflows over time. The annual hydroperiod presents data on water level changes during a year, including flood depth and duration and the amount of soil saturation, but does not tell us explicitly about the topographic and climatological factors that cause the changes.

A hydrologic, or water, budget is the total of water flows into and out of a site. It is an important tool because it reveals the relative importance of each hydrologic process for a given wetland. Water budgets, along with information about the local soils and surficial geology, can provide an understanding of the hydrologic processes and water chemistry, help explain the diversity and distribution of species in the plant community, and provide insight into the changes that might result from hydrologic disturbance.

Water inflows are generally driven by climate and include precipitation, surface runoff, groundwater inflows, and, in coastal systems, tidal ebb and flow. Mass balance equations are often used to describe the flows of water into and out of a wetland (Huff and Young 1980), and are generally calculated to solve for volume such that:

$$\Delta V/\Delta t = \text{water inputs} - \text{water outputs} \quad (3.1)$$

or more specifically:

$$\Delta V/\Delta t = S_i + G_i + P_n - ET - S_o - G_o \pm T \quad (3.2)$$

where

$\Delta V/\Delta t$ = change in volume of water (storage) per unit time, t
S_i = surface inflow
G_i = groundwater inflow
P_n = direct precipitation
ET = evapotranspiration
S_o = surface outflow
G_o = groundwater outflow
T = tidal inflow (+) or outflow (–); not present in inland wetlands

TABLE 3.1
Examples of Water Budgets for a Variety of Wetland Types

Wetland Type and Location	Inflows			Outflows			Tides	($\Delta V/\Delta t$)
	S_i	G_i	P_n	ET	S_o	G_o		
Great Lakes coastal marsh, Ohio	576	131[a]	38	67	653		[a]	+25
Mangrove swamp, Florida			121	108	90	28	in = 1228 out = 1177	–54
Prairie pothole, North Dakota	40		37	64		18		–5
Okefenokee Swamp, Georgia	39[b]		131	93	73	4		0
Fen, North Wales	38[b]		102	49	100			–9
Green Swamp, central Florida			89–180	86–99	5–79	5–6		–10 to 0
Bog, Massachusetts			145	102	24[c]			+19
Pocosin swamp, North Carolina			117	67	49	1		0

Note: The units are cm yr^{-1}. Not all of the terms in the water budget equation apply in each type of wetland, and in some, not every part of the budget was measured (S_i = surface inflow, G_i = groundwater inflow, P_n = Precipitation, ET = evapotranspiration, S_o = surface outflow, G_o = groundwater outflow).

[a] Groundwater inflow is combined with tidal inflows and outflows.

[b] Surface inflow is combined with groundwater inflow.

[c] Surface outflow is combined with groundwater outflow.

Data compiled by Mitsch and Gosselink 2000.

A wetland's annual water budget may change from one year to the next because of climatic variability, which in turn may result in a change in the magnitude of different components of the budget. A comparison of water budgets for several North American wetlands is shown in Table 3.1. The terms in the water budget vary in importance depending on the type of wetland, and not all terms apply to all types of wetlands. In the examples in which the change in volume is small or zero, the water level at the end of the study period was close to the water level at the beginning of the study period (Mitsch and Gosselink 2000).

1. Transpiration and Evaporation

Transpiration (water that passes through vascular plants to the atmosphere) is an important parameter in wetland plant studies because it represents the interaction between a wetland's hydrologic regime and its vegetation. Transpiration is the only component of the water budget that is dependent entirely upon plants. Estimates of transpiration are often combined with evaporation (water that vaporizes directly from the water or soil); this measure is known as evapotranspiration (ET). When water supplies are not limiting, meteorological factors tend to control rates of ET. The rate of evapotranspiration is affected by solar radiation, wind speed and turbulence, available soil moisture, and relative humidity. Rates vary with the difference in vapor pressure at the water surface or leaf surface and the vapor pressure of the atmosphere. As the vapor pressure of the water or leaf surface increases relative to the atmosphere (due to solar energy or increases in temperature, for example), ET rates increase. When differences in vapor pressure decrease, for example when humidity increases or wind speeds decrease, ET rates decrease in response (Mitsch and Gosselink 2000).

On an ecosystem level, water outputs due to ET are largely controlled by vegetation (both the species present and their areal extent) and the supply of water (Lafleur 1990a, b; Gilman 1994). In many cases, ET is the largest loss term in the water balance equation (Hollis et al. 1993; Gilman 1994; Owen 1998). For example, Verhoeven and others (1988), working in mesotrophic and eutrophic fens, found ET rates of 482 mm yr^{-1}. This accounted for 60% of total annual precipitation.

Additionally, soil porosity may affect ET by limiting or facilitating the movement of water in the soil to roots or to the soil surface. Mann and Wetzel (1999) demonstrated this in a mesocosm study using *Juncus effusus* (soft rush). When grown in clay, *J. effusus* did not cause a decrease in soil water levels. However, in more porous sandy soils, where water movement in the soil is relatively quick, *J. effusus* caused a decline in the water level. Table 3.2 summarizes the results of some studies comparing the rates of ET from different vegetation stands.

TABLE 3.2
Mean Daily Summer ET Rates for Wetlands in Different Regions

Vegetation Type	Location	ET (mm/d)	Ref.
Reed swamp	Czechoslovakia	6.9	Smid 1975[a]
	Czechoslovakia	3.2	Priban and Ondok 1985[a]
Freshwater marsh	Florida	5.1	Dolan et al. 1984[a]
Low arctic bog	Canada	4.5	Roulet and Woo 1986[a]
Quaking fen	Netherlands	2.5	Koerselman and Beltman 1988[a]
Coastal marsh (wet)	Ontario, Canada	3.1	Lafleur 1990b
Coastal marsh (dry)	Ontario, Canada	2.6	Lafleur 1990b
Reed swamp	North Germany	~10	Herbst and Kappen 1999

[a] Data compiled in Lafleur 1990b.

2. Measuring Transpiration and Evaporation

Two general approaches to the quantification of ET are direct field measurements and estimates calculated using models of atmospheric conditions. Under certain conditions it is possible to isolate and measure transpiration rates directly (i.e., to measure transpiration separately from evaporation). For example, transpiration can be measured directly by taking readings of stomatal conductance of water vapor. Water lost through stomata is a function of atmospheric conditions but it is also influenced by stomata density (number per unit leaf area) and the diameter of the stomatal openings. Measurements are taken with leaf photosynthesis meters (infrared gas analyzers) that provide data on both inorganic carbon uptake and water loss through the stomata (e.g., Mann and Wetzel 1999). Results are generally reported in mol H_2O m^{-2} s^{-1}. Measures of stomatal conductance, while precise, present difficulties when one attempts to apply the results to the population or community level. Using this method, Mann and Wetzel (1999) found transpiration rates in a population of *Juncus effusus* ranged from 0.16 to 0.43 mol H_2O m^{-2} s^{-1}. The highest seasonal rates were found during the summer and autumn, and the highest daily rates in the early evening.

In wetlands without standing water, diurnal fluctuations in the groundwater table (during periods of no precipitation) can be used to estimate ET. Groundwater levels typically fluctuate by several millimeters over a 24-h period. Water levels decline during the day and remain stable or increase slightly overnight. Any increase in water level during the night is due to groundwater influx, which is assumed to occur at a constant rate over a 24-h period. Rates of evapotranspiration can be calculated as follows (Gilman 1994):

$$ET = S/100\ (24r \pm s) \tag{3.3}$$

where

ET = evapotranspiration
S = the specific yield of the soil (i.e., how quickly water can move through the soil profile)
r = the hourly rate of change in the water table during the night (generally taken from the night before, calculated from 8 P.M. to 6 A.M. or midnight to 4 A.M.)
s = the net increase or decrease in the water table over the 24-h period

Diurnal fluctuations can be attributed solely to transpiration and not evaporation by comparing changes in groundwater levels in areas that are vegetated with areas that are not (for instance, where plants have been cleared). Transpiration was measured in this way using continuous water level recorders at Wicken Fen in the United Kingdom (Gilman 1994). Groundwater levels showed diurnal fluctuations from mid-June to late September. Early in the growing season, rapid growth rates and high temperatures led to high transpiration rates. As the growing season progressed and both the water table and water demand by the plants declined, transpiration rates declined as well (as evidenced by the decreasing amplitude in diurnal groundwater level changes; Figure 3.1). The accumulation of plant litter can also affect transpiration. Using this method, lower rates of transpiration were found in natural stands of herbaceous plants where standing crop and accumulated litter reduced water loss. By comparison, transpiration rates were higher in grazed or mowed areas with little accumulated litter.

One technique to measure evapotranspiration is to convert data from pan evaporation. Data from a Class A evaporation pan, which provide an estimate of open water evaporation

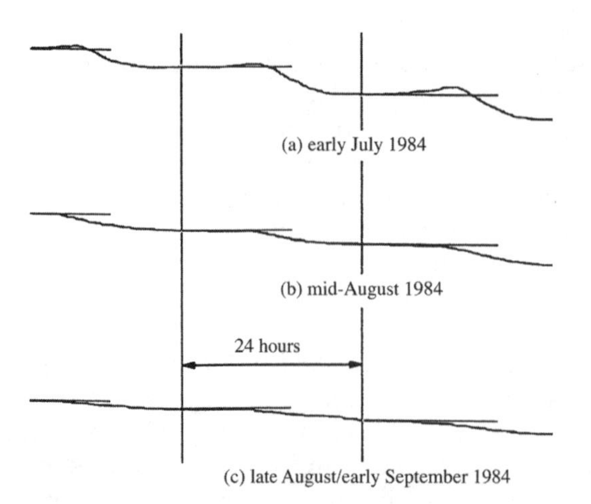

FIGURE 3.1
Diurnal groundwater level fluctuations due to transpiration on three dates recorded at Wicken Fen in the United Kingdom. The fluctuations have an amplitude of several millimeters which declines as the growing season progresses. In each case the fluctuations were recorded over a 3-day period: (a) July 6–8, 1984, (b) August 17–19, 1984, (c) August 31–September 2, 1984. (From Gilman, 1994. *Hydrology and Wetland Conservation*. Chichester. John Wiley & Sons. Reprinted with permission.)

(E_0), are converted to ET of a vegetated area by multiplying by an empirically derived coefficient. ET is assumed to be less than E_0 and the coefficient 0.7 is often used (Chow 1964). Coefficients vary however, depending on environmental conditions and species. In a study of *Typha domingensis*, for example, coefficients varied from 0.7 to 1.3, depending on salinity levels (Glenn et al. 1995).

Indirect estimates of ET are based on physical variables. These methods tend to ignore the influence that plant species composition may have. For example, Thornthwaite's equation (Chow 1964) to calculate potential evapotranspiration (i.e., the maximum rate possible when water is not limiting) requires only the input of mean monthly temperature:

$$\text{PET (mm mo}^{-1}\text{)} = 16\,(10\,T_i\,/I)^a \tag{3.4}$$

where

PET = potential evapotranspiration
T_i = mean monthly temperature, °C

$$I = \text{local heat index} = \sum_{i=1}^{12} (T_i\,/5)^{1.514}$$

a = location dependent coefficient =
$(0.675 \times I^3 - 77.1 \times I^2 + 17{,}920 \times I + 492{,}390) \times 10^{-6}$

Penman (1948) also developed a model to calculate ET. Temperature is also used as a factor in this model, but other meteorological data are also used, including wind speed, solar radiation, elevation, and vapor pressure. This model allows the calculation of daily ET rates. Monteith (1965) modified the Penman equation (commonly referred to as the Penman–Monteith equation) to more clearly take into account the effects of stomatal resistance and wind. In essence, the Penman–Monteith model incorporates all parameters that govern energy exchange and the corresponding latent heat flux (i.e., evapotranspiration) from a uniform bed of vegetation. These parameters are either measured directly or calculated from weather data. Souch and others (1998) investigated the effects of disturbance histories (ditching and drainage) on evapotranspiration rates in wetlands in the Indiana

Dunes using this method. Measuring ET as its energy equivalent, the latent heat flux, they found that ET losses were approximately 3.5 mm d^{-1} whether standing water was present (undisturbed sites) or absent (disturbed sites).

B. The Effects of Hydrology on Wetland Plant Communities

Wetland plant communities have been shown to respond to different hydrologic regimes in several ways including differences in primary productivity, species diversity, and the distribution of species within the ecosystem.

1. Hydrology and Primary Productivity

The duration and frequency of flooding may reduce or enhance primary productivity, depending upon the physiological benefit or stress that is created. Increased water inflows to wetlands carry additional nutrients and facilitate the exchange of dissolved elements (e.g., phosphorus nitrogen, oxygen, and carbon) by decreasing the thickness of the boundary layer at the plant surface, thus enhancing primary productivity (Odum 1956; Brown 1981; Madsen and Adams 1988; Carr et al. 1997). However, in some wetland types, prolonged inundation can cause stress if the soils become anoxic (Mitsch and Ewel 1979; Odum et al. 1979; Brinson et al. 1981; Conner and Day 1982).

Several studies concerning the influence of hydrology on primary productivity have been performed in forested wetlands. Studies in Florida (Carter et al. 1973; Mitsch and Ewel 1979), Louisiana (Conner and Day 1976, 1982; Conner et al. 1981), and Kentucky (Mitsch et al. 1991) have shown that stagnant, continuously flooded forested wetlands have lower primary productivity than sites open to flow and with a more pulsing hydrology. In a review of this relationship, Mitsch (1988) used a parabolic curve to describe primary productivity as a function of water flow (Figure 3.2). His model of forested wetlands designed to investigate this relationship showed that primary productivity is highest when hydrologic inputs are "pulsing." The high primary productivity of wetlands with pulsing hydrology has been attributed to higher nutrient loads.

Brown (1981) found a similar pattern when comparing flow-through, sluggish flow, and stagnant cypress wetlands in Florida. She concluded that phosphorus inflow, which is coupled with hydrologic flow, was the critical variable in determining primary productivity. Brinson, Lugo, and Brown (1981) characterized the link between hydrology and primary productivity in wetlands in order of greatest to least productivity as:

flowing water wetlands > sluggish flow wetlands > stillwater (stagnant) wetlands

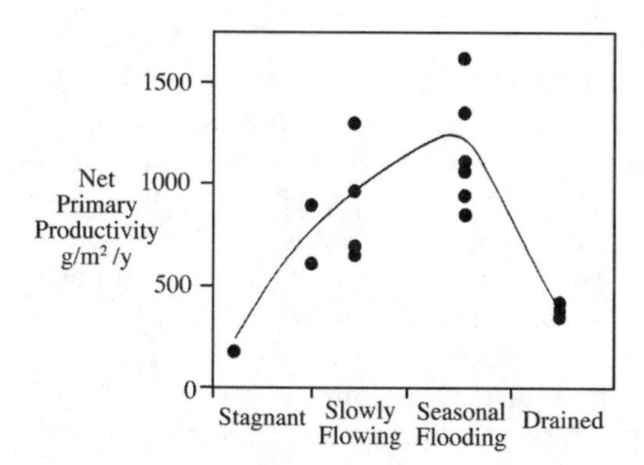

FIGURE 3.2
The relationship between hydrology and net primary productivity in forested wetlands. Productivity is highest when wetlands have "pulsing" hydrology, shown here as a seasonal pattern of flooding. (From Conner, W.H. and Day, J.W., Jr. 1982. *Wetlands: Ecology and Management*. B. Gopal, R.E. Turner, R.G. Wetzel, and D.F. Whigham, Eds. Jaipur, India. National Institute of Ecology and International Scientific Publications. Reprinted with permission.)

The data summarized in their review show that stillwater forested wetlands averaged 707 g dry weight m^{-2} yr^{-1}; systems with sluggish flows averaged 1090 g dry weight m^{-2} yr^{-1}, and flowing water wetlands (excluding data on shrub wetlands) averaged 1498 g dry weight m^{-2} yr^{-1}. In another review of forested wetlands, Lugo, Brown, and Brinson. (1988) stated that ecosystem complexity (i.e., structural and functional characteristics) and primary productivity are correlated with both higher hydrologic energy and higher nutrient supply. They called these the "core factors" that govern plant community response.

The "fertilizer effect" from hydrologic subsidies may elicit other responses from the plant community. Many studies have shown that inputs of water that contain nutrients not only result in higher biomass production but also higher tissue concentrations of these elements (Barko and Smart 1978, 1979; Jordan et al. 1990; Neill 1990). In Florida, wetland plots receiving high rates of wastewater effluent had increased net biomass production (including roots, shoots, and rhizomes) and higher phosphorus concentrations in plant tissues when compared to control plots (Dolan et al. 1981). Similarly, at a site in Michigan, Tilton and Kadlec (1979) found higher biomass production in a zone nearest the point of wastewater discharge. Tissue concentrations of phosphorus were significantly higher in this zone when compared to areas farther from the discharge point.

Bayley and others (1985) found that primary productivity in a freshwater marsh was more dependent on the simple presence of standing water than nutrient subsidies. In their study, emergent vegetation in peat-accumulating marshes showed no difference in primary productivity when nutrient-enriched wastewater was applied as opposed to unenriched water. In this case, standing water (in spite of the difference in nutrient status) led to anoxic conditions in the peat and the release of dissolved phosphorus to the overlying water. This internal nutrient input, while a result of hydrology, outweighed any differences from hydrologic inputs.

Current velocities have been linked to increased primary productivity in submerged plants. Westlake (1967) found that photosynthesis and respiration rates increased in the submerged species, *Ranunculus peltatus* and *Potamogeton pectinatus*, as current velocities increased from 0 to 5 mm s^{-1}. Over this range, the photosynthetic rate of *R. peltatus* increased by a factor of 6. This response was attributed to increased exchange rates of gases and solutes as faster flows decreased the boundary layer around plants. Similarly, Madsen and Sondergaard (1983) found that the growth of *Callitriche cophocarpa* increased as flow rates increased up to 1.5 cm s^{-1}. In their study, photosynthesis rates increased by 20 to 28% with increasing current velocity after a 30-min incubation period. However, if current velocities exceed an optimal level, primary productivity can be reduced. Madsen and others (1993) found that for eight species of macrophytes, primary productivity was reduced as flow rates increased from 1 to 8 cm s^{-1}. Chambers and others (1991) also found that the biomass of submerged plants decreased in the Bow River, Canada as current velocities increased from 10 to 100 cm s^{-1}.

A long-term study in constructed marshes in Illinois was designed to test how different hydrologic regimes influenced plant community development, including primary productivity. Phytoplankton and periphyton net primary productivity was greater in two marshes with high hydrologic inflow (48 cm wk^{-1}) than in two marshes with low inflow rates (8 cm wk^{-1}; Cronk and Mitsch 1994a, b). However, in the first two growing seasons following construction, the macrophyte community did not respond to the different water regimes (Fennessy et al. 1994a). Differences in mean water depths in the four basins may have confounded the results. The discrepancy in the results of the algal community vs. the macrophyte community may also be a function of the response time of the different communities. Given enough time, macrophyte primary productivity may become greater in the high flow wetlands.

2. Hydrologic Controls on Wetland Plant Distribution

Plant species zonation occurs in response to variations in environmental conditions, particularly water depth. A species' habitat along a water depth gradient is a result of its individual adaptations. The shoreline of many wetlands, where hydrological conditions change with elevation and where water levels fluctuate over the long term, supports different zones of vegetation (Figure 3.3, top). For example, lacustrine wetlands have submerged vegetation where the water is deepest, floating-leaved plants at higher elevations and emergent species along the water's edge. In coastal wetlands, both tidal and freshwater inputs influence plant zonation. The most salt-tolerant species are found closest to tidal inputs or where salt water collects. Often, salt-tolerant plants are excluded from less saline areas of the wetland because they are unable to compete with other plants there (see Chapter 2, Section III.A.1, Coastal Marshes; and Section III.B.1, Coastal Forested Wetlands: Mangrove Swamps).

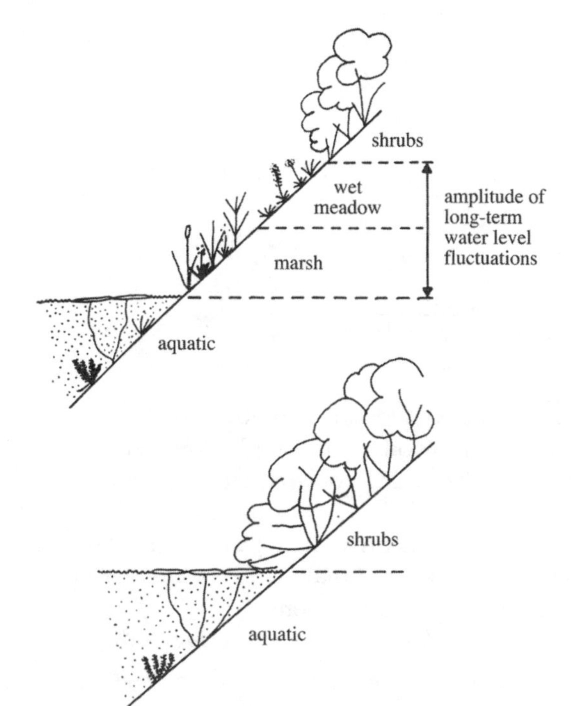

FIGURE 3.3
A conceptual diagram showing how stabilizing water levels can compress the zonation of wetlands species from four zones (top) to two zones (bottom). Overall species diversity in the community declines as a result. (From Keddy, P.A. 2000. *Cambridge Studies in Ecology*. H.J.B. Birks and J.A. Weins, Eds. Cambridge. Cambridge University Press. Reprinted with permission).

Hydrology not only structures plant communities in space, but also in time. For example, flood duration exerts control on the type of community present in a given location as well as species distribution within the community. Keddy (2000) summarized the relationship between community type and hydroperiod for inland wetlands. He organized inland wetlands into four community types defined by the length of time they are flooded each year:

- *Forested wetlands* (swamps, bottomland forests, riparian, or floodplain forests). These areas are only periodically flooded. Where elevations rise they grade into upland species and where elevations fall they give way to more flood-tolerant species. Lugo (1990) described forested wetlands as areas wet enough to exclude upland species but not wet enough to kill trees. The survival time for selected wetland trees in flooded conditions is shown in Table 3.3.

TABLE 3.3

Estimated Survival Time When Inundated for Selected Species of Flood-Tolerant Trees

Species	Survival Time (years)
Quercus lyrata	3
Q. nuttalii	3
Q. nigra	2
Q. palustris	2
Q. macrocarpa	2
Acer saccharinum	2
A. rubrum	2
Fraxinus pennsylvanica	2
Gleditsia triacanthos	2
Populus deltoides	2
Carya aquatica	2
Salix interior	2
Nyssa aquatica	2
Taxodium distichum	2
Celtis laevigata	2
A. negundo	0.5
Platanus occidentalis	0.5
Pinus contorta	0.3

After Keddy 2000, data from Crawford 1982.

- *Wet meadows.* These tend to replace forested wetlands at lower elevations. Occasional flooding in this zone tends to kill woody plants and allow germination of wet meadow species from the seed bank. If flood frequency is reduced, woody species tend to move in.
- *Marshes.* Marshes tend to be flooded for the majority of the growing season. Species here can tolerate long periods of flooding, but many still require drawn-down conditions for germination and seedling establishment.
- *Deepwater aquatic sites.* These occur at the lowest elevations where flooding is essentially continuous.

Kushlan (1990) also described the distribution of plant associations in the Florida Everglades in terms of duration of flooding (Table 3.4). In the Everglades, plant communities change both in composition and growth form as hydroperiods shorten (from greater than 9 months to less than 6 months of flooding) and as fire frequency increases.

3. The Effects of Water Level Fluctuation on Wetland Plant Diversity

One of the major controls on the diversity of any plant community is the ability of each species to become established and persist under existing environmental conditions. The establishment phase is critical, and the conditions that a given species requires to germinate and become established might differ markedly from the conditions to which they are adapted when mature. This set of requirements for germination and establishment has been dubbed the "regeneration niche" by Grubb (1977). Subsequent reproduction by the individual is often vegetative. Many wetland plant seeds and seedlings require drawn-

TABLE 3.4
Environmental Characteristics of Marsh Communities in the Everglades

Vegetation Type	Hydroperiod (time flooded)	Fire Frequency	Organic Matter Accumulation Rates
Water lily	<9 months	<once/decade	high
Submerged	<9 months	<once/decade	high
Cattail	6–9 months	once/decade	high
Flag	6–9 months	once/decade	moderate to high
Sawgrass	6–9 months	once/decade	moderate to high
Wet prairie	<6 months	>once/decade	low

From Kushlan, J.A. 1990. *Ecosystems of Florida.* R.L. Myers and J.J. Ewel, Eds. Orlando, FL. University of Central Florida Press. Reprinted with permission.

down conditions for germination and establishment (see Chapter 5, Section II.C.1, Seedling Dispersal and Establishment).

When water levels are stabilized, a reduction in both plant species diversity and the diversity of vegetation types often results (Figure 3.3; Keddy and Reznicek 1986; Kushlan 1990; van der Valk et al. 1994; Shay et al. 1999). Hydrologic variabilities, such as short-term drawdowns (i.e., those lasting 1 to 3 years followed by extended inundation), maintain species over time. Different species are adapted to the different hydrologic conditions, thus allowing for greater diversity over time. For example, when water level is stabilized in prairie potholes, the wet meadow and marsh zones tend to disappear and overall diversity is reduced (Galatowitsch and van der Valk 1995). As diversity declines, species such as *Typha glauca* tend to encroach and become dominant (Shay et al. 1999). Rare plant communities are particularly vulnerable to extirpation when water levels are stabilized (Schneider 1994). Variable water levels are important in offering propagules the opportunity to establish, making community composition temporally variable.

In an ecosystem level experiment at the Delta Marsh, in Manitoba, van der Valk and others (1994) investigated the effects of stabilizing water levels at higher than normal levels on freshwater marsh vegetation. The site consisted of ten wetland cells (6 to 8 ha in size) in which the water levels could be manipulated. Three water level treatments were applied including normal (levels maintained at the level of the surrounding natural marsh), and 30 and 60 cm above normal. As water levels increased during a 5-year study period, the area of open water increased, and the number of vegetation types and species richness declined. The cover of emergent species declined by an average of 40% in the high water treatments. This decline was not apparent until the third year of the study, a fact that corroborates previous research showing that 2 or 3 years of higher water levels are needed to reduce emergent vegetation (e.g., van der Valk and Davis 1978; Farney and Bookhout 1982).

Wetlands dominated by woody species often show much slower responses to elevated water levels. For instance, Malecki and others (1983) found that when bottomland hardwood forests in New York were subjected to spring flooding (to a depth of 30 cm from March to late June), the composition of the major tree species did not change significantly, even after 12 years of flooding. However, tree growth rates were lower in the flooded forests than in similar bottomland hardwood wetlands with natural hydrology. Declines in mean annual increment amounted to an annual mean of approximately –0.3 cm for *Acer rubrum* ($p = 0.05$), –0.7 cm for *Fraxinus pennsylvanica* ($p = 0.005$), and –0.4 cm for *Quercus bicolor* ($p = 0.10$). The herbaceous understory community showed shifts in species composition over this time period. For example, the density and cover of *Peltandra virginica* (arrow arum) and *Decodon verticillatus* (swamp loosestrife) increased significantly. Greenhouse experiments have also shown that inundation during the growing season can result in high seedling mortality. *Acer rubrum* (red maple) has been found to be more tolerant of inundation than several other wetland trees such as *Ulmus americana* (American elm), *Betula nigra* (river birch) and *Platanus occidentalis* (sycamore; Jones et al. 1989).

4. Riparian Wetland Vegetation and Stream Flow

Rivers show distinct seasonality in discharge rates and water levels. In temperate rivers, floods tend to occur in the spring due to rains and snowmelt. In tropical rivers, flooding results from seasonal changes in precipitation (Junk and Welcomme 1990). The hydrology of many wetlands is linked to adjacent streams and rivers. As a result, wetland hydroperiods also exhibit seasonal changes. The amplitude and frequency of water level fluctuations control wetland characteristics, for instance by eliminating existing vegetation and creating space for the reestablishment of species from the seed bank (Keddy 2000).

The composition of riparian plant communities along rivers and streams is both diverse and dynamic in response to the high degree of physical disturbance. Much of the disturbance is related to current velocity (Nilsson 1987). Increased flows tend to favor higher levels of species richness up to some optimal level. Beyond that, scouring becomes severe and species richness declines, thus supporting the *intermediate disturbance hypothesis,* which states that diversity is maximized at intermediate levels of disturbance (i.e., where disturbance levels are low, competitive exclusion tends to reduce diversity, and where disturbance is high, only highly tolerant species are able to persist). Plant diversity also has been shown to be a function of river discharge, stream order, soils, and microtopographic relief.

Nilsson and others (1994) investigated the relationship between the environmental characteristics of a riparian zone and diversity of stream edge vegetation in a large river and its tributaries. Species richness varied with mean annual discharge of the river, substrate type (cover of peat and silt), altitude, and exposure to waves and flow. They hypothesized that riparian species richness results from propagule transport in the water (*hydrochory*) such that the number of plant species per area of riverbank increases with mean annual discharge. Their data provide evidence that the size of the species pool in upstream areas places a limit on species richness downstream. As flow rates increase, the downstream distance affected by upstream species richness increases as well. Bornette and others (1998) confirmed this relationship in a test of the hypothesis that intermediate levels of hydrologic connectivity (floods) between rivers and riparian wetlands result in maximum propagule inputs, and so maximum species richness. They found that intermediate floods not only bring in propagules, but also scour some areas, allowing for the germination and establishment of new species. However, excess connectivity (flooding) tends to impede recruitment while low connectivity can result in lower species richness due to competitive exclusion (see Chapter 7, Section V.B.1, Floods).

C. Hydrological and Mineral Interactions and Their Effect on Species Distribution

Wetland water chemistry is a function of hydrologic links between the landscape and the wetland ecosystem itself. For example, the mineral composition of the bedrock and the soils helps control hydrology and water chemistry and so controls the formation of specific wetland types. The position of the wetland in the landscape also dictates the amount and quality of surface runoff to the site. The balance of landscape position and surficial geology creates hydrologic and geochemical gradients that are strongly correlated to plant species distributions (Wilcox et al. 1986; van der Valk et al. 1987; Bedford 1999). Building on this idea, Bedford (1996) states that "plant communities may be the best indicators of the many possible unique combinations of climatic and hydrogeologic factors in a landscape."

The chemical composition of groundwater that discharges into wetlands has a strong bearing on community structure (Siegel and Glaser 1987; Rey Benayas et al. 1990; Rey Benayas and Scheiner 1993; Shedlock et al. 1993; Johnson and Leopold 1994; Grootjans et al. 1998). Plant species diversity has been shown to vary with many chemical factors including pH, conductivity, and dissolved cations (Johnson and Leopold 1994). As the value of these parameters increases, species richness has been shown to increase as well; this is attributed to higher fertility of the site. Compounds can accumulate and become toxic in some wetlands, however. Rey Benayas and Scheiner (1993) studied the relationship between soil water chemistry and diversity in wet meadows receiving groundwater discharge in central Spain, and found that species numbers declined as the concentration of ions (e.g., Ca^{2+}, Mg^{2+}, SO_4^-, CO_3^{2-}, HCO_3^-, Cl^-, Na^+) increased. They attributed the lower species richness to toxicity from the accumulation of some of these compounds.

Besides groundwater inputs, a wetland may also receive surface water inflows, tidal inputs, and precipitation. The relative contribution of these sources influences the type of wetland that will develop (Figure 3.4). The links between hydrogeologic setting and vegetation have been extensively studied in peatlands (Wilcox et al. 1986; Thompson et al. 1992; Bridgham and Richardson 1993; Halsey et al. 1997). Fens, for example, depend on groundwater discharge for their nutrient supply. Fen water tends to be high in calcium and magnesium bicarbonates with circum-neutral pH (Wilcox et al. 1986). Oligotrophic bogs, on the other hand, are dependent solely on precipitation for their water supply.

As described in Chapter 2, fen nomenclature is based in part on the minerotrophic status of the system. Vitt and Chee (1990) described the relationship between vegetation, surface water, and peat chemistry in 23 Canadian fens that were grouped into categories, each of which had its own characteristic species. The topography of each fen and the inputs of groundwater as a function of that topography affected species distribution within each site. Wilcox and others (1986) performed a comprehensive analysis of a fen in Indiana. The position of species within the wetland was highly correlated with groundwater flows and water chemistry. These in turn were affected by a 4.1-ha convex peat mound in the center of the wetland. The mound rises as much as 1.4 m above the surrounding wetland. The authors mapped eight distinct vegetation communities including a *Quercus velutina* (black oak) woodland on upland dunes, an *Acer rubrum* (red maple) swamp adjacent to the

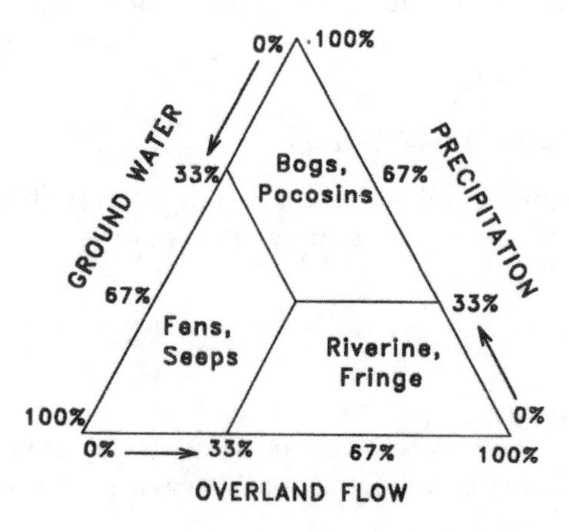

FIGURE 3.4
The relative contributions of water sources including groundwater, surface water, and precipitation determine the type of wetland that will form. (From Brinson, M.M. 1993b. *Wetlands* 13: 65–74. Reprinted with permission.)

upland dunes, intergrading *Typha* marshes and scrub-shrub communities, a *Carex-Calamagrostis* marsh, and a *Phragmites-Typha* marsh on a drier site not on the mound itself. *Thuja occidentalis* (northern white cedar) occurs on the mound where soils are drier. *Larix laricina* occurs adjacent to the mound, where water chemistry is rich in calcium and magnesium (90 ± 12 and 50 ± 5 mg l^{-1}, respectively) and where the soils are saturated year-round.

In coastal wetlands, community composition is affected by salinity gradients, and any hydrologic changes that also alter salinity may change vegetation composition. Visser and others (1999) investigated this phenomenon in the Louisiana coastal zone where wetland losses have been severe. They found that as long as freshwater inputs from the Atchafalaya River were maintained, freshwater species continued to dominate. In their study area, the dominant species in the freshwater marshes changed between 1968 and 1992. In 1968, *Panicum hemitomon* (maidencane) was the dominant plant, covering 51% of the study area. By 1992, *P. hemitomon* covered only 14% of the site. It was replaced by *Eleocharis baldwinii* (roadgrass), which increased from 3% cover in 1968 to 41% in 1992. The authors attribute the change in part to higher water levels that have resulted from flow manipulations to the Atchafalaya River.

Species diversity has declined in the Tijuana Estuary, just north of the U.S.–Mexico border as a result of closures to the estuary inlet. When the inlet closed for a full growing season in 1984, surface soils in the marshes became dessicated and hypersaline (Zedler 2000a). The presence of the salt-tolerant plant, *Salicornia virginica* (glasswort), increased significantly: before the closure none of the study plots contained *S. virginica*, but during the closure, 45% of them did. Shallow-rooting species such as *Salicornia bigelovii* (glasswort) and *Suaeda esteroa* (estuary seablite) were nearly eliminated, presumably because their roots were not able to reach groundwater. The inlet was opened by dredging 8 months later, and after a full decade of return to normal hydrological conditions, *Salicornia bigelovii* and *Suaeda esteroa* had failed to recover. In addition, the marsh as a whole had not recovered its pre-closure level of species richness, dropping from a mean of 5.2 species per 0.25 m^2 originally, to 1.7 per 0.25 m^2 while the inlet was closed, to only 2.6 per 0.25 m^2 a decade after it was re-opened. The failure of the marsh to recover is blamed on the loss of the seed bank during the closure.

In an important feedback loop, it should be noted that wetland plants have important effects on water quality. Their influence can either increase or decrease waterborne materials. For instance, plant root systems help to stabilize sediments and their tissues may accumulate nutrients or metals, eliminating them from the water column. Nutrient "pumping" may also occur. In this case, macrophytes take up minerals or nutrients from the sediments and release them to the water column. Mechanisms for release include tissue senescence and subsequent decomposition, or the loss of organic compounds from plant tissues (Wiegleb 1988).

III. Growth Conditions in Wetlands

Aside from hydrologic forces, a range of environmental factors create unique growth conditions in each wetland. We describe some of the major commonalties among wetland types in the following discussion.

A. Anaerobic Sediments

Under upland soil moisture conditions, plant roots experience the same oxygen level as that of the atmosphere (about 21%). When soils are flooded, the pore spaces are filled with water rather than gas, and the rate of oxygen diffusion is reduced by a factor of 10,000. This slow rate of diffusion combined with the respiratory demands of plant roots and soil microorganisms results in little or no oxygen available externally to plant roots. Reduced inorganic and organic compounds accumulate in the soil. Within hours to days of being flooded, soil becomes hypoxic (with less than 2 mg O_2 l^{-1}) and eventually anoxic.

Wetland soils are either perpetually flooded, and therefore anoxic throughout much of the soil profile, or they are flooded only part of the time. When wetland soils are drained during part of the year, the upper part of the soil profile may become oxidized, enabling seed germination, and in some cases, the invasion of upland plants. Even in soils that are always flooded, a thin oxidized upper layer usually persists, made possible by benthic algal production of oxygen as well as gas exchange with the atmosphere (Gambrell and Patrick 1978; Ponnamperuma 1984; Ernst 1990; Koch et al. 1990; Pezeshki 1994). Since all wetlands are flooded at least part of the time and plants require oxygen for respiration, rooted wetland plants are adapted to the lack of oxygen (see Chapter 4, Section II, Adaptations to Hypoxia and Anoxia).

1. Reduced Forms of Elements

In drained soils, microorganisms use oxygen as a terminal electron acceptor during respiration. When soils are flooded, anaerobic soil microorganisms (both facultative and obligate anaerobes) are able to thrive because they use other terminal electron acceptors in respiration, namely, nitrate (NO_3^-), manganic ions (Mn^{4+}), ferric ions (Fe^{3+}), sulfate (SO_4^{2-}), carbon dioxide (CO_2), and some organic compounds. As a result of anaerobic microbial respiration and chemical oxidation-reduction reactions, reduced forms of these elements are produced. As the time of flooding increases, the soil *oxidation-reduction potential* (or redox potential) decreases. The redox potential (expressed in millivolts, mv) is a measure of the willingness of an electron carrier to donate electrons. The more negative the redox potential, the more the carrier will act as a reducing agent (i.e., it will acquire electrons). A drained soil typically has a redox level of about + 400 to +700 mv, a moderately reduced soil has a redox of about +100 mv, and a strongly reduced soil has a redox of about –300 mv.

As the redox potential decreases, a sequence of redox reactions ensues (Table 3.5). Oxygen is depleted at about +330 mv. The reduction of nitrate begins when redox drops to about +250 mv. Manganic ions are reduced at +225 mv; ferric ions, at +120 mv; sulfate, at about –75 to –150 mv; and carbon dioxide at still lower redox values, from –250 to –350 mv. Not all of the reduction processes are strictly separated at these redox values; some adjoining reactions can occur simultaneously. In each reduction process, bacteria gain energy from the reaction, although the amount of energy derived decreases from the reduction of oxygen to the reduction of carbon dioxide. The reduced forms of the elements create a habitat that is stressful or even toxic to plants (Ponnamperuma 1984; Laanbroek 1990; Pezeshki 1994).

a. Nitrogen

Nitrogen is a vital nutrient for plant growth. In aerated soils, with redox potentials in the range of +400 to +700 mv, nitrogen is found in its most oxidized form, nitrate (NO_3^-). In flooded soils, once most or all of the oxygen has been depleted and soil redox is reduced to about +250 mv, nitrate levels decline to zero within about 3 days (Ernst 1990).

In reduced soils, nitrate may be transformed in two different reduction processes: *denitrification* and *nitrate ammonification.* In the first, nitrate is reduced by anaerobic bacteria in a series of redox reactions to nitrite (NO_2^-), then to N_2O, and ultimately to dinitrogen gas (N_2), which is released to the atmosphere. Denitrification is dependent on the

TABLE 3.5

The Oxidized and Reduced Forms of Nitrogen, Manganese, Iron, Sulfur, and Carbon, the Redox Value at Which They Are Reduced in Flooded Soils, and the Effect the Reduced Form Has on Plants

Oxidized Form	Reduced Form(s)	Redox (mv)	Effect on Plants
NO_3^-	NH_4^+, N_2O, N_2	+250	Denitrification removes N from soil
Mn^{4+}	Mn^{2+}	+225	High levels of reduced Mn interfere with enzyme structure and nutrient consumption
Fe^{3+}	Fe^{2+}	+120	High levels of reduced Fe interfere with Mg during chlorophyll formation and cause discolored leaves, diminished photosynthetic activity, and decreased root respiration
SO_4^{2-}	S^{2-}, HS^-, H_2S	–75 to –150	Reduced sulfur inhibits enzymes involved in photosynthesis and reduces the capacity of the roots to respire both aerobically and anaerobically
CO_2	CH_4	–250 to –350	Methane is transported through wetland plants' internal gas spaces and released to the atmosphere

presence of NO_3^-, which is itself produced by the oxidation of ammonia and nitrite. When NO_3^- is limited, denitrification is also limited (D'Angelo and Reddy 1993).

In nitrate ammonification, a second type of nitrate-reducing bacteria, called nitrate ammonifying bacteria, reduces NO_3^- to NH_4^+. Nitrate ammonification uses eight electrons in the reduction of one molecule of nitrate while denitrification consumes five electrons. The higher number of reactions in nitrate ammonification means that more organic matter is oxidized per molecule of nitrate by nitrate ammonifying bacteria than by denitrifying bacteria. Due to the greater consumption of organic carbon in nitrate ammonification, it tends to dominate when the ratio of organic carbon to nitrate is high (such as in marine sediments). Denitrification tends to dominate where relatively large amounts of nitrate are available.

Whether denitrifying bacteria or nitrate ammonifying bacteria dominate in a soil determines the nitrogen level in the soil. Nitrate ammonification conserves total nitrogen in the soil while denitrification removes nitrogen from the soil (Laanbroek 1990).

b. Manganese

Manganese, a micronutrient, is reduced from Mn^{4+} to Mn^{2+} at a redox potential of about +225 mv. The transformation is a microbial process (Laanbroek 1990). In the reduced form

(Mn^{2+}) manganese is slightly more available to plants than in the oxidized form (Mn^{4+}). While plants require a small amount of manganese, high levels of Mn^{2+} interfere with enzyme structure and nutrient consumption (Laanbroek 1990; Pezeshki 1994). Manganese toxicity occurs at lower concentrations than those typically found under waterlogged conditions (about 3.2 mmol Mn). Some wetland plants have higher manganese resistance compared to upland populations of the same species or upland species of the same genus (Ernst 1990).

c. Iron

At a redox potential of about +120 mv, the insoluble oxidized form of iron, called ferric iron (Fe^{3+}) is reduced to the ferrous form (Fe^{2+}). Iron can be reduced both chemically and microbially. Microbial involvement has been indicated by at least three observations: (1) ferric oxide reduction can be inhibited by bacterial inhibitors, (2) it occurs at a distinct temperature optimum (30°C), and (3) sterilizing the medium prevents iron reduction. Some of the microorganisms involved in iron reduction can also reduce nitrate. They favor nitrate, so the reduction of ferric oxide starts once nitrate has been depleted (Laanbroek 1990).

In the reduced form, iron becomes soluble and is more readily bioavailable. Under reduced conditions, soluble iron (Fe^{+2}) can increase up to a few mmol l^{-1} and its uptake is not controlled metabolically. Consequently, the iron concentration in wetland plant tissue is often greater than that of upland plants. Iron toxicity is one result, occurring in some plants at levels as low as 0.07 mmol l^{-1}, while other plants can withstand concentrations greater than 1 mmol l^{-1} with no ill effects. Iron-affected plants have discolored leaves, diminished photosynthetic activity, and decreased root respiration. Iron interferes with magnesium during chlorophyll formation (Talbot and Etherington 1987; van Wijck et al. 1992). If iron is oxidized within the plant's oxidized rhizosphere, an insoluble iron plaque (Fe^{3+}) can form, coating the roots and blocking nutrient uptake (Ernst 1990; Pezeshki 1994; Christensen and Wigand 1998).

d. Sulfur

Sulfur is an essential micronutrient for plant growth, and it is normally taken up in the form of sulfate (SO_4^{2-}). Its concentration within plants varies but it is commonly about 0.1% of a plant's dry weight. Sulfur is used in the synthesis of some amino acids and proteins and in coenzyme A, which plays a role in aerobic respiration.

Two groups of sulfate-reducing bacteria have been distinguished: one which oxidizes organic compounds to the level of fatty acids, and the other which oxidizes organic matter completely to CO_2. The first group of sulfate-reducing bacteria is found in freshwater systems while the latter is only found in saline environments (Laanbroek 1990). Once sulfate is reduced to sulfide, it may be chelated by metal ions such as iron, copper, and manganese, forming insoluble inorganic sulfides. Some of the sulfide remains soluble in the acid ion form (HS^-) or as a highly soluble gas (H_2S; Gambrell and Patrick 1978; Ingold and Havill 1984; Ernst 1990; Pezeshki 1994).

Sulfate is an abundant ion in seawater: the average sulfate concentration is 0.3 m*M*, as compared to the world average for freshwater rivers, which is 1.2 μ*M* (Lugo et al. 1988). Sulfate reduction can be quite high in salt marshes and mangroves (Howarth and Teal 1979). In the soil water of salt marshes of England and Holland, soluble forms of sulfide have been measured up to 0.5 m*M* (Ernst 1990), and in an Alabama salt marsh, sulfide levels ranged from 1.3 to 8.1 m*M* (Lee et al. 1999). In mangrove forests, sulfide concentrations in unvegetated areas reach over 2 m*M* (McKee et al. 1988).

While the uptake of sulfate is metabolically controlled, sulfide can enter plants without control. Plants growing in high sulfide soils have been shown to contain high concentrations of sulfur (Van Diggelen et al. 1987; Pearson and Havill 1988; Koch and Mendelssohn 1989). When sulfide is taken up by plants it inhibits enzymes involved in photosynthesis and reduces the capacity of the roots to respire aerobically. Sulfide may limit the generation of energy through anaerobic metabolism by inhibiting alcohol dehydrogenase activity (an enzyme involved in ethanol production; see Chapter 4, Section II.B, Metabolic Responses). If roots are unable to maintain either aerobic respiration or anaerobic metabolism, important metabolic functions, such as the uptake of nitrogen, decrease (Havill et al. 1985; Koch and Mendelssohn 1989; Koch et al. 1990; Pezeshki 1994; Lee et al. 1999). As a result, sulfide has a negative effect on the primary productivity of plant communities, reducing the biomass of even flood-tolerant species such as *Atriplex patula, Festuca rubra, Puccinellia maritima* (Ingold and Havill 1984), *Oryza sativa* (Pearson and Havill 1988), and *Potamogeton pectinatus* (van Wijck et al. 1992).

e. Carbon

At extremely low redox levels (–250 to –350 mv), carbon dioxide (CO_2) and organic carbon (methyl compounds) are reduced to methane (CH_4) in a microbial process known as *methanogenesis*. This usually occurs once all of the sulfur compounds have been reduced to sulfide although some sulfate reduction has been observed to occur after peak CH_4 production (Crozier et al. 1995). Decomposing plant matter is the source of carbon for methanogenesis. Methane is not retained by plants, but diffused from the root zone through the plants' internal gas spaces and released to the atmosphere (Howes et al. 1985; Sebacher et al. 1985; Brix et al. 1996; Jespersen et al. 1998). Average rates of methane emission are greater from freshwater wetlands (which release up to 500 mg C m^{-2} d^{-1}) than from saltwater wetlands (up to 100 mg C m^{-2} d^{-1}). Where sulfate concentrations are high, such as in saltwater wetlands, methane production is low, possibly because (1) methane is oxidized to CO_2 by sulfate reducers, (2) sulfur and methane bacteria compete for substrates, (3) sulfate or sulfide inhibits methane bacteria, (4) methane bacteria may depend on the products of sulfur-reducing bacteria, or (5) the redox potential does not drop low enough to reduce CO_2 because of an ample supply of sulfate (Valiela 1984; Mitsch and Wu 1995).

2. Nutrient Availability under Reduced Conditions

Nutrient availability in wetlands is determined by sediment and watershed characteristics as well as hydrology. Wetlands with higher water throughflows tend to have higher levels of available nutrients. Nutrient availability is also affected by the redox potential though not all nutrients are reduced or altered. Essential plant nutrients such as phosphorus, potassium, magnesium, and calcium are not reduced, but the reduction of other elements can change their availability.

For example, in oxidized soils, less phosphorus is available to plants than in reduced soils. Under oxidized conditions, phosphate (PO_4^{3-}) is adsorbed onto iron oxyhydroxides, thus removing phosphorus from circulation. Under reducing conditions, the conversion of iron from Fe^{3+} to Fe^{2+} releases phosphate, making it more available. Potassium, magnesium, and calcium are all positively charged ions that are more available in reduced conditions; however, toxic cations such as copper, zinc, and manganese are also more available (Laanbroek 1990).

3. The Presence of Toxins under Reduced Conditions

A wide range of soluble organic compounds found in wetland soils are toxic to plants. Some toxins are from the anaerobic decomposition of cellulose and lignin. Microbial metabolism brings about a potentially toxic accumulation of acetic and butyric acids, and anaerobic metabolism results in the accumulation of ethanol. These compounds, along with the other phytotoxins mentioned above (reduced iron and manganese, hydrogen sulfide, and dissolved sulfides), produce a hostile environment for plant growth (Barko and Smart 1983; Crawford 1993; Pezeshki 1994). Plants withstand the accumulation of toxic compounds by diffusion of oxygen from the roots into the adjacent soil (see Chapter 4, Section II.A.5, Radial Oxygen Loss).

B. Substrate Conditions in Saltwater Wetlands

A wetland is considered to be brackish or saline if the salt concentration is greater than 0.5 ppt (or about 1.4% the concentration of seawater at an average of 35 ppt NaCl). The average concentrations of the most abundant ions in ocean water are 460 m*M* sodium (Na^+), 50 m*M* magnesium (Mg^{2+}) and 540 m*M* chloride (Cl^-); other ions are relatively scarce. The salinity of soil pore water in saltwater wetlands may be higher or lower than seawater depending on proximity to the tides and on the ratio of evaporation to precipitation. Rainwater or freshwater inputs from inland may dilute seawater while evaporation can cause pore water salinity to be greater than that of the ocean (Flowers et al. 1986; Fitter and Hay 1987). Salinity may also vary with season. For example, in a southern California salt marsh, salinity is low in winter and spring and high in summer and fall when there is less precipitation (Callaway et al. 1990).

The ions in seawater are toxic to most plants and even salt-tolerant plants succumb to very high levels of salt. Salt limits plant growth at about 100 m*M* in the soil solution. The salt concentration of seawater is around 500 m*M* and it commonly rises to concentrations of about 1 *M* in saltwater wetlands (Flowers et al. 1986). The problems faced by plants in highly saline environments are the following:

- It is difficult to acquire water under high salt conditions. Under non-saline conditions, water moves into plants because the external water potential is greater than within the plant. Solutes such as salt decrease the water potential so the plant's internal water potential must be even lower in order for the water to move into the plant. Plants that can acquire water under high salt concentrations are able to decrease their internal water potential below that of salt water (see Chapter 4, Section III.A.1, Water Acquisition).
- There is a different ionic mix in salt water than in fresh water, making the uptake of beneficial ions difficult. For example, where Na^+ is in high concentrations, the uptake of K^+ is inhibited because the two ions are chemically similar and plants take up Na^+ in place of K^+ (Tomlinson 1986; Fitter and Hay 1987).
- High salinity interferes with the uptake of carbon dioxide. As rhizosphere salinity increases, the net rate of carbon dioxide uptake declines. Emergent plants open their stomata to take in carbon dioxide, but lose water at the same time. Since the plants are under water stress and cannot afford to lose water through transpiration, taking in carbon dioxide becomes problematic (Pomeroy and Wiegert 1981; Longstreth et al. 1984; Bradley and Morris 1991b).

When salt concentrations rise to extremes, the primary productivity of saline wetlands decreases (Lugo et al. 1988; Srivastava and Jefferies 1996; Teal and Howes 1996). In mangroves, the shortest trees (<1 m) are reported in areas of high salinity (Pool et al. 1977) compared to over 50 m in areas where growth conditions are favorable.

The plants of saline wetlands are also subjected to high sulfide concentrations, which are stressful and potentially toxic (see Section III.A.1.d, Sulfur). Plants of saline wetlands have a number of adaptations which allow them to persist in a high salt, high sulfide environment (see Chapter 4, Section III, Adaptations in Saltwater Wetlands).

Plants in saline wetlands have been shown to be nitrogen-limited in a number of studies (Valiela and Teal 1974; Sullivan and Daiber 1974; Gallagher 1975; Mendelssohn 1979). Nitrogen limitation in marine plants increases with distance from the shore (Ryther and Dunstan 1971) because the major source of nitrogen is from land runoff. In freshwater systems, cyanobacterial fixation of nitrogen provides another important source; however, its contribution in saltwater systems is minimal (Valiela 1984).

C. Substrate Conditions in Nutrient-Poor Peatlands

Nutrient-poor peatlands (referred to here as bogs) are hydrologically isolated; they receive mineral and nutrient inputs only from precipitation and dry matter deposition. Plant growth in these environments is often limited by one or several nutrients (N, P, or K^+). Nutrient inputs are low and nutrients from the breakdown of organic matter are not readily available due to low decomposition rates and the incomplete decay of organic compounds (Moore and Bellamy 1974). The chemical composition of water in pools on the bog surface closely resembles that of local rainwater, although it tends to be more concentrated due to evaporation. Bog plants tend to show little response to nutrient additions probably due to their inherently slow growth rate as well as their adaptations to a nutrient-poor environment (Proctor 1995). Plants adapted to bog conditions often exhibit a number of nutrient conservation or acquisition mechanisms such as evergreen leaves, carnivory, and nitrogen fixation (see Chapter 4, Section IV, Adaptations to Limited Nutrients).

Bog waters tend to be acidic (pH < 5.0) with a low calcium concentration (<2 mg Ca l^{-1}). This is opposed to mineral-rich fens which have a pH greater than 7 and a calcium concentration above 30 mg Ca l^{-1} (Glaser 1992). Low pH precludes many plant species and affects the availability of plant nutrients. The most important influences on the pH in nutrient-poor peatlands are (Clymo and Hayward 1982; Kilham 1982; Gorham et al. 1985):

- *The production and accumulation of organic acids:* Bogs often have water rich in organic acids, known as humic acids, due to the incomplete decomposition of organic matter. If the concentration of humic acids is high, they are important in the acid base balance of a bog because the dissociation of humic acids can be a major source of H^+ ions. Polygalacturonic acid (PGA) makes up the cell walls of *Sphagnum* and constitutes up to 21% of its dry weight. In a study of 28 North American bogs, the dissociation of PGA and other organic acids was found to be the primary cause of low pH (where acid deposition was not a factor; Gorham et al. 1985).
- *Cation exchange by Sphagnum moss:* Cation exchange occurs when the hydrogen ions of an organic acid are exchanged for other cations in the environment such as Ca^{2+}, Mg^{2+}, K^+, and Na^+. A solution containing Ca^{2+} and SO_4^{2-} before cation exchange would become a solution of H_2SO_4 after cation exchange, thus decreasing the pH. The cation exchange capacity (CEC) of living *Sphagnum* moss is

about 1 meq g^{-1} dry weight which is similar to that of some clays (montmorillonite and vermiculite).

- *Atmospheric deposition of ions:* In the eastern half of North America the minimum pH of precipitation has decreased from 4.4 to 4.0 since the 1950s, due to urban and industrial emissions of sulfur and nitrogen oxides (Bischoff et al. 1984; Laws 1993). Because precipitation is often the major source of external cations, nutrients, and hydrogen ions in bogs, acid precipitation is of significance in their water chemistry. Other sources of ions are the surrounding land and water. Where bogs are near the ocean, sea spray is a major input, supplying high levels of Na^+, Cl^-, and Mg^{2+}. Near prairies and cultivated fields, dustfall tends to be high in Ca^{2+} and Mg^{2+} (Gorham et al. 1985).
- *Biological uptake of nutrients:* The uptake of plant nutrients (Ca^{2+}, K^+, NH_4^+, and NO_3^-) affects the acid–base balance of a bog. When a plant takes up a positively charged ion a positively charged ion such as H^+ is released into the environment. Similarly, when a negatively charged ion is taken up, a negatively charged ion like OH^- is produced. In a balanced system the uptake and release of ions from plants would be equal. However, bog ecosystems are not necessarily balanced because cations are concentrated by evaporation and differentially taken up by mosses.
- *Oxidation and reduction of sulfur compounds:* The substrate of bogs is typically aerobic near the surface and anaerobic with increasing depth. Under anaerobic conditions, sulfate reduction increases the alkalinity by producing OH^- ions, while the oxidation of sulfur compounds such as pyrite increases acidity by producing H^+ ions in the form of sulfuric acid. The pH of the bog may be influenced by both of these reactions.
- *Hydrologic regime:* The hydrologic regime determines the concentration of ions and organic compounds dissolved in the bog water. Atmospheric and groundwater inputs are concentrated by evaporation and evapotranspiration, causing high ion concentrations which are available for exchange with the moss. In addition, surface or groundwater inputs as well as rain can dilute ionic concentrations and change the pH.

The acidity, the nutrient-poor status, and the high water table of nutrient-poor peatlands excludes many plant species and inhibits high rates of primary productivity (Proctor 1995). The primary productivity of ombrotrophic bogs is generally the lowest among wetland types (see Chapter 6, Table 6.2).

D. Growth Conditions for Submerged Plants

Submerged plants are exposed to different light and carbon dioxide regimes than emergent plants. Light and carbon dioxide availability change throughout the course of a day and with the seasons.

1. Light Availability

Light availability is probably the most important regulator of submerged plants' distribution within water bodies (Sand-Jensen and Borum 1991). The spectral composition and amount of light in the submerged environment are influenced by season and latitude. As light passes through the atmosphere, it is selectively absorbed by oxygen, ozone, carbon dioxide, and water vapor, which removes high frequency wavelengths. Once light reaches the surface of the water, a portion is reflected away. Greater reflectance occurs under the

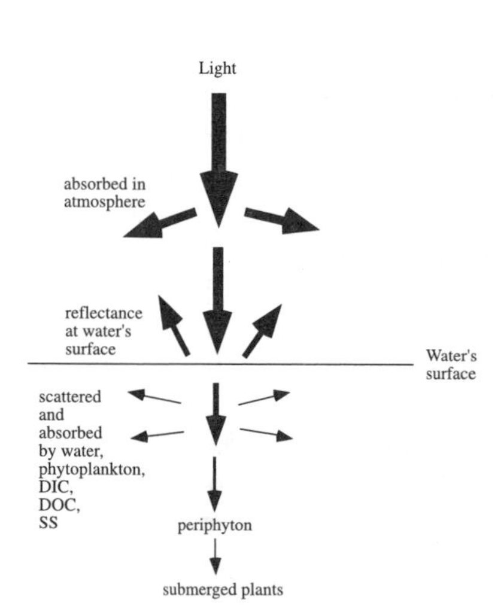

FIGURE 3.5
Light availability for submerged plants is less than that of emergent or upland plants. Light is reflected at the water's surface; it is scattered and absorbed by water, phytoplankton, and other particles (DIC = dissolved inorganic carbon; DOC = dissolved organic carbon; SS = suspended solids) in the water. It is intercepted by periphyton on the surface of submerged macrophytes.

circumstances in which light is less available, i.e., in winter, at the beginning or end of the day, and at high latitudes. Reflectance also increases when the surface of the water is disturbed by wave turbulence or is covered with snow. The portion of light that enters the water column is scattered and absorbed by water molecules, by dissolved inorganic and organic compounds, and by suspended particulate matter. With increasing water depth, radiant energy is attenuated. The depth at which light attenuation occurs varies throughout the year in relation to changes in external inputs to the water body and internal changes due to productivity and decomposition (Figure 3.5; Wetzel 1988; Dennison et al. 1993; Kirk 1994).

The region of the water column in which plants can photosynthesize is referred to as the *euphotic zone.* Submerged plants usually require from 4 to 29% of the incident light measured just below the water surface (Dennison et al. 1993). Vascular macrophytes are usually limited by light to a maximum of 12 to 17 m depth in most freshwater lakes (Chambers and Kalff 1985). Water color (caused by suspended solids resulting from tributary inputs, sedimentation, and resuspension) directly affects the depth at which submerged plants can grow. In lakes with dark color, submerged plants do not colonize to as great a depth as in lakes with less color or high water clarity (Chambers and Prepas 1988; Lauridsen et al. 1994).

The structure and productivity of submerged communities are affected by activities that alter the light regime by depositing or resuspending sediments, such as boat traffic, shoreline erosion, and bioturbation caused by the bottom feeding activities of species such as the common carp (*Cyprinus carpio*). Periphyton growing on submerged macrophytes' surfaces have been shown to limit light and subsequently reduce macrophyte productivity in lakes in Austria (Schiemer and Prosser 1976), England (Philips et al. 1978), Denmark (Kiorboe 1980), and Sweden (Strand and Weisner 1996). In the Chesapeake Bay, where excessive nutrients from human activities stimulated high periphyton productivity, whole communities of submerged plants declined dramatically in the 1960s and 1970s (Orth and Moore 1983, 1984; Twilley et al. 1985). With the restoration of cleaner water, submerged plant communities have reappeared, although with altered species composition (Davis 1985; Carter and Rybicki 1986; Stevenson et al. 1993).

2. Carbon Dioxide Availability

Carbon dioxide is the most readily used form of inorganic carbon for photosynthesis. Like oxygen, carbon dioxide diffuses about 10,000 times more slowly in water than in air. If other nutrients and light are in sufficient amounts for productivity, the rate of CO_2 transport from the atmosphere can become limiting to plant growth, especially in quiescent or highly productive waters (Stengel and Soeder 1975; Beer and Wetzel 1981; Stevenson 1988; Wetzel 1988). In water, carbon dioxide is in equilibrium with carbonic acid (H_2CO_3), bicarbonate (HCO_3^-), and carbonate (CO_3^{2-}). When CO_2 enters water, a portion of it (<1%) hydrates to form carbonic acid:

$$CO_2 + H_2O \leftrightarrow H_2CO_3 \tag{3.5}$$

Because the concentration of carbonic acid is low compared to that of CO_2, the two are often considered together to be the concentration of CO_2. Some of the carbonic acid dissociates to form bicarbonate and hydrogen ions:

$$H_2CO_3 \leftrightarrow HCO_3^- + H^+ \tag{3.6}$$

At high pH (>8.3), bicarbonate dissociates to form carbonate and hydrogen:

$$HCO_3^- \leftrightarrow CO_3^{2-} + H^+ \tag{3.7}$$

During photosynthesis, plants take up CO_2 from the water column. The uptake of CO_2 results in an increase in pH because less carbonic acid is in solution. On days when the water column plant community is highly productive, the pH of fresh water can increase 3 to 4 units (from 6 to 10, for example). At night, CO_2 is released from plants and other organisms and the pH decreases again.

In waters where the pH is 5 or below, dissolved carbon dioxide is the predominant form of inorganic carbon. Above 9.5, carbonate (CO_3^{2-}) is found in significant quantities. At the pH levels that are the most prevalent in fresh surface waters around the world (between 7 and 10), bicarbonate (HCO_3^-) is the most plentiful form of inorganic carbon in the water. At pH 8.4, the bicarbonate concentration in water that is in equilibrium with the air is 1 m*M*, whereas the concentration of CO_2 is only 10 µ*M*. In marine waters, the inorganic carbon supply is fairly consistent with a concentration of about 2.5 m*M* C, largely in the form of bicarbonate. In fresh waters, the inorganic carbon concentration varies much more (from 50 µ*M* to 10 m*M*) due to the differing alkalinity of water bodies and high photosynthetic carbon demands (Prins et al. 1982a; Wetzel 1988). Submerged plants exhibit a number of adaptations to low carbon dioxide levels, including the ability to incorporate bicarbonate ions for use in photosynthesis (see Chapter 4, Section V.B, Submerged Plant Adaptations to Limited Carbon Dioxide).

Summary

The hydrologic regime of a wetland has a major influence on the composition, distribution, and diversity of wetland plants. Species richness, successional trends, primary productivity, and organic matter accumulation are influenced by hydrological factors such as water depth, water chemistry, and flow rates. The construction of a water budget is one technique used to understand the relative importance of different hydrologic processes, to help explain the diversity and distribution of species in the plant community, and to provide insight into the changes that might result from hydrologic disturbance. Evapotranspiration (ET) is a fundamental component of a wetland's hydrologic budget that results from the interaction of water availability and vegetation, and it is often the largest loss term in the water balance equation.

Wetland soils are anaerobic during part or all of the growing season due to the presence of water. In anaerobic soils, N, Mn, Fe, S, and C are found in reduced forms. The reduced forms of these elements create a habitat that is stressful or even toxic to plants. Nutrient availability is also affected by low redox potentials. Essential plant nutrients such as phosphorus, potassium, magnesium, and calcium are not themselves reduced, but the reduction of other elements can change their availability.

In saltwater wetlands, salinity and high sulfide levels are stressful to plants. The ions in seawater are toxic to many species and high salinity levels can be fatal even to salt-tolerant plants. In some peatlands (bogs), nutrient levels are low and the pH of the substrate is acidic. In the underwater environment, both light and carbon dioxide may be limiting factors.

Part II

Wetland Plants: Adaptations and Reproduction

4

Adaptations to Growth Conditions in Wetlands

I. Introduction

The greatest difference between wetland and upland plants is the ability of rooted wetland plants to survive in saturated soil. In addition, submerged plants grow with little or no exposure to the atmosphere, and exhibit adaptations to low light and low carbon dioxide levels in the water column. Free floating plants, able to absorb dissolved nutrients directly from the water, thrive without anchoring roots. While many wetland plant adaptations are unique to the wetland habitat, some are also found in upland plants, such as the enhancement of nutrient uptake through nitrogen fixation, or various defenses against herbivores.

In this chapter, we describe the fate of an upland plant when subjected to anoxic sediments, as well as the many adaptations that have evolved in wetland plants as a result of anoxia. We also discuss plant adaptations to high salt and sulfide concentrations in salt marshes and mangrove forests. We give examples of adaptations that allow for improved nutrient uptake or nutrient conservation. We describe adaptations of submerged plants to life underwater, the defenses some wetland plants have developed against herbivory, and finally, wetland plants' adaptations to water shortages.

A. Aerobic Respiration and Anaerobic Metabolism

Every plant cell requires oxygen for *aerobic respiration*. A green plant produces more oxygen than it needs during daylight hours; however, the oxygen produced during photosynthesis diffuses away from the plant and very little of it is transported to the root tips. As a consequence, the foliage of plants must take in oxygen from the atmosphere, and the roots of plants in drained soils must take in oxygen from the soil pore spaces.

During aerobic respiration, the 6-carbon glucose molecule produced during photosynthesis is broken down to a pair of 3-carbon molecules of pyruvate in *glycolysis*. When oxygen is available, pyruvate is completely oxidized to carbon dioxide. In this process, ATP is formed from ADP and phosphate. In aerobic respiration, the oxidation of one molecule of glucose results in the optimal net yield of 36 ATP molecules. An active cell requires more than 2 million molecules of ATP per second to drive its biochemical machinery. If the production of ATP completely shuts down, the cell, and eventually the plant, will die.

In the absence of oxygen, plant cells undergo *anaerobic metabolism*, or alcohol fermentation. Glycolysis occurs as in aerobic respiration, but the resulting pyruvate molecules are broken down first into acetaldehyde and then into ethanol and CO_2. Thus, the chain of major products of anaerobic metabolism is glucose → pyruvate → acetaldehyde → ethanol.

In anaerobic metabolism, only two molecules of ATP are produced per molecule of glucose, and cell activities such as cell extension, cell division, and nutrient absorption decrease or stop altogether (Raven et al. 1999). Plants that cannot tolerate long periods of flooding-induced anaerobiosis usually die due to insufficient energy (ATP) generation to sustain cell integrity (Vartapetian and Jackson 1997).

B. Upland Plant Responses to Flooding

Much of the research on plants under the stresses of anaerobiosis has been done using crop plants, especially tomatoes, maize, and rice. In tomatoes, maize, and other upland crop plants, some of the signs of stress due to waterlogged sediments begin to appear within minutes to hours. When the roots lack oxygen, the plant's ability to transport water decreases, leading to a decrease in water uptake and a wilted appearance. The stomata close to decrease water loss and, subsequently, photosynthetic activity decreases. In some species, the plant hormone ethylene stimulates *hypertrophy*, or swelling at the stem base. Hypertrophy expands the gas spaces in the stem base and may aid in the diffusion of gases to the roots. Another sign of stress is *epinasty*, or non-uniform elongation of cells, in which the cells on the upper side of a leaf petiole elongate at a faster rate than the cells on the lower side. Epinasty may provide an advantage in water conservation, as it tends to decrease direct insolation of leaf surfaces.

Plant cells deprived of oxygen convert to anaerobic metabolism. Ethanol is the main product of anaerobic metabolism, with lactic acid and alanine produced to a lesser extent. During anaerobic metabolism, ATP production decreases, leaving less energy available for the maintenance of cellular pH and the transport of ions. The optimum pH for the activity of many plant enzymes is 7, so as the pH declines (due to processes discussed in Section II.B.2.b, Davies' Hypothesis), cell metabolism is disturbed. This condition, called *cytoplasmic acidosis*, is a secondary effect of the absence of oxygen in root cells (Roberts et al. 1984). The ultimate cause of plant death in flooded soils is drastically reduced ATP production which shuts down the cell's metabolism (Crawford 1993; Jackson 1994; Lambers et al. 1998).

II. Adaptations to Hypoxia and Anoxia

A number of adaptations allow wetland plants to sequester oxygen or tolerate the consequences of low oxygen levels. We start our discussion with the structural adaptations that affect wetland plants' oxygen supply. The most common adaptation is the formation of *aerenchyma* (porous tissue) in the shoots and roots. We also discuss other root and shoot adaptations, as well as the mechanisms by which oxygen moves through the plants and into the root zone. Following our discussion of structural adaptations, we cover plant metabolic responses to anoxia and some of the research in this field.

A. Structural Adaptations

1. Aerenchyma

Virtually all rooted wetland plants form internal gas-transport systems made up of large gas-filled spaces called *lacunae* (Crawford 1993). The lacunae are held together in a porous tissue referred to as *aerenchyma* (the most commonly used term), *aerenchymous tissue*, or

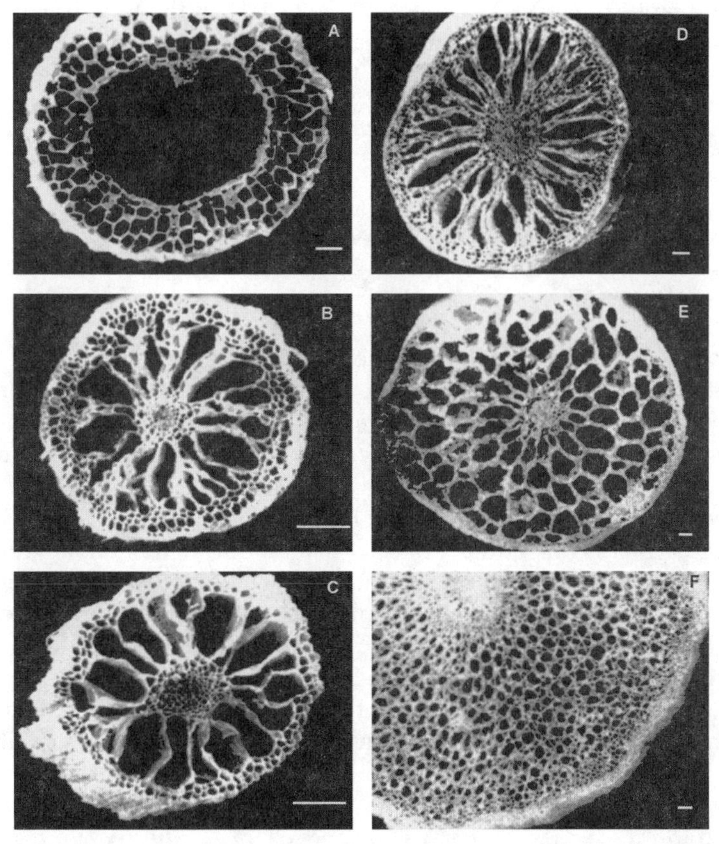

FIGURE 4.1
Cross-sectional electron scanning micrographs of the roots of six wetland macrophytes showing large air spaces, or aerenchyma. (A) *Isoetes lacustris,* (B) *Littorella uniflora,* (C) *Luronium natans,* (D) *Nymphoides peltata,* (E) *Nymphaea alba,* (F) *Nuphar lutea*. Bars represent 100 μm. (From Smits et al. 1990. *Aquatic Botany* 38: 3–17. Reprinted with permission; photos courtesy of G. van der Velde.)

aerenchymatous tissue. Gases are transported throughout the plant along the channels formed by aerenchyma and there is little or no resistance to gas movement (Figure 4.1; Laing 1940; Armstrong 1978). In emergent wetland plants, oxygen enters the aerial parts of the plant via stomata in leaves, and via *lenticels* in stem or woody tissue. It travels toward the roots through aerenchyma, usually via diffusion. Carbon dioxide follows the opposite route, moving upward from the roots where it is produced as a by-product of respiration, through the aerial portion of the plant, where it is released into the atmosphere through the stomata (Armstrong 1978; Topa and McLeod 1986). Aerenchyma forms in both new and old tissue in the roots, rhizomes, stems, petioles, and leaves of both woody and herbaceous wetland plants (Jackson 1989; Arteca 1997). In some species, such as *Cladium mariscus* (twig rush) and *Spartina alterniflora* (cordgrass), a continuous air space extends from the leaves to the roots (Teal and Kanwisher 1966; Smirnoff and Crawford 1983).

a. Aerenchyma Formation

Aerenchyma forms in flood-tolerant species and, to a lesser extent, in many flood-intolerant species (Armstrong 1978, 1979; Crawford 1982; Justin and Armstrong 1987). In flood-intolerant plants, the spaces may occupy 10 to 12% of the total root cross-sectional area, but in flood-tolerant plants, the total area of gas spaces may be over 50 to 60% of the root area (Smirnoff and Crawford 1983; Smits et al. 1990). The volume of aerenchyma varies considerably among species, but porosity is generally greater in emergent than in submerged plants (Sculthorpe 1967).

Aerenchyma forms in two ways:

1. By cell wall separation and the collapse of cells, known as *lysigeny*
2. By the enlargement and separation of cells (without collapse), known as *schizogeny*

In lysigeny, the cells disintegrate and the total number of cells in the air spaces is reduced. Lysigeny is more common than schizogeny (Arteca 1997). Smirnoff and Crawford (1983) noted lysigeny in *Mentha aquatica, Ranunculus flammula, Potentilla palustris, Juncus effusus, Narthecium ossifragum, Glyceria maxima,* and *G. stricta* and in some members of the Cyperaceae (sedge family), including *Eriophorum vaginatum, E. angustifolium, Carex curta,* and *Trichophorum cespitosum*.

In schizogenous plants, the number of cells is not reduced, but a honeycomb structure is produced by the enlargement of intercellular spaces. The cells move farther from one another, thus creating space, but do not disintegrate (Arteca 1997). Schizogeny has been observed in *Caltha palustris, Filipendula ulmaria* (Armstrong 1978; Smirnoff and Crawford 1983) and *Rumex maritimus* (Laan et al. 1989).

The precise mechanism of aerenchyma formation is not entirely defined, but the gaseous plant hormone, ethylene, is clearly involved. When a chemical inhibitor is used to stop ethylene production, aerenchyma formation also stops (Arteca 1997). Low oxygen levels stimulate the production of the enzyme, 1-aminocyclopropane-1-carboxylate (ACC) synthase, which in turn brings about increased levels of another enzyme, ACC oxidase. ACC oxidase is directly responsible for ethylene production, which requires oxygen. ACC oxidase diffuses throughout the plant and ethylene is produced in the aerated plant parts (Jackson 1994; Arteca 1997). Ethylene normally diffuses away from plants, but diffusion is inhibited when the plant is surrounded by water. As ethylene accumulates, it stimulates cell rupture, cell wall degeneration, and an increase in the activity of compounds that degrade cell walls (Vartapetian and Jackson 1997).

The amount of porosity in plant tissues increases with increasingly reduced conditions. Smirnoff and Crawford (1983) noted that several flood-tolerant species formed aerenchyma at the onset of waterlogging, and that porosity increased as the soil redox potential decreased. Plants taken from fens, bogs, and a reed swamp had from 1.2 to 33.6% porosity after 11 weeks of waterlogging, and as waterlogging time increased to 32 weeks, the percent porosity increased up to 50% in some species (*Eriophorum vaginatum* and *E. angustifolium*). As the soil water content increased from 70 to 90%, the root porosity of *Senecio aquaticus* increased from 10 to 35% (Figure 4.2). Lacunal space increases with increasing sediment anaerobiosis in other herbaceous plants as well, notably in the seagrass, *Zostera marina* (Penhale and Wetzel 1983), and in *Salicornia virginica* (Seliskar 1987), *Oryza sativa* (deep water rice; Kludze et al. 1993), *Spartina patens* (salt marsh hay; Burdick and Mendelssohn 1990; Kludze and DeLaune 1994), *Cladium jamaicense* (sawgrass) and *Typha domingensis* (cattail; Kludze and DeLaune 1996). *Taxodium distichum* (bald cypress) also forms more aerenchyma under increasingly reduced conditions (Kludze et al. 1994).

Some plants, such as *Oryza sativa, Schoenus nigricans,* and some *Juncus* (rush) species, form aerenchyma even in well-aerated soils as a part of ordinary root development. This suggests that the formation of aerenchyma in these plants has a genetic component and

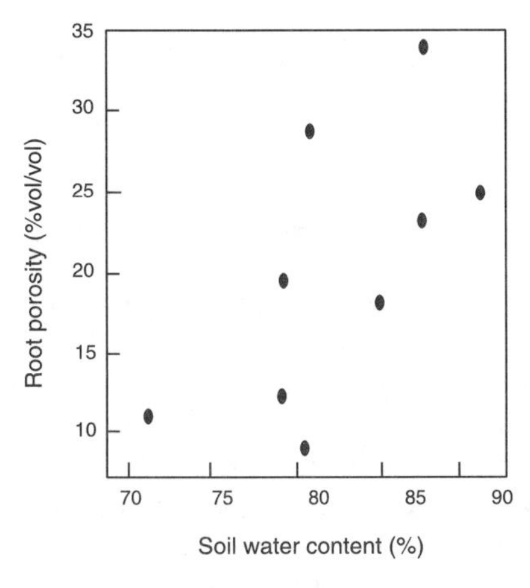

FIGURE 4.2
The relationship between soil water content and the porosity of the root systems of *Senecio aquaticus* plants growing in a peatland of the Orkney Valley, United Kingdom. (From Smirnoff, N. and Crawford, R.M.M. 1983. *Annals of Botany* 51: 237–249. Redrawn with permission.)

does not require ethylene accumulation (John et al. 1974; Jackson et al. 1985; Justin and Armstrong 1987; Jackson 1990).

b. Aerenchyma Function

Aerenchyma decreases the resistance to flow encountered by oxygen and other gases in plant tissue, allowing oxygen to reach the buried portions of the plant relatively unimpeded (Vartapetian and Jackson 1997). Aerenchyma also allows plant-produced gases such as carbon dioxide and ethylene to escape into the atmosphere (Visser et al. 1997). Aerenchyma is effective in aerating the roots and rhizomes of wetland plants; however the aeration is often incomplete. Even with extensive aerenchyma, roots may suffer some degree of anoxic stress and shift to anaerobic metabolism (Saglio et al. 1983). At the beginning of the growing season, the roots and rhizomes of some emergent species experience oxygen deficiency until their new shoots arise and connect them to the atmosphere (Burdick and Mendelssohn 1990; Koncalova 1990; Naidoo et al. 1992; Weber and Brandle 1996).

Aerenchyma also provides storage for gases. The gas storage capacity of herbaceous plants is limited, however, and can be depleted in minutes to hours. New inputs from the atmosphere are required to sustain the plant's oxygen needs. In general, the more air space within the plant, the greater its storage capacity, and monocots tend to have greater porosity and storage capacity than eudicots (Crawford 1993). In *Typha latifolia* (broad-leaved cattail), approximately half of the total leaf volume is occupied by gas spaces and the internal leaf concentration of CO_2 is up to 18 times greater than atmospheric levels (Constable et al. 1992). The internal CO_2 is assimilated by the leaves and provides the plant with a significant carbon supplement (Constable and Longstreth 1994).

2. Root Adaptations

Besides the formation of aerenchyma, wetland plants may undergo other root changes in response to flooded conditions. Among these are the development of *adventitious roots* (roots that arise from other than root tissues) and *shallow rooting* (Laan et al. 1989; Koncalova 1990). In woody plants, other root adaptations include *pneumatophores, prop roots,* and *drop roots.*

a. Adventitious Roots

Within a few days of flooding, some plants form adventitious roots that grow laterally from the base of the main stem. They spread into the surface layers of the soil or grow above the soil surface. In standing water, adventitious roots are in contact with oxygen-containing water, while in areas of saturated soil with no standing water, adventitious roots are in contact with the air. Adventitious roots replace the roots of deeper soil layers that have died due to anoxia. With fewer roots belowground, less root biomass needs to be aerated (Ernst 1990; Arteca 1997; Vartapetian and Jackson 1997).

Adventitious roots form in many herbaceous wetland plants such as *Rorippa nasturtium-aquaticum* (=*Nasturtium officinale*; water cress; Sculthorpe 1967), *Cladium jamaicense, Typha domingensis* (Kludze and DeLaune 1996), and various species of *Rumex* (Laurentius et al. 1996). Adventitious roots have aerenchyma, and the entire stem/root system forms a highly porous continuum (Vartapetian and Jackson 1997). Adventitious roots form in many flood-tolerant tree and shrub species, including *Salix* species, *Alnus glutinosa, Cephalanthus occidentalis, Pinus contorta, Thuja picata, Tsuga heterophylla,* and *Ulmus americana* (Crawford 1993).

The plant hormone, auxin, is involved in the formation of adventitious roots. In flood-tolerant *Rumex* species, the diffusion of auxin into oxygen-deficient roots is slowed and auxin accumulates at the root-shoot junction where adventitious roots form (Laurentius et al. 1996). Some studies have implicated ethylene in the formation of adventitious roots as well (Kawase 1971; Jackson et al. 1981), although results are contradictory (Jackson 1990; Arteca 1997; Vartapetian and Jackson 1997). Unlike aerenchyma, adventitious roots do not increase if the substrate becomes increasingly anoxic. They are simply triggered by an initial flooding (Kludze and DeLaune 1996).

Adventitious roots aid in water and nutrient uptake in flood-tolerant plants. They enhance nitrate availability to plants under anoxic stress because they are in contact with oxygenated soil, air, or water. When adventitious roots are cut daily as they emerge, leaf senescence and dehydration are accelerated and survival rates are decreased (Jackson 1990). In a number of monocots, the large surface area of adventitious roots enhances the

FIGURE 4.3
The shallow roots of a tree growing in saturated soil. (Photo by H. Crowell.)

FIGURE 4.4
An uprooted tree, or "tip-up," indicating shallow rooting and saturated soil conditions. (Photo by H. Crowell.)

rapid absorption of nutrients (Koncalova 1990). Adventitious roots also allow the end product of alcoholic fermentation, ethanol, to diffuse from the plant more easily, rather than accumulating in and near the plant (Crawford 1993).

b. Shallow Rooting

Both herbaceous and woody species tend to have shallower root systems when in flooded conditions (Figure 4.3). Surface or sub-surface roots are above the soil or in the oxygenated portion of the soil profile, thereby alleviating the problem of oxygen shortages in the roots. In a German salt marsh dominated by *Aster tripolii* and *Agropyretum repentis,* the highest root density was found in the soil sub-surface (0 to 8 cm; Steinke et al. 1996). *Phragmites australis* (common reed) also concentrates root growth at or near the soil surface when in flooded sediments (Weisner and Strand 1996). In a study in which *Taxodium distichum* saplings were continuously flooded, only 6% of their total root mass was found below 30 cm in the soil profile. Periodically flooded saplings had 30% of their root biomass below 30 cm. The relatively shallow rooting zone of the continuously flooded plants allows the roots access to nitrate and oxygen (Megonigal and Day 1992). Trees with shallow roots are sometimes felled by high winds and such uprooted trees ("tip-ups") are an indicator of continuous soil saturation (Figure 4.4).

c. Pneumatophores

Pneumatophores are modified erect roots that grow upward from the roots of *Taxodium distichum* and some mangrove species. In *T. distichum,* pneumatophores are commonly called "knees." Cypress knees rise out of the soil wherever water covers the soil surface for extended periods (Figure 4.5). Their height often corresponds to the mean high water level and the highest part of the knee is exposed to air much of the time. Most of the oxygen brought into the plant from the knees is consumed within the knee itself. There is little oxygen exchange between the knee and the roots so they do not aerate the subsurface roots. Cypress knees do have a role in gas exchange, however, since they release 3 to 22 times more carbon dioxide per unit area than an equivalent area of trunk surface and account for 6 to 21% of stem respiration (Brown 1981).

FIGURE 4.5
A "knee" of a *Taxodium distichum* (bald cypress) in the Florida Everglades (approximately 60 cm high). The height of cypress knees usually corresponds to the mean high water level. (Photo by H. Crowell.)

In mangroves, there are several different types of pneumatophores, variously called pneumatophores, root knees, and plank roots (Figure 4.6; Tomlinson 1986):

- The pneumatophores of *Avicennia* and *Sonneratia* species are erect lateral branches of the horizontal roots. They appear at regular intervals along the root (in *Avicennia*, every 15 to 30 cm). A single *Avicennia* tree may have up to 10,000 pneumatophores. In *Avicennia*, pneumatophores are usually less than 30 cm high while in *Sonneratia* they can be up to 3 m. The pneumatophores of *Laguncularia* spp. are erect and blunt-tipped and rarely exceed 20 cm in height. They do not grow in all *Laguncularia* populations and appear to be facultative.
- The root knees of *Bruguiera* and *Ceriops* are actually horizontal roots which periodically re-orient and grow upward through the substrate. The tip of the upward growth forms a loop and then growth continues horizontally so that the root appears to curl its way in and out of the substrate. Branching occurs at the knees and new horizontal anchoring roots are formed. Some *Xylocarpus* species also have root knees that are localized erect growths on the upper surface of horizontal roots that can grow up to 50 cm.
- The plank roots of *Xylocarpus* granatum are horizontal roots that become extended vertically and appear to be shallow roots half in and half out of the substrate. The roots curve laterally back and forth on the soil surface in a series of S-shaped loops.

In all of these root systems, the aboveground component of the root ventilates the buried portion. The entire root system is permeable to the mass flow of gases, with atmospheric exchange occurring through lenticels in the aboveground portions of the roots. About 40% of the root system is gas space, so gases brought in through the lenticels move

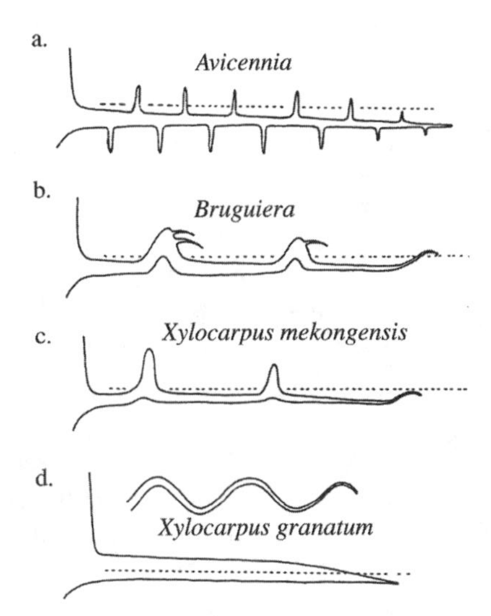

FIGURE 4.6
The aerial roots of mangroves: (a) *Avicennia, Sonneratia,* and *Laguncularia* have horizontal roots buried in the substrate and from them arise erect lateral branches called pneumatophores. (b) *Bruguiera* and *Ceriops* have root knees that are upward growths from the horizontal roots. Branching occurs at the root knees and new horizontal anchoring roots form. (c) Some *Xylocarpus* species also have root knees but without the growth of new anchoring roots. (d) *Xylocarpus granatum* forms horizontal roots, called plank roots, that lie half in and half out of the sediments. The plank roots curve back and forth forming S-shaped curves; this is shown as if from above. (From Tomlinson, P.B. 1986. *The Botany of Mangroves*. London. Cambridge University Press. Reprinted with permission.)

freely. If the lenticels are blocked, the level of oxygen in the submerged roots falls and the roots become asphyxiated (Scholander et al. 1955; Tomlinson 1986; Crawford 1993).

Upward growth from underground roots is reported in other woody wetland species, notably the shrubs *Myrica gale, Viminaria juncea,* and *Melaleuca quinquenervia.* When flooded, the roots reverse direction and grow upward, toward the soil surface, rather than away from it. Although these roots do not emerge from the soil surface like pneumatophores, they have aerenchyma and allow the deeper roots to be aerated (Jackson 1990).

d. Prop Roots and Drop Roots

Rhizophora species (mangroves) form prop roots that develop from the lower part of stems and branch toward the substrate and drop roots that drop from branches and upper parts of the stem into the soil (Figure 2.8). Prop roots and drop roots are covered with lenticels that allow oxygen to diffuse into the plant and carbon dioxide and other gases to diffuse out. Both drop and prop roots branch and form feeding and anchoring roots. Feeding roots are shallow and fine with many root hairs that expand the surface area of the roots. Anchoring roots are thicker with a protective cork layer and extend as deep as 1 m into the substrate (Odum and McIvor 1990). Prop and drop roots give the plant stability, particularly in the face of tides and shifting substrates, and they increase the root surface area, thus improving aeration (Crawford 1993).

3. Stem Adaptations

In addition to the ability of wetland plant stems to form aerenchyma, they exhibit other adaptations to avoid oxygen deprivation. Total submergence stimulates the stems of some wetland plants to grow rapidly toward the water surface in order to reach atmospheric oxygen. The stems of both herbaceous and woody plants sometimes swell at the base due to increased porosity (hypertrophy). The aerenchyma within the stems of many submerged and floating-leaved plants allows them to float near or at the water's surface where they have access to oxygen, light, and carbon dioxide.

a. Rapid Underwater Shoot Extension

Rapid underwater shoot extension, or stem elongation, has been observed in many wetland plants including *Sagittaria pygmaea, S. latifolia, Nymphaea alba, Nymphoides peltata,*

Oryza sativa, Potamogeton distinctus, Victoria amazonica, and species of *Ranunculus* and *Rumex* (Ridge 1987; Arteca 1997; Vartapetian and Jackson 1997). Shoot elongation brings the plant near or above the water's surface, where it has greater access to light, oxygen, and carbon dioxide (Jackson 1990; Laurentius et al. 1996). Rapid shoot elongation is most prevalent when normally emergent or floating-leaved species are submerged (Ridge 1987). Jackson (1982) showed that the shoots of *Callitriche platycarpa* grew an average of 7.3 cm in 4 days when submerged (as opposed to 2.5 cm for floating shoots). Petioles of *Regnellidium diphyllum* (a floating or emergent fern ally) increased petiole length by an average of 4.5 cm d^{-1} when submerged and by only 0.6 cm d^{-1} when emergent (Ridge 1987). Shoot extension usually starts within about 30 min of flooding. It may be stimulated by an accumulation of ethylene which causes the shoot's cells to elongate (Jackson 1990, 1994; Vartapetian and Jackson 1997).

Rapid elongation of stems from overwintered roots, rhizomes, and tubers has been observed at the beginning of the growing season in *Scirpus lacustris, Scirpus maritimus, Typha latifolia, Acorus calamus,* and *Potamogeton pectinatus.* Since stem growth is rapid, the plant comes into contact with oxygen and carbon dioxide before the plant's winter reserves are exhausted (Vartapetian and Jackson 1997; Summers et al. 2000). The mechanism of rapid shoot growth from underground plant parts may be explained by enhanced rates of glycolysis under anaerobiosis (known as the *Pasteur effect*: see Section II.B, Metabolic Responses; Summers et al. 2000).

b. Hypertrophy

Hypertrophy is swelling of the stem base that occurs in response to flooding in both herbaceous and woody plants. It has been observed in a number of flood-intolerant crop plants as well as in many flood-tolerant plants. The swelling is due to accelerated cell expansion caused by cell separation and rupture that occur as aerenchyma forms, and it is stimulated by ethylene (Jackson 1990). Hypertrophy increases the porosity of the stem base and enhances aeration. In some species, hypertrophy enables the development of adventitious roots (Arteca 1997).

Wetland trees often exhibit swelling at the trunk base, called buttressing. *Taxodium distichum* trees, for example, are often buttressed (Figure 4.7). Buttressing increases the plant's stability in water by broadening its base.

c. Stem Buoyancy

The aerenchyma of submerged plants serves not only as a channel for the diffusion of gases, but also provides buoyancy so the stems remain upright and in optimum position for taking in oxygen and carbon dioxide at the water's surface (Kemp et al. 1986). Submerged plants like *Myriophyllum spicatum* (Eurasian watermilfoil; Grace and Wetzel 1978), *Hydrilla verticillata* (hydrilla; Yeo et al. 1984), and *Lagarosiphon major* (African elodea; Howard-Williams 1993) form buoyant canopies of stems, concentrating much of their biomass at the water's surface.

Floating-leaved plants have elongated stems that are also buoyant and help keep the leaves afloat. The stems are supported by the surrounding water and thereby relieved of the burden of holding the plant's leaves erect. The long stems allow the leaves to spread out on the water's surface. Floating leaves may be seen as an adaptation to low oxygen, light, and carbon dioxide levels within the water column (Sculthorpe 1967).

FIGURE 4.7
The thickly buttressed base of *Taxodium distichum* (bald cypress). (Photo by H. Crowell.)

4. Gas Transport Mechanisms in Wetland Plants

Aerenchyma enables gases to move relatively easily between the aerial and subterranean portions of wetland plants. The actual mechanism that drives the movement of gas through most plants is passive molecular diffusion, although other gas movement mechanisms exist in some wetland species. These are *pressurized ventilation, underwater gas exchange,* and *Venturi-induced convection.* These mechanisms enhance gas diffusion and further enable wetland plants to persist in anoxic substrates (Dacey 1980, 1981; Brix 1993).

a. Passive Molecular Diffusion

Passive molecular diffusion is the most prevalent process by which gases move through plants of all kinds. Diffusion is a physical process in which a substance moves from sites of higher concentrations (or partial pressures) to sites with lower concentrations. Gas diffusion rates vary as a function of the medium in which the diffusion occurs, the molecular weight of the gas, and the temperature. Oxygen diffuses freely into the aerial parts of a plant through stomata or lenticels, and then diffuses through gas spaces toward the buried portions of the plant. Within wetland plants, oxygen is usually found in greater concentrations in the aerial parts of the plant than in the belowground parts (Figure 4.8). In *Phragmites australis,* for example, the oxygen concentration has been measured at close to atmospheric levels (20.7%) in the aerial stems and at only 3.6% in the rhizomes. The reverse is true for carbon dioxide and methane. Carbon dioxide in *P. australis* decreases from 7.3% in the rhizomes to 0.07% in the stems. The gradient of concentration of gases within wetland plants indicates that diffusion is the major means of gas transport (Armstrong 1978; Brix 1993).

b. Pressurized Ventilation

Diffusion may be augmented by other mechanisms of gas movement. In some plants, a mechanism variously called *pressurized ventilation, mass flow, bulk flow* (Dacey 1981), or *convective throughflow* (Brix 1993) plays a significant role in the aeration of the plant's belowground parts. In pressurized ventilation, air moves into the plant through the stomata of younger leaves, down the stem to the rhizomes, and then up the stems of the older leaves and back out

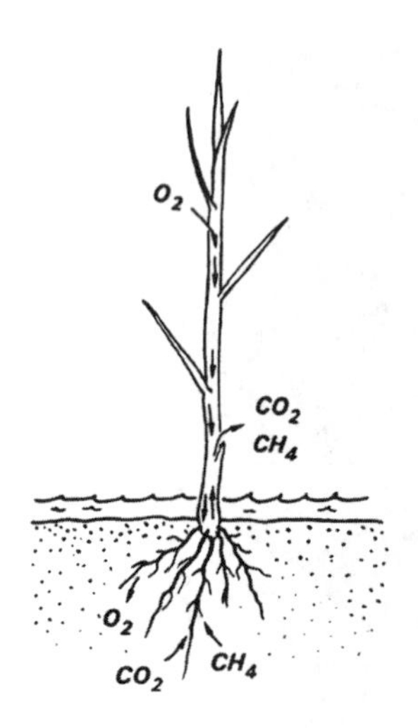

FIGURE 4.8
Passive diffusion of gases in wetland plants. Oxygen diffuses along a concentration gradient from the atmosphere into the aerial plant parts and down the internal gas spaces to the rhizomes and roots. Carbon dioxide produced by root respiration and methane produced in the sediment diffuses along reverse concentration gradients in the opposite direction. (From Brix, H.T. 1993. *Constructed Wetlands for Water Quality Improvement.* G.A. Moshiri, Ed. Boca Raton, FL. Lewis Publishers. Reprinted with permission.)

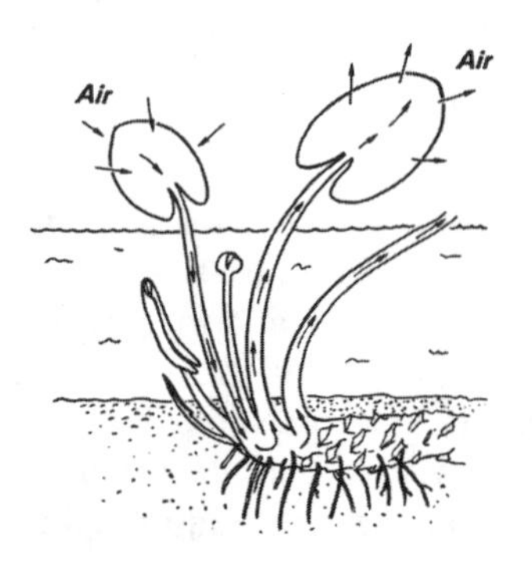

FIGURE 4.9
Pressurized ventilation (or convective throughflow) in *Nuphar lutea* (yellow water lily) as described by Dacey (1981). Air enters the youngest emergent leaves against a small pressure gradient as a consequence of humidity-induced pressurization and thermal transpiration, passes down their petioles to the rhizomes, and up the petioles of older emergent leaves back to the atmosphere. (From Brix, H.T. 1993. *Constructed Wetlands for Water Quality Improvement.* G.A. Moshiri, Ed. Boca Raton, FL. Lewis Publishers. Reprinted with permission.)

to the atmosphere (Figure 4.9). The process is driven by temperature and water vapor pressure differences between the inside of the leaves and the surrounding air (Brix 1993).

The first plant in which this system of ventilation was described in detail was *Nuphar lutea* (yellow water lily or spatterdock; Dacey 1980, 1981). The yellow water lily's rhizome can be 10 cm in diameter and several meters long, and make up 80% of the plant's biomass during the growing season (Figure 4.10). Long petioles support rosettes of floating and emergent leaves that arise from the rhizomes each spring. New leaves continue to emerge throughout the growing season. A large proportion of the plant's volume is aerenchyma (60% in the petiole and 40% in the roots and rhizomes).

Dacey (1981) established the path of gas flow in *N. lutea* using a gas tracer. When the tracer was injected into the upper end of the young leaves' petioles, the tracer moved quickly to the lower end of the petiole, through the rhizome, and then up through the petioles of the older emergent leaves. None of the tracer escaped from the younger leaves; it left the plant through the older leaves. Dacey measured up to 22 l of air entering the youngest leaves of a single plant each day and moving down the petioles to the rhizome at a rate of up to 50 cm/min. The incoming air forced carbon dioxide-rich gas from the roots and rhizomes upward through the petioles of the older leaves and out to the atmosphere.

The youngest leaves have the smallest pores (<0.1 μm in diameter) and due to temperature and water vapor differences between the interior and exterior of the leaves, the gas pressures in the youngest leaves increase to a greater level than in the older leaves. As the leaves grow and mature, the size of their pores increases, and older leaves become leaky,

FIGURE 4.10
An exposed rhizome of *Nuphar lutea* (yellow water lily), measuring approximately 5 cm in diameter. (Photo by M.S. Fennessy.)

losing their capacity to support pressure gradients. Since the lacunae of the older leaves are continuous with those of younger leaves, the older leaves vent the pressure generated in the youngest leaves. As a result, large volumes of oxygen are transported to the buried rhizomes. This method is so effective that the oxygen in the rhizomes is at ambient levels during daylight (21%) and less than 10% at night, with most of the oxygen coming from the atmosphere rather than as a by-product of the plant's own photosynthesis (Dacey and Klug 1982). Dacey called the system a pump because it brings in air against a pressure gradient (Dacey 1980, 1981; Grosse and Bauch 1991; Brix 1993; Vartapetian and Jackson 1997).

Young leaves maintain high pressures via two strategies: both are physical and do not depend on plant metabolism. The first, called *thermal transpiration*, requires a porous partition within the leaf (made up of lacunae), plus heat from the sun. When the interior of the leaf is warmer than ambient temperature, gas moves into and through the porous partition. With higher temperatures, gas expands and the pressure increases within the young leaves. The gas pressure in the young leaves is highest at midday and declines at night (Figure 4.11). The ability of young leaves to trap heat appears to be maximized by the red pigment, anthocyanin, which may increase their absorption of light. As the leaves mature, they lose their reddish color.

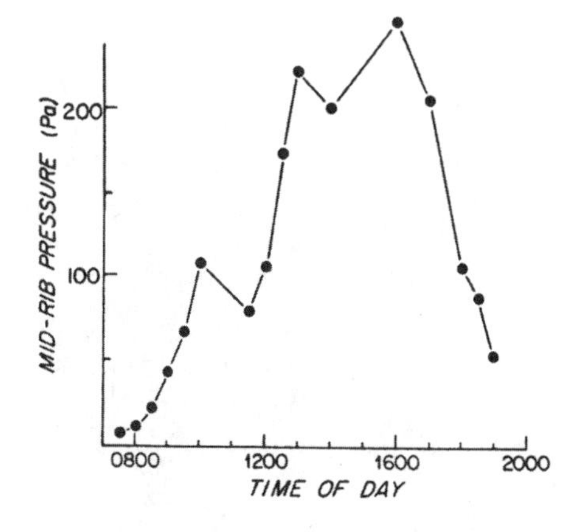

FIGURE 4.11
Representative daily pressurization time-course for an influx leaf of *Nuphar lutea*. The gas pressure in the young leaves is highest at midday and declines at night. (From Dacey, J.W.H. 1981. *Ecology*, 62: 1137–1147. Reprinted with permission.)

The second strategy, called *humidity-induced pressurization* or *hygrometric pressure*, also requires a porous partition and heat. In addition, a constant supply of water within the plant is needed. When there is a difference in water vapor pressure across the partition, then the total pressure is higher on the more humid side. The vapor pressure is greater on the warmer, more humid side of the partition (i.e., inside the young leaf), so the young leaf maintains a greater air pressure than the rest of the plant (Dacey 1981). The slightly increased pressures in the young leaves cause gases to flow through the leaf petioles, through the buried rhizomes, and back to the atmosphere via the petioles and leaf blades of the older leaves. Some researchers have found that the role of gradients in water vapor pressure is negligible and that thermal transpiration is the only significant strategy at work in pressurized ventilation (Armstrong and Armstrong 1991; Grosse 1996).

Pressurized ventilation has been described in a number of other species besides *Nuphar lutea*, including *Euryale ferox, Hydrocleys nymphoides, Nelumbo nucifera, Nymphoides peltata, N. indica, Victoria amazonica,* and some species of *Nymphaea* (Figure 4.12; Grosse 1996). All of these have floating or emergent round leaves like the yellow water lily, although monocots (e.g., *Phragmites australis*, and species of *Eleocharis, Schoenoplectus,* and *Typha*) with linear leaves have also been found to move gases via pressurized ventilation (Armstrong and Armstrong 1991; Brix 1993). This gas-flow mechanism provides the plant with substantial benefits since it helps aerate the roots and rhizomes and thereby alleviates oxygen stress without incurring any metabolic cost. It has apparently evolved several times since it is not restricted to closely related species (Grosse 1996).

FIGURE 4.12
Species that have been found to aerate their roots and rhizomes via pressurized ventilation include (a) *Hydrocleys nymphoides* (water poppy) of the Limnocharitaceae of South and Central America (bar = 1 cm), and two members of the Nymphaeaceae: (b) *Euryale ferox*, found in south and east Asia (bar = 1 cm), and (c) *Victoria amazonica* of South America (bar = 3 cm). (From Cook, C.D.K. 1996. *Aquatic Plant Book.* The Hague. SPB Academic Publishing/Backhuys Publishers. Reprinted with permission.)

c. Underwater Gas Exchange

Underwater gas exchange, or *non-throughflow convection,* is based on the exchange of gases between submerged plant tissues and the surrounding water. In a coastal plant such as *Avicennia germinans* (black mangrove), the pneumatophores are submerged during high tide (Figure 4.13). During submergence, the partial pressure of oxygen decreases within the roots because it is consumed by respiration. Carbon dioxide is produced in respiration, but it does not fill the void left by the decrease in oxygen. Rather, it diffuses from the plant roots and is dissolved in water. Since both gases are depleted within the roots, the total gas pressure is decreased during the period of submergence, creating a vacuum. When the tide goes back out, air is drawn into the first exposed pneumatophore. From there it moves into the rest of the root system, restoring the balance of gas pressures between the atmosphere and the plant's roots (Scholander et al. 1955; Tomlinson 1986; Brix 1993).

A similar mechanism is at work in the sedge, *Carex gracilis* (Koncalova et al. 1988), and in *Oryza sativa* (Raskin and Kende 1985). When their roots are submerged, carbon dioxide is released and dissolved in the surrounding water. The gas pressure within the plants' internal gas spaces decreases, causing a mass flow of air into the aerated portion of the plant.

FIGURE 4.13
Underwater gas exchange (or non-throughflow convection) in *Avicennia germinans* (black mangrove). *A. germinans* has pneumatophores with numerous lenticels. When the tide covers the lenticels, the partial pressure of oxygen in the roots' aerenchyma decreases because it is consumed by respiration. When the tide falls and the lenticels are again exposed to the air, atmospheric air is drawn into the root system through the first emerging pneumatophore (From Brix, H. 1993. *Constructed Wetlands for Water Quality Improvement.* G.A. Moshiri, Ed. Boca Raton, FL. Lewis Publishers. Reprinted with permission.)

d. Venturi-Induced Convection

A fourth mechanism of gas movement has been described for *Phragmites australis* (Armstrong et al. 1992). This mechanism, called *Venturi-induced convection,* is based on gradients in wind velocity. The dead, hollow, broken shoots and stubbles of *P. australis* may remain attached to the rhizome for 2 to 3 years. They are closer to the ground than the taller live shoots. The tall shoots are exposed to higher wind velocities and therefore lower external air pressures. Gas concentrations within the tall shoots are lower than within the broken shoots. This creates a pressure gradient in which gases are driven from the area of higher concentration (the broken shoots) into the area of lower concentration (the taller shoots). In effect, air is pulled through the whole plant, including the underground portions, by the deficit in gas pressure in the wind-exposed taller shoots. The pull of air is balanced by air inputs into the broken shoots (Figure 4.14).

Models of Venturi-induced convection predict that a constant wind speed of 3 m s^{-1} blowing across a single culm would produce an influx of 0.3×10^{-8} m^3 s^{-1} of air, raising the rhizome oxygen concentration to 79% of its potential maximum (rhizome size = 0.3 to 0.4 m in length). If the wind speed doubles to 6 m s^{-1}, the rhizome oxygen concentration increases to 90% of its potential maximum. The proportion of oxygen that enters the rhizome via Venturi-induced convection may be quite significant in high winds or when the number of dead and broken shoots per unit length of rhizome is high (Armstrong et al. 1992).

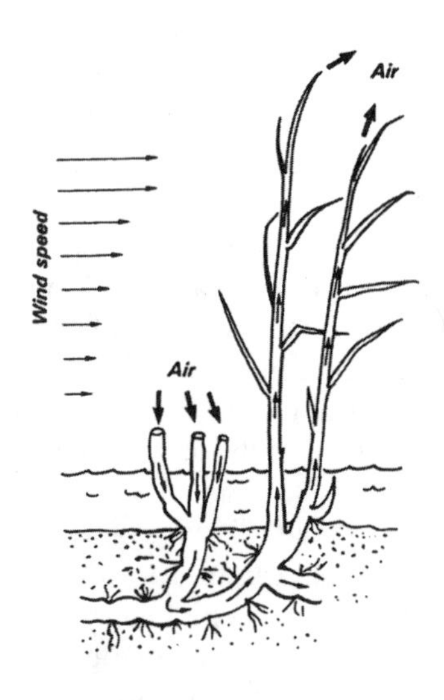

FIGURE 4.14
Venturi-induced throughflow in *Phragmites australis* (common reed). The taller shoots are exposed to higher wind velocities than broken shoots and stubbles close to ground level. This induces a pressure differential that draws atmospheric air into the underground root system. The air is released through the taller shoots. (From Brix, H. 1993. *Constructed Wetlands for Water Quality Improvement*. G.A. Moshiri, Ed. Boca Raton, FL. Lewis Publishers. Reprinted with permission.)

The system is analogous to that which ventilates prairie dog tunnels. The openings of prairie dog tunnels are maintained at various elevations above the soil surface. The taller openings are exposed to higher wind velocities, and therefore lower pressures. This pressure differential draws atmospheric air into the lower openings, through the tunnels, and out of the taller openings (Brix 1993).

5. Radial Oxygen Loss

The oxygen channeled through the plant's aerenchyma is depleted by root and rhizome respiration and by *radial oxygen loss* from the plant roots to the surrounding substrate. Plant roots leak oxygen into the surrounding substrate by diffusion. Radial oxygen loss usually results in the oxygenation of the area immediately adjacent to the plant roots and thereby an increase in the sediment redox potential (Armstrong 1978; Koncalova 1990). The ability of plants to oxygenate the rhizosphere varies with the plant's root oxygen levels, with the size of the plant's root mass, and with the permeability of the roots. Many species of submerged, emergent, floating-leaved plants, and trees, have been observed to oxidize the rhizosphere via radial oxygen loss (Teal and Kanwisher 1966; Barko and Smart 1983; Chen and Barko 1988; Kludze et al. 1993; Kludze et al. 1994; Moore et al. 1994; Grosse 1996; Vartapetian and Jackson 1997). Radial oxygen loss often exhibits diurnal variation, with the greatest oxygen loss to the sediments occurring during the daytime (Grosse 1996). Radial oxygen loss occurs along the entire length of the roots of some plants (e.g., *Isoetes lacustris, Littorella uniflora,* and *Luronium natans*) and only at the root apex in some species (e.g., *Nymphoides peltata, Nymphaea alba,* and *Nuphar lutea;* Smits et al. 1990).

Radial oxygen loss is driven by diffusion, so the greater the oxygen concentration in the adjacent soil, the less oxygen diffuses out of the plant (Reddy et al. 1989). At low redox (–300 mv), *Oryza sativa* roots released 35 µmol O_2 per plant per day. As redox was increased to –200 mv, the roots released about 27 µmol O_2 day^{-1} and at +200 mv, the roots released only about 20 µmol O_2 day^{-1} (Figure 4.15; Kludze et al. 1993). Similarly, in *Taxodium distichum* seedlings, radial oxygen loss is greater under flooded conditions than drained. Kludze and others (1994) measured the loss of oxygen from *T. distichum* roots as 4.6 mmol O_2 g dry $weight^{-1}$ day^{-1} in flooded plants and 1.4 mmol O_2 g dry $weight^{-1}$ day^{-1} in drained plants.

In general, submerged plants have less extensive aerenchyma than emergents, and they oxygenate the rhizosphere to a lesser extent (Barko and Smart 1983). For submerged

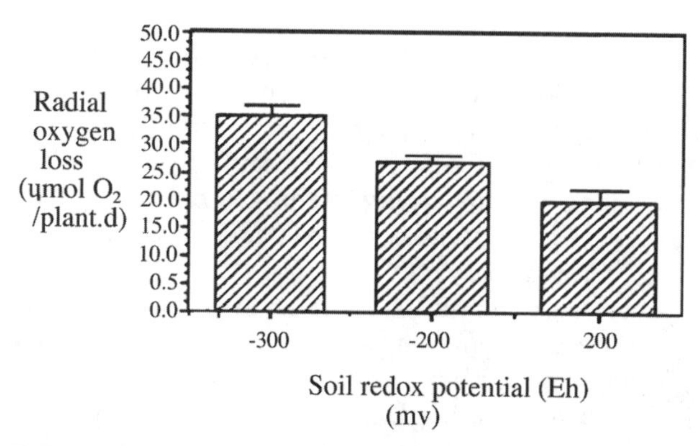

FIGURE 4.15
Radial oxygen loss from rice roots as a function of soil redox potential (Eh). Bars represent ±1 standard deviation. Radial oxygen loss decreases with increasing Eh. (From Kludze et al. 1993. *Soil Science Society of America Journal* 57: 386–391. Reprinted with permission.)

plants, the range of oxygen release per unit root mass is from 0.08 to 5.4 µg O_2 mg^{-1} h^{-1}. In emergent macrophytes the range is higher, from 0.8 to 9.8 µg O_2 mg^{-1} h^{-1} (Carpenter et al. 1983). In a comparison of the emergent plant, *Sagittaria latifolia* (arrowhead), and the submerged plant, *Hydrilla verticillata,* Chen and Barko (1988) found that *S. latifolia* radial oxygen loss affected the soil redox and changed the conditions from reduced to oxidized within 6 weeks. *H. verticillata,* on the other hand, did not noticeably alter the sediment redox, perhaps due to its smaller root system.

Radial oxygen loss also varies considerably among species due to morphological differences such as the root-to-shoot ratio, the canopy type, and growth form. Kludze and DeLaune (1996) measured radial oxygen loss in *Cladium jamaicense* and *Typha domingensis* and found that the radial oxygen loss of *T. domingensis* was greater than twice that of *C. jamaicense.* In a comparison of submerged species, Wigand and others (1997) found that the redox potential in the root zone of *Vallisneria americana* (wild celery) was significantly higher than in the root zone of *Hydrilla verticillata* (+125 mv vs. –5 mv at 4 cm depth).

Although radial oxygen loss depletes the root oxygen supply, it may benefit plants by oxidizing potentially toxic compounds in the rhizosphere, such as reduced metals and gases, dissolved sulfides, and soluble organic compounds (Barko and Smart 1983). Radial oxygen loss often supplies enough oxygen so that nitrifying bacteria, which require oxygen, can transform ammonia to nitrate (Tolley et al. 1986). It also brings about the precipitation of manganese hydroxides and oxides on the root surface, thus preventing the uptake of manganese (Ernst 1990). Reduced iron uptake is also avoided by the oxidation of iron outside of the root via radial oxygen loss (Ernst 1990). Oxidized iron appears as rust-colored spots in the substrate and such plaques are often found in the vicinity of plant roots (Crowder and Macfie 1986; Howes and Teal 1994; Wigand and Stevenson 1994).

Radial oxygen loss may not be sufficient in most herbaceous wetland plants to oxidize sulfide, which is found at very low redox levels (–75 to –150 mv). Sulfide diffuses into the root tissue and exposed plants must be able to tolerate high sulfur concentrations (Havill et al. 1985; Koch and Mendelssohn 1989; Ernst 1990). Some mangrove species (e.g., *Avicennia germinans* and *Rhizophora mangle*) oxidize the substrate sufficiently to reduce sulfide levels (Thibodeau and Nickerson 1986; McKee et al. 1988; see Section III.B, Adaptations to High Sulfide Levels).

6. Avoidance of Anoxia in Time and Space

When flooding is seasonal, some plants' active growth or sensitive periods such as seedling establishment coincide with the dry season. Flood-tolerant trees tend to concentrate their active growth during the late spring and summer when dry conditions are likely. Many flood-tolerant trees are unflooded for 55 to 60% of the growing season. *Liriodendron tulipifera* (tulip tree) can survive prolonged flooding but dies after only a few days of flooding in May or June because the demand for oxygen is greater during the period of active growth (Crawford 1993).

Most wetland plants are perennials and they overwinter as rootstocks, rhizomes, tubers, turions, bulbs, or other perennating structures. Perennating plant parts are usually exposed to anoxic sediments with no connection to atmospheric oxygen. Perennial plants such as species of *Typha, Nymphaea, Nuphar,* and many others avoid oxygen stress in winter by entering a period of low metabolic activity in which there is little demand for oxygen. At the onset of the growing season, their shoots grow rapidly, using stored carbon and nutrients for energy (Ernst 1990; Crawford 1993; Vartapetian and Jackson 1997).

The seeds of many wetland plants only germinate when water levels are low and the substrates are exposed. By germinating only in drier places, the young plant avoids exposure to anoxic stress. Many wetland plants have buoyant seeds that float away from the parent plant. The seeds that arrive at the wetland edges or in areas of shallow water have the best chance of germinating and surviving (see Chapter 5, Section II.B.2, Seed and Fruit Dispersal).

7. Development of Carbohydrate Storage Structures

The length of time plants can survive anoxia varies widely. Most flood-intolerant plants are unable to survive anoxia for more than 3 days. Flood-tolerant plants show a range of survival times, from 4 to more than 90 days (Table 4.1). In a study of plant rhizomes, the largest rhizomes, from species of *Iris, Phragmites, Scirpus, Spartina,* and *Typha,* were able to survive for longer periods of time than small, thin rhizomes of *Carex, Juncus, Ranunculus,* and *Mentha* species (Barclay and Crawford 1982; Braendle and Crawford 1987). Under anaerobic metabolism, the production of sufficient ATP to continue cell metabolism requires a greater amount of glucose than under aerobic respiration. Therefore, plants with a greater stock of fermentable compounds, such as the carbohydrate stores in large rhizomes, are generally able to survive anoxia for longer periods (Studer and Braendle 1987).

The condition of the rhizomes and the season also affect flood-tolerant plants' survival under anoxia. When plants have large carbohydrate reserves at the beginning of the growing season, they can be kept alive under anoxia for longer periods than later in the summer when carbohydrate supplies have been reduced. For example, *Glyceria maxima* (manna grass) rhizomes can survive 7 to 14 days under anoxia in the early spring, but are killed by 7 days' anoxia in mid-summer (Barclay and Crawford 1982).

B. Metabolic Processes

While the development of structural tissues in response to anaerobiosis may take days, plant cells display metabolic responses to anoxia within minutes to hours (Xia and Saglio 1992; Ricard et al. 1994). Most of the research concerning plants' metabolic responses to anoxia has been conducted in laboratories, usually with one of four crop plants: *Oryza sativa* (rice; e.g., Pearce and Jackson 1991; Gibbs et al. 2000), *Zea mays* (maize; e.g., Saglio et al. 1983; Roberts and et al. 1989, 1992; Xia and Saglio 1992; Xia et al. 1995), *Lycopersicon esculentum* (tomatoes; e.g., Germain et al. 1997), or *Triticum aestivum* (wheat; e.g., Menegus

TABLE 4.1
Length of Anaerobic Incubation That Can Be Endured in Detached Rhizomes of Flood-Tolerant Plants without Loss of Regenerative Power

Species	Anoxia Endurance (days)[a]	Shoot Elongation
Carex rostrata	4	None
Mentha aquatica	4	None
Juncus effusus	4–7	None
J. conglomeratus	4–7	None
Ranunculus lingua	7–9	None
R. repens	7–9	None
Eleocharis palustris	7–12	None
Fililpendula ulmaria	7–14	None
Carex papyrus	7–14	None
C. alternifolius	7–14	None
Glyceria maxima	7–21	Occasional
Spartina anglica	>28	None
Iris pseudacorus	>28	None
Phragmites australis	>28	None
Typha latifolia	>28	Frequent
T. angustifolia	>28	Frequent
Scirpus americanus	>28	Frequent
S. fluviatilis	>90	Frequent
S. tabernaemontani	>90	Frequent
S. lacustris	>90	Frequent

Note: Species with large rhizomes survive longer than those with thin rhizomes.

[a] These figures represent the minimum time that the species were able to survive anoxia; longer periods of anoxia survival may be possible in those species that survived 90 days or more.

From Braendle, R. and Crawford, R.M.M. 1987. *Plant Life in Aquatic and Amphibious Habitats.* R.M.M. Crawford, Ed. Oxford. Blackwell Scientific Publications. Reprinted with permission.

et al. 1991; Waters et al. 1991). In most of the studies, the compounds (e.g., ethanol) that are produced by plant parts (often maize root tips, rice coleoptiles, and various seeds) are measured. The plant parts under study are usually moved abruptly from aerobic conditions into anoxia. In some studies, plants are acclimated to low oxygen levels for several hours or days before being plunged into anoxia (e.g., Xia et al. 1995; Germain et al. 1997). Acclimated plants tend to survive longer periods of anoxia than nonacclimated plants (Xia and Saglio 1992; Xia et al. 1995; Raymond et al. 1995; Germain et al. 1997). Some researchers have examined the metabolic responses of wetland plants (other than rice; e.g., Rumpho and Kennedy 1981; Mendelssohn et al. 1981; Mendelssohn and McKee 1987; Summers et al. 2000). In all cases, the ability to survive anoxia requires both the availability of a fermentable substrate (e.g., sucrose) and the avoidance of excessive cell acidification (Raymond et al. 1995).

In wetlands under natural conditions, anoxia may not be complete, although sediment oxygen levels are generally low enough to cause plant root stress. In addition, wetland plant parts are not moved abruptly from aerobic into anaerobic conditions as they are in the laboratory. Nonetheless, the metabolic responses of wetland plants have been found to be similar in many ways to those of study plants, whether the study plants are categorized as flood-tolerant (i.e., wetland species) or not. The major mechanism of survival in anoxic conditions is a conversion to anaerobic metabolism. We discuss some of the findings regarding anaerobic metabolism and some of the hypotheses that have been the basis of many of the studies of flood tolerance in plants.

1. Anaerobic Metabolism and the Pasteur Effect

When deprived of oxygen, plant cells convert from aerobic to anaerobic metabolism. Anaerobic metabolism is considered to be an adaptation to anoxia since it allows ATP production to continue, although usually at a much lower rate than under aerobic respiration. Anaerobic metabolism allows the plant to withstand brief periods of anoxia (hours to a few days; Studer and Braendle 1987). If oxygen is re-introduced to the plant by the de-submergence of the plant's roots or the development of aerenchyma or other oxygen-carrying structures, then the plant cells convert to aerobic respiration. A number of chemical changes occur within plant cells during anaerobic metabolism (many of them during only the first minutes or hours). These include the accumulation of ethanol and organic acids and a pH reduction in plant cells. If anoxia is prolonged, plants must be able to withstand these changes.

Carbon dioxide is produced in both aerobic respiration and alcoholic fermentation. At equal rates of glycolysis, the ratio of anaerobic CO_2 production to aerobic CO_2 production is 1:3. When anaerobic CO_2 production exceeds this ratio, it is known as the *Pasteur effect.* The Pasteur effect is caused by an increased rate of sugar oxidation through glycolysis. Rapid glycolysis offsets the decreased rate of ATP production in anaerobic metabolism (Summers et al. 2000). In an example of an unusually enhanced Pasteur effect, Summers and others (2000) showed that the rate of glycolysis in *Potamogeton pectinatus* tubers was roughly six times faster in anaerobic conditions than in air. The increased rate of glycolysis resulted in rapid stem growth from the tubers. Overwintering tubers are rich in carbohydrates, and the breakdown of these probably fuels rapid glycolysis. The Pasteur effect has also been observed in rice coleoptiles. In a study of two cultivars of rice, the more flood-tolerant of the two exhibited a pronounced Pasteur effect and rapid shoot growth (Gibbs et al. 2000). The ability of plants to increase the rate of anaerobic metabolism enables them to sustain ATP production for growth. Rapid growth of stems allows the plant to move into more oxygenated conditions closer to the water's surface.

2. Hypotheses Concerning Metabolic Responses to Anaerobiosis

Two major hypotheses have been the basis of much of the research on metabolic tolerance of anaerobiosis. The first, proposed by McManmon and Crawford in 1971, is based on the idea that ethanol, the end product of anaerobic metabolism, is toxic. They hypothesized that flood-tolerant plants must have metabolic adaptations that allow them to avoid ethanol toxicity. The second major hypothesis is that flood-tolerant plants are able to avoid the cytoplasmic acidosis brought about by the accumulation of organic acids (Davies 1980).

a. McManmon and Crawford's Hypotheses

McManmon and Crawford (1971) suggested that flood-tolerant plants must have ways of surviving the accumulation of ethanol, a compound that was widely considered to be toxic. They proposed that while flood-intolerant plants suffer an acceleration of the production of ethanol during anaerobic metabolism, flood-tolerant plants avoid this acceleration and also undergo a metabolic switch from ethanol to malate production.

ADH activity — Anaerobic metabolism is driven by a number of enzymes synthesized in anoxic plant tissues. The most studied of these is *alcohol dehydrogenase*, or *ADH*. ADH catalyzes the final step in the synthesis of ethanol. A measurement of ADH activity provides an assessment of the plant's capacity to produce ethanol. High ADH activity indicates that the plant's respiration is suboptimal, i.e., at least partially anaerobic.

ADH activity increases very soon after flooding. When plants develop adaptive tissues or structures that allow for the diffusion of oxygen to the roots, ADH activity subsequently declines. In a study of *Spartina patens*, root ADH levels increased within 3 days of flooding, then declined as root aeration increased (aerenchyma expanded to 50% of the root volume after 29 days of flooding). After 2 months of flooding the ADH activity decreased to levels equivalent to drained control plants (Burdick and Mendelssohn 1990).

McManmon and Crawford (1971) proposed that flood-tolerant plants have a lower ADH activity (and thereby produce less ethanol) than flood-intolerant plants. Less ethanol production would allow them to avoid ethanol toxicity. They observed that ten flood-tolerant species had lower ADH activity when deprived of oxygen than nine flood-intolerant plants. They surmised that flood-tolerant plants were able to switch from ADH activity to the enzyme that catalyzes malic acid production, MDH. Subsequent research has not upheld their theory. Other researchers have found that both flood-intolerant and flood-tolerant plants activate ADH as soon as the oxygen supply is removed. Lower ADH activity has not been observed consistently in flood-tolerant plants and flood tolerance does not correlate with the level of ADH activity (Kennedy et al. 1987; Studer and Braendle 1987; Kennedy et al. 1992; Vartapetian and Jackson 1997).

Alternative end products — McManmon and Crawford also hypothesized that flood-tolerant plants can switch from ethanol production during anaerobic metabolism to the formation of less toxic alternative end products, which would generate energy for the plant. While ethanol is the main end product of anaerobic metabolism, various organic acids do accumulate in flooded plants including malic acid, shikimic acid, oxalic acid, glycolic acid, lactic acid, and pyruvic acid. McManmon and Crawford's 'alternative end products hypothesis' has been the basis for many studies on the tolerance for low oxygen levels and on the alternative end products of fermentation. The tenet that alternative end products allow wetland plants to survive anoxia has been widely accepted and taught; however, a number of studies have shown that alternative end products of fermentation do not explain flood tolerance.

For example, malate was proposed as an alternative end product of fermentation that is less damaging than ethanol (McManmon and Crawford 1971), and some studies have shown that flooded plants do accumulate malate (Crawford and Tyler 1969; Linhart and Baker 1973; Keeley 1979; Rumpho and Kennedy 1981; Ap Rees and Wilson 1984), while others have shown that the level of malate does not increase, but slowly decreases under anoxia (Saglio et al. 1980; Fan et al. 1988; Menegus et al. 1989). No ATP is produced by the malate pathway and therefore no energy is provided to the plant. For this reason, malate would not be a viable alternative to ethanol production (Vartapetian and Jackson 1997).

In addition, there has been no convincing evidence that alternative end products are synthesized in preference to ethanol in flood-tolerant species. A study of the genus *Rumex*, which has both flood-tolerant and flood-intolerant species, shows that the most flood-tolerant species form the most ethanol and do not convert to the production of other end products. This trend is the reverse of that hypothesized by McManmon and Crawford (as reviewed by Davies 1980; Ernst 1990; Kennedy et al. 1992; Crawford 1993; Vartapetian and Jackson 1997). Ethanol is the main product of fermentation in higher plants, whether they are flood-tolerant or not (Ricard et al. 1994). The hypothesis that flood-tolerant species possess alternative energy-generating pathways has been largely dispelled. Rather, responses to anoxia appear to be part of metabolic regulation processes that are common to both flood-tolerant and flood-intolerant species (Henzi and Braendle 1993).

Is ethanol toxic? — Ethanol may not be as toxic to plants as previously thought. It may not inhibit plant growth until concentrations are reached that exceed those found in flooded plants. When ethanol (at a concentration close to that found in flooded soil, 3.9 m*M*) was supplied to *Pisum sativum* (garden pea) roots in both aerobic and anaerobic nutrient solutions, growth of both roots and shoots was essentially the same under all treatments. In addition, both *Oryza sativa* and *Echinochloa crus-galli* (barnyard grass) are tolerant of high ethanol levels (Rumpho and Kennedy 1981; Jackson et al. 1982).

Despite increased ethanol concentrations under flooded conditions, ethanol does not necessarily accumulate in plant tissue. In many flooded plants, such as flood-tolerant *Spartina alterniflora* (Mendelssohn et al. 1981; Mendelssohn and McKee 1987) and flood-intolerant crop plants (maize, tomato, and pea), ethanol diffuses from the roots to the external medium (Davies 1980). In some *Salix* and *Oryza* species, and in *Nyssa sylvatica* var. *biflora*, the production of ethanol is increased under flooded conditions. However, the additional ethanol is diffused to the atmosphere or water through the plants' adventitious roots. In rice, up to 97% of the ethanol produced in oxygen-deprived roots is vented through adventitious roots (as reviewed by Crawford 1993). In some plants, such as *Echinochloa crus-galli*, ethanol is transported from poorly aerated tissues belowground to well-aerated tissues aboveground, where it is metabolized (Rumpho and Kennedy 1981; Jackson et al. 1982).

While ethanol does not appear to inhibit plant growth at the levels usually found in flooded conditions, the precursor to ethanol, acetaldehyde, is toxic to plants (Perata and Alpi 1991). When plants are re-exposed to well-oxygenated conditions, ethanol is oxidized and becomes acetaldehyde, with potentially fatal consequences for the plant (Monk et al. 1987; Crawford 1992).

b. Davies' Hypothesis

Short-term tolerance of anoxia may involve the tight regulation of cellular pH to prevent cytoplasmic acidosis (Davies 1980). Under anaerobiosis, pyruvate is initially converted to lactic acid, which reduces cytoplasmic pH. As the pH decreases, the lactate-activating enzyme, LDH, is inhibited, thus decreasing the production of lactic acid. This occurs within minutes of the onset of anoxia. After LDH levels decrease, ethanol production dominates (Roberts 1989). In work on maize root tips, Roberts (1989) showed that the cytoplasmic pH decreased from 7.3 to 6.8 within 20 min of the onset of anoxia. The pH then stabilized, perhaps because lactate was transported into the vacuole, thus isolating it from the rest of the cytoplasm. Roberts (1989) suggested that after prolonged anoxia (>10 h), the transfer of protons into the vacuole ceases to function. Acid leaks from the vacuole into the rest of the cytoplasm causing cytoplasmic acidosis. The proton gradient between the vacuole and the rest of the cytoplasm collapses. The inability of the cells to maintain a near-neutral pH may be due, at least in part, to insufficient ATP to maintain the proton gradient between the vacuoles and the rest of the cytoplasm (Roberts et al. 1984). On the other hand, the pH may become stable because the production of lactate decreases after about 1 h of anoxia and is followed by increased ethanol production (Ricard et al. 1994).

Some research in this area has indicated that lactic acid may not be the cause of decreased cytoplasmic pH after flooding. In maize root tips, the changes in cytoplasmic pH were much more rapid than changes in the level of lactic acid. Instead, the change in pH followed the time course of a decrease in ATP (Saint-Ges et al. 1991). This study suggested that the decrease in ATP was the main cause for the rapid decline in pH. Acidification may result from insufficient ATP for proton pumping, as suggested by Roberts et al. (1984), and from proton release through ATP hydrolysis (Ricard et al. 1994).

In a study in which maize root tips were slowly acclimated to low oxygen levels (they were exposed to about 14% of ambient oxygen levels for up to 48 h before being deprived of oxygen), the root tips produced less lactic acid than nonacclimated root tips and also excreted it into the medium. As a result, cytoplasmic pH was higher in acclimated root tips than in nonacclimated root tips (Xia and Saglio 1992). In a subsequent study, acclimated maize root tips were shown to have higher levels of ATP and a pH that was maintained near neutral (Xia et al. 1995). Similarly, in tomato roots, a period of acclimation resulted in less lactic acid production at the onset of anoxia than in nonacclimated roots (Germain et al. 1997).

The research concerning pH regulation and avoidance of cytoplasmic acidosis has involved mostly flood-intolerant crop plants. It is not clear whether flood-tolerant plants are better able to regulate cellular pH than flood-intolerant ones. Results from studies of some flood-tolerant plants indicate an ability to avoid acidosis. For example, *Oryza sativa* var. *arborio* showed a slight alkalinization during the first 8 h of anoxia (changing from pH 6.0 to 6.2; Menegus et al. 1989, 1991). *Echinochloa phyllopogon* showed no change in pH following flooding (Kennedy et al. 1992). In *Potamogeton pectinatus*, the pH fell by ≤0.2 units immediately following flooding (Summers et al. 2000), a decrease that is smaller than that seen in maize (0.5 to 0.6 units; Roberts 1989).

The mechanism for pH maintenance is not clearly defined (Kennedy et al. 1992; Vartapetian and Jackson 1997; Summers et al. 2000). However, a lack of detectable lactate was observed in the growth medium of *P. pectinatus* plants. It is possible that lactate production is only a minor pathway in *P. pectinatus* (Summers et al. 2000). Other flood-tolerant plants such as *Trapa natans* and *O. sativa* var. *arborio* have also been shown to produce little lactate (Menegus et al. 1989, 1991).

3. Other Metabolic Responses to Anoxia

Research on metabolic responses to anoxia has centered on the changes brought about as a result of anaerobic metabolism (the accumulation of ethanol, the increase in ADH activity, and the decrease in cellular pH). Other categories of study may eventually provide additional insight into the ability of flood-tolerant plants to survive long periods of anoxia.

For example, metabolic responses to anoxia are reflected in protein metabolism and in the repression or expression of genes under different levels of oxygen availability. For example, some of the proteins produced under anaerobic conditions are those involved in ethanol fermentation. These proteins are involved in the pathways that mobilize sucrose or starch for ethanol fermentation and they are necessary to maintain energy production under anaerobic conditions. In addition to these proteins, others have been noted in some plants, for example, proteins that induce the production of alanine and lactate (Ricard et al. 1994). *Echinochloa crus-galli*, a flood-tolerant grass, produces anaerobic proteins during the first 24 h of flooding, but resumes aerobic protein synthesis thereafter (Kennedy et al. 1992). Further discovery and detailing of altered gene expression under anoxia may indicate ways in which flood-tolerant plants are metabolically adapted to anoxia (Kennedy et al. 1992; Ricard et al. 1994; Bouny and Saglio 1996; Setter et al. 1997; Vartapetian and Jackson 1997).

Mitochondrial adaptations may also play a role in flood tolerance. Mitochondria develop abnormally without oxygen in many plants (i.e., polypeptides synthesized in anoxic mitochondria differ qualitatively and quantitatively from those produced when oxygen is available), including flood-tolerant *Oryza sativa* (Vartapetian et al. 1976; Couée et al. 1992; Ricard et al. 1994). However, the mitochondria of flood-tolerant *Echinochloa phyllopogon* develop normally whether exposed to oxygen or not (Kennedy et al. 1992). When glucose is supplied to mitochondria that are developing abnormally under anaerobiosis, their structure is preserved and they resemble mitochondria that develop in the presence of oxygen. It may be that mitochondrial tolerance to anoxia is enhanced when sufficient glucose is available (Davies 1980). The study of mitochondrial adaptations may provide insight into whole-plant adaptations to anoxia.

III. Adaptations in Saltwater Wetlands

A. Adaptations to High Salt Concentrations

Apart from some algal species, nearly all salt-tolerant plants are angiosperms. Salt tolerance occurs in about one third of the angiosperm families, with somewhat different adaptations among the monocots and the eudicots. Plants adapted to high levels of salinity are known as *halophytes;* those that are not adapted to salinity are called *glycophytes.* To successfully grow in a saline environment, halophytes must be able to acquire water and avoid accumulating excess salt. Halophytes do not require salt; however, the growth of some eudicot halophytes is optimal at moderate concentrations of salt (50 to 250 m*M* NaCl). Halophytes accumulate salt and maintain a higher ion content than glycophytes can withstand (Flowers et al. 1977, 1986; Partridge and Wilson 1987).

1. Water Acquisition

The greatest problem faced by plants exposed to high levels of salt is the acquisition of water. In general, water moves along a gradient from areas of higher water potential to lower water potential. Water potential is the free energy content of water per unit volume, expressed in the same units used to express pressure (energy per unit volume, called megapascals, or MPa). The water potential of pure water is assumed to be zero at ambient temperature and atmospheric pressure. Under non-saline conditions, the water potential of soil water is greater than the water potential within a plant. The range in water potential of herbs of moist forests is from –0.6 to –1.4 MPa, while the soil water potential is generally greater than –0.1. Since water flows from higher to lower water potentials, external water enters the plant. Plant roots tend to have a higher water potential than plant shoots or leaves allowing water to flow upward from the roots to the shoots.

The addition of a solute, such as salt, causes the water potential to decrease. Salt water has a water potential of –2.7 MPa, and plants growing in salt water must maintain an even lower water potential in order to acquire water. When a non-halophyte is placed in a saltwater solution it loses water since the water moves from the higher water potential inside the plant to the lower water potential outside of the plant. In the short term, the plant wilts, and if the plant is unable to adjust to the lowered external water potential, it dies (Queen 1974; Salisbury and Ross 1985; Fitter and Hay 1987).

Plants that are able to take in water despite low external water potentials do so by a process called *osmotic adjustment* or *osmo-regulation.* The plant increases its internal solute concentration with NaCl or other compounds, known as *compatible solutes.* Examples of compatible solutes are glycine betaine (Cavalieri and Huang 1981; Marcum 1999; Mulholland and Otte 2000), proline (Stewart and Lee 1974), mannitol (Yasumoto et al. 1999), and dimethylsulphonioproprionate (DMSP; Stefels 2000). It should be noted that these compounds are sometimes found in quantities too low to affect osmo-regulation. They may play a different role in some plants, such as carbon or nitrogen conservation (Stewart and Lee 1974) or cell protection (e.g., proline; Soeda et al. 2000).

The increased solutes within the plant cause the plant's water potential to fall lower than that of the external medium. Because high salt levels are potentially toxic and can threaten cell processes, increased internal solute concentrations are damaging to most plants. Halophytes are able to tolerate high internal solute concentrations and withstand higher external levels of salt than glycophytes (Queen 1974; Flowers et al. 1986; Fitter and Hay 1987).

2. Salt Avoidance

Halophytes avoid or tolerate high salt levels through *exclusion, secretion, shedding,* and *succulence*. Usually several salt avoidance mechanisms function within a single plant.

a. Exclusion

Salt exclusion is the most important means of surviving high salt concentrations and all halophytes exclude most of the salt in their growth medium (Waisel et al. 1986). In several mangrove genera (*Bruguiera, Lumnitzera, Rhizophora, Sonneratia*) nearly 99% of the salt in the surrounding seawater is excluded at the roots (Tomlinson 1986). *Spartina alterniflora* excludes from 91 to 97% of the ions in salt water (Bradley and Morris 1991a). Salt may be inhibited from entering the entire plant, or only sensitive tissues. Exclusion on the whole plant level occurs by inhibition at the roots. In mangrove trees of the genera *Rhizophora, Laguncularia,* and *Conocarpus,* the roots perform ultrafiltration at plasma membranes. The process is driven by tension in the xylem resulting from low xylem pressures (–3 to –6 MPa, considerably lower than the water potential of seawater at –2.7 MPa). The low pressure from within pulls water into the roots and salt is filtered out at root cell membranes.

For most species, salt exclusion at the roots is not entirely sufficient, and other mechanisms are employed. Some plants sequester salt ions in specialized tissues that prevent them from reaching sensitive tissues. For example, some species of the families Leguminosae and Chenopodiaceae can absorb Na^+ in mature parts of the roots, blocking them from advancing into the shoots. The Na^+ accumulation capacity of these cells is limited so this process is effective only at low levels of salinity (Hagemeyer 1997).

Casparian strips may also play a role in excluding salt from the inner root tissues. Casparian strips are bands of tissue containing suberin (fatty tissue) and lignin. They block the passage of substances through the *apoplast* (the cell wall continuum of a plant or organ) thereby excluding materials that cannot be transported within the *protoplasts* (living substance of the cell). Casparian strips have been found in the root hypodermis of macrophytes, notably in salt-tolerant plants such as *Ruppia maritima* and *Potamogeton pectinatus* and in seagrasses such as *Zostera marina, Z. japonica, Z. capensis,* and *Halophila ovalis* (Flowers et al. 1986; Barnabas 1996).

Another means of excluding salt is to recognize the ions, Na^+ and Cl^-, and to prevent their uptake. The absorption of ions from the external medium is regulated by active transport mechanisms located in cell membranes. In the case of Na^+, exclusion is difficult since Na^+ is chemically similar to K^+. In excluding Na^+, the plant may also exclude K^+, which is an essential plant nutrient (Queen 1974; Fitter and Hay 1987). *Spartina alterniflora* is capable of preferentially absorbing K^+ and excluding Na^+ (Bradley and Morris 1991a).

b. Secretion

Salt glands on the leaves of many halophyte species secrete salt. The voided salt is in solution, and the liquid evaporates leaving salt crystals on the leaf exterior that are blown or washed away by wind and rain. Salt secretion is also called *excretion* and *recretion* (Waisel et al. 1986).

Salt glands have been observed in a number of salt marsh species including *Distichlis spicata, Spartina alterniflora, S. patens, S. foliosa, S. townsendii,* and *Limonium* species (Anderson 1974). In *S. alterniflora,* salt glands selectively secrete Na^+ relative to K^+ (Bradley and Morris 1991a). Salt glands are abundant on the leaves of some mangrove

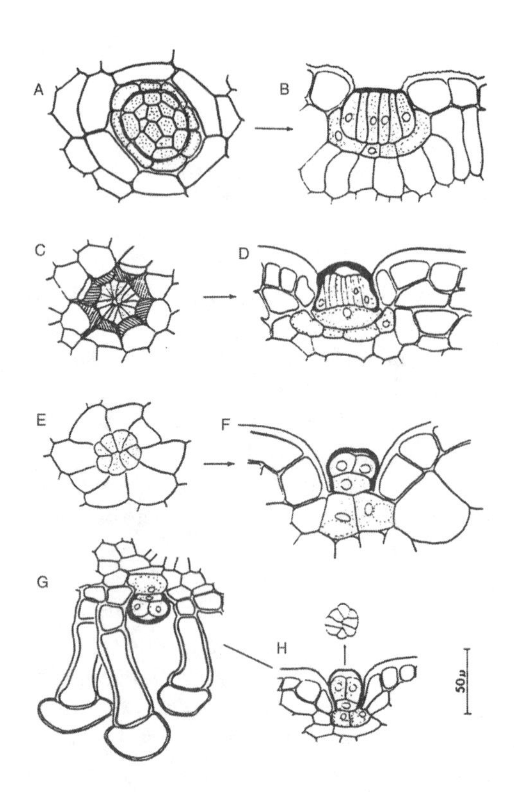

FIGURE 4.16
Salt glands in mangroves. A and B are *Aegialitis annulata,* C and D are *Aegiceras corniculata,* E and F are *Acanthus ilicifolius,* G and H are *Avicennia marina.* All are on the upper surface of the leaves except G, which is on the lower surface. (From Tomlinson, P.B. 1986. *The Botany of Mangroves.* London. Cambridge University Press. Reprinted with permission.)

FIGURE 4.17
Salicornia sp. (glasswort), a halophyte with succulent shoots. (Photo by H. Crowell.)

genera (*Acanthus, Aegialitis, Aegiceras, Atriplex, Avicennia,* and *Halimione*), although they are sometimes obscured by the presence of hairs (Figure 4.16). The structure of salt glands in all of the salt-secreting mangroves is quite similar despite the fact that they are only remotely related, an example of evolutionary convergence. Mangrove salt glands are highly selective, secreting Na^+, Cl^-, and HCO_3^- against a concentration gradient while Ca^{2+}, NO_3^-, SO_4^{2-}, and $H_2PO_4^-$ are retained (Tomlinson 1986).

Some halophytes, such as *Salicornia virginica*, and several mangrove genera (*Bruguiera, Lumnitzera, Rhizophora,* and *Sonneratia*) do not secrete salt (Anderson 1974; Tomlinson 1986). In general, halophytes that do not secrete salt tend to be more efficient at salt exclusion. In mangroves, the xylem sap of salt secreters has an average salt concentration that is 10% that of salt water. In non-secreting mangrove species, on the other hand, the salt concentration of the xylem sap is only 1% that of salt water, indicating that the non-secreting mangroves exclude more salt (Tomlinson 1986).

c. Shedding

Salt is lost from some plants by the loss of plant parts, usually leaves, in which salt has accumulated. If the salt-containing leaf or shoot falls directly below the plant, salt can accumulate in the plant's root zone unless the plant parts are carried away by tides or other sources of water (Waisel et al. 1986). Mangroves shed leaves as the leaves age, but salt is not actively transported to senescing leaves (Tomlinson 1986).

d. Succulence

Succulence is an increase in water content per unit area of leaf. When succulence occurs, each cell increases in size, the leaves or shoots become thicker, and the number of leaves per plant decreases. The increased succulence in halophytes dilutes the internal salt water and thereby lessens salt's negative effects (Flowers et al. 1986). Succulence occurs in the leaves of some eudicot halophyte genera such as *Atriplex* and *Suaeda*, and in the shoots of *Salicornia* and *Arthrocnemum* (Figure 4.17). It is also observed in non-halophytes in arid regions. Succulence may be a response to the difficulty in obtaining water under high salt conditions rather than a response to salt. When it is difficult to acquire water, plants respond by closing their stomata to conserve water. Succulent plants often close their stomata during the day and open them at night, thereby minimizing daytime water loss (Fitter and Hay 1987).

Succulence occurs in many mangrove species and increases in occurrence as the plant ages. The leaves become more fleshy in texture and leaf thickness increases. In *Laguncularia racemosa* (white mangrove), leaf thickness increases fourfold from the youngest to the oldest leaves on a shoot. In *Rhizophora mangle* (red mangrove), leaf thickness increases with increasing soil salinity (Tomlinson 1986).

B. Adaptations to High Sulfide Levels

Despite the toxicity of sulfide, salt marsh and mangrove plants survive chronic sulfide exposure. The mechanisms of sulfide tolerance are a matter of current study and are not yet completely described. Some adaptations to anoxia help plants avoid exposure to high levels of sulfide. For example, both adventitious and shallow rooting concentrate roots in oxidized areas. Radial oxygen loss, in which oxygen diffuses from the roots into the rhizosphere, provides plants with a means to detoxify the soil environment. However, the oxidation of sulfide requires a greater amount of oxygen than most herbaceous plants lose through radial oxygen loss. While some sulfide may be oxidized in this way, sulfide still enters plants in high sulfide environments (Crawford 1982; Havill et al. 1985; Koch et al. 1990; see Case Study 4.A, Factors Controlling the Growth Form of *Spartina alterniflora*).

Mangroves, on the other hand, may release sufficient oxygen through their roots to oxidize sulfide. In *Avicennia germinans* (black mangrove), the majority of the roots extend horizontally away from the trunk, near the soil surface. At intervals of about 25 cm, pneumatophores extend above the soil. The pneumatophores are covered with lenticels that allow air to enter the air spaces within the roots. The soil surrounding *A. germinans* pneumatophores is consistently more oxidized and sulfide levels are up to three times lower than in nearby unvegetated soil (Thibodeau and Nickerson 1986). Sulfide levels near *Rhizophora mangle* roots have also been found to be less than in adjacent unvegetated areas (0.33 vs. 1.63 m*M*; McKee et al. 1988).

Plants emit a number of sulfur compounds such as dimethylsulfide, hydrogen sulfide, carbonyl sulfide, carbon disulfide, and dimethyl disulfide. The release of these compounds may reduce sulfide toxicity (Ernst 1990).

Some sulfide-tolerant plants may have the capacity to oxidize sulfide within the root tips. Sulfide oxidation has been observed in the root tips of *Spartina alterniflora*. The oxidation may occur because of the presence of sulfate-oxidizing bacteria on the root surface or it may be due to as yet undescribed enzymes that are catalysts for sulfide oxidation (Lee et al. 1999).

IV. Adaptations to Limited Nutrients

Nutrients come from precipitation and dry atmospheric deposition as well as the weathering of rocks and soil minerals and the decomposition of organic matter. In wetlands, decomposition is slow and nutrients tend to be bound in organic matter rather than mineralized. If little surface drainage enters a wetland from surrounding uplands, the plants can be completely dependent on atmospheric sources of nutrients. Wetlands with low nutrient status include raised peatlands, cypress domes, and basin mangrove forests.

The ability of some wetland plants to procure nutrients is enhanced by *mycorrhizal associations, nitrogen fixation,* and *carnivory*. Some exhibit strategies to conserve nutrients including *nutrient translocation* and *evergreen leaves*.

A. Mychorrhizal Associations

Mycorrhizae are symbiotic fungi associated with plant roots. They benefit the host plants by increasing the plant's ability to capture water, phosphorus, and other plant nutrients, such as nitrogen and potassium. The mycorrhizae benefit from the association because the plant roots provide carbohydrates. Mycorrhizae are common in upland plants and have been found to be associated with many wetland plants as well (Sondergaard and Laegaard 1977).

There are two major types of mycorrhizae: *endomycorrhizae* and *ectomycorrhizae* (Fitter and Hay 1987; Crum 1992; Raven et al. 1999). Endomycorrhizae, also called vesicular-arbuscular mycorrhizae, or VAM, are by far the most common type of mycorrhizae. They are found in 80% of angiosperms as well as some bryophytes (liverworts but not mosses) and pteridophytes (ferns and fern allies).

VAM produce two types of structures, called *arbuscules* and *vesicles*. Arbuscules are highly invaginated branching structures that are probably the site of nutrient exchange. Vesicles are storage bodies. VAM infect the roots of wetland plants and have been found in many submerged, free-floating, floating, and emergent species, including members of the Juncaceae (rushes) and Cyperaceae (sedges), two families that had previously been thought to be non-mycorrhizal (Sondergaard and Laegaard 1977; Clayton and Bagyaraj 1984; Ragupathy et al. 1990; Wigand and Stevenson 1994; Rickerl et al. 1994; Wetzel and van der Valk 1996; Christensen and Wigand 1998; Cooke and Lefor 1998; Turner et al. 2000).

Ectomycorrhizae, which are usually associated with trees, form a mantle, or sheath, around a plant's roots. There are no intercellular connections between the fungus and the roots, which are usually stunted. Ectomycorrhizae occur in only 3% of plants, some of which grow in northern peatlands, such as *Larix laricina* (tamarack), *Picea mariana* (black spruce), *Alnus incana* (speckled alder), *Betula glandulosa* (dwarf birch), and *B. pumila* (low birch).

Two additional types of mycorrhizae are found among the Ericaceae (heath family) and the Orchidaceae (orchid family). These are sometimes classified as endomycorrhizae (Fitter and Hay 1987). In the Ericaceae, which commonly grow in peatlands, hyphae form an extensive web over the root surface. The principal role of the fungus is to release enzymes into the soil that break down organic compounds, making nitrogen available to the plant and allowing the Ericaceae to inhabit nitrogen-poor peatlands.

In the Orchidaceae, mycorrhizae are associated with the seeds and seedlings. Without the appropriate mycorrhizae, the orchid seed will not germinate since it has no endosperm and depends on the fungus as a carbohydrate source for germination and seedling growth. Many orchids including species of the genera *Cypripedium* (lady-slipper), *Orchis* (orchis), *Habenaria* (rein orchid), *Listera* (twayblade), and *Spiranthes* (ladies' tresses), as well as *Isotria verticillata* (whorled pogonia), *Arethusa bulbosa* (dragon's mouth), *Pogonia ophioglossoides* (rose pogonia), and *Calopogon tuberosus* (grass-pink), can be found in bogs, often on raised hummocks out of the saturated zone.

In phosphorus-deficient soils, mycorrhizal plants grow better than non-mycorrhizal ones. In laboratory studies with upland plants, mycorrhizae have been shown to improve plant growth by enhancing phosphorus uptake. More phosphorus diffuses from the soil into mycorrhizal hyphae than into plant roots because the hyphae have a greater surface area and thus increase the potential for absorption of water, phosphorus, and other nutrients. Nitrogen uptake is also enhanced in plants with VAM associations.

Keeley (1980) compared the growth of mycorrhizal and non-mycorrhizal seedlings of *Nyssa sylvatica* (water tupelo) and found that those with VAM associations had significantly higher biomass than those without. In a Chesapeake Bay population of the submerged plant, *Vallisneria americana*, the uptake of both nitrogen and phosphorus was reduced when the mycorrhizae were removed with a fungicide (Wigand and Stevenson 1994).

Some plants have greater degrees of mycorrhizal infection than others. Plants that lack root hairs or have coarse root systems, such as those in the Magnoliaceae (magnolia family), tend to have a high dependence on mycorrhizae. Plants on the other extreme with finely branched roots and dense root hairs, such as the Poaceae (grasses), are often non-mycorrhizal except in phosphorus-poor soils. Clayton and Bagyaraj (1984) examined submerged species from several New Zealand lakes and found that 22 of them had mycorrhizal associations. As in upland plants, the presence or absence of root hairs was one of the determinants of the degree of VAM infection. None of the 15 species with abundant root hairs had median infection levels above 5%, while 13 of the 14 species with few or no root hairs had median infections above 20%.

Plants growing in drier or more oxidized soils tend to have a greater degree of mycorrhizal infection than those growing in reduced soils. Mycorrhizae require oxygen and may more readily infect roots in oxidized zones because of oxygen availability there. In New Zealand lakes, the infection level of 12 VAM-associated species declined with increasing water depth. The plants with a high degree of infection typically only grew in oxidized soils in the wave zone of lakes (Clayton and Bagyaraj 1984). In South Dakota freshwater marshes, the roots of six emergent species showed greater mycorrhizal infection levels in drier soils (54% soil water content) than in wet soils (75% soil water content; Rickerl et al. 1994). The wetland tree, *Chamaecyparis thyoides* (Atlantic white cedar), grows on the tops of raised hummocks, and its roots extend along the sides of the hummocks into saturated hollows. Roots throughout the hummocks are colonized by VAM, but the frequency of VAM occurrence is significantly lower at the bottoms of the hummocks where oxygen is less plentiful (Cantelmo and Ehrenfeld 1999).

As in upland habitats, the level of mycorrhizal infection in wetland plants appears to correspond roughly to the amount of available phosphorus in the soil. Wetzel and van der Valk (1996) examined 19 emergents in North Dakota and Iowa prairie potholes and found VAM associated with all of them, including the Cyperaceae. The degree of VAM infection was greatest in areas of low phosphorus availability. In Iowa wetlands, with 6 to 52 μg available phosphorus per gram of soil, the VAM infection levels were lower (0.2 to 52.1%) than in North Dakota (7.3 to 71.8%), where the available phosphorus was only 0.1 to 5 $\mu g\ g^{-1}$.

B. Nitrogen Fixation

Nitrogen fixation is the process by which the gaseous form of nitrogen (N_2) is made available to plants and other organisms. Some types of soil bacteria are capable of fixing N_2 and many exist in symbiosis with plants. Nitrogen-fixing bacteria develop in root nodules formed by the host plant's vascular system and derive energy from the plant while the plant benefits from the additional nitrogen. Nitrogen fixation is not particular to wetland plant species, but it appears in a few wetland plants and probably affords the plants supplementary nitrogen, just as it does in upland environments.

The major group of plants that form root nodules housing nitrogen-fixing bacteria is the Leguminosae (legumes). The legumes form a large family of 657 genera, at least four of which have aquatic species (*Neptunia, Discolobium, Aeschynomene,* and *Sesbania*) that grow as emergents or floating plants in the tropics or subtropics. The Australian flood-tolerant shrub, *Viminaria juncea,* is also in the Leguminosae (Tjepkema 1977; Walker et al. 1983; Cook 1996). Nitrogen fixation occurs in several genera outside of the legumes, including the bog-inhabiting *Alnus* (alder) and *Myrica* (gale, myrtle, bayberry). *Myrica gale* (sweet gale) is found in open peatlands and along lake and stream shores in northern North America and Europe. In a Massachusetts bog, *M. gale* nitrogen fixation added five to six times more nitrogen to the site than was added in bulk precipitation. Nitrogen fixation provided 43% of the estimated annual nitrogen requirement for *M. gale*. The leaves of *M. gale* contained over 2% nitrogen, more than the average nitrogen content (about 1.7%) in leaves of nearby shrubs of the Ericaceae (Schwintzer 1983).

In mangrove forests, N-fixing cyanobacteria have been found on the aerial roots and on the sediments surrounding *Avicennia marina* (grey mangrove; Potts 1979), *A. germinans* (Zuberer and Silver 1978; Toledo et al. 1995), *Rhizophora mangle,* and *Laguncularia racemosa* (white mangrove; Sheridan 1991). Like other saline wetlands, mangrove forests tend to have low nitrogen availability, yet the trees manage to grow, presumably due in part to their association with cyanobacteria (although the transfer of nitrogen from the cyanobacteria to the plant has not been confirmed with tracer studies). Nitrogen fixation occurs continuously, regardless of tidal submergence, with peaks at the highest temperatures and during the daylight hours. The association between the cyanobacteria and the mangroves may be of mutual benefit, with the cyanobacteria deriving carbon or other resources from the mangroves while the mangroves gain nitrogen (Potts 1979; Hussain and Khoja 1993; Toledo et al. 1995).

C. Carnivory

Carnivorous plants have the ability to entrap animal prey ranging in size from zooplankton to insects and, rarely, to small frogs and birds. The leaves of carnivorous plants have adaptations associated with the attraction, retention, trapping, killing, and digestion of animals and the absorption of their nutrients. The absorption of insects' nutrients is assumed to enhance the plant's fitness in terms of increased growth, chances of survival, pollen production, or seed set in areas that are low in nitrogen and/or phosphorus. Many carnivorous plants are found in nutrient-poor peatlands (Albert et al. 1992; Stewart and Nilsen 1992, 1993; Lowrie 1998).

1. Habitat and Range of Carnivorous Plants

There are six families of carnivorous plants with 13 to 15 genera (Table 4.2). The two genera with the greatest number of species, *Drosera* (sundew) and *Utricularia* (bladderwort), grow in wetlands and waters around the world, while genera with fewer species often have very localized habitats. Many of the carnivorous plants are found in the tropics or subtropics in South America and Africa. Australia's flora is particularly rich in carnivorous plants with about 35% of the estimated 500 species of carnivorous plants, 65% of which grow in the southwestern region of Western Australia. Despite a long dry season there, some low-lying areas in shallow depressions, spring-fed watersheds, floodplains, and peatlands retain sufficient moisture to support carnivorous plants. Many endure the dry season in much the same way as temperate carnivorous plants endure the winter: as dormant tubers, roots, buds, or seeds. *Nepenthes, Aldrovanda, Byblis,* and *Cephalotus* are all found in Western Australia as well as over 70 species and a number of subspecies of *Drosera* and about 20 species of *Utricularia* (Lowrie 1998).

In the northeastern U.S. and eastern Canada, *Sarracenia purpurea* (northern pitcher plant), several species of *Drosera* (sundew; *Drosera rotundifolia, D. linearis, D. anglica,* and *D. intermedia*), and *Utricularia* are common in *Sphagnum* bogs. *Pinguicula vulgaris* and *P. villosa* (butterwort) are also found in this area on the sandy shores of large lakes. Their range extends north and west into the subarctic regions of Canada and Alaska and south into northern California. The same species of pitcher plants, sundews, and bladderworts are found in bogs within the Appalachian Mountains of the eastern U.S. and as far south as Alabama (Schnell 1976).

In the southeastern coastal plain of the U.S., along the Gulf of Mexico from the panhandle of Florida to Texas, are the Gulf Coast pitcher plant bogs, which share a number of carnivorous species with northern peatlands. Seven of the eight eastern *Sarracenia* species (other than *S. purpurea*) as well as several sundews (*Drosera intermedia, D. filiformis, D. capillaris,* and *D. brevifolia*) and butterworts (*Pinguicula lutea, P. cearulea, P. planifolia, P. primuliflora* and *P. ionantha*) are found in this area. Perhaps the most striking feature of these bogs is that their original distribution in pre-settlement times of 2935 km^2 has shrunk to a preserve of only 12 km^2, with most of the remaining area threatened by altered drainage and lack of fire (Folkerts 1982).

The only western hemisphere Venus' flytrap (*Dionaea muscipula*) is found in savannahs and peatlands in southeast North Carolina and the east coast of South Carolina (Schnell 1976; Slack 1979). The California pitcher plant (*Darlingtonia californica*) occurs only in Pacific coastal bogs and mountain slopes from Oregon to northern California.

TABLE 4.2

The Families and Genera of Carnivorous Plants and Their Geographic Distribution

Family and Genus	No. of Species	Geographic Distribution
Sarraceniaceae		
Heliamphora	5	Guyana; Venezuela
Sarracenia	9	Eastern North America
Darlingtonia	1	N. California and S. Oregon
Nepenthaceae		
Nepenthes	65	Eastern Tropics to Ceylon and Madagascar
Droseraceae		
Dionaea	1	North and South Carolina, U.S.
Aldrovanda	1	Europe, India, Japan, Africa, Australia
Drosophyllum	1	S. Portugal, S.W. Spain, Morocco
Drosera	90	Ubiquitous
Byblidaceae		
Byblis	5 to 6	Western Australia
Cephalotaceae		
Cephalotus	1	Australia, extreme S.W.
Lentibulariaceae		
Pinguicula	48	Northern Hemisphere, Old and New Worlds; 3 species in South America
Utricularia[a]	275	Ubiquitous
Polypompholyx[b]	2 to 4	Southern Australia
Genlisea	16	W. Africa, E. South American tropics

[a] *Utricularia* includes two species that were formerly classified within a separate genus, *Biovularia*. Both species are found in Cuba and eastern South America.

[b] *Polypompholyx* is included as a subgenus of *Utricularia* by some authors.

From Lloyd 1942; Slack 1979; Cook 1996; and Lowrie 1998.

2. Types of Traps

The traps of carnivorous plants provide the main means of distinguishing the genera. The traps are described here in order from the most passive types of trap, in which plant movement is not a part of the trapping process, to the most active, in which plant parts move or snap shut (Table 4.3; Schnell 1976; Lowrie 1998).

a. Pitfall Trap

Pitfall traps are tube-shaped leaves, or pitchers, with various modifications. The prey is lured to the trap (which is sometimes red, resembling a flower) by nectar, enters it or falls in, and is unable to escape. Modifications such as downward pointing hairs within the traps make crawling out difficult (Figure 4.B.2). The pitchers of many species are covered by a lid, or hood, which shades the fluid and inhibits its evaporation or dilution by rainwater. The pitchers are usually filled with fluid in which the prey drowns. Enzymes are secreted by the plant that act with bacteria and resident insects to break down the prey into usable forms of nutrients (Fish and Hall 1978). Pitchers are inhabited by a number of organisms that resist digestion including algae, fungi, bacteria, protozoa, and various

TABLE 4.3

Types of Traps and Their Distribution among the Genera of Carnivorous Plants

Type of Trap	Genus
Pitfall	*Heliamphora*
	Sarracenia
	Darlingtonia
	Nepenthes
	Cephalotus
Lobster pot	*Genlisea*
Passive adhesive	*Byblis*
	Drosophyllum
Active adhesive	*Pinguicula*
	Drosera
Bladder	*Utricularia*
	Polypompholyx
Snap-trap	*Dionaea*
	Aldrovanda

From Lloyd 1942 and Lowrie 1998.

insect larvae. Spiders frequently spin webs across pitcher openings and intercept the plants' potential prey (Creswell 1991).

In the U.S., the pitfall trap is present in the pitcher plants, *Sarracenia* (see Case Study 4.B, Carnivory in *Sarracenia purpurea*) and *Darlingtonia californica. D. californica* has a tubular pitcher that is narrow at the bottom and widens to 12 to 15 cm at the top. *D. californica* has no digestive glands; its prey is decomposed by resident microorganisms and the nutrients are absorbed by the cells that line the pitcher (Schnell 1976).

Cephalotus follicularis (Albany pitcher plant) grows only in Australia. The outside of the pitchers of *C. follicularis* have ladder-like ridges covered with long stiff hairs that give crawling insects a foothold as they climb to the pitcher's mouth, attracted by the nectar (Figure 4.18). A rim of inward pointing teeth allows insects to enter, but not crawl out of the pitcher. The insect is also lured downward because the nectar is sweeter lower on the inside walls. Where the nectar becomes sweeter, a waxy zone causes insects to slip into the pitcher's fluid. The prey drowns and is slowly dissolved with the aid of enzymes and bacteria.

Nepenthes has 65 species in the tropics and subtropics of Asia and in Australia. *N. mirabilis* (tropical pitcher plant) is a climbing plant of Australia, Papua New Guinea, and southern China. It has tendrils and two types of pitchers. In the plant's rosette stage, the pitchers have two serated wings at the front with stiff hairs at the apex of each serration. The wings seem to act as ladders for insects climbing from the ground to the pitcher's mouth. When the plant forms a vine, the upper pitchers have no wings or hairs, which would have no function since these pitchers are suspended in the air (Figures 4.19a and b; Lowrie 1998).

b. Lobster Pot Trap

The lobster pot trap is found only in the genus *Genlisea*, which grows in the tropics of Africa and South America. All 16 species are small rootless plants that are seasonally submerged, with only the inflorescence emerging above the water's surface. They form rosettes with two kinds of leaves: foliage and trapping. The trapping leaves start with a

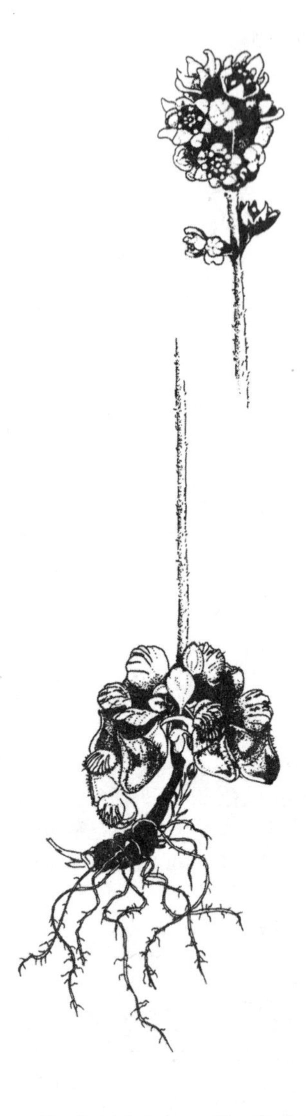

FIGURE 4.18
Cephalotus follicularis (Albany pitcher plant), a perennial that forms compact colonies of leaves and pitchers (pitchers are approximately 2 to 3 cm at the widest point. (From Pietropaolo, J. and Pietropaolo, P. 1986. *Carnivorous Plants of the World*. Portland, OR. Timber Press. Reprinted with permission.)

cylindrical footstalk, then develop a hollow bulb. From the bulb grows a long tubular neck that has a slit-like opening at the end. The neck splits into two branches that twist and look like spirals (Figure 4.20). Inside the tubular neck are ridges with inward pointing hairs. The prey, which consists of several zooplankton species, enters at the mouth and cannot retreat due to the hairs. Some insects die and are broken down within the tubular neck while some move into the bulb where the nutrients are absorbed. The spiral appendages anchor the plants in the sand when they are left exposed during dry periods (Lloyd 1942; Slack 1979; Cook 1996).

c. Passive Adhesive Trap

Adhesive, or flypaper, traps are leaves covered by sticky glands in which prey become stuck. In passive adhesive traps, there is no plant movement in response to prey capture: the prey is broken down where it becomes stuck. Passive adhesive traps are found in *Byblis*, and in *Drosophyllum lusitanicum*.

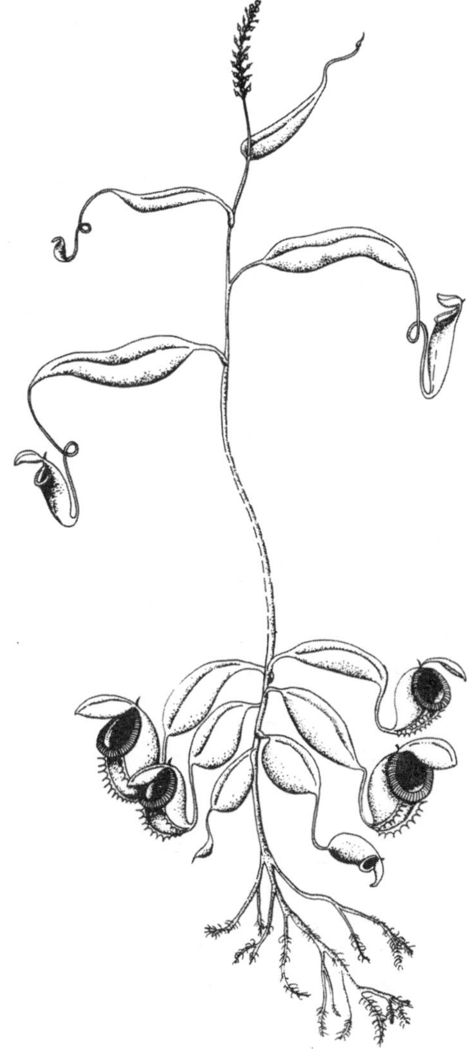

FIGURE 4.19a
Nepenthes mirabilis (tropical pitcher plant), a perennial with leafy basal rosettes, each leaf with a tendril bearing an insect-trapping pitcher at the end. Vine-like stems up to 10 m long arise from the rosettes. The upper pitchers lack the bristles seen on the basal pitchers (pitchers are approximately 5 cm at the widest point. (From Pietropaolo, J. and Pietropaolo, P. 1986. *Carnivorous Plants of the World.* Portland, OR. Timber Press. Reprinted with permission.)

FIGURE 4.19b
The pitchers of another species of tropical pitcher plant, *Nepenthes alata.* (Photo by J. Cronk.)

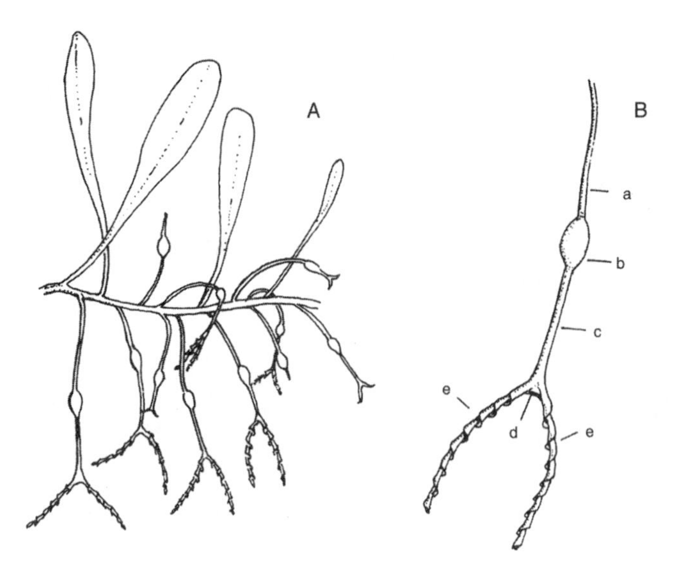

FIGURE 4.20
Genlisea sp. (A) part of a typical plant showing foliage and trap leaves. The length of the trapping leaf varies between 2.5 and 15 cm. (B) An enlarged trap leaf showing [a] a cylindrical footstalk, [b] a hollow bulb, [c] a cylindrical neck, [d] a slit-like mouth, [e] the two branches that extend from the mouth in a spiral. (From Slack, A. 1979. *Carnivorous Plants.* Cambridge, MA. MIT Press. Reprinted with permission.)

The five or six species of *Byblis* grow in Western Australia. They have long narrow leaves covered with both long- and short-stalked glands. The glands are shaped like straight pins and the flattened head produces a viscous fluid. Insects that land on the leaves are trapped by the sticky fluid on the long glands. As the insect struggles, it becomes more entrapped as more glands adhere to it. The fluid from all of the glands involved form a larger pool of liquid that engulfs the prey. The insect is decomposed within the fluid and the nutrients are absorbed by the smaller glands (Figure 4.21; Lowrie 1998).

Drosophyllum lusitanicum grows only in Spain, Portugal, and Morocco and differs from most carnivorous plants in that it is usually found in dry, often alkaline soil. Like *Byblis,* it forms long tapering leaves about 20 cm in length. The leaves are covered with two types of glands. Red mucilage-secreting glands secrete clear drops of a glutinous liquid. The liquid is not as viscous as that secreted by other adhesive trap species and the insect is not tightly held where it lands. It can move up and down the leaf, collecting beads of liquid from each gland it passes. The liquid eventually engulfs the insect and drowns it. A second set of glands, called the digestive glands, then start to secrete an enzyme-filled fluid that breaks down the prey in as little as 24 h (Slack 1979).

d. Active Adhesive Trap

Plants with active adhesive traps secure their prey with a glandular secretion just as plants with passive adhesive traps do. Once the prey is stuck, the leaves or glands move around it, curling at the margins, or slowly closing completely. *Pinguicula* leaves have stalked sticky glands that capture and detain prey as well as stalkless glands that aid in digestion and absorption of nutrients. The movements of a struggling insect stimulate the margins of the leaves to roll inward so even more glands touch the prey. If an insect lands in the center of the leaf, the leaf forms a "dish" around it, by curling a small leaf area into a tiny bowl. Many

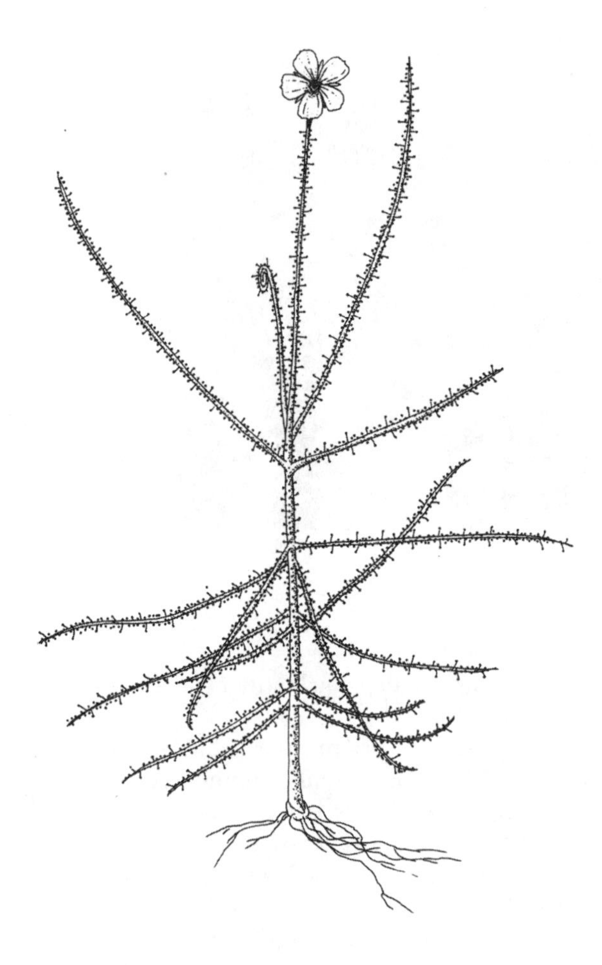

FIGURE 4.21
Byblis filifolia is an annual erect plant that grows up to 40 cm tall. Its leaves are covered with sticky glands that entrap insects. (From Pietropaolo, J. and Pietropaolo, P. 1986. *Carnivorous Plants of the World*. Portland, OR. Timber Press. Reprinted with permission.)

Pinguicula species grow in acidic bogs, but some grow in alkaline conditions on limestone rocks.

In *Drosera*, the glands secrete a sticky fluid in which insects become trapped. The sundews move their glands toward the struggling captured prey and position the prey to be in contact with a greater number of glands for effective digestion. If an insect lands on glands at the edge of the leaf, the glands bend inward, moving the prey to the center of the leaf where glands are more numerous (Figure 4.22; Slack 1979; Lowrie 1998) Larger insects are able to break the threads of mucilage on adhesive traps and escape. Insects that are over 1 cm long can generally escape *Drosera* traps and those over 0.5 cm can escape *Pinguicula* traps (Gibson 1991).

e. Bladder Trap

Bladder traps are found in the genera *Utricularia* (bladderwort) and *Polypompholyx*. Many *Utricularia* species are submerged; all are rootless (Figure 4.23). Their traps are positioned along the stems among the leaves. The traps are somewhat bulbous and have an inward opening flap over an entrance at one end (Figure 4.24). Sensitive external hairs next to the flap are stimulated by movement in the water. In response, the flap opens inward and the prey (usually zooplankton) and the surrounding water are engulfed in the bladder trap and the flap closes. The plant then actively pumps ions and water out of the bladder and transports nutrients from the digested prey into the rest of the plant. A trap can fire, reset,

FIGURE 4.22
Drosera rotundifolia (sundew) with small insects trapped on some of the leaves. (Photo by H. Crowell.)

and fire again every 20 min. The prey is broken down within the bladder traps by bacteria and possibly by enzyme activity.

While many of the carnivorous plants are found in nutrient-poor habitats, *Utricularia* species are found in a wide range of nutrient regimes. Because the submerged species of *Utricularia* are rootless, all of their nutrients must be derived from the water column. Carnivory enhances shoot nutrient uptake and probably allows these plants to thrive without contact with sediment nutrients (Schnell 1976; Knight and Frost 1991; Knight 1992; Lowrie 1998). Some species of *Utricularia* grow in wet soils and some South American ones are epiphytic, growing in the moss and rotting bark on rain forest trees. Some of the submerged species grow only in the water bowls formed by the leaves of other epiphytes.

Polypompholyx is sometimes included as a subgenus of *Utricularia*. Its rosettes and flowers grow without roots, but it is anchored to the sediments by bladder traps that form on the tips of its leaves. It is seasonally submerged, but can grow on dry soils. Its two to four species are found only in Australia. The traps work in a similar manner to the bladder traps of *Utricularia* (Figure 4. 25; Slack 1979; Lowrie 1998).

f. Snap-Trap

The snap-trap, or spring trap, is bivalved (two similar halves connected by a midrib) and the two halves close around prey. Only two species, both in the Droseraceae, have snap-traps, *Dionaea muscipula* (Venus' flytrap) and *Aldrovanda vesiculosa* (waterwheel plant).

Dionaea muscipula, the well-known Venus' flytrap, has leaves that end in two lobes, the margins of which have 15 to 20 tough pointed bristles. In the center of each lobe are two to four trigger hairs which, when touched by prey, stimulate the closing of the two lobes. Each trap is rather short-lived and can normally catch only three insects before becoming inactive. *D. muscipula* grows in both dry and wet soils and is not necessarily a wetland plant.

Aldrovanda vesiculosa is a submerged rootless plant that floats just below the water's surface in acidic waters of Europe, Asia, and Australia. The plant's major stem is up to 20 cm long with whorls of leaves at short intervals along its length. Each whorl has five to

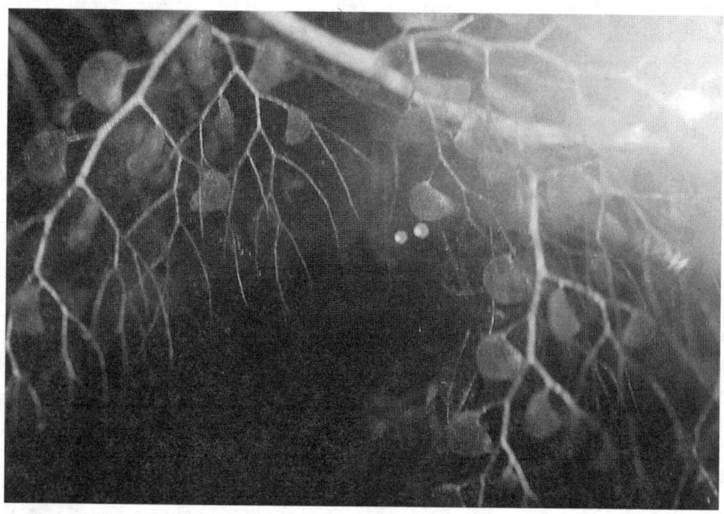

FIGURE 4.23
Utricularia macrorhiza (bladderwort), showing the round bladder traps among the leaves. (Photo by H. Crowell.)

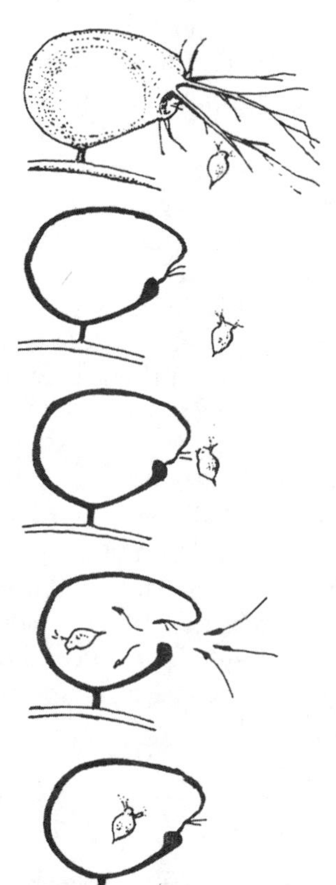

FIGURE 4.24
A bladder trap of *Utricularia macrorhiza* showing prey-guiding antennae; section of bladder with door closed and partial vacuum within; trigger hairs touched by swimming *Daphnia*; door immediately opening, releasing partial vacuum, prey being sucked in by inrush of water; door closing, imprisoning prey. Partial vaccum gradually restored (trap diameter is approximately 50 mm). (From Slack, A. 1979. *Carnivorous Plants.* Cambridge, MA. MIT Press. Reprinted with permission.)

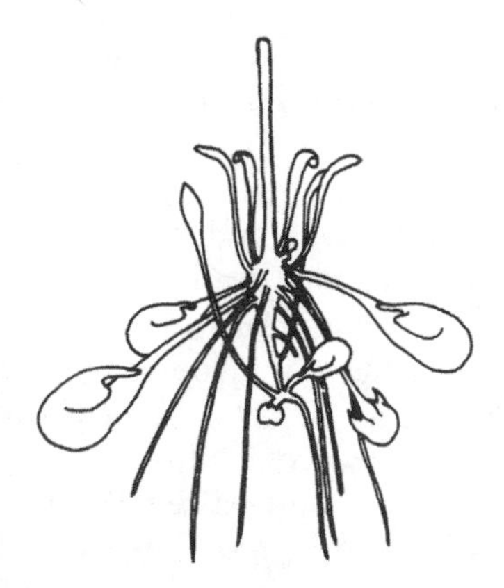

FIGURE 4.25
Polypompholyx tenella (also called *Utricularia tenella*), an annual plant with a compact basal rosette of leaves 10 mm in diameter with bladder traps among the leaves. The traps and leaves are covered by a film of water at the time of flowering. (From Cook, C.D.K. 1996. *Aquatic Plant Book.* The Hague. SPB Academic Publishing/Backhuys Publishers. Reprinted with permission.)

nine leaves, and each leaf ends in a bivalved trap. When prey enters the open trap, the two sides rapidly snap shut. The struggling prey causes the trap to close more tightly and to seal. The prey is decomposed within the trap (Figure 4.26; Lowrie 1998).

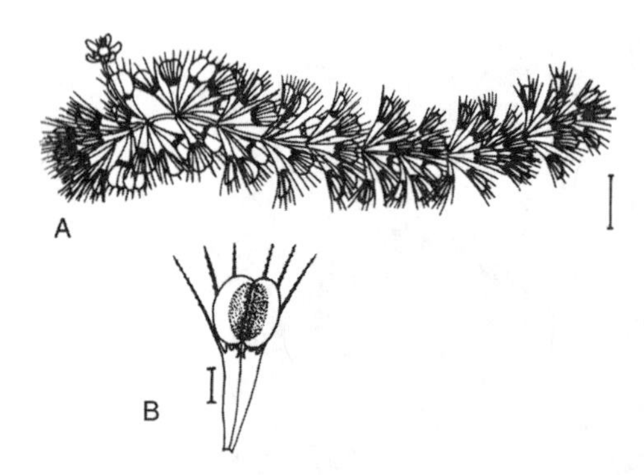

FIGURE 4.26
Aldrovanda vesiculosa (waterwheel plant), a submerged rootless plant with bivalved "snap-traps" at the end of each leaf. (a) stem (bar = 1 cm) and (b) leaf with trap (bar = 2 mm). (From Cook, C.D.K. 1996. *Aquatic Plant Book.* The Hague. SPB Academic Publishing/Backhuys Publishers. Reprinted with permission.)

3. Benefits and Costs of Carnivory

Carnivorous plants are assumed to have a growth and fitness advantage in nutrient-poor habitats. Indeed, some studies have shown that carnivorous plants fare better with prey than without. A study of two sundews, *Drosera intermedia* and *D. rotundifolia,* showed that when supplied with fruit flies, both species reacted with increased plant size and more flowers per plant. In *D. intermedia,* the frequency of vegetative reproduction also increased compared to control plants that depended solely on soil and rainwater nutrients (Thum 1989a, b). Another study showed the importance of prey sources of nitrogen. The bladderwort, *Utricularia macrorhiza,* derives up to 75% of its nitrogen needs from prey (Knight and Frost 1991). In *Pinguicula vulgaris,* the uptake of soil nitrogen is enhanced by increased prey capture. Prey capture may stimulate root growth and thus increase plant uptake capacity (Hanslin and Karlsson 1996).

The sundews, *Drosera binata* var. *multifida* (native to Australian and New Zealand bogs) and *D. capensis* (found in South African bogs), appear to be facultative carnivores. In a study in which some plants were kept in exclosures to prevent prey capture while controls were not, none of the species benefited from insect capture in either nutrient-poor or -rich soils. In fact, plants that were not able to capture prey were slightly larger than those that could (Stewart and Nilsen 1993).

The benefits of carnivory must be balanced with the energy cost of maintaining the traps. While the relative balance of benefits and costs changes under different nutrient regimes, benefits may outweigh the costs during stressful periods (Karlsson et al. 1991; Knight and Frost 1991). Some carnivorous plants appear to be facultative carnivores, so carnivory may provide them with an alternative source of nutrients in nutritionally hard times (Stewart and Nilsen 1993). For example, when the nutrient status of a peatland diminishes during a prolonged period without fire, carnivory may become a significant source of nutrients and vital to survival and competition (Folkerts 1982).

D. Nutrient Translocation

Most wetland plants are perennials that conserve nutrients through nutrient translocation. Herbaceous perennial species, such as members of the Poaceae and Cyperaceae, translocate nutrients and carbohydrates from aboveground tissues to belowground ones such as roots, rhizomes, tubers, and bulbs. The stored material allows the plant to overwinter and is a source of energy for the following growing season's initial growth. This strategy allows plants to conserve nutrients from one growing season to the next (Crawford 1978; Grace 1993).

Trees in temperate areas retain nutrients in woody tissue during the winter and translocate them to the foliage in the spring. In a study of nutrient translocation in two wetland trees, *Taxodium ascendens* (pond cypress) and *Nyssa sylvatica* var. *biflora* (swamp black gum), the foliar nitrogen concentration was about three times greater in the spring (nearly 3% nitrogen content) than in the fall (1%). The foliar phosphorus concentration decreased by about half over the course of the growing season (from 0.15 to about 0.08%). In November and December the foliar nutrients were translocated into the twigs. After December the nutrients diffused from the twigs to the branches, the trunk, and the roots. In the spring, the opposite movement of nutrients was observed. This conservation and recycling of nutrients are especially beneficial in cypress domes where decomposition is slow and a large portion of the nutrients is locked up in undecayed litter and humus. The ability of these trees to redistribute nutrients throughout the year enables them to grow where nutrients seem limiting (Dierberg et al. 1986).

E. Evergreen Leaves

Evergreens are found in many habitats, including wetlands. A number of peatland species such as the shrubs of the Ericaceae (Figures 2.23 and 2.24) and *Picea mariana* (black spruce) are evergreen. Evergreen leaves allow plants to maintain foliar nutrients longer than a single growing season and may be particularly adaptive in basin forests such as ombrotrophic bogs with few external nutrient inputs (Schlesinger 1978; Crawford 1993). The Ericaceae retain their leaves for 2 years, which may help them compete with another peatland shrub, the nitrogen-fixing *Myrica gale* (sweet gale; Schwintzer 1983).

V. Adaptations to Submergence

Submerged plants face different growth constraints than upland, emergent, floating, and floating-leaved plants since within the water column they are exposed to lower oxygen and carbon dioxide concentrations and to a lower light regime.

A. Submerged Plant Adaptations to Limited Light

While all plants can suffer a lack of light by shading, upland as well as emergent and floating-leaved plants often cope with shade by growing fast, growing toward the source of light, and growing where there is an opening. Submerged plants have similar shade strategies as well as structural adaptations involving leaf design, shape, and thickness (Table 4.4).

While chloroplasts in land plants are found largely in the mesophyll (internal tissue of a leaf), the chloroplasts of freshwater submerged species of *Ceratophyllum*, *Myriophyllum*, and *Potamogeton* as well as several marine angiosperms are concentrated in the epidermis. The surface of the submerged leaf is the site of most of the plant's photosynthesis. The mesophyll serves mainly for the storage of starch or oils (Sculthorpe 1967).

Submerged leaves are often ribbon-like or highly dissected with a high surface area-to-volume ratio which facilitates both the penetration of light and the diffusion of dissolved gases to the plant's inner tissues. Both dissected and ribbon-like leaves offer little resistance to currents and can trail freely in water (Sculthorpe 1967).

TABLE 4.4

Similarities and Changes in Basic Plant Structures That Occurred in the Evolution from Land to Submerged Plants

Plant Part	Function in Land Plants	Function in Submerged Plants
Roots	Roots anchor the plant in soil, and with root hairs and mycorrhizae, absorb water and nutrients from soil; large quantities of water are required (100 g water g^{-1} dry wt. gained by the plant)	Roots anchor the plant and absorb soil nutrients, but shoots are also capable of absorbing water; submerged shoots can absorb nutrients from the water column, but the majority of submerged plants' nutrients come from sediments;[a] rootless plants such as *Utricularia* and *Ceratophyllum* absorb nutrients from the water column
Leaves	Leaves are many layers thick; photosynthetically active cells are just below the epidermis, or surface layer, in the mesophyll	Submerged leaves are 1 to 3 cell layers thick and photosynthetically active cells are concentrated in the epidermis, which maximizes their proximity to light, dissolved gases, and dissolved nutrients
Xylem	Xylem provides a pathway for moving water and nutrients to the shoot, and lignified cell walls are important structural elements; most water is lost to transpiration	The quantity of xylem and its degree of lignification are much reduced; since water surrounds the shoot, transpirational loss does not occur
Cuticle	The cuticle (a waterproof/water-repellent layer) occurs at all cell wall/gas phase interfaces, especially the outer surface of the shoot; the cuticle restricts water loss through the plant surface	The cuticle is usually thin and not a significant barrier to nutrient and water uptake
Stomata	Stomata (epidermal pores whose openings can be varied in response to environmental and endogenous signals) control gas exchange	Stomata are infrequently found on submerged shoots; generally, submerged leaves do not have stomata; dissolved gases enter and exit the plant through diffusion
Intercellular	Gas spaces permit carbon dioxide distribution in the plant; they permit the growth of tall plants, affording them advantages in competing for light	Air spaces take up much more volume than in terrestrial plants, providing buoyancy and allowing increased gas transport within the plant

[a]Bristow and Whitcombe 1971; Toetz 1974; Carignan and Kalff 1980; Denny 1980; Barko and Smart 1980; 1981.

From Sculthorpe 1967 and Raven 1984.

In some species with both submerged and emergent foliage, the chlorophyll concentration is higher in the submerged leaves. For example, in several mangrove species (*Acanthus ilicifolius, Avicennia officinalis,* and *Bruguiera gymnorhiza*), the leaves that are periodically submerged by tides have a greater chlorophyll content than other leaves. The higher chlorophyll content may reflect the lower light conditions in the turbid estuarine water (Misra et al. 1984). Increased chlorophyll content has also been noted in the submerged leaves of *Potamogeton polygonifolius* and *P. perfoliatus*. The increased chlorophyll leads to an increase in the photosynthetic rate at low light (Kirk 1994).

Submerged species with *apical growth* (growing upward from the tips of the stalks) grow toward the water surface and can partly compensate for light attenuation in the water by concentrating their leaf biomass close to the water surface (Barko and Smart 1981a; Sand-Jensen and Borum 1991). Some submerged species with apical growth, notably those in the Hydrocharitaceae (frogbit) and some *Potamogeton* species (pondweed), form lush canopies with most of their foliage near the water's surface (Dibble et al. 1996). Canopy formers such as *Myriophyllum spicatum* (Eurasian watermilfoil) and *Hydrilla verticillata* (hydrilla) shade other submerged plants below. This capacity has led to their success as invasive plants in many water bodies of the U.S., Canada, and elsewhere (Stevenson 1988; Madsen et al. 1991; Howard-Williams 1993; see Chapter 8, Invasive Plants in Wetlands).

B. Submerged Plant Adaptations to Limited Carbon Dioxide

Submerged plants exhibit a number of adaptations to limited or variable CO_2 levels. Some of the adaptations to low carbon dioxide levels are the same as for low light and oxygen (Table 4.4). For example, aerenchyma promotes buoyancy, enabling the plant to grow close to the water's surface where it has greater access to light and atmospheric CO_2. Dissected or ribbon-like leaves, a thin cuticle, and the presence of chloroplasts in the epidermis seem to be adaptations to both low light levels and limited CO_2 since they increase the surface-to-volume ratio and decrease the distance that inorganic carbon must travel in order to be used (Sculthorpe 1967).

Submerged plants have several other adaptions to low CO_2:

- Many submerged plants are able to assimilate bicarbonate ions for photosynthesis. This is perhaps the most critical mechanism that enables submerged plants to inhabit the water column (Prins et al. 1982a; Lucas 1983).
- Through a mechanism called *aquatic acid metabolism,* some submerged plants are able to assimilate CO_2 at night, when it is more plentiful (Cockburn 1985).
- Some submerged species can recycle respired CO_2 within their aerenchyma and assimilate it in photosynthesis. They maintain internal CO_2 pressures that are higher than external pressures (Madsen and Sand-Jensen 1991).
- High soil respiration rates create a pool of available CO_2 that some submerged plants are able to use (Bowes and Salvucci 1989).

1. Use of Bicarbonate

Many submerged plants are able to use bicarbonate (HCO_3^-) for photosynthesis. Uptake of HCO_3^- has been observed in Myriophyllum spicatum (Grace and Wetzel 1978), Scirpus subterminalis (water bulrush; Beer and Wetzel 1981), Hydrilla verticillata, Egeria densa (egeria), Elodea canadensis (elodea), and Potamogeton species (pondweed), as well as in non–angiosperms such as the Characeae (stoneworts) and many algal species (Prins et al. 1982a). In general, HCO_3^- use is seen more often in monocots than in eudicots.

There is a continuum of bicarbonate use that ranges from plants that use CO_2 exclusively to those that are able to make effective use of HCO_3^- (Allen and Spence 1981; Prins and Elzenga 1989). Species that use CO_2 exclusively include *Myriophyllum aquaticum* (formerly called *M. brasiliense*; parrot feather), *M. verticillatum* (green milfoil), *M. hippuroides, Ludwigia natans, Echinodorus tenellus* (burhead), and *E. paniculatus* (Prins et al. 1982a). Most plants' capacities to use HCO_3^- are flexible and they convert to HCO_3^- use when CO_2 concentrations become limiting. For the genus *Potamogeton,* HCO_3^- use appears to be correlated to the abundance of HCO_3^- in the native habitat of each species. *Potamogeton* species that normally grow in waters with high HCO_3^- concentrations are better able to exploit HCO_3^- than species from waters of low HCO_3^- (Prins and Elzenga 1989).

A mechanism called the *polar model* has been proposed to explain the capacity of some plants to incorporate HCO_3^- and convert it to CO_2. In the polar model, the use of HCO_3^- results in the production of one hydroxide (OH^-) molecule for every molecule of CO_2 assimilated. To control the pH within the cells, OH^- is excreted from the plant, so the net effect for the plant is the same as CO_2 fixation (Prins et al. 1982a). The mechanism is called polar because HCO_3^- is taken up on the lower side of the leaf and OH^- is excreted on the upper side; the pH is higher on the upper side of the leaf due to the excreted OH^-. The conversion from HCO_3^- to CO_2 may take place within an invagination in the cell wall. Potassium (K^+) moves along with the OH^- and balances the charge within the cells. This model has been confirmed in *Elodea canadensis, Hydrilla verticillata,* and *Potamogeton lucens* (Figure 4.27; Prins et al. 1982b; Lucas 1983; Sondergaard 1988; Krabel et al. 1995).

Submerged plants that are able to use both CO_2 and HCO_3^- may have a wider range of habitats available to them. Since the use of inorganic carbon throughout the day results in raising the pH and raising the ratio of HCO_3^- to CO_2 molecules, HCO_3^- assimilators would have an advantage over plants that can only use CO_2, especially in water bodies with slow mixing. Ultimately, however, plants use HCO_3^- only when CO_2 is not available. The use of HCO_3^- is less efficient than CO_2 use and results in lower photosynthetic rates (Prins and Elzenga 1989).

2. Aquatic Acid Metabolism

Aquatic acid metabolism (AAM) is similar to Crassulacean acid metabolism (CAM) which is usually seen in plants of xeric landscapes. In CAM, the stomata are closed during the day to prevent water loss and opened at night to allow for the uptake of CO_2. AAM occurs in submerged plants that have no stomata. In AAM, CO_2 uptake occurs by diffusion during the night. Both systems are referred to as diel photosynthetic acid metabolism (DPAM). The basic features of DPAM are (Cockburn 1985):

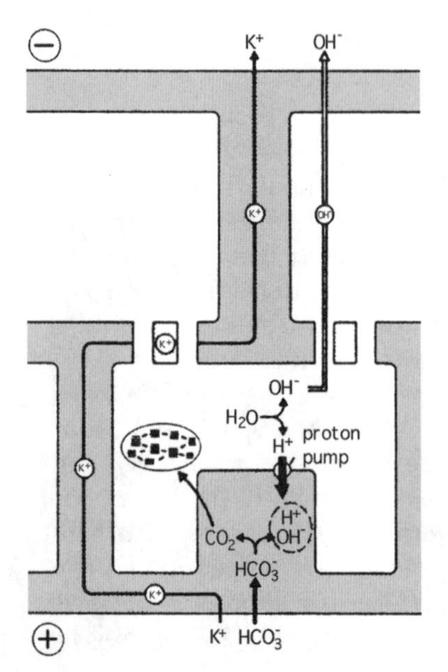

FIGURE 4.27
Proposed scheme for bicarbonate conversion into carbon dioxide by means of light-dependent proton pumps. Two cell layers, the upper and lower epidermis, are shown (the leaves of *Elodea* spp. and some other submerged species typically have only two to three cell layers). The cells of both layers are connected by plasmodesmata (narrow strands of cytoplasm that pass through pores in plant cell walls and join the cells to one another). An invagination in the plasmalemma (the cell membrane that lines the connecting plasmodesmata between cells) is schematically depicted as an ingrowth of the cell wall. The HCO_3^- ion enters the invagination and splits into CO_2 and OH^-. The CO_2 is used in photosynthesis and the OH^- is excreted at the upper side of the leaf. K^+ transport proceeds via the cell wall, indicated by the shaded area, and balances the charge within the cells. (From Prins et al. 1982b. *Studies on Aquatic Vascular Plants.* J.J. Symoens, S.S. Hooper, and P. Compere, Eds. Brussels. Royal Botanical Society of Belgium. Reprinted with permission.)

- CO_2 is acquired from the atmosphere or water in darkness.
- CO_2 is incorporated into the carboxyl group of an organic acid (usually malic acid) which accumulates in the cell in which it is synthesized. During the following day, the acid is decarboxylated, meaning that the CO_2 is released and used in photosynthesis.

AAM has been observed in several submerged angiosperms such as *Hydrilla verticillata, Littorella uniflora,* and *Scirpus subterminalis,* as well as in a number of submerged Isoetaceae (pteridophytes). In the submerged plant's environment, CO_2 can become limiting during the day and its concentration is at a maximum at night. AAM plants are able to take up CO_2 at night without competition from non–AAM plants (Cockburn 1985).

3. Lacunal Transport

Submerged plants have extensive aerenchyma in which gases are stored and transported. Carbon dioxide produced by respiration in the roots and rhizomes is transported to the leaves and used in photosynthesis in some plants. This recycling of CO_2 has been observed in *Juncus bulbosus* and species of *Isoetes* in soft water lakes with low carbon content (Stevenson 1988). *Hydrilla verticillata* and *Elodea nuttalli* both store CO_2 in internal gas spaces. In *E. nuttalli,* the internal CO_2 level has been measured at 100 to 500 times the external CO_2 level (Madsen and Sand-Jensen 1991).

4. Sediment-Derived CO_2

The sediment is a source of inorganic carbon for some submerged plants. Due to both root and animal respiration, soil water CO_2 concentrations can be about twice as high as in the water column. In experiments in which labeled CO_2 was supplied to the roots of *Lobelia dortmanna, Isoetes lacustris,* and *Littorella uniflora,* the carbon was fixed in the plants' leaves. Sediment-derived CO_2 can be up to 90% of carbon uptake in these species; however, not all submerged plants derive carbon from the sediments. Species of *Myriophyllum, Vallisneria, Heteranthera,* and *Hydrilla* take less than 1.5% of their inorganic carbon needs from the sediments (as reviewed by Bowes and Salvucci 1989). Conversely, in the emergent species, *Scirpus lacustris* and *Cyperus papyrus,* sediment CO_2 was found to be the only source of inorganic CO_2 for photosynthesis in submerged young green shoots (Singer et al. 1994).

C. Adaptations to Fluctuating Water Levels

In zones along the edges of lakes, streams, or wetlands where flooding is regularly alternated with periods of dessication, many plants are able to grow as both submerged and emergent plants. Some of these form differently shaped leaves when submerged than when they are emergent. This strategy, called *heterophylly,* allows plants to survive under both dry and submerged conditions and may give them a competitive edge over submerged plants that cannot survive outside of water and emergent plants with little tolerance for continual submergence.

Many heterophyllous plants have ovate, elliptic, or rounded emergent leaves, while their submerged leaves are longer and ribbon-like with little or no differentiation of a blade. These include several species of *Sagittaria* as well as *Rotala indica, R. rotundifolia, Callitriche palustris, Cryptocoryne beckettii, C. ciliata, C. thwaitesii, C. wendtii, Didiplis diandra, Echinodorus brevipedicellatus, E. grisebachii, E. tenellus, Ludwigia arcuata, L. repens, L. palustris,* and *Butomus umbellatus.* The submerged leaves of some plants, such as

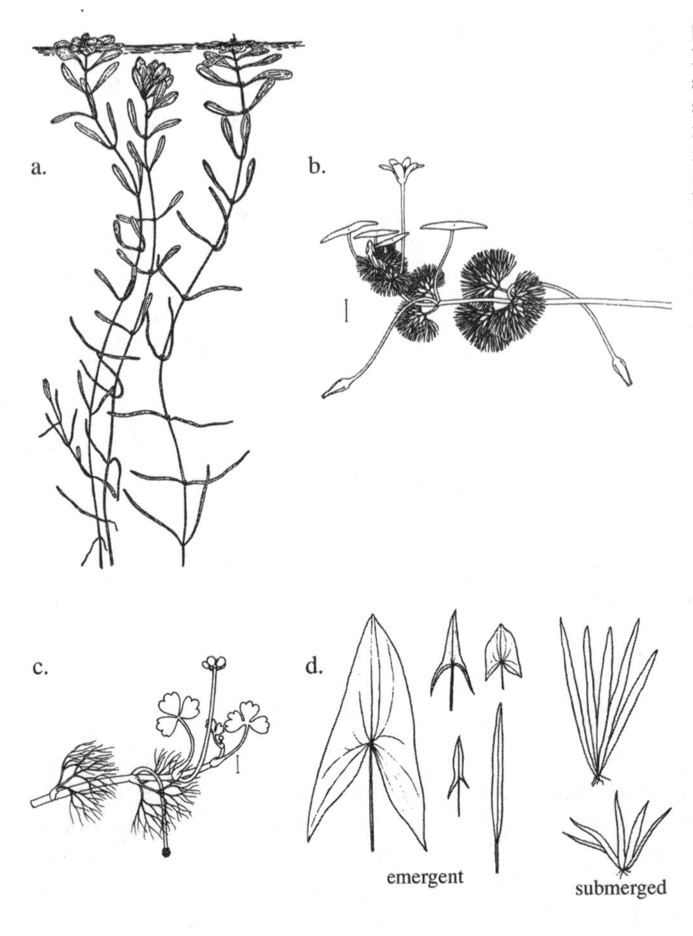

FIGURE 4.28
Plants showing heterophylly with submerged leaves that differ in shape from those that are either near, on, or above the water's surface. (a) *Callitriche palustris* (water starwort) with thinner, more ribbon-like leaves farther below the surface and rounder, spatulate leaves near the water's surface (leaves are 1.5 to 2 cm long). (From Fassett, N.C. 1957. *A Manual of Aquatic Plants.* Madison, WI. University of Wisconsin Press. Reprinted with permission.) (b) *Cabomba furcata* with fan-like dissected leaves below the surface and entire leaves floating on the water's surface (bar = 1 cm). (c) *Ranunculus peltatus* subsp. *baudotii* (bar = 1 cm) with dissected leaves below surface (b and c from Cook, C.D.K. 1996. *Aquatic Plant Book.* The Hague. SPB Academic Publishing/ Backhuys Publishers. Reprinted with permission.) (d) *Sagittaria cuneata* (northern arrowhead) with ribbon-like leaves below the water and sagittate, or arrow-shaped leaves emergent (submerged leaves are from 6 to 20 cm long). (From Hotchkiss, N. 1972. *Common Marsh, Underwater and Floating-Leaved Plants of the United States and Canada.* New York. Dover Publications, Inc. Reprinted with permission.)

Proserpinaca palustris and some *Ranunculus* species, are more highly dissected with longer lobes than their emergent leaves (Figures 4.28 and 4.29; Sculthorpe 1967; Kaul 1976).

Highly dissected or ribbon-like leaves increase the surface area-to-volume ratio and are thought to be adaptations to enhance light and nutrient absorption and the uptake of CO_2 (Wetzel 1983a). The elongated underwater leaf shape is brought about by cell elongation rather than cell division. In some species, increased temperatures or a longer photoperiod stimulate the formation of aerial leaves. In others, the submerged leaf form develops when the CO_2 level decreases to 5% of ambient levels (Maberly and Spence 1989). In some species, the plant hormone, giberellic acid, induces the formation of submerged leaf forms while abscisic acid induces aerial leaf morphology (Jackson 1990).

Species of the genus Ranunculus (buttercup) inhabit a range of moisture conditions from upland to submergence. Those species that inhabit either wet terrestrial or dry terrestrial habitats with stable water levels are inflexible in leaf shape. However, R. flammula, which inhabits the fringe of water bodies where water levels fluctuate, is flexible in leaf shape and displays the greatest level of heterophylly within the genus (Figure 4.30; Cook and Johnson 1968; Barrett et al. 1993).

Littorella uniflora, another heterophyllous species, can grow either as a submerged or emergent plant, depending on the water level. The submerged leaves have lacunae; their epidermis is thinner than that of the emergent leaves, and they have few stomata. The submerged leaves die after a day of emergence. After 2 to 5 days aerial leaves grow from the

FIGURE 4.29
Proserpinaca palustris (mermaidweed) with highly dissected leaves below the water's surface and leaves with toothed margins above the water's surface. (Photo by H. Crowell.)

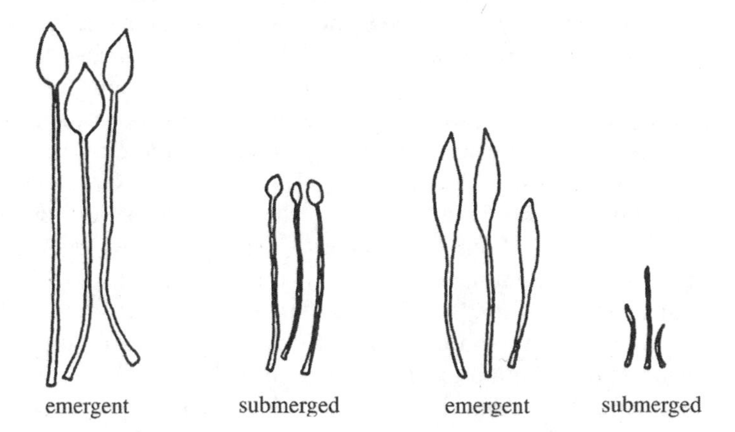

FIGURE 4.30
Leaf silhouettes of *Ranunculus flammula* showing the different shapes and sizes of leaves found under emergent and submerged conditions in two different lakes. (From Cook, S.A. and Johnson, M.P. 1968. *Evolution* 22: 496–516. Reprinted with permission.)

same rosette. These can survive flooding and change gradually into submerged leaves (Hostrup and Wieglieb 1991).

In an interesting display of heterophylly, two *Eleocharis* species (of the Cyperaceae) have been shown to use two different modes of photosynthesis, depending on whether their stems are submerged or emergent. In *Eleocharis vivipara* (Ueno et al. 1988) and *E. baldwinii* (Uchino et al. 1995), emergent stems assimilate atmospheric carbon dioxide via the C_4 photosynthetic pathway. When *E. vivipara* is submerged, it uses the C_3 photosynthetic mode (Ueno et al. 1988). In *E. baldwinii,* submerged plants fix inorganic carbon via a system that appears to be an intermediate between the C_3 and C_4 pathways. In addition, submerged *E. baldwinii* plants are able to assimilate carbon at night via aquatic acid metabolism (see Section V.B.2, Aquatic Acid Metabolism; Uchino et al. 1995).

The photosynthetic plant parts of *E. vivipara* are able to differentiate into the C_4 and C_3 modes under emergent and submerged conditions. In other words, the emergent plant parts develop a Kranz-type anatomy and well-developed bundle-sheath cells with numerous large choroplasts, features that are typical of C_4 plants. Submerged plant parts display reduced bundle-sheath cells with only a few small chloroplasts (Ueno et al. 1988). The development of C_4 features is inducible by exposure to the air. The changes are reversible, that is, when aerial plant parts are re-submerged, they lose the C_4 anatomy and metabolic pathway. When re-exposed to the air, they re-develop the C_4 anatomy (Ueno et al. 1988). The development of C_4 anatomy may be stimulated by the plant stress hormone, abscisic acid. When submerged plants are grown in water containing high levels of abscisic acid, they develop C_4 anatomy (Ueno 1998).

Both species grow at the fringes of warm water bodies (both were found in Florida) where periodic wetting and drying occur. In both species, the two forms are visibly dissimilar. In *E. vivipara,* emergent plants are composed of a rosette of long, slender, leafless stems, typical of the genus. The submerged form has whorls of hair-like leaves at nodes along the stem (Ueno et al. 1988). In *E. baldwinii,* the submerged stems are softer and more flexible than the emergent stems (Uchino et al. 1995). Softer, more pliable stems provide less resistance to underwater currents. The whorls of leaves in *E. vivipara* provide a greater surface area-to-volume ratio than the leafless emergent stems, and may allow greater absorption of light and carbon dioxide in the underwater environment. While these visible features of heterophylly seem to follow the pattern seen in other heterophyllous species, the adaptive significance of inducible changes in photosynthetic mode is not clear.

The C_4 photosynthetic pathway has been shown to be advantageous in tropical and subtropical areas where plants experience high irradiance, high temperatures, and intermittent water stress (Ehleringer and Monson 1993). The presence of C_4 plants in wetlands would seem to be a contradiction of this generally accepted explanation for the adaptive significance of C_4 photosynthesis. Some researchers have suggested that C_4 photosynthesis may confer a competitive advantage in areas of low nitrogen such as salt marshes, sandy soils, and other nutrient-poor settings (Jones 1987, 1988; Li et al. 1999) because plants with C_4 photosynthesis have been shown to have higher nitrogen use efficiency than C_3 plants (Wilson 1975; Bolton and Brown 1980; Jones 1987, 1988; Anten et al. 1995; McJannet et al. 1995; Sage et al. 1999). However, no clear pattern of C_4 prevalence or of greater productivity in C_4 plants has emerged in low-nitrogen settings, including wetlands (Sage and Pearcy 1987; Mozeto et al. 1996; Li et al. 1999). The reasons for the expression of C_3 metabolism underwater and C_4 above the water's surface remain to be explained (Ueno et al. 1988; Ueno 1998).

VI. Adaptations to Herbivory

Macrophytes are an important trophic link in wetland food webs. Many wetland plants have developed defenses to deter herbivory which include both chemical defenses, or *secondary metabolites,* and structural defenses such as thorns or tough leaves. Wetland plant defenses against herbivory are similar to those of upland plants.

A. Chemical Defenses

Secondary metabolites include a number of compounds, such as alkaloids, phenolics, quinones, essential oils, glycosides, and raphides. Plants that produce secondary metabolites are unpalatable to most herbivores. Many wetland plants produce secondary metabolites that deter herbivores and in some cases make the plants dangerous to humans.

Plants with long life spans tend to invest more in chemical defenses than do shorter-lived plants, perhaps because they are more likely to eventually become prey for herbivores. While longer-lived species endure the cost of producing secondary metabolites, they benefit by being able to reproduce during more than one growing season. Short-lived species, on the other hand, shunt resources into reproduction rather than defenses against herbivory. The relatively few studies of secondary metabolites in wetland plants confirm this trend and indicate that the rank of mean phenolic content in wetland plants is trees > floating-leaved plants > emergents > submerged ≥ algae (Lodge 1991).

Alkaloids are produced by a number of floating-leaved genera of the Nymphaeaceae including *Nuphar, Nymphaea,* and *Nelumbo.* Several submerged species have also been found to produce alkaloids in sufficient amounts to render plant tissue unpalatable or even toxic to invertebrates. These plants may also produce other chemical herbivore deterrents as well; only the alkaloids have been recorded. Some *Myriophyllum* species produce a cyanogenic compound that deters herbivores (Ostrofsky and Zettler 1986).

Arundo donax (giant reed) is an emergent grass of Eurasia and an invasive plant in the U.S. that produces steroids and alkaloids that deter herbivory. These compounds have been extracted and used to inhibit herbivory on agricultural plants (Miles et al. 1993). Like other members of the Asclepiadaceae (milkweed family), *Asclepias incarnata* (swamp milkweed; Figure 4.31) produces alkaloids and glycosides and only insects that have evolved to withstand these consume the plants. Monarch butterflies and certain other insects can accumulate the plant toxins and become unpalatable or poisonous to birds and other insect consumers (Raven et al. 1999).

Toxicodendron vernix (poison sumac; Figure 4.32) grows in peatlands, often in close proximity to *Larix laricina. T. radicans* (poison ivy; Figure 4.33) grows in both wet and upland areas as a shrub or a vine. Both species are in the Anacardiaceae (cashew family), a largely tropical family with both edible and poisonous plants. Both *T. vernix* and *T. radicans* produce an oily sap called urushiol that, when released from the ruptured epidermis of stems, leaves, roots, and fruits, deters herbivory and causes an irritating skin rash in humans (Voss 1985).

Several members of the Apiaceae (=Umbelliferae; parsley family) grow in wet woods and marshes, such as *Cicuta maculata* (water hemlock or spotted cowbane), *C. bulbifera* (bulb-bearing water hemlock), *Conium maculatum* (poison hemlock), and *Oxypolis rigidior* (cowbane). They all produce alkaloids and other toxins and are highly poisonous to animal herbivores as well as to humans (Voss 1996).

B. Structural Defenses

Structural defenses are more commonly found among upland plants than wetland ones (Lodge 1991). Most wetland plants with structural defenses usually do not grow in

FIGURE 4.31
Asclepias incarnata (swamp milkweed) produces alkaloids and glycosides which deter many insects. (Photo by H. Crowell.)

standing water, but are more commonly found in peatlands or wet forests. Some wetland plants, such as *Rosa palustris* (swamp rose), have thorns that dissuade large herbivores. Leathery leaves are common in peatland shrubs such as *Chamaedaphne calyculata* (leatherleaf; Figure 2.23), *Kalmia polifolia* (swamp laurel), *Ledum groenlandicum* (labrador tea), and *Andromeda glaucophylla* (bog rosemary; Figure 2.24), and they may deter herbivory as well as aid in water retention (Crum 1992). *Pubescent,* or hairy leaves, such as those found in some *Salix* (willow) species, may deter herbivory while also aiding in water retention.

VII. Adaptations to Water Shortages

Most wetland plants either do not have adaptations to water stress or show only a weak expression of them. Some wetland plants that grow where dry periods are predictable, or in cold climates, exhibit adaptations to water shortages. For example, southeastern U.S. cypress swamps often experience dry periods in the spring and summer. In Florida, cypress domes have a perched water table caused by underlying hardpans and clay layers. Clay layers inhibit root penetration to groundwater sources so the plants' water supply is limited to the water stored within the dome basin. Cypress trees exhibit a number of water conservation traits. In Florida, the transpiration ratio (the ratio of the amount of water lost through transpiration to the amount of organic matter produced by photosynthesis during the photoperiod) of *Taxodium distichum* was measured as 156 to 220 g water lost g^{-1} organic matter produced. When compared to Florida marshes (transpiration ratio = 414 to 1820), corn (400), grain crops (650 to 750), and nonsucculent plants in a subtropical dry forest (mean 310), the water use efficiency of cypress forests appears to be high (ratios from Brown 1981). *T. distichum* also has small vertically oriented needle-shaped leaves that minimize heating loads and maximize cooling by convection, thus reducing the water lost through cooling by transpiration. *T. distichum* has leaves with thick cuticles and deeply sunken, low-density stomata, which also help prevent water loss (Brown 1981).

The shrubs of the Ericaceae exhibit a decumbent growth habit and tough, leathery leaves with heavy cuticles and sunken stomata, which may help them avoid water loss and

FIGURE 4.32
Toxicodendron vernix (poison sumac) produces oils that deter herbivory and are a harsh skin irritant. (Photo by H. Crowell.)

FIGURE 4.33
Toxicodendron radicans (poison ivy) produces an oil that deters herbivory and causes an itchy rash in humans. (Photo by H. Crowell.)

protect them from frost. However, peatland plants with xeromorphic adaptations have not been shown to retain water better than peatland species without such adaptations. They may have retained family characteristics that were in existence prior to their adaptation to the peatland environment (Crum 1992). In subarctic coastal wetlands of the Hudson Bay, five shrub species and one sedge (*Salix planifolia, S. reticulata, Betula glandulosa, Myrica gale,* and *Carex aquatilis*) were found to have midday stomatal depression. With midday stomatal depression, the plants had a decrease in transpiration rate when the air temperatures were high and the atmospheric humidity was low. The habitat did not appear to be water-limited so stomatal depression may serve another purpose such as resistance to dehydration by freezing in winter (Blanken and Rouse 1996).

Summary

All plant cells require oxygen for aerobic respiration. When the sediments are flooded, very little or no oxygen is available to plant roots. Wetland plants have developed a number of adaptations to the lack of oxygen in the soil environment. These adaptations include the development of air spaces, called aerenchyma, that allow oxygen to move from aerial parts to belowground parts of the plant. Other adaptations include adventitious rooting, shallow rooting, and a variety of root structures known as pneumatophores (found on trees). Some mangrove species have aerial roots known as drop and prop roots. Stem adaptations include rapid underwater shoot extension, hypertrophy, and stem buoyancy.

Gases (oxygen, carbon dioxide, methane, and others) move through plants via diffusion. Some wetland plants also exhibit the capacity to move gases via pressurized ventilation, underwater gas exchange, and Venturi-induced convection. Oxygen diffuses from plant roots into the surrounding sediments (called radial oxygen loss) and the resulting oxygenated rhizosphere provides a habitat for aerobic microbes and an area within the otherwise saturated soil in which elements may become oxidized.

When plant cells are deprived of oxygen, anaerobic metabolism begins. With anaerobic metabolism, ATP production continues, although at a much decreased rate. An indicator that plant cells are undergoing anaerobic metabolism is increased ADH activity. Ethanol, the main product of alcoholic fermentation, may not be as toxic as originally thought. During the first minutes of anoxic conditions, the cytoplasmic pH decreases in most plants. This may be caused by the production of lactic acid or by the decrease in the amount of ATP to regulate pH. Some flood-tolerant plants appear to be able to avoid the decrease in pH. Metabolic responses to anoxia are also reflected in protein metabolism under different levels of oxygen availability. Mitochondrial adaptations may also play a role in flood tolerance.

Plants growing in saline environments must be able to acquire water without accumulating excess salt. The water potential of halophytes must be lower than the water potential of the surrounding medium. Salt-tolerant plants are able to increase their internal solute concentration by osmotic adjustment and thereby lower their water potential. Some halophytes are able to avoid salt toxicity through salt exclusion and excretion, by shedding salt-laden leaves, or by succulence. High sulfide levels in salt marshes and mangroves create a stressful environment for plant growth. Some of the adaptations for low oxygen levels also help plants avoid sulfide toxicity, such as adventitious and shallow rooting and radial oxygen loss.

In some wetlands, nutrients are in limited supply. Wetland plant adaptations to low nutrient levels include mycorhizzal associations, nitrogen fixation, and carnivory. Some plants exhibit strategies to conserve nutrients including nutrient translocation and evergreen leaves.

Submerged plants are subject to low carbon dioxide levels and low light. Several structural adaptations such as leaf design and shape aid in sequestering light underwater. Some submerged plants are able to use bicarbonate (a form of inorganic carbon that is often more available underwater than carbon dioxide) in photosynthesis. Some are able to assimilate carbon dioxide at night, when it is more plentiful, in a process called aquatic acid metabolism. Some can recycle respired carbon dioxide within their aerenchyma and assimilate it in photosynthesis. Some submerged plants are able to acquire carbon dioxide from the sediments. Heterophylly is observed in some submerged plants and is thought to be an adaptation to fluctuating water levels.

Wetland plant adaptations to herbivory are similar to those of upland plants. They include both chemical and structural defenses such as thorns or tough leaves. Some wetland plants that are routinely exposed to dry periods have adaptations to water shortages. For example, cypress trees have high water use efficiency, and vertically oriented needle-shaped leaves that minimize heating loads and maximize cooling by convection.

Case Studies

4.A. Factors Controlling the Growth Form of *Spartina alterniflora*

Monospecific stands of *Spartina alterniflora* (cordgrass) stretch across the tidal salt marshes of the Atlantic and Gulf of Mexico coasts of the U.S. (Figure 2.2). Within the marshes, *S. alterniflora* grows in two distinct forms, called tall and short *Spartina* (an intermediate form sometimes coexists with the tall and short forms). Tall *Spartina* inhabits the banks of tidal creeks while the short form is inland from the creeks where tidal flushing occurs only infrequently and freshwater inputs other than rain are few. The tall form appears greener, more robust, and it can reach 3 m in height, while the short form is often only 10 to 40 cm tall (Table 4.A.1). The short form grows more densely than the tall form, perhaps because the taller plants shade the sediments and inhibit the growth of new shoots (Valiela et al. 1978). The net aboveground primary productivity of the tall form is greater than that of the short form, although the short form invests more in belowground primary productivity (Table 4.A.1). Average total (above- and belowground) net primary productivity for the tall form is 3900 g dry weight m^{-2} yr^{-2}, or about 500 g dry weight greater than for the short form.

The different forms occur across a broad latitudinal area, so gradients in temperature, photoperiod, and rainfall do not appear to cause the differences in growth form (Valiela et al. 1978). Several causes for the differences in the two growth forms have been suggested including genetic differences between the two forms and increased stresses where the short form grows (i.e., nitrogen limitation, higher salinity levels, lower sediment redox potential, and higher sulfide levels).

Genetic Studies

Electrophoretic studies of the two forms indicate that the chromosomes of the two forms are identical (Shea et al. 1975; Valiela et al. 1978; Anderson and Treshow 1980), but there may be differences among the genes of the two forms that have not been detected. Transplant studies provide conflicting results. When tall *Spartina* was transplanted into short *Spartina* areas in a Connecticut salt marsh, it grew at the same rate and to the same height as the surrounding short plants (60 cm). The short plants transplanted into the tall areas grew to be about 1.5 times as tall as the short plants in the short form area, but they did not attain the height of the surrounding tall plants during the first growing season. By the end of the third growing season, the transplanted plants were indistinguishable from surrounding undisturbed populations, indicating that environmental differences brought about the two height forms (Shea et al. 1975). In Louisiana salt marshes, Mendelssohn and McKee (1988) also found that transplanted tall *Spartina* had reduced standing crops when planted among the short form. The reverse experiment, with short plants grown among tall ones, brought about an increase in standing crop.

In another transplant experiment, tall and short plants were removed from a Delaware salt marsh and planted under the same conditions in a garden plot. After 9 years of identical

TABLE 4.A.1

Characteristics of the Tall and Short Forms of *Spartina alterniflora* in Salt Marshes of Georgia, South Carolina, North Carolina, New Jersey, and Massachusetts

Characteristics	Tall	Short
Habitat	Tidal creek banks	Flat marsh, landward of tall form
Habitat hydrology	Tidal flushing, some fresh or brackish water inputs	Often stagnant, low or no freshwater inputs
Height	1 to 3 m	<1 m
Aspect	Dark green, robust	Often chlorotic, leaf tips burned
Stem diameter	2–9 mm	1–3 mm
Stem density (stems m–2)	80–230	400–1100
Leaf longevity	72 days	49 days
Ramet longevity	231 days	204 days
Flowering	Common	Absent or infrequent
Clonal propagation	Common	Common
NPP: aboveground (g m^{-2} yr^{-1})	2245	900
NPP: belowground (g m^{-2} yr^{-1})	1660	2560
Total NPP (g m^{-2} yr^{-1})	3905	3460
Belowground proportion of NPP	40%	84%

Descriptions from Valiela et al. 1978; Anderson and Treshow 1980; Wiegert et al. 1983; and Dai and Wiegert 1996. Average NPP calculated from data in Gallagher and Plumley 1979; Smith et al. 1979; Dame and Kenny 1986; Morris and Haskin 1990; and Dai and Wiegert 1996.

irrigation, fertilization, and salinity regimes, the two forms retained their gross morphological distinctions, and plant biomass, culm height, stem density, stem diameter, and flowering frequency remained significantly different. A genetic difference may have caused the differences in growth form or, if the two growth forms have the same genetic composition, certain genetic characteristics may be "turned on" in the seedling stage and persist despite long-term exposure to another environment (Gallagher et al. 1988).

Nitrogen Limitation

In salt marshes, nitrogen levels have been shown to be low and salt marsh plants are often nitrogen-limited (Hopkinson and Schubauer 1984). A number of researchers have found that short *Spartina* responds positively to nitrogen inputs (Sullivan and Daiber 1974; Gallagher 1975; Valiela et al. 1976; Linthurst and Seneca 1981; Broome et al. 1983); however, even under prolonged nitrogen fertilization treatments, the net primary productivity of the short form remains less than that of the tall form (Howes et al. 1986). Some have found that the short form grows in areas of higher interstitial ammonium than the tall form, but nonetheless remains stunted (Mendelssohn 1979). So, while the short form may be nitrogen-limited, environmental conditions such as high salinity, high sulfide concentrations, and low soil oxygen levels may prevent nitrogen uptake and thereby keep the short form short (Bradley and Morris 1990; Mendelssohn and Morris 2000).

Salinity

In tidal salt marshes, the soil salinity is lowest near sources of estuarine or fresh water and highest in areas removed from those sources. Plant community zonation appears to be related to salinity in many salt marshes, brackish swamps, and mangrove forests, with the most salt-tolerant species inhabiting the areas of highest salinity (Haller et al. 1974; Mahall and Park 1976; Cooper 1982; Snow and Vince 1984; Burchett et al. 1989; Allen et al. 1994). The growth forms of *S. alterniflora* appear to be influenced by salinity since the short forms usually grow where the salinity is highest (Nestler 1977; Linthurst and Seneca 1981). Tall *Spartina* usually grows in salinities from 10 to 30 ppt with maximum growth at low salinity levels (i.e., 10 to 15 ppt; Longstreth and Strain 1977; Linthurst and Blum 1981; Bradley and Morris 1991a), while the short form may be exposed to salinities up to 60 ppt due to accumulation of salt, evaporation between tidal cycles, and distance from freshwater sources (Webb 1983). High salinity (in excess of 50 ppt) inhibits the uptake of ammonium, which may be the reason the short form persists even where soil ammonium levels are high (Bradley and Morris 1991b). In many salt marshes, however, salinity is not consistently higher in short form stands, so other causes, such as soil redox levels, must ultimately bring about the differences in height forms, at least in some marshes (Howes et al. 1981; Howes et al. 1986; Teal and Howes 1996).

Sediment Oxidation

Wherever the water table is depressed and the soil is oxidized, the tall form of *S. alterniflora* can be found (Howes et al. 1981). Biomass in the field and in controlled studies is positively correlated to sediment oxidation (Linthurst and Seneca 1981; Wiegert et al. 1983; Howes et al. 1986; Pezeshki et al. 1988). Drainage and the maintenance of an aerated root zone are aided by animal burrows such as those of the mud fiddler crab (*Uca pugnax;* Bertness 1985).

Vegetated areas of the salt marsh have a higher redox level than unvegetated areas and the redox level in the root zone of the tall form is higher than beneath the short form (Howes et al. 1981; Mendelssohn and McKee 1988). Higher redox levels in tall form zones may be because more of the tall plants' area is allocated to lacunae than in the short form, so more oxygen diffuses from the tall plants' roots (Arenovski and Howes 1992). Below the root zone the redox level is quite low and not significantly different from unvegetated areas of the salt marsh (Figure 4.A.1). Where nutrient concentration is increased and/or sediment saturation is decreased, plant growth increases, thereby setting up a feedback loop in which root zone oxidation is increased further and nutrient uptake is enhanced (Howes et al. 1981).

Sulfide Concentration

Under reduced conditions, sulfate is reduced to hydrogen sulfide, which, in high concentrations, has been found to inhibit nutrient uptake in wetland plants (Howes et al. 1981; Havill et al. 1985; Bradley and Morris 1990; Koch et al. 1990). Near tidal creeks, where the tall form thrives, sulfide concentrations are lower than in the short form areas of salt marshes (King et al. 1982; Bradley and Dunn 1989). Sulfide concentrations in the root zone of short *Spartina* and sulfide accumulation within the plants are greater in the short form than in the tall form (Ornes and Kaplan 1989). Increased sediment oxidation removes hydrogen sulfide and other reduced elements by oxidation, thereby enhancing plant nutrient uptake (Howes et al. 1986).

Summary

Tall *Spartina* grows in areas with more oxidized sediments and greater subsurface water movement than in the areas where short *Spartina* grows. The greater water flow-through results in an environment with less sulfide, less salt, and less ammonium. Tall *Spartina* is able to maintain aerobic root respiration and the uptake of ammonium, and consequently produce large amounts of biomass. The reducing conditions surrounding the short form

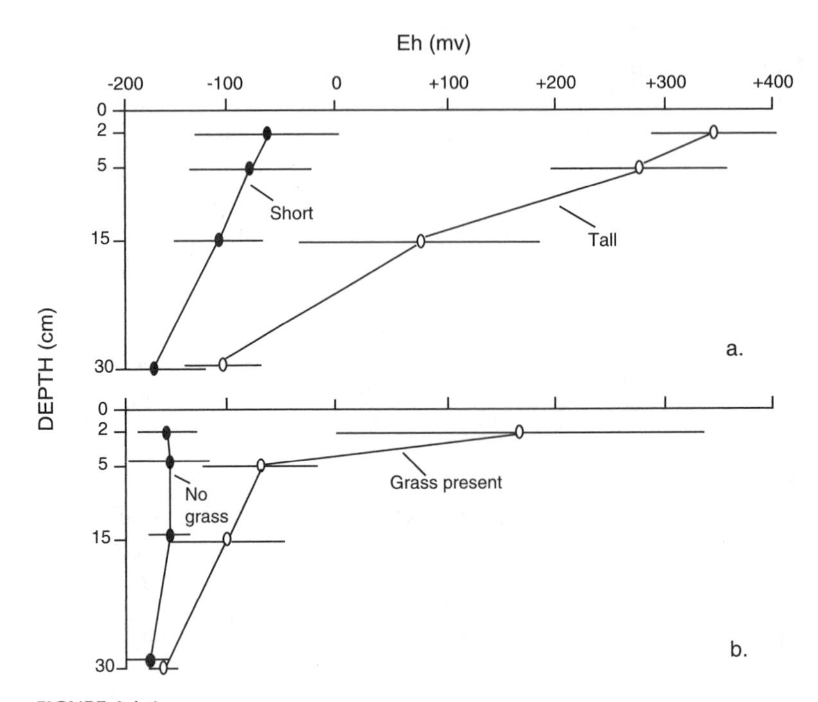

FIGURE 4.A.1
(a) Profiles of soil redox (Eh) in sediment with tall and short stands of *Spartina alterniflora*. (b) Profiles of Eh in an area devoid of vegetation and in nearby clumps of *S. alterniflora* expanding into the bare area. (From Howes et al. 1981. *Limnology and Oceanography* 26: 350–360. Redrawn with permission from the American Society of Limnology and Oceanography.)

lead to anaerobic root metabolism and inhibited ammonium uptake due to high sulfide levels. Under these environmental conditions, short *Spartina* remains stunted (Teal and Howes 1996; Mendelssohn and Morris 2000).

4.B. Carnivory in *Sarracenia purpurea* (Northern Pitcher Plant)

Sarracenia purpurea (northern pitcher plant) grows in peatlands of the northeastern U.S. and Canada. The plant forms a rosette of modified leaves, called pitchers, and spreads vegetatively along rhizomes that grow through and on top of the surrounding *Sphagnum* moss. *S. purpurea* blooms in the late spring or early summer with one insect-pollinated flower atop a 10- to 15-cm stalk (Figure 4.B.1). The pitchers secrete nectar and contain water. Insects and other prey are attracted to the pitchers by their reddish color and the nectar. Many cannot escape once they have entered because of the downward pointing hairs that line the pitcher's hood (Figure 4.B.2). The captured prey are broken down by insects and bacteria, and by digestive enzymes produced by the plant (Fish and Hall 1978).

The pitchers originate in the center of the rosette and form a whorl. Each new pitcher arises nearly opposite the previous pitcher (Figure 4.B.3). As new pitchers are produced, the older pitchers are displaced outward. Eventually they die and become detached. The newest pitchers are paler in color than the older pitchers whose outer margins eventually fade and turn brown. In a Massachusetts population of *S. purpurea*, new pitchers were produced about every 20 days with an average of 5.5 pitchers per plant per growing season (Fish and Hall 1978).

FIGURE 4.B.1
Sarracenia purpurea (northern pitcher plant) showing several flowers and pitchers. (Photo by D. Greenberg.)

FIGURE 4.B.2
Sarracenia purpurea pitcher showing the downward pointing hairs within the pitcher that keep insects from crawling out. (Photo by H. Crowell.)

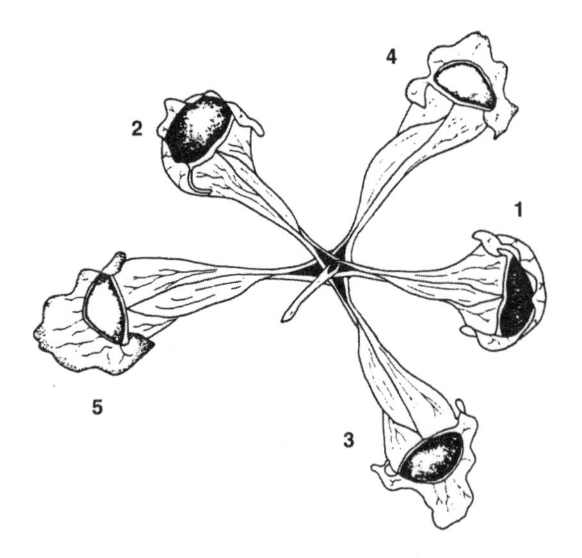

FIGURE 4.B.3
Chronological arrangement of leaves on a midseason pitcher plant, with the youngest leaf marked "1." The petioles of the older leaves nearly surround those of younger leaves and new leaves arise from the center of the whorl. (From Fish, D. and Hall, D.W. 1978. *The American Midland Naturalist* 99: 172–183. Reprinted with permission.)

The function of the pitchers changes as they age. The youngest pitchers capture more prey than the older ones. A plant with several pitchers of different ages may have only one or two that are actively capturing insects at any given time (Fish and Hall 1978). Pitchers with the highest degree of pigmentation and nectar attract the greatest numbers of prey. The youngest pitchers' nectar has measurable amounts of total carbohydrates, while the nectar of the older pitchers does not (Cipollini et al. 1994). In some cases it is difficult to determine the age of the leaves and therefore their function because some pitchers do not grow in rosettes, but arise independently on extended petioles at some distance from the original plant (Creswell 1993).

A number of organisms, including bacteria, protozoa, rotifers, nematodes, copopods, mites, and three insect larvae, inhabit *S. purpurea* pitchers without becoming digested. These organisms are resistant to enzymes produced by the plant and they subsist on the prey insects. The three insect larvae are a fly (*Blaesoxipha fletcheri*), a mosquito (*Wyeomyia smithii*), and a chironomid (*Metriocnemus knabi*). They partition the resources within the pitchers and each inhabits the pitcher at a different fluid depth. The depth where each is found seems to be determined by the insects' oxygen needs and respiratory anatomy. Nearest the fluid's surface is the fly larva, which depends on atmospheric oxygen. The fly larvae inhabit the newest pitchers that are at the peak of their prey capture ability. Females lay eggs within the pitchers and the larvae go through the first three larval instars within the pitchers. The larvae feed at the surface of the pitcher plant fluid and retreat into the pitcher only when disturbed. They consume recently captured floating insects. Pupation occurs outside the pitchers in the *Sphagnum* moss (Fish and Hall 1978).

The mosquito larvae occur in several of the newest pitchers. In the newest pitcher, eggs and first instars are found, in the second youngest pitcher there are large numbers of second and third instars, and the third youngest pitcher contains mostly fourth instars and pupae. Each pitcher is occupied by only one generation of mosquitos as the eggs are laid only in the newest leaf. The mosquito larvae swim thoughout the pitcher plant fluid and return to the surface only periodically for gas exchange through posterior siphons. They feed on suspended particulate matter made up of captured insects that bacteria have begun to break down (Fish and Hall 1978).

The chironomid larvae are found at the bottom of pitchers of all ages. They do not need to rise to the surface because they take in dissolved oxygen directly from the pitcher fluid.

The chironomid larvae eat the insect remains that have settled on the bottom of the pitchers (Fish and Hall 1978). As the chironomid larvae break down settled insects, some of the insect parts become resuspended and feed the mosquito larvae. The chironomid larvae accelerate the prey breakdown and increase the food supply for the mosquito. Both the chironomid and the mosquito larvae are found exclusively within *S. purpurea* pitchers (Heard 1994). The chironomid larvae are generally more abundant than the mosquito larvae; in one study an average of 56 chironomid larvae per pitcher were counted, while an average of 17 mosquito larvae per pitcher were found only in the younger and larger pitchers (Nastase et al. 1995).

The insect and other animal inhabitants of the pitchers, known collectively as the *inquilines*, may benefit the plants by breaking down prey and making nutrients available for plant absorption. The pitchers are capable of taking up the end products (ammonia) of inquiline metabolism from the pitcher fluid. Both the pitchers and the inhabitants may also benefit from gas exchange processes. The pitchers take up carbon dioxide produced from inquiline metabolism in the pitcher fluid while the inquilines take up oxygen produced by the plant (Figure 4.B.4; Bradshaw and Creelman 1984).

Pitcher plant fluid sometimes becomes anaerobic and takes on a red color due to the presence of anaerobic bacteria. The bacteria occur elsewhere in peatlands and probably arrive in the pitcher on inquilines. The bacteria are able to exploit anaerobic conditions in pitchers and even when other inquilines cannot survive, anaerobic bacteria continue to degrade prey (Bradshaw and Creelman 1984).

Pitchers also house nitrogen-fixing bacteria. The nitrogenase activity in pitchers may provide ample nitrogen for the plants, and along with sediment and prey nitrogen, the nitrogen supplied is in excess of the plant's needs. Tracer studies of nitrogen uptake indicate that the plant takes up nitrogen from all of these sources (Prankevicius and Cameron 1991).

Carnivory has long been thought to be an adaptation to nutrient-poor conditions; however, it may be that prey capture simply enhances nitrogen availability rather than supplying all of the plant's needs. A nutrient budget for pitcher plants reveals that the nitrogen from captured insects provides about 10% of the plant's annual nitrogen requirements. The soil nitrogen is often more than adequate to meet the plant's needs (Chapin and Pastor 1995). It appears that carnivory is not obligatory in the pitcher plant and it may reflect past environmental conditions more than today's. Pitcher plant carnivory may provide a "back-up plan" that would supply nutrients in case they are in short supply.

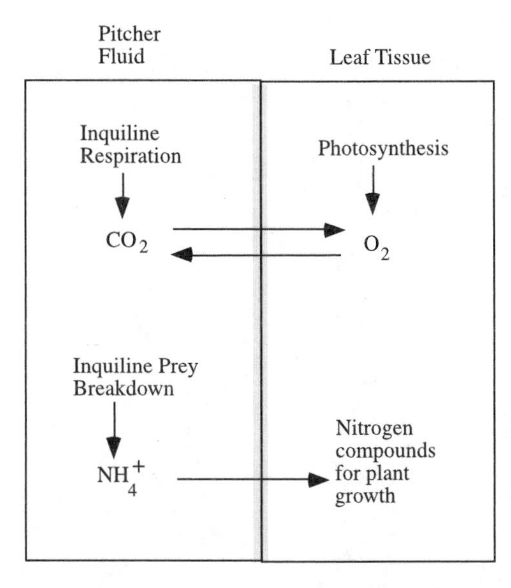

FIGURE 4.B.4
Gas and nitrogen flow between the pitcher fluid of *Sarracenia purpurea* and the plant's leaf tissue under aerobic conditions. Respiration of the inquiline population (including bacteria, protozoa, rotifers, and insects) produces carbon dioxide which is taken up by the plant in photosynthesis. The oxygen produced by the plant is used by the inquilines. As the prey organisms are broken down by the inquilines they are made available to the plant. (Adapted from Bradshaw and Creelman 1984 and Crooks 1997.)

5

Reproduction of Wetland Angiosperms

I. Introduction

Wetland angiosperms reproduce both sexually, by the development of seeds that grow into new plants, and asexually, by the growth of clonal structures that separate and become independent plants. For the majority of wetland angiosperms, sexual reproduction is essentially the same as for terrestrial angiosperms. In order to reproduce sexually, wetland flowers depend on one of several pollination mechanisms followed by the effective dispersal and germination of seeds. Sexual reproduction presents a challenge for submerged, floating, and floating-leaved plants because, with few exceptions, the plants must produce an aerial flower in order for pollination to occur. Sexual reproduction brings about new genotypes because it involves the fusion of two genetically unique nuclei.

Asexual reproduction preserves the parent's genotype. Wetland plants show a high propensity for asexual, or vegetative, reproduction. Nearly all wetland plants form *propagules,* which are clonal structures that can separate from the parent plant and develop into an independent plant. Clonal structures may have evolved for functional reasons other than reproduction, such as perennation, anchorage, and carbohydrate storage. Most wetland plant species are capable of both sexual and asexual reproduction. Many species, such as those in the floating family, Lemnaceae, flower only very rarely. Others, such as the submerged annual, *Najas* (bushy pondweed), rely almost entirely on sexual reproduction.

A. A Brief Review of Floral Structures Involved in Reproduction

The sexual reproductive cycle in all angiosperms begins with the flower. The flower is the site of *gamete* (haploid reproductive cell) formation and fusion. The sexual cycle involves (1) the production of special reproductive cells following meiosis, (2) pollination, (3) fertilization, (4) fruit and seed development, (5) seed or fruit dispersal, and (6) seed germination.

The floral parts involved in sexual reproduction are the male structures, or *androecium* (*stamens*), and the female structures, or *gynoecium* (*pistil*; Figure 5.1). Within the androecium, each stamen has an *anther* supported by a *filament.* Inside the gynoecium, the number of *carpels*, or "vessels," and their arrangement vary greatly among species. A single carpel consists of an *ovary* (basal portion), a *style* (a stalk connecting the ovary and the stigma), and a *stigma* (tip). In flowers of only one sex, the female is called the *pistillate* flower, and the male is called the *staminate* flower. Flowers with both male and female parts are called *hermaphroditic* or *perfect.*

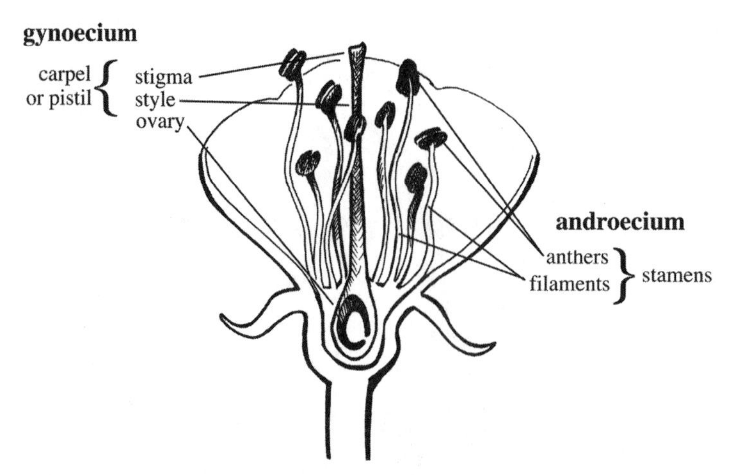

FIGURE 5.1
A diagram of a *perfect*, or *hermaphroditic*, flower. The *gynoecium*, or female part of the flower, consists of the *ovaries*, *styles*, and *stigmas*. Each set of these is called either a *pistil* or a *carpel*. A gynoecium may have several carpels, or a single one. The *androecium*, or male part of the flower, consists of the *anthers* and the *filaments*; each set of these is called a *stamen*. When flowers are of only one sex, they are called either *pistillate* (female) or *staminate* (male) flowers. (Drawing by B. Zalokar.)

Pollination occurs when the pollen is transported from the anther to the stigma of the same or adjacent flowers, by insects, birds, bats, wind, gravity, or water. Seeds are the result of fertilization and once mature, they are dispersed singly or within a fruit by gravity, wind, animals, or water. Seeds germinate when they are in favorable moisture, oxygen, and temperature conditions. Some seeds only germinate after a period of dormancy. Dormancy is often broken after a period of cold, or high heat (fire), or once the seed coat has been scratched or damaged (Rost et al. 1984).

B. Challenges to Sexual Reproduction in Wetland Habitats

Emergent plants have some or all of their aboveground parts above the surface of the water. Their flowers also grow above the water surface and pollination occurs via mechanisms typical of terrestrial plants. In submerged, floating-leaved, and floating plants, the main obstacles to sexual reproduction are floral induction and pollen transport (Titus and Hoover 1991). A major obstacle to pollination in many wetland plants is that the pollen of all but water-pollinated species cannot withstand wetting. If pollen gets wet, it can sprout a pollen tube (*precocious germination*) or rupture before it has reached the stigma. In most cases, precocious germination means that pollination does not occur. Wet stigmas can be non-receptive to pollen and can decompose before pollination occurs.

Many wetland plants have retained upland flowering strategies, perhaps because of the multiple adaptations necessary to overcome water disruption of pollination. Upland plants avoid water by, for example, hanging flowers facing downward, closing flowers to exclude water (many close at the approach of inclement weather), or protecting pollen by a cover of hairs or scales that prevent water from entering the flower (Faegri and van der Pijl 1979). The flowers of wetland plants avoid contact with water in a variety of ways (Sculthorpe 1967; Philbrick and Les 1996):

- *Air bubbles*. Some floating flowers close and entrap an air bubble when pulled below the water's surface as seen in the pistillate flower of *Hydrilla verticillata* (hydrilla; Yeo et al. 1984).
- *Elongated peduncles.* Some submerged and floating-leaved plants develop long flower stalks, or *peduncles,* that are longer, more rigid, or more buoyant than other stems. Species of *Nymphaea* and *Ranunculus* produce flowers that float at the end of long resilient stalks and sway with currents, thus decreasing the likelihood of the submergence of flowers by waves. *Lobelia dortmanna* (water lobelia), the only submerged species of the Lobeliaceae, has retained the flowering strategy of the upland members of the family by extending long peduncles through the water column (Figure 5.2a). These elongated structures are easily damaged and the deeper the plant, the more resources are necessary to elongate the peduncle (Philbrick and Les 1996). Peduncle elongation may be triggered by water temperature in some species. For example, a population of *L. dortmanna* initiated peduncle elongation at a similar water temperature (19 ± 1.5°C) for 5 consecutive years. Other plants may be stimulated by the accumulation of sufficient biomass. *Vallisneria americana* (tape grass; wild celery), a submerged plant that forms elongated peduncles, fails to flower below 0.75 g dry weight. The challenge of peduncle elongation may preclude sexual reproduction for plants growing in deep water. *V. americana* rarely flowers at depths greater than 2 m even though it can grow at depths up to 5 m (Titus and Hoover 1991).
- *Modified structures*. Many species prevent contact of aerial flowers with the water through the use of modified leaves, branches, and floral axes. In *Cabomba, Brasenia, Nymphoides, Potamogeton,* and some species of *Polygonum* and *Ranunculus,* groups of floating leaves support aerial floral axes. In *Hottonia*

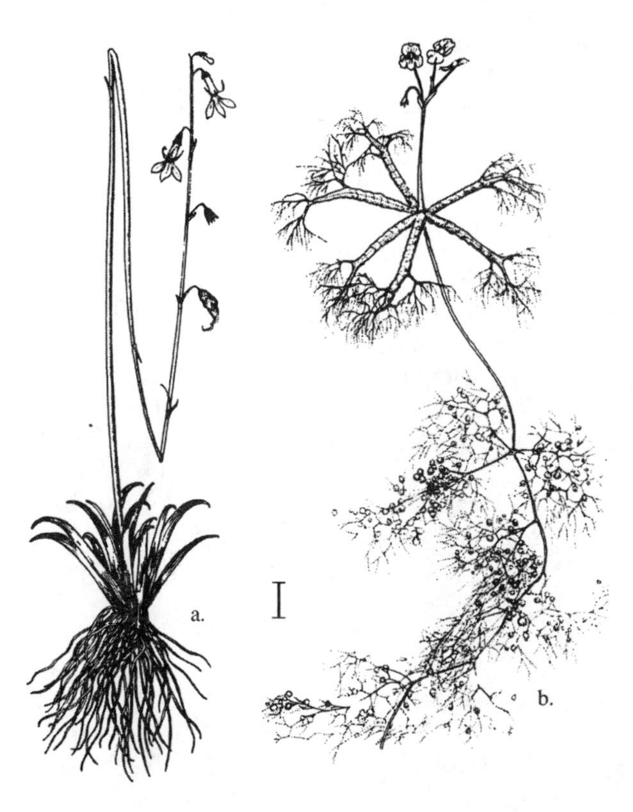

FIGURE 5.2
Submerged plants often possess highly modified vegetative organs but retain flowers with features similar to their terrestrial ancestors. Detrimental contact of flowers with water is prevented by a variety of adaptations. (a) *Lobelia dortmanna* (water lobelia) grows as a submerged basal rosette. At flowering, it extends a long peduncle through the water column. Its aerial insect-pollinated flowers closely resemble the terrestrial members of its genus. (b) *Utricularia radiata* (bladderwort) is a submerged species that develops a set of air-filled floats that keep its flowers afloat. The floats are modified portions of the highly dissected submerged stems. *Utricularia* flowers resemble those of the closely related terrestrial family, Scrophulariaceae (snapdragons; bar = 1 cm). (From Philbrick, C.T. and Les, D.H. 1996. *BioScience* 46: 813–826. Reprinted with permission © 1996 American Institute of Biological Sciences.)

inflata the flower stalk itself is inflated. The submerged *Utricularia radiata* (bladderwort) develops a set of floats at the base of the peduncle which keeps its flowers out of the water (Figure 5.2b; Cook 1996; Philbrick and Les 1996).

- *Self-pollination.* Submerged hermaphroditic flowers of several genera including *Myriophyllum, Podostemum, Ruppia,* and *Utricularia,* are able to self-pollinate without opening their flowers (*cleistogamy*). In the unisexual submerged flowers of some species of *Callitriche,* pollen tubes grow internally through vegetative tissue from staminate to pistillate flowers of the same plant. When flowers self-pollinate in either of these ways, pollen does not come in contact with water.

A few submerged wetland plants have pollen that can withstand wetting. These plants, described as *hydrophilous,* have developed pollination systems that are a significant evolutionary leap from those of terrestrial species (see Section II.A.3, Water Pollination).

II. Sexual Reproduction of Wetland Angiosperms

In general, sexual reproduction in plants is thought to be advantageous in changing or heterogeneous environments, while asexual reproduction is favored in stable or uniform habitats. Many wetland habitats are generally thought to be relatively stable due to the buffering capacity of water and predictable daily and seasonal changes. Nearly all wetland plants reproduce asexually and in many, sexual reproduction is rare.

Because wetland habitats are often stable, why should sexual reproduction exist there at all? In fact, wetland environments can be unstable. Short-term changes in water level due to sudden flooding or droughts can bring about drastic variations in the growth conditions of wetland plants. Within the time frame of plant evolutionary processes, such habitat-altering and land-forming forces as glaciation and continental drift have brought about changes in wetland habitats. Vascular wetland plants adapt to these changes through genetic selection that occurs with sexual reproduction. Sexual reproduction also allows for genetic responses to pathogen outbreaks, which are unpredictable and can negatively affect a population at any time (Ehrlich and Raven 1969; Philbrick and Les 1996).

A. Pollination Mechanisms

Wetland plant pollination strategies include those used by terrestrial angiosperms as well as water pollination, a phenomenon unique to the wet habitat (Table 5.1). Aerial flowers use insects (*entomophily*) and the wind (*anemophily*) as pollen vectors. Flowers of some submerged plants use water to transport pollen (*hydrophily*). In hydrophily, pollen is transferred on the water's surface (*epihydrophily*) or below the surface (*hypohydrophily*). Some aerial and submerged species are also capable of self-pollination. Pollination by bats has been observed in one genus of mangrove trees, *Sonneratia,* and by birds in the large-flowered mangrove genus, *Bruguiera,* both of Southeast Asia (Tomlinson 1986). The earliest flowering plants were pollinated by animals, and wind pollination is thought to have derived from these early biotic systems. Water pollinators probably evolved from wind-pollinated plants.

1. Insect Pollination

Insect-pollinated plants usually possess a colorful *perianth* (sepals and petals) and produce nectar and volatile compounds with distinctive odors. Flowers produce an attractant that satisfies a need in the pollinator, usually for food, which may be in the form of pollen, nectar, oil,

TABLE 5.1

Pollination Mechanisms in Wetland Plants

Pollination Mechanisms	Flower Position
1. Insect (entomophily)	Aerial and water's surface
2. Wind (anemophily)	Aerial and water's surface
3. Water pollination (hydrophily)	
a. Surface pollination (epihydrophily)	Water's surface
1. Dry epihydrophily	
2. Wet epihydrophily	
b. Submerged pollination (hypohydrophily)	Submerged
4. Self-pollination	
a. Aerial self-pollination	Aerial and water's surface
1. Within the same flower (autogamy)	
2. Different flowers on the same plant (geitonogamy)	
3. Within closed flowers (cleistogamy)	
b. Submerged self-pollination (hydroautogamy)	Submerged
1. Pollen on the surface of a bubble (bubble autogamy)	
2. Different flowers on the same plant (geitonogamy)	
3. Within closed flowers (cleistogamy)	

or water. In some cases, insects use flowers for shelter, protection from predators, or as a trap for capturing prey (Faegri and van der Pijl 1979).

The majority of all angiosperms are insect-pollinated, as are most wetland angiosperms (about 60 to 66%; Cook 1988; Titus and Hoover 1991; Cook 1996). Thousands of emergent species, particularly the eudicots, use a vast array of insect pollinators. Among the herbaceous emergent wetland plants, at least 52 families of wetland plants worldwide are predominantly entomophilous (Cook 1996). Many wetland shrubs as well as most mangrove species are also insect-pollinated. The most common pollinators are bees and flies, and the majority of plants are served by a variety of insect pollinators. In the case of the emergent *Pontederia cordata* (pickerelweed), there are several insect pollinators, but one bee, *Dufourea novae-angliae,* apparently visits no other plant and its annual emergence coincides with the onset of the plant's flowering (Sculthorpe 1967). Some examples of floating-leaved plants and their insect pollinators are *Nuphar* species (spatterdock) which are pollinated by

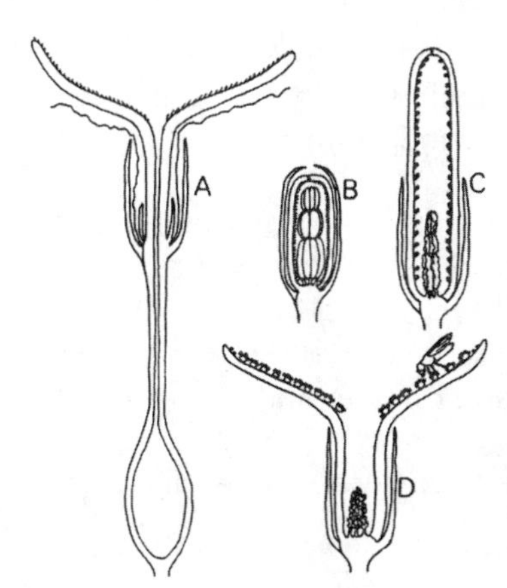

FIGURE 5.3
Longitudinal sections of the insect-pollinated flowers of *Blyxa octandra*, which offer no nectar or other reward. The staminate and pistillate flowers resemble each other and it may be that they only offer insects a resting place. (A) Pistillate flower: note hair-like petals and petal-like stigmas; (B) staminate flower in bud; (C) staminate flower showing pollen being carried on the elongating petals; (D) staminate flower with spreading petals, fluid droplets, and pollen, being visited by a fly (not drawn to scale; staminate petals and pistillate stigmas are 5 to 15 mm long). (From Cook, C.D.K. 1982. *Studies on Aquatic Vascular Plants.* J.J. Symoens, S.S. Hooper, and P. Compere, Eds. Brussels. Royal Botanical Society of Belgium. Reprinted with permission.)

aquatic beetles and flies. In *Nymphoides* species (floating heart) and *Cabomba caroliniana* (fanwort), flies and bees are pollinators. Many submerged plants are also insect-pollinated and so must elevate their flowers above the water's surface. Among these are a number of species of *Utricularia*, with their brightly colored pink, purple, or yellow flowers.

Cook (1982) describes two types of insect pollination in the aquatic family, Hydrocharitaceae (frogbit): insect pollination with enticement and without enticement. Species in the genera *Ottelia, Stratiotes, Hydrocharis,* and *Egeria* all attract pollinators with their white petals, yellow infertile stamens, strong scent, and nectar reward. They do not all attract the same pollinators, however. *Ottelia* attracts bees with its sweet, pleasant scent, obvious on sunny afternoons; the pistillate flowers of *Hydrocharis* bear nectaries that are lacking in the staminate flowers (the staminate flowers closely resemble the pistillate ones); *Egeria* flowers secrete nectar; and *Stratiotes* releases a strong carrion-like odor on sunny mornings to which flies are attracted.

Species of the tropical genus, *Blyxa*, also of the Hydrocharitaceae, do not entice and offer no reward to insect visitors. The staminate and pistillate flowers of this *dioecious* genus (pistillate and staminate flowers are on separate plants) resemble one another (Figure 5.3). The staminate flower spreads its petals to display pollen. The pistillate flower's petals are thread-like, but its stigmas look like the staminate petals. At midmorning, the petals of the staminate flower and the stigmas of the pistillate flowers spread and exude water-like droplets. The pollen is wettable (i.e., it can get wet without precociously germinating) and forms clumps on the male petals. The flowers attract various species of flies that move from flower to flower with pollen on their feet. The flowers may only offer resting places for the insects. Pollination is efficient despite the lack of pollinator reward perhaps because insects are so plentiful in the tropical *Blyxa* habitat. Because the pollen is wettable, *Blyxa* may represent an intermediate stage in the evolution to water pollination.

2. Wind Pollination

Overall, about two thirds of wetland plants are biotically pollinated, while one third is abiotically pollinated. These proportions hold true around the world at all latitudes. Wind pollination, or anemophily, is the dominant form of abiotic pollination comprising about 98% of known examples. It is prevalent in several families that are well represented among wetland plants, particularly the monocots: Poaceae (grasses), Cyperaceae (sedges), and Juncaceae (rushes; Faegri and van der Pijl 1979). Certain kinds of wetlands like reedswamps and sedge-dominated communities are dominated by wind-pollinated genera (Cook 1988).

Anemophily evolved from entomophily several times in many plant families. Anemophilous wetland plants flower and pollinate like their terrestrial counterparts. Virtually all anemophilous wetland plants were already wind-pollinated before they adapted to the wetland habitat.

The flowers and pollen of anemophilous species differ from entomophilous plants. Several adaptations characterize wind pollination (Whitehead 1983; Cook 1988):

- Pollen production is increased, usually by an increase in anther size and often associated with a decrease in the number of anthers per flower and an increase in the number of staminate flowers.
- The flowers are usually unisexual: the staminate flower dispenses pollen and the pistillate flower receives it.

- The perianth and bracts are reduced, which increases aerodynamic efficiency of dispersal.
- There are sometimes special arresting or explosive mechanisms that ensure that pollen gets into the airstream.
- The pollen is released at certain times of the season or the day to maximize the chance that it will be transferred to downwind stigmas.
- The number of ovules in each flower is usually reduced; the reason for this is not clear.
- The structure of the inflorescence and its location on the plant tend to maximize the likelihood that pollen will be carried away by the wind. Inflorescences are located on exposed stalks, often separate from leaves, and high above the ground or water surface.

Because pollen begins to fall as soon as it is blown off the male flower, it is advantageous for wind pollinators to have staminate flowers or inflorescences located higher than pistillate ones (Whitehead 1983). Such is the case in many wetland plants, particularly many *Carex* species (sedges), in which the staminate portion of the inflorescence is superior to the pistillate portion (Figure 5.4a), and in *Typha* (cattail; Figure 5.4b). In *Typha* the pollen count is assumed to be quite high since seed production has been estimated at an average of 220,000 seeds per spike (Voss 1972). Tall conspicuous flowers are evident in many wetland grasses and sedges such as *Phragmites australis* (common reed), and several species of *Scirpus* (bulrush).

Pollen grains of anemophilous plants also show adaptations to being transported by the wind. They are usually small (20 to 40 μm in diameter) and light and are readily spread

FIGURE 5.4a
The pistillate flowers are below the staminate flowers in many wind-pollinated species, which may enhance pollen capture efficiency. *Carex* sp. with separate staminate and pistillate flowers on the same stalk. (Photo by H. Crowell.)

FIGURE 5.4b
The wind-pollinated *Typha angustifolia* (cattail) with staminate flowers in the upper portion of the inflorescence and pistillate flowers below. (Photo by H. Crowell.)

in high winds. They tend to be powdery, smooth, and not sticky. They are resistant to changes in temperature and to desiccation, probably because the size and/or the number of apertures is reduced. Often, pollen is released at times of low humidity and when there is a low probability of rainfall. This timing helps keep the pollen dry (Whitehead 1983; Cook 1988).

On a landscape scale, wind pollination often dominates in areas where the vegetational structure is relatively open. In open areas pollen grains have a greater chance of arriving on stigmas, rather than solely on nonstigmatic surfaces. Compatible plants are usually relatively closely spaced (Whitehead 1983). These conditions are met in several types of open canopy wetlands such as freshwater marshes, sedge fens, and salt marshes. In addition, many wetland plants grow in monotypic stands which enhance wind pollination effectiveness.

3. Water Pollination

In water pollination, or hydrophily, water is a vector in the transportation of pollen. While some authors restrict hydrophily to plants whose pollen moves strictly underwater or comes into contact with the water (Les 1988; Philbrick 1991; Les et al. 1997), we use a more inclusive definition in order to describe the variety of strategies involved (Table 5.2). Cox (1988) states that, "[in hydrophily], it is not necessary for the pollen itself to come in contact with water because water borne conveyances such as anthers or flowers may carry dry pollen across the water surface." Hydrophily occurs in both temperate and tropical marine and freshwater systems. Over half of the hydrophilous plants are seagrasses whose pollination is affected by the tide (Les et al. 1997). Seagrasses are aquatic monocots (not actually grasses of the Poaceae) that are capable of surviving in the marine environment. We include them here in order to cover the complete range of hydrophily. Many hydrophilous species are quite widely distributed and span temperate and tropical regions. This may be due to the thermal buffering properties of water which enable the plants to thrive and flower even where air temperatures are extreme (Philbrick 1991).

Hydrophilous taxa represent a tiny fraction of angiosperms (125 to 150 of about 250,000 described angiosperm species; Philbrick 1991). Most genera that contain hydrophilous species are small, with fewer than 10 species. Low speciation among the hydrophiles may be due to the homogeneity of the wet habitat combined with infrequent sexual reproduction. An exception is *Najas* which has from 35 to 50 species, perhaps because most *Najas* species are annuals with high rates of sexual reproduction (Les 1988; Cook 1996).

a. Planes of Water Pollination

Hydrophily can be divided into four categories based on whether pollination occurs above, on, or below the water surface or a combination of these (Cox 1988). We describe the four categories as (1) pollen transported above the water surface (*dry epihydrophily*), (2) pollen transported on the water surface (*wet epihydrophily*), (3) pollen transported beneath the water surface (*hypohydrophily*), and (4) surface and underwater pollination (*combination*).

Pollen transported above the water surface — During pollination above the water surface (*dry epihydrophily*), the reproductive structures remain dry. Dry epihydrophily requires the fewest adaptations to the wet habitat since it is essentially an aerial strategy (Philbrick 1988). The predominant mechanism of above-surface pollination is the collision of floating staminate and pistillate flowers. The pollen is transported by floating staminate flowers that are released underwater and float to the surface where they dehisce. The pollen is dry, spherical, and sometimes agglutinated into sticky masses. The floating pistillate flowers are attached to the shoots by long peduncles. The pistillate sepals and petals are partly unwettable. The pistillate flower often forms a depression in the water's surface (Cox 1993). All the taxa that rely on the collision of flowers are in the same family, the Hydrocharitaceae.

The species of the submerged dioecious monocot, *Vallisneria*, are pollinated above the water's surface (Figure 5.5). At flowering, the solitary pistillate flower is carried to the water's surface on an elongated peduncle. The male plants produce a large number of staminate flowers on a short peduncle. Batches of staminate flowers are released underwater with the perianth enclosing a gaseous bubble. At the surface, the perianth opens and two sterile anthers serve as a tiny sail while two fertile anthers bear the pollen. The staminate flower floats until it reaches the small depression in the surface formed by the larger pistillate flower. The staminate flower literally falls into the dimple and pollination occurs, barely above the surface of the water (Voss 1972).

The African pondweed genus, *Lagarosiphon*, has staminate flowers that are released underwater and float to the surface as buds. When the buds open, the sepals and petals bend backward and form the flower's vessel. Three fertile stamens extend horizontally while three infertile stamens grow upright and function as sails. The male flowers link together in large rafts and collide with the floating pistillate flowers, depositing the sticky pollen on the stigma (Figure 5.6; Cook 1982; Cox 1993).

In *Hydrilla verticillata*, the staminate flowers are released underwater and float to the surface. Pollen is catapulted into the air and a small percentage of pollen grains land in floating pistillate flowers (Figure 5.7). In addition, staminate flowers sail upright on the water surface and bump into pistillate flowers spilling more pollen onto stigmas. If the water level rises and the stigmas sink, they are protected by an air bubble formed by the inward bending sepals and petals (Yeo et al. 1984). The first strategy, in which pollen is sent into the air, is arguably anemophilous, since the pollen is airborne before pollination occurs. However, the second strategy requires water to carry the staminate flower to the pistillate flower, so it resembles *Vallisneria* and *Lagarosiphon* as well as a number of other Hydrocharitaceae genera: *Appertiella, Enhalus, Maidenia,* and *Nechamandra*.

TABLE 5.2
The Hydrophilous Orders, Families, Genera, and Their Traits

Family	Distribution	No. of Hydrophilous spp.	Habitat	Life Form	Sexual Reproduction	Sexual Condition	Type of Hydrophily	Pollen Apparatus
Monocotyledons								
Alismatales								
Hydrocharitaceae	Cosmopolitan, mainly warm							
Appertiella		1	F	A	Common	D	Dry epi-	Male flower
Elodea		5	F	A and P	Rare	M and D	Wet epi-	Pollen
Enhalus		1	M	P		D	Wet, dry epi- and hypo-	Male flower
Halophila		10	M	A and P	Rare to common	M and D	Wet epi- and hypo-	Mucilaginous chains and pollen tube
Hydrilla		1	F	A and P	Rare to common	M and D	Dry epi-	Pollen or male flower
Lagarosiphon		9	F	P		M and D	Dry epi-	Male flower
Maidenia		1	F	A and P		M and D	Dry epi-	Male flower
Nechamandra		1	F	A and P		D	Dry epi-	Male flower
Thalassia		2	M	P	Rare to common	D	Hypo-	Slime mat chains and pollen tube
Vallisneria		2–4	F	A and P	Rare to common	D	Dry epi-	Male flower
Najadaceae	Cosmopolitan							
Najas		About 40	B, F	A and P	Common	M and D	Hypo-	Pollen tube + pollen
Posidoniaceae	Mediterranean, S.W. Asia, Australia							
Posidonia		5	M	P	Rare to common	H	Hypo-	Thread-like pollen
Potamogetonaceae[a]	Cosmopolitan							
Potamogeton		4–5	F	A and P	Rare to common	H	Wet epi- and hypo-	Pollen
Ruppia		2–10	B, F, M	A and P	Common	H	Wet epi- and hypo-	Pollen chains
Zosteraceae	All coasts							
Heterozostera		1	M	P	Common	M	Hypo-	Pollen

Phyllospadix		5	M	P	Common	D	Wet epi- and hypo-	Pollen
Zostera		12	M	A and P	Rare to common	M	Wet epi- and hypo-	Pollen chains
Zannichelliaceae	Cosmopolitan							
Althenia		1 or 3[b]	B, F	A and P	Common	D	Wet epi- (or autogamous)	
Lepilaena		5	B, F	A and P	Common	M and D	Wet epi- and hypo-	Pollen chains
Pseudalthenia		1	B, F	P		M	Hypo-	
Zannichellia		1 or 6[b]	B, F	A and P	Common	H, M, D	Hypo-	Pollen tube + mucus
Cymodoceaceae	Tropical and subtropical							
Amphibolis		2	M	P	Common	D	Wet epi- and hypo-	Filiform pollen/chains
Cymodocea		4	M	P	Rare	D	Wet epi-	Pollen
Halodule		6	M	P	Rare	D	Wet epi-	Pollen
Syringodium		2	M	P	Common	D	Hypo-	Pollen
Thalassodendron		2	M	P	Rare	D	Hypo-	Pollen/pollen chains
Non-Monocotyledon								
Ceratophyllales[c]								
Ceratophyllaceae	Cosmopolitan							
Ceratophyllum		4–6	B, F	P	Rare to common	M	Hypo-	Anthers or pollen grains with pollen tubes

Note: Habitat categories are B = brackish, F = freshwater, M = marine. Life form categories are A = annual, P = perennial. Sexual condition categories are D = dioecious, M = monoecious, H = hermaphroditic

[a]Potamogetonaceae includes *Ruppia* according to Cook (1996). Some place *Ruppia* in a separate family, Ruppiaceae, due to genetic differences.

[b]Either 1 polymorphic species or the higher number.

[c]The order, Ceratophyllales, has one family, Ceratophyllaceae, and one genus, *Ceratophyllum*. It is classified outside of both the monocotyledons and the eudicotyledons in a separate ancestral angiosperm line (Les et al. 1991).

Adapted from Les 1988 and Philbrick 1991, with additional information from Cox 1983, 1988, 1993; Philbrick 1988; Cook 1996; Titus and Hoover 1991; and Steward 1993.

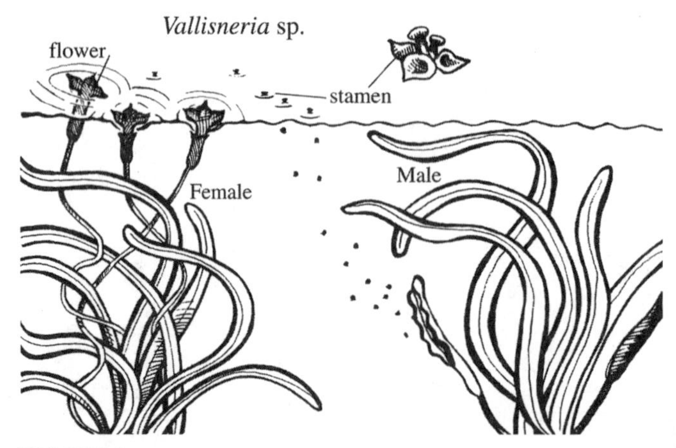

FIGURE 5.5
Water pollination in *Vallisneria* sp. occurs on the water's surface and is called *dry epihydrophily*. The staminate flowers abscise from the male plant and float to the surface. The floating pistillate flower creates a surface depression and the staminate flowers float. Pollination occurs when the staminate flowers collide with the pistillate flower. The free-floating staminate flower is about 1 mm in diameter; it is shown enlarged at the upper right. (From Cox, P.A. 1993. *Scientific American* 269: 68–74. Redrawn, with permission, by B. Zalokar.)

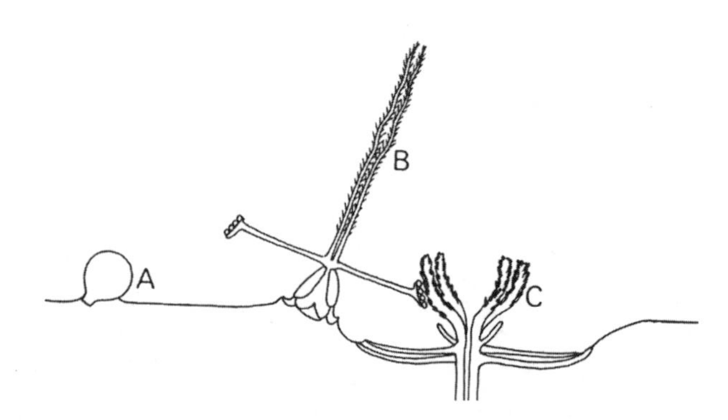

FIGURE 5.6
The staminate flowers of *Lagarosiphon* are released underwater and float to the surface as buds where they open and form large "rafts" that collide with the pistillate flowers without wetting the pollen (*dry epihydrophily*). (A) Bud of staminate flower floating on water; (B) staminate flower at anthesis (bud opening), with an open anther touching the stigma of a pistillate flower; (C) pistillate flower, one stigma in contact with the open anther of a staminate flower (the total height of the staminate flower is about 4 mm). (From Cook, C.D.K. 1982. *Studies on Aquatic Vascular Plants.* J.J. Symoens, S.S. Hooper, and P. Compere, Eds. Brussels. Royal Botanical Society of Belgium. Reprinted with permission.)

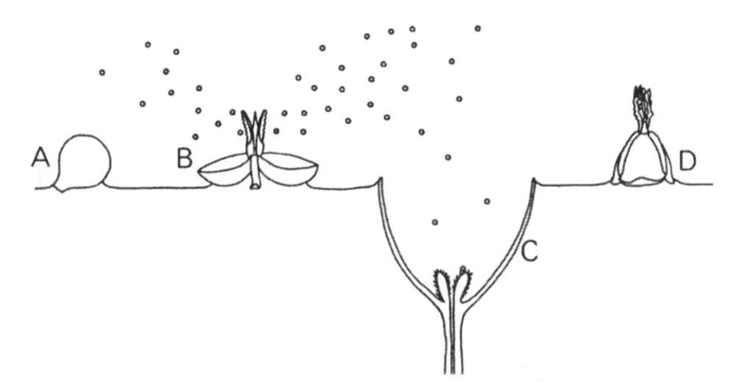

FIGURE 5.7
The flowers of *Hydrilla verticillata,* which are pollinated at the water's surface without wetting the pollen (*dry epihydrophily*). (A) Bud of staminate flower floating on water; (B) staminate flower explosively releasing pollen; (C) pistillate flower receiving pollen; (D) spent staminate flower (the opening of the pistillate flower is about 2 mm deep and the staminate flower is about 1.5 mm tall) From Cook, C.D.K. 1982. *Studies on Aquatic Vascular Plants.* J.J. Symoens, S.S. Hooper, and P. Compere, Eds. Brussels. Royal Botanical Society of Belgium. Reprinted with permission.)

Pollen transported on the water surface — When pollination occurs on the water surface (*wet epihydrophily*) the pollen is in direct contact with the water (Philbrick 1988). Surface pollinators use various strategies to keep pollen dry and afloat. The pollen grains are often long, thread-like, and sticky, and they aggregate into clumps. For example, in some species of *Halodule, Halophila, Ruppia, Lepilaena,* and *Amphibolis,* the pollen grains form chains that float inside a mat of mucilaginous slime. The slime protects the pollen from wetting and enlarges the effective size of the pollen, thus increasing its chances of reaching a stigma (Cox 1993). The floating stigmas of epihydrophilous plants create depressions in the water surface (freshwater taxa) or synchronize flowering with tidal phenomena (marine taxa; Cox 1988).

The pistillate flower of the freshwater genus, *Elodea,* is attached to the plant by a long peduncle and the staminate flowers are released as buds that float to the surface (Figure 5.8). The staminate flower dehisces and sprays pollen onto the water's surface where it floats and drifts into the pistillate flower. The chances of the pollen reaching the pistillate flower seem relatively slim, but the pistillate flower has an adaptation that attracts the pollen grains. Its sepals are water repellent and they form a small depression around the flower. If a pollen grain comes within about 1 cm of the pistillate flower, it is drawn toward the stigma, probably because of changes in surface tension (Voss 1972; Cook 1982). Similarly, in the Australian freshwater *Lepilaena cylindrocarpa,* the pistillate flower forms a depression that captures pollen released as a floating slime mat (Cox 1993).

Rather than forming depressions in the water's surface, marine species tend to use the movement of the tide to carry pollen to the stigma. In the Fiji Islands, low spring tides expose entire populations of the seagrass, *Halodule pinifolia.* When it is exposed, long filaments bearing anthers become erect and release cottony masses of filiform pollen onto the water surface (Figure 5.9). These masses float together and form millions of large pollen "rafts." As the tide returns, the smooth filamentous stigmas float on the water surface where they collide with the pollen rafts (Cox 1993). Like *H. pinifolia,* the male flowers of *Amphibolis antarctica* float to the surface and dehisce during extremely low spring tides. At the surface, the filamentous pollen (3,000 to 5,000 μm long) aggregates to form thousands

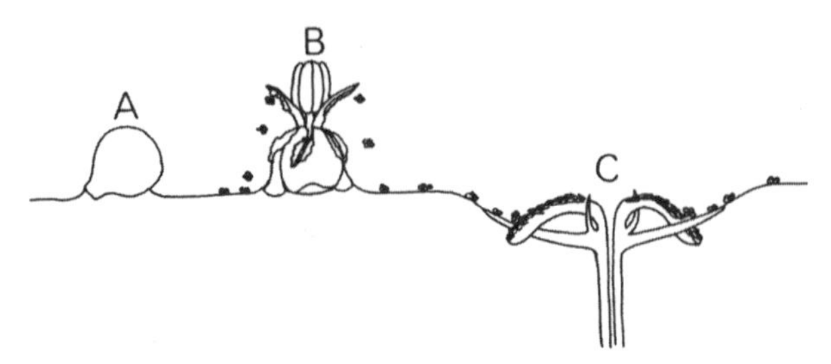

FIGURE 5.8
The flowers of *Elodea nuttalli* are pollinated at the water's surface and the pollen is wettable (*wet eiphydrophily*). (A) Staminate flower bud floating on water; (B) staminate flower releasing pollen; (C) pistillate flower collecting pollen (the staminate bud is about 1 mm in diameter). (From Cook, C.D.K. 1982. *Studies on Aquatic Vascular Plants*. J.J. Symoens, S.S. Hooper, and P. Compere, Eds. Brussels. Royal Botanical Society of Belgium. Reprinted with permission.)

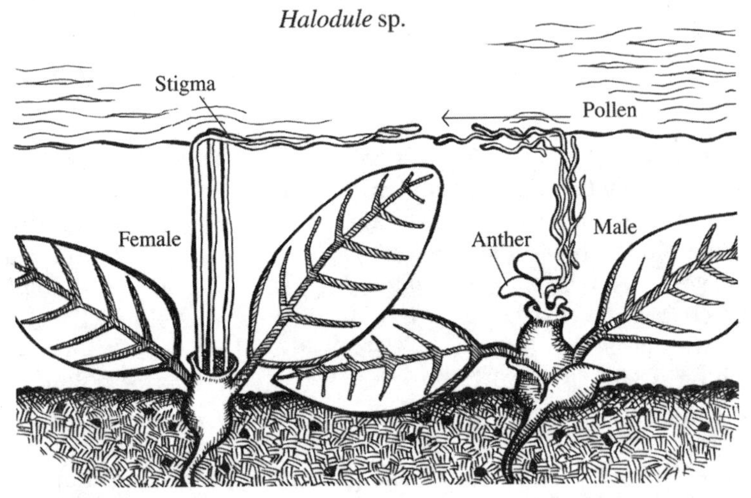

FIGURE 5.9
Halodule is a marine genus that produces filamentous pollen and thread-like stigmas. The stigmas float on the water's surface and collide with the stigmas. The pollen is wettable and pollination occurs at the water's surface (*wet epihydrophily*; the body of the flowers, not including the pollen or stigmas, is about 1 cm long). (From Cox, P.A. 1993. *Scientific American* 269: 68–74. Redrawn, with permission, by B. Zalokar.)

of pollen rafts about 8,500 μm in diameter. The tops of the pistillate plants are exposed to the air and pollination occurs through collision of the floating rafts with the stigmas (Cox and Knox 1988; Cox 1993).

Pollen transported beneath the water surface — Pollination below the water surface (*hypohydrophily*) involves transferring pollen within the water column. It has evolved in plants that are always submerged and have little opportunity for pollination at the water's surface (Cox 1993). In hypohydrophily, the pollen is released underwater, and both pollen and stigma are wet during pollination. Hypohydrophily occurs in both marine and freshwater plants. In some cases, hypohydrophily has not been observed, but deduced because the flowers are submerged. Plants that pollinate underwater have negatively buoyant pollen that is filamentous or spherical or it is released in mucilaginous strands. In marine species, the stigmas are stiff and occasionally *papillate* (with projections on the surface), borne on rigid, flattened inflorescences that are held at a right angle to the marine substrate. In freshwater taxa, pollen production is copious and the stigmas tend to be small and filiform or feathery, borne on flexible three-dimensional inflorescences (Cox 1988).

The most common form of hypohydrophily is the "pollen shower" in which pollen is released from elevated but often submerged staminate flowers. It showers down to submerged stigmas that are, in some cases, quite large and specialized, as in the funnel-shaped stigmas of *Zannichellia palustris* (horned pondweed; Ackerman 1995; Figure 5.10). *Z. palustris* is *monoecious* (staminate and pistillate flowers are separate, but on the same plant), with the staminate flower superior to the pistillate flower. The pollen is released as floating clouds that slowly drop round pollen grains onto the stigmas below. One plant can produce over 2 million seeds in 5 months (Voss 1972; Van Vierssen 1982; Cox 1993).

FIGURE 5.10
Some species that are pollinated underwater (*hypohydrophily*) have large pistillate flowers that capture floating clouds of pollen that are released from staminate flowers located above. The submerged plant, *Zannichellia palustris* (horned pondweed), has funnel-shaped pistillate flowers in the leaf axils that may enhance pollen capture. It is shown here with the staminate flower above (the height of the pistillate flower is about 2 mm). (From Voss, E. 1972. *Michigan Flora Part I Gymnosperms and Monocots.* Bloomfield Hills, MI. Cranbrook Institute of Science. Reprinted with permission.)

A common genus of freshwater wetlands, *Ceratophyllum* (hornwort), contains the only hydrophilous species that is not a monocot (neither is it a eudicot; Ceratophyllaceae are classified as an ancestral angiosperm line; Les et al. 1991; Philbrick and Les 1996). In *Ceratophyllum*, the stamens abscise underwater and float to the surface where they dehisce. Then the pollen grains slowly sink. They have to fall directly on the stigmatic grooves of pistillate flowers in order for pollination to occur. A similar process occurs in the freshwater monocot, *Lepilaena bilocularis*, but the sinking pollen grains are dispersed in strings of mucilaginous slime until captured by feathery stigmas. The freshwater monocot genus, *Najas*, has pollen grains that germinate before being released into the water column. The grain plus the growing pollen tube drift onto the stigma. *Najas* produces seeds in abundance that are an important waterfowl food (Fassett 1957; Voss 1972).

Pollen can also be transported by water currents. The reliance on currents is found primarily in the coastal marine habitat, probably because these taxa are adapted to the more energetic and regular water motion of tides (Ackerman 1995). In the seagrass, *Thalassia testudinum* (turtle grass), flowering coincides with spring tides. Flowers are raised a few centimeters above the substrate surface. The staminate flowers open at night with anthers dehiscing by means of longitudinal slits. The round pollen is dispersed in strings of mucilaginous slime that sink and glide along the substrate. Pollination occurs by collision with the stiff, bristled stigmas of the pistillate flowers (Cook 1982; Cox 1993). *Syringodium* species also release pollen that drifts toward the stigma on water currents.

Surface and underwater pollination — Some species combine epihydrophily and hypohydrophily. Generally these are seagrasses with both intertidal populations that release pollen on the water's surface, and subtidal populations that are never exposed. The staminate flower of the seagrass, *Phyllospadix scouleri*, releases filamentous pollen that aggregates into snowflake-like rafts. Pollen is released both on and below the water's surface. Water currents carry the pollen rafts to floating or submerged pistillate flowers. *Zostera marina*, which grows in coastal regions throughout the Atlantic and Pacific, releases oily, water-repellent, elongated pollen that forms rafts and drifts with the currents onto the floating stigmas. It also releases linear bundles of long, thread-like pollen that are intercepted by stigmas beneath the water's surface (Cox 1993; Laushman 1993; Ackerman 1997a).

b. Hydrophilous Pollen Adaptations

The pollen of hydrophilous species has undergone a number of adaptations to become wettable. In many, the *exine* (the outer coat of a pollen grain) is reduced or lacking. Only a rudimentary exine exists in *Ceratophyllum* and *Thalassia* and it is lacking in some species of *Najas* and *Amphibolis*. In terrestrial species, the exine prevents desiccation. Since this function is unnecessary underwater, there may be little selection pressure to retain the exine in hydrophilous plants. The exine also contains substances that allow the stigma to accept or reject unsuitable pollen. These substances may be dissolved or leached out in water (Philbrick and Les 1996).

The size, shape, and dispersal mechanisms of hydrophilous pollen show evolutionary convergence. Hydrophilous pollen grains often elongate to remarkable lengths, up to 5 mm. Instead of producing the typical spherical pollen grains of most angiosperms, many hydrophilous species produce thread-like, or *filiform*, pollen, a shape distinctive to hydrophily. Filiform pollen is found in all the hydrophilous seagrasses of the Zosterales and may have evolved only once, in an ancestor plant to this order (Cox 1993). The unrelated seagrasses of the Hydrocharitales (*Halophila, Thalassia,* and *Enhalus*) have spherical and ellipsoidal pollen. Round pollen grains of *Thalassia* and *Najas* can germinate precociously and with the pollen tube added to their length, they become filiform in shape. The ellipsoidal pollen of *Halophila* aggregates into long mucilaginous groups (Ackerman 1995). Filiform pollen and the elongated shapes of clumped pollen grains are found only in seagrasses and not in related hydrophilous freshwater genera. Freshwater genera are more frequently surface-pollinated and have spherical pollen (Ackerman 1995). The various means by which hydrophilous pollen becomes larger or moves toward the pistillate flower (i.e., pollen + pollen tube, pollen groups of mats and chains, or the entire staminate flower) are referred to as the *pollen apparatus* in Table 5.2.

Cox (1983, 1988, and 1993) explains the evolution of the increased length of seagrass pollen grains using a military analogy. Search theory, developed during World War II, shows that if a search vehicle traces a random path in two dimensions, its probability of meeting a fixed target increases with the width of the path it traces. In the military realm, the theory showed that even a slight increase in the range of ship-borne radar would increase the chances of detecting a fixed target. Cox and Knox (1988) applied the theory to epihydrophily and showed that the larger the "search vehicle" (pollen apparatus), the more likely it is to reach the "target" (stigma). The model seems to explain the selective advantage of the elongated or aggregated floating pollen as well as the even larger "search vehicles" used by several epihydrophilous species of the Hydrocharitaceae: the entire male flower.

In hypohydrophily, in which the search takes place in three dimensions, search theory does not necessarily apply. In two dimensions, a random search is more efficient because the search path is recurrent and any target on the plane will eventually be hit after enough time. In three dimensions, however, even an infinite amount of time does not guarantee that search vehicle and target will meet (Cox 1993). In three dimensions, pollen and stigmas are dispersed throughout a volume rather than concentrated on a plane, so more pollen is needed. Ackerman (1995, 1997a, b) proposed a model for the evolution of filiform pollen in seagrasses that is based on physical properties of both the water and the pistillate flower. He asserted that filiform pollen shapes evolved because of their fluid dynamic advantage in the velocity gradients around submerged flowers. Filiform pollen has increased rotation (as compared to round pollen) forcing it to tumble and turn in areas of greater water velocity. The direction and velocity of water movement surrounding the pistillate flower causes the elongated pollen grains to rotate and tumble toward the stigma, thereby increasing the incidence of pollination. Rotational motion increases with length so an increase in pollen length is selected for because the grains with greater movement are more likely to reach the stigma.

c. Hydrophilous Stigma Adaptations

Stigmas also show adaptations which heighten the chances of pollination. Some species have a relatively wide "target" that increases the likelihood of encounter. The stigmas of the seagrass genera, *Halodule, Halophila,* and *Thalassodendron,* are long and thread-like and are more likely to come in contact with pollen than circular stigmas (Figures 5.9 and 5.11; Cox and Knox 1988). *Vallisneria* and *Enhalus* have floating pistillate flowers that are much larger than the staminate flowers. In addition, they create a depression in the water surface, which draws the staminate flowers toward them. The Australian *Lepilaena cylindrocarpa* uses the same strategy. Its three stigmas are held together by a bract that forms a dip in the surface, thus enlarging the target area. The pistillate flowers of *Zannichellia* are funnel-shaped and lower on the plant than the stamens. Since pollen is released higher in the water column, the pistillate flowers' position and shape may improve the likelihood of pollen capture (Figure 5.11). Some plants use an oscillatory motion that can effectively widen the pollen's target. In slow-moving streams, the pistillate flowers of *Ruppia marina* move back and forth with the current and thereby collect the floating pollen (Verhoeven 1982). Some flowers appear to have localized effects. The submerged flowers of *Zostera* cluster together and affect underwater flow patterns, thus concentrating pollen near stigmas (Cox 1993).

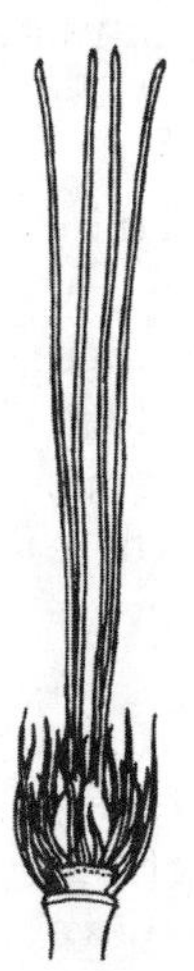

FIGURE 5.11
Some seagrasses have long thread-like stigmas that are more likely to come in contact with pollen than round stigmas. The seagrass, *Thalassodendron pachyrhizum,* has elongated stigmas (stigmas about 30 mm). (From Cook, C.D.K. 1996. *Aquatic Plant Book.* The Hague. SPB Academic Publishing/Backhuys Publishers. Reprinted with permission.)

d. The Evolution of Hydrophily

Most angiosperm flowers are aerial and require dry conditions for pollination and fertilization. That a clear majority of submerged angiosperms retain aerial flowers indicates a strong selective pressure against flowering underwater. A major obstacle to the evolution of water pollination is the series of adaptations necessary for seed set with wetted pollen and stigmas. It is not clear how selection for wettability occurred and acted on aerial flowers since partially wet intermediate steps are not obvious (Philbrick 1988).

Water pollination shares a number of characteristics with wind pollination. Both anemophily and hydrophily are random abiotic strategies that lack the specificity of biotic pollination. Both have high pollen production and much of the pollen does not contact a stigma. Both anemophilous and hydrophilous species have reduced perianths and high pollen/ovule ratios. The pollen exine is reduced and the receptive surface, or stigma, is often enlarged. Anemophily and hydrophily differ in pollen dispersal distances. The pollen dispersal distance of anemophiles is limited by wind speed and the pollen grain's ability to remain airborne. Hydrophiles' pollen dispersal is limited to the water body they inhabit. Pollen that is dispersed to deep or disturbed areas where plants do not live is lost (Les 1988).

Because several of the modifications for the evolution to hydrophily are similar to those that occurred in evolving toward anemophily, it is thought that hydrophily evolved from anemophily (Philbrick 1988). Hydrophiles in the families Cymodoceaceae, Posidoniaceae, Ruppiaceae, Zannichelliaceae, and Zosteraceae all have anemophilous evolutionary antecedents (Philbrick and Les 1996). In the Hydrocharitaceae, hydrophily may have evolved from entomophily with flowers of the genus *Blyxa* still displaying transitional features (see Section II.A.1, Insect Pollination; Cook 1982; Philbrick and Les 1996).

Anemophilous and hydrophilous angiosperms display frequent *dicliny*. Dicliny occurs when the staminate and pistillate flowers are separated either on the same plant (monoecy) or on different plants (dioecy). Hydrophiles are overwhelmingly diclinous. Only 4 of 27 genera have hermaphroditic flowers. Among hydrophiles, dioecy is more frequent than monoecy with only three of the genera exclusively monoecious. In fact, the predominance of dioecy is one of the most striking features of water pollination. As a comparison, only about 3 to 4% of terrestrial angiosperms are dioecious (Cox 1993). Dioecy may be present in hydrophiles because it ensures outcrossing (Cox 1993). It may have been a feature of upland ancestors of hydrophilous species that predisposed them to the development of hydrophily.

For the pollination strategies involving wet pollen and stigmas (wet epihydrophily and hypohydrophily), three evolutionary models have been proposed (Philbrick and Les 1996):

- *The gradual model:* Wet epiydrophily and hypohydrophily may have evolved by gradual selection on aerial floral systems that led to the accumulation of hydrophilous characters. Some of the intermediate steps may have included steps seen in dry epihydrophily: pistillate flowers pollinated by detached floating staminate flowers where the anthers come in direct contact with the stigma, or the discharge of pollen that reaches the stigma by falling through the air. These strategies would lead to repeated contact of pollen with water and might eventually select for wettable pollen.
- *The selfing intermediate model:* This is an elaboration of the gradual model. Self-pollination occupies a key intermediate position that allows hydrophilous features to accumulate while maintaining seed production. Submersion of the flower does not require wettable pollen and stigma. Such flowers may have

experienced increased fitness from self-pollination, not hydrophily, and this may have facilitated the abandonment of aerial flowers. Philbrick (1988) suggests that bubble self-fertilization (*autogamy*), as seen in the submerged flowers of *Potamogeton pusillus* (small pondweed), may be an intermediate step between aerial and underwater pollination (Figure 5.12). In several *Potamogeton* species, air spaces are present in the flowers. Gases within the air spaces build up in the anther and as the flower opens, a carbon dioxide bubble is formed as the anthers dehisce. Pollen is transported on the outer surface of the bubble. The bubble increases in size until it extends from the anther to the stigma. Pollen is then deposited onto the stigma from the bubble surface. The bubble eventually floats to the surface and other bubbles form as each anther opens. Earlier in the plant's evolution, the pollen may have been shed dry and stayed inside the bubble where there was high humidity. Pollen and stigmas that could withstand the high humidity were selected for. In addition, pollen that could survive wetting could rest on the outside of the bubble and perhaps have a greater chance of reaching the stigma. In this way there could have been a gradual selection for both wettable pollen and wettable stigmas. However, since most hypohydrophilous plants have unisexual flowers, bubble autogamy may not have led to the modifications required for hydrophily (Philbrick 1988; Ackerman 1995).

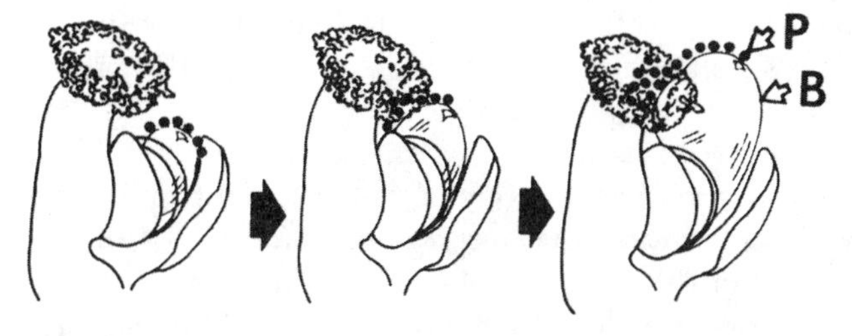

FIGURE 5.12
Idealized diagrams of a portion of a *Potamogeton pusillus* flower that illustrate the progressive enlargement of a bubble (B) during anthesis and the deposition of pollen (P) onto the stigma from the bubble surface. The pollen is moved toward the stigma on the surface of the bubble and the flower is self-fertilized (bubble autogamy; the diameter of the stigma is about 1 mm). (From Philbrick, C.T. 1988. *Annals of the Missouri Botanical Gardens* 75: 836–841. Reprinted with permission.)

- *The punctuated model:* Hydrophily may have evolved in one large step, with most features arising simultaneously. This model is the least likely since so many flower features had to adapt to the water habitat. Nonetheless, the development of hydrophily in Ceratophyllaceae may provide an example. The Ceratophyllaceae have reproductive features found in the gymnosperms (ancestors to flowering plants). These features include branched pollen tubes, exineless pollen, monoecy, and lack of stigma. These primitive characteristics are similar to features that have evolved in other hydrophiles, but *Ceratophyllum* species may have been pre-adapted to this environment and exploited these features.

While all three of these evolutionary models may have been at work in some species during the evolution of hydrophily, the gradual model is the most likely for the greatest number of species.

4. Self-Pollination

Self-pollination occurs when pollen grains travel from the anthers to the stigmas of the same flower or other flowers on the same plant. Self-fertilization occurs unless the plant's own pollen is incompatible with its stigmas. This method of fertilization is beneficial when other plants are not available for mating. Self-pollinators do not rely on animals or other pollen vectors, and self-pollinated plants are often present where flower-visiting animals are rare, such as on high mountains or in extremely cold climates (Raven et al. 1999). Some plants always self-pollinate, while in others self-pollination is rare. Self-pollination can occur in plants that also have other pollination strategies such as wind or insect pollination. Plants that always or usually self-fertilize have low genetic variation and "blur the line between sexual and asexual reproduction" (Philbrick and Les 1996).

Several flowering strategies seem to inhibit self-pollination (Rost et al. 1984):

- *The dioecious condition*: Separate male and female plants force outcrossing.
- *The monoecious condition:* With staminate and pistillate flowers on different parts of a plant, selfing can occur but is less likely than in plants with hermaphroditic flowers.
- *Self-incompatibility:* Stigmas have biochemical substances that detect pollen from the same flower and inhibit its germination or pollen tube growth.
- *Heterostyly:* Heterostylous plants have two or more flower forms and the pollen is only compatible with stigmas of unlike flowers.
- *Timing of maturation:* Anthers may mature more slowly than stigmas (or vice versa) and cross-pollination takes place before or after the flower's own pollen grains are shed.

Self-pollination occurs in aerial flowers of wetland plants in the same way as for terrestrial plants. Both insects and wind can carry pollen from the anther to the stigma of the same flower (*autogamy*) or different flowers on the same plant (*geitonogamy*), and closed flowers can develop that self-pollinate without opening (*cleistogamy*).

In some submerged hydrophilous taxa, pollination can occur on the surface of bubbles that form around flowers or inflorescences (Figure 5.12). The hydrophilous genera that have hermaphroditic flowers, *Posidonia, Potamogeton, Ruppia,* and *Zannichellia,* may all be capable of bubble autogamy (Les 1988). Submerged self-pollination (*hydroautogamy*) may also occur as for aerial flowers, through geitonogamy and cleistogamy.

Cleistogamy, or pollination within closed flowers, may occur when flowers that are normally aerial remain submerged (Sculthorpe 1967). Cleistogamy appears to be an alternative to cross-pollination in several submerged genera, such as *Myriophyllum, Podostemum, Ruppia,* and *Utricularia* (Philbrick and Les 1996). *Ottelia ovalifolia,* a submerged tropical plant, produces aerial flowers in mid-summer. In the spring or autumn or if the plants are crowded, it produces small, submerged cleistogamous flowers. In some submerged plants, notably *Subularia aquatilis, Blyxa alternifolia,* and *Ottelia alismoides,* submerged flowers can self-pollinate if they fail to emerge from the water's surface. Unlike most cleistogamous flowers, they subsequently open and are similar in size to the species' normal aerial flowers (Sculthorpe 1967).

The annual submerged species, *Callitriche heterophylla* and *C. verna,* are self-fertilized in a slightly different way. Each flower is unisexual, so self-pollination within the same flower is impossible. Instead, the pollen grains germinate within undehisced anthers and pollen tubes grow through vegetative tissue to the pistillate flowers of the same plant, thus effecting an internal form of geitonogamy (Philbrick and Les 1996).

B. Fruits and Seeds

1. Types of Fruits Produced by Wetland Plants

Many wetland plants produce one-seeded dry indehiscent or dehiscent fruit, and some produce fleshy fruit (Table 5.3). The fruit types can all be found in upland plants as well. Fruit terminology varies somewhat among the plant keys, and we present here alternative fruit names as well as those most often used.

TABLE 5.3

Types of Fruits Produced by Wetland Plants, Their Description, and Examples of Families with Wetland Species That Bear Each Type of Fruit

Type of Fruit	Description	Families
Dry Indehiscent		
Achene	One-seeded fruit from a single ovary; pericarp can be easily detached; seed within attached to placenta by a stalk (*nut* is sometimes used as a synonym)	Cyperaceae Ceratophyllaceae Cymodoceaceae Polygonaceae
Caryopsis	One-seeded fruit with seed coat firmly united with pericarp (synonymous with *grain*)	Poaceae
Cypsela	An achene-like fruit from an inferior ovary with a circle of hairs (pappus) formed from a modified group of sepals	Asteraceae
Mericarp	Nutlet-like, one seeded portions of a schizocarp that are split off from a syncarpous ovary (carpels are fused)	Apiaceae Callitrichaceae
Nutlet	One-seeded fruit derived from one of several carpels from an apocarpous ovary (carpels are not fused)	Alismataceae Lamiaceae Potamogetonaceae
Samara	One- or two-seeded, with an outgrowth of the ovary wall that forms a wing	One-seeded: Ulmaceae Two-seeded: Aceraceae
Dry Dehiscent		
Capsule	Many-seeded fruit from a compound ovary of two or more carpels)	Hydrocharitaceae Iridaceae Juncaceae Lentibulariaceae Lythraceae Menyanthaceae Primulaceae
Follicle	Usually dehiscent fruit from a single carpel that splits along one suture	Aponogetonaceae Butomaceae Cabombaceae Ranunculaceae
Pod	Fruit from a single carpel that splits along two sutures or into one-seeded portions	Leguminosae
Fleshy		
Berry	Derived from a compound ovary; usually many seeded	Araceae Nymphaeaceae
Drupe	Fruit with a stony endocarp usually containing one seed	Nyssaceae

From Sculthorpe 1967; Voss 1972; Rost et al. 1984; Cook 1996.

Dry indehiscent fruits do not split open to release the seeds. They are distinguished by the way in which the seed is or is not attached to the surrounding *pericarp* (fruit wall), by the type of ovary from which the fruit is derived, and by appendages to the fruit, as in the *cypsela* and the *samara*. The *achenes* of the Cyperaceae and other families are usually small fruits that can be mistaken for the seed. However, the pericarp can be detached easily (Figure 5.13a). Achenes are also called nuts by some authors (i.e., Fassett 1957; Cook 1996). The *caryopsis*, or *grain*, of the Poaceae (grasses) is usually enclosed in two bracts characteristic of grasses, the *lemma* and *palea* (Figure 5.13b).

Many species of Asteraceae (a very large family of about 1,300 genera, at least 15 of which have submerged or emergent species) produce a *cypsela*, a fruit that is similar to an achene, but with an appendage called a *pappus* formed by a group of sepals (Figure 5.13c). A well-known example of a cypsela with a pappus is the wind-dispersed fruit of the upland species, *Taraxacum officinale* (common dandelion). Among wetland plants the cypsela is seen in *Hydropectis aquatica* and other aquatic members of the Asteraceae. *Mericarps* and *nutlets* are both dry indehiscent fruits that derive from ovaries with more

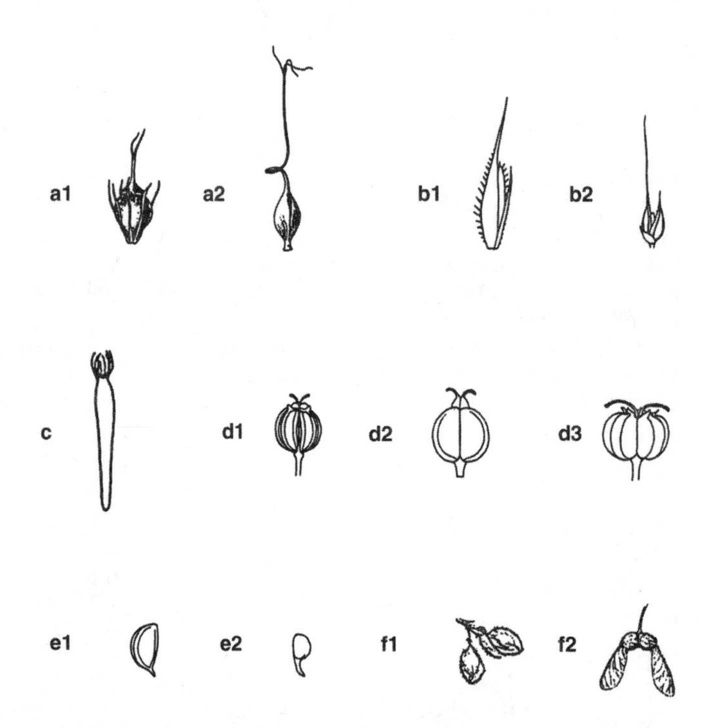

FIGURE 5.13
Examples of dry indehiscent fruits produced by wetland plants.
(a) **achene** — (1) *Eleocharis obtusa*, fruit diameter at widest point = 0.6–1 mm; (2) *Carex lupulina*, fruit diameter at widest point = 1.5 mm. (From Voss, E. 1972. *Michigan Flora Part I Gymnosperms and Monocots.* Bloomfield Hills, MI. Cranbrook Institute of Science. Reprinted with permission.)
(b) **caryopsis** — (1) *Spartina pectinata*, fruit length = 1–1.5 cm; (2) *Echinochloa stagnina*, fruit length = 1–1.5 cm.
(c) **cypsela** *Hydropectis aquatica*, fruit length = 2 cm.
(d) **mericarp** — (1) *Carum verticillatum*, fruit diameter at widest point = 5 mm; (2) *Berula erecta*, fruit diameter at widest point = 1 mm; (3) *Cicuta virosa*, fruit diameter at widest point = 3 mm.
(e) **nutlet** — (1) *Baldellia ranunculoides*, fruit length = 2 mm; (2) *Potamogeton crispus*, fruit length = 2–3 mm.
([b–e] From Cook, C.D.K. 1996. *Aquatic Plant Book.* The Hague. SPB Academic Publishing/ Backhuys Publishers. Reprinted with permission.)
(f) **samara** — (1) *Ulmus crassifolia*, fruit length = 1.0–1.2 cm; (2) *Acer saccharum*, fruit length = 2.5–3 cm. (From Little, E. 1980. *The Audubon Society Field Guide to North American Trees.* New York. Alfred A. Knopf, Inc. Reprinted with permission.)

than one carpel. When the carpels are fused, the ovary is called *syncarpous*, and the pericarp and seed within can break off from the others in a one-seeded *mericarp*. The fruit bearing the group of mericarps is called a *schizocarp*. The Apiaceae (=Umbelliferae), a family of many emergent, submerged, and floating genera, such as *Cicuta, Hydrocotyle,* and *Sium,* produce two one-seeded mericarps (Figure 5.13d). The submerged Callitrichaceae produce a four-lobed schizocarp, which splits into four flattened mericarps.

When the carpels of an ovary are not fused, the resulting fruit is called a *nutlet*. Nutlets are common among wetland plants, notably in the entirely aquatic families, Alismataceae and Potamogetonaceae, and a family with many emergents, Lamiaceae (=Labatiae, mint; Figure 5.13e). The *samara,* sometimes called a *key,* is the fruit of the Aceraceae (maple) and some members of the Ulmaceae (elm). In elm trees, the samara is one-seeded, while in maples it is two-seeded (Figure 5.13f).

Dry dehiscent fruits split open to release the seeds. They are distinguished by the number of seeds they contain and the ways in which they dehisce (Table 5.3). *Capsules* are derived from a compound ovary and are many-seeded. They are seen in a number of wetland plant families (Figure 5.14[a]). Dehiscent capsules ripen above water in the Eriocaulaceae, Mayacaceae, Xyridaceae and Monochoria, Elatinaceae, Podostemaceae, Menyanthaceae, and in the aquatic members of Lythraceae. Capsules developing underwater are rare. They resemble aerial capsules but they dehisce differently. In aerial capsules,

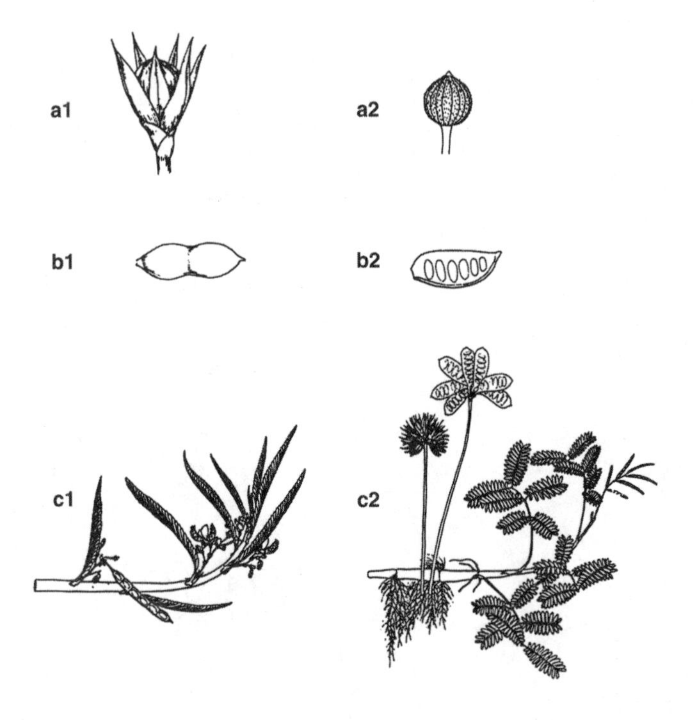

FIGURE 5.14
Examples of dry dehiscent fruits produced by wetland plants.
(a) **capsule** — (1) *Juncus bufonius,* mature perianth, bracteoles, and capsule, fruit diameter at widest point = 3 mm; (2) *Thalassia testudinum,* fruit diameter at widest point = 2.5 cm. (From Voss, E. 1972. *Michigan Flora Part I Gymnosperms and Monocots.* Bloomfield Hills, MI. Cranbrook Institute of Science. Reprinted with permission.)
(b) **follicle** — (1) *Brasenia schreberi,* fruit length = 7 mm; (2) *Caltha natans,* fruit length = 3–4 mm.
(c) **pod** — (1) *Aeschynomene aspera,* floating shoot with pod and flowers, fruit length = 8–10 mm; (2) *Neptunia oleracea,* inflorescence and pods, fruit length = 2 cm. ([b and c] From Cook, C.D.K. 1996. *Aquatic Plant Book.* The Hague. SPB Academic Publishing/Backhuys Publishers. Reprinted with permission.)

dehiscence is brought about by desiccation, but in submerged capsules, pressure from within the fruit causes it to split open. The Juncaceae are distinguished by the fact that their capsules are subtended by three sepals and three petals (Figure 5.14[a1]).

Two other dry dehiscent fruits, *follicles* and *pods*, are similar to one another and both usually contain more than one seed. Follicles dehisce along one suture (though sometimes they are indehiscent as in *Brasenia schreberi*), while pods split along two sutures (as in the Leguminosae; Figures 5.14[b] and [c]).

Some wetland plants produce *fleshy fruits* called *berries* and *drupes*. Berries usually have many seeds and are derived from a compound ovary. Examples of berry-producing plants are those in the Araceae, which has a number of wetland plants; some of the most common are in the genera *Acorus, Calla,* and *Peltandra* (Figure 5.15[a]). In some species of Nymphaeaceae, the berry dehisces and releases its seeds, some of which have small appendages marking the point of attachment to the fruit (*arils*). Because the berry of *Nymphaea* is dehiscent, it is also sometimes called a capsule (Sculthorpe 1967).

A *drupe* has a fleshy *exocarp* (outermost layer of the pericarp) and a stony *endocarp* (inner layer of the pericarp) as in a peach or cherry. Trees of the Nyssaceae (tupelo) that grow in bottomland hardwood forests produce drupes (Figure 5.15[b]). Cook (1996) calls the fruits of the Sparganiaceae and the genus *Zanichellia* drupes, although others call them nutlets (Fassett 1957) or achenes (Voss 1972).

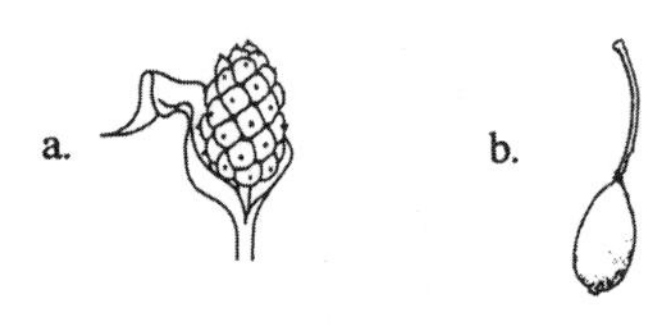

FIGURE 5.15
Examples of fleshy fruits produced by wetland plants. (a) Berry: *Calla palustris,* fruit length = 2 to 4 cm (From Cook, C.D.K. 1996. *Aquatic Plant Book.* The Hague. SPB Academic Publishing/ Backhuys Publishers. Reprinted with permission.) (b) Drupe: *Nyssa aquatica,* fruit length = 2.5 cm. (From Little, E. 1980. *The Audubon Society Field Guide to North American Trees.* New York. Alfred A. Knopf, Inc. Reprinted with permission.)

FIGURE 5.16
Underwater coiling of the peduncle of *Enhalus acoroides* at the beginning of fruit development. Some plants retract their fruit from the water's surface by coiling the peduncle and the fruit matures underwater (the fruit is approximately 3 to 4 cm in diameter.) (From Cook, C.D.K. 1996. *Aquatic Plant Book.* The Hague. SPB Academic Publishing/Backhuys Publishers. Reprinted with permission.)

2. Seed and Fruit Dispersal

Seed or fruit dispersal is a critical stage in the life cycle of angiosperms. In some cases, the seedling is dispersed, rather than the seed or fruit. All of these forms, as well as clonal structures from a plant, are referred to as *diaspores*. For seed dispersal to be effective, the seeds must be carried to favorable sites. The seed's resting place ultimately determines whether it will germinate and if it does, whether it will be able to survive.

Many wetland plant seeds are dispersed by gravity, dropping to the sediments below the parent plant soon after release, thus allowing little chance for long-range transport unless they are ingested by animals. Some submerged, floating-leaved, and floating genera, such as *Hottonia, Nuphar, Utricularia,* and *Lobelia,* maintain their fruit above the water until they mature. When mature, the seeds usually sink into the sediment near the parent. Some plants (including *Enhalus, Ruppia,* and *Vallisneria*) retract their aerial fruits from above the water on a coiling peduncle and the fruits mature underwater (Figure 5.16). In *Vallisneria,* the fruits detach from the peduncles and rise to the surface. The seeds sink when the fruit decays (Titus and Hoover 1991). In many other genera (*Aldrovanda, Aponogeton, Brasenia, Cabomba, Hydrocharis, Limnobium, Nymphaea, Nymphoides, Ottelia, Ranunculus, Trapa,* and *Victoria,* and most of the genera in the Pontederiaceae and Potamogetonaceae) the aerial fruit becomes submerged because the peduncle bends toward the water. The fate of the seeds is generally to drop into the adjacent sediment and germinate there if conditions are favorable (i.e., enough oxygen, light, and space for growth; Sculthorpe 1967).

Seeds that are dispersed farther from the parent plant are spread by wind, water, animals, and people. Wind-dispersed diaspores tend to be the light seeds and fruits of aerial flowers such as the caryopses of grasses or the samaras of maple trees. Wind can blow otherwise water-transported seeds farther away from the parent plant before they drop in the water.

Seed dispersal by water, called *hydrochory,* is of particular interest in wetlands. Seeds of some hydrochorous plants are dispersed by rain wash, others by floods, currents, and tides. Most dispersal by water is downstream. Water dispersal is especially important for wetland plants with buoyant diaspores. In general, the seeds and fruits of submerged plants do not float, while emergents tend to have fruits with more adaptations for remaining buoyant (Sculthorpe 1967).

Some plants' diaspores float simply because they are small and light and the water's surface tension allows them to stay afloat for short periods of time. Once submerged, they sink. Examples of these are the seeds of *Ranunculus repens* (creeping buttercup), and *Myosotis scorpioides* (forget-me-not). Other plants' seeds float because they are attached to a floating plant part. In *Limnocharis,* a floating plant of warm areas, the mericarps float briefly and as their suture widens the sinking seeds are released. In *Alisma* the seeds float on the split pericarp (Sculthorpe 1967).

Some fruits have special adaptations that allow them to remain buoyant. They have a low specific weight caused by air spaces or corky tissues. Large intercellular spaces in the pericarp keep the seeds afloat (as seen in the fruits of *Alisma, Cladium, Limnocharis, Orontium, Pontederia, Potamogeton, Sparganium,* and some species of *Mentha, Scirpus,* and *Scutellaria*). In other genera the air spaces are in the testa, which is a protective coating formed from the integuments of the ovule (*Aponogeton, Calla, Iris, Menyanthes,* and *Scheuchzeria*). In *Nelumbo lutea,* the entire light-weight receptacle for the seeds breaks off and floats upside down, releasing the loosened seeds that sink, but emerge again as floating seedlings (Sculthorpe 1967).

Some fruits and seeds remain afloat due to enlargement by appendages or hairs. The mericarps of species of *Oenanthe* and *Sium* are buoyant because they have lateral masses of waterproof corky tissue. The achenes of *Sagittaria* and *Limnophyton* also have lateral wings, but theirs contain parenchyma with large air spaces. The caryopses of some wetland grasses and the achenes of sedges sometimes float on water because they are enclosed by one or more bracts that form an envelope of air around them (Sculthorpe 1967). In some mangrove genera, the seed or fruit is modified to facilitate buoyancy. In *Cerbera, Cynometra, Heritiera, Laguncularia, Lumnitzera, Nypa, Pelliciera, Scyphiphora,* and *Terminalia*, the fruit wall is fibrous and in *Xylocarpus* the testa is thick and corky. In mangrove genera that also have terrestrial species (*Heritiera* and *Terminalia*), the terrestrial taxa differ in dispersal. Their seeds are winged and suited for wind dispersal (Tomlinson 1986).

The period of buoyancy is important since, at least in theory, the longer the seed can float, the farther it can travel from the parent plant. Buoyant seeds eventually become waterlogged and sink. For some plants the period of buoyancy is just a few hours to a day (e.g., *Eichhornia, Lemna, Myriophyllum, Ranunculus, Trapa,* and *Wolffia*), for others a few days (*Oenanthe* and *Orontium*), and for still others it can be up to 2 months (*Alisma* and *Menyanthes*). The seeds of some species (*Baldellia ranunculoides, Lythrum salicaria, Scrophularia aquatica,* and some species of *Juncus*) sink as soon as they fall into the water, but their seedlings rise to the surface and float for several weeks before becoming permanently rooted (Sculthorpe 1967).

Hydrochory is important in riparian wetlands where its main effect is to redistribute seeds from the sites where gravity has deposited them. Schneider and Sharitz (1988) studied seed fall and dispersal in a forested floodplain of the Savannah River in South Carolina. The two dominant species, *Taxodium distichum* (bald cypress) and *Nyssa aquatica* (water tupelo), depended heavily on water for the dispersal of their seeds. About 86% of *T. distichum* seeds and over 98% of *N. aquatica* seeds were dispersed by water. The remaining seeds simply fell below the tree. The seeds accumulated where emergent substrates (logs, cypress knees, etc.) blocked their paths. The average periods of buoyancy were 42 days for *T. distichum* and 85 days for *N. aquatica.* Most seeds were released in the late fall when water levels were high. The presence or absence of high water determined the distribution of the seeds.

Many wetland plants rely on animals (usually birds and fishes) for dispersal of all types of diaspores. They are transported within animals as well as on their bodies (Sculthorpe 1967; van der Pijl 1982). The fruits and seeds of many freshwater plants are eaten by the abundant wildlife typically seen in wetlands. Hundreds of species of water fowl, marsh birds, shore birds, and game birds as well as mammals and fishes consume the seeds, fruits, and vegetative matter of wetland plants (Fassett 1957). Migrating birds may carry seeds long distances. The passage of food can take 7 to 12 h in ducks, swans, and geese (perhaps less in migrating birds whose rates of metabolism are faster), so presumably seeds that are not digested could be eliminated at a considerable distance from where they were ingested. In this way animals have a direct influence on the distribution of the plant species whose seeds they consume (Sculthorpe 1967; Agami and Waisel 1986, 1988).

Humans are also important dispersal agents of wetland plants. Seeds of wetland species are spread among agricultural crops such as rice. All types of diaspores are transported on the hulls of boats and on agricultural vehicles. Some exotics, such as *Eichhornia crassipes* (water hyacinth) and *Lythrum salicaria* (purple loosestrife), have spread into the wild from gardens. Humans have significantly increased the rate of introduction of species to new habitats, sometimes with deleterious effects (see Chapter 8, Invasive Plants in Wetlands).

3. Seed Dormancy and Germination

Seeds often enter a period of dormancy once they are released from the parent plant. Dormancy is a method by which seeds cope with variation in climate and resource availability. Dormancy gives the seed a means of surviving prolonged unfavorable periods (Bliss and Zedler 1998). Many seeds of temperate species require a cold and/or wet period before germination. Tropical and marine species often do not have a dormancy period, but germinate as soon as the seed is mature. Some species do not germinate unless the seed coat is broken, usually by passing through the digestive tract of an animal. Wetland plants share these requirements with upland plants. Some, however, have special requirements. For some, a period of time when the water level is at or below the soil surface is necessary, presumably to ensure an oxygenated root zone. In a few species, the reverse is true, and germination is enhanced by anaerobic conditions.

A majority of wetland angiosperms exhibit prolonged dormancy. Many have hard seed coats with heavy cuticles that prevent both water loss and water entry (e.g., *Alisma, Halophila,* and *Zostera*). Many seeds germinate after the first winter or period of dry conditions, while others are dormant for several years. In some, a period of drying induces the rupture of the testa or endocarp. Once ruptured, water can be absorbed and the seed can germinate. This phenomenon was observed in *Mayaca fluviatilis* (pool moss), a submerged and sometimes emergent angiosperm of warm areas of the U.S. Seeds that were dried for 6 weeks and then returned to water germinated, but seeds that were kept wet did not. In some species, however, such as *Zizania aquatica* (wild rice), drying kills the seeds (Sculthorpe 1967). The seeds of *Nymphoides peltata* can withstand drying for 2 1/2 years and *Sparganium erectum* seeds remain viable after 4 years whether kept cold or at room temperature, wet or dry. The Asiatic *Nelumbo nucifera* produces very long-lived seeds. Sculthorpe (1967) recounts that a 237-year-old *N. nucifera* seed germinated on an herbarium sheet at the British Museum of Natural History when the museum was accidentally flooded after bombing in 1940.

Many submerged plants can germinate in a certain depth of water or on moist or flooded soil if the temperature is right and there is sufficient oxygen (Sculthorpe 1967). However, many emergent species cannot germinate or survive as seedlings underwater. They rely on vegetative reproduction during flooded conditions. Seeds of these species germinate only when and where conditions are favorable (Titus and Hoover 1991). Once the seedlings become well established they are better able to withstand the low oxygen levels brought on by flooding (Kushlan 1990). For example, although *Taxodium distichum* (cypress) seeds cannot germinate when soils are flooded, their seedlings grow best in saturated soils (Ewel 1990b).

Under flooded and anaerobic conditions, most seeds absorb water and the seed coat is ruptured, as at the onset of germination. Continued soaking in the absence of oxygen results in seed death. However, in some species, germination occurs without oxygen. In both *Oryza sativa* (deep water rice) and *Echinochloa crus-galli* (barnyard grass), the seeds can germinate under both aerobic and anaerobic conditions (Rumpho and Kennedy 1981). In ephemerally wet ecosystems, the ability to germinate in anaerobic conditions may afford a selective advantage. California's vernal pools have summer droughts with rain only in some years. Germination of the vernal pool grass *Tuctoria greenia* occurs almost exclusively under anaerobic conditions; therefore it germinates only in years when there is sufficient rainfall for its subsequent growth. Seeds of the vernal pool grass *Orcuttia californica* were collected and exposed to various treatments in a laboratory. About 25% of the seeds germinated under aerobic conditions with a 12-h photoperiod, but germination increased to 50% when the seeds were held under anaerobic conditions. Such responses to anaerobic conditions favor germination in years when the vernal pools are filled with water (Keeley 1988).

The substrate type can strongly influence germination. For example, *Vallisneria americana* and *Najas* species germinated more readily in a lake with alkaline, silty sediments than in a bog with high organic matter and acidity (Titus and Hoover 1991). Salinity of the substrate also plays an important role. In an inland saline marsh in Ohio, soil salinity was found to be the primary factor controlling seed germination of *Hordeum jubatum, Atriplex triangularis,* and *Salicornia europaea*. Seedling survival and distribution were strongly correlated to the salinity gradient and the plants' tolerance of salinity (Badger and Ungar 1994). In many salt marshes of the eastern U.S., *Spartina alterniflora* (cordgrass) coexists with the invasive species, *Phragmites australis* (common reed). Both can germinate under saline conditions, though *S. alterniflora* has a higher tolerance for salinity and can germinate even at concentrations as high as 40 g NaCl l^{-1}. *P. australis* germination is inhibited above 25 g NaCl l^{-1}. This difference may have significant ramifications in the control of *P. australis* in eastern U.S. salt marshes (Wijte and Gallagher 1996a, b; see Chapter 8, Section V.E, *Phragmites australis*).

Some seeds require that their outer coat be scratched, broken, or that protective chemicals be leached out (*scarification*). Such seeds usually do not germinate unless they have passed through the digestive tract of an animal. Many wetland plant seeds are ingested and excreted by animals and this process is thought to enhance their germination rates.

Evidence suggests that some aquatic plants may produce soft and easily digested seeds as a bait, and harder, digestion-resistant seeds that are the actual dispersal agents of the plant. For example, *Najas marina*, a submerged plant of brackish to salty water in temperate and warm regions around the world, produces at least two kinds of seeds. Some are soft and easily cracked, and some have an impermeable seed coat. A study of *N. marina* germination showed that with no treatment of the seeds, germination was about 11%, but rose to 62% after the seed coats were mechanically cracked (Agami and Waisel 1984). Agami and Waisel thought that an agent that ruptures *N. marina* seed coats must exist in nature. They tested the theory by feeding both types of seeds to mallard ducks (*Anas platyrhynchos*), a species whose flyway corresponds to the distribution of certain species of *Najas* (Agami and Waisel 1986). About 70% of the seeds consumed by the birds were digested, and therefore no longer available for germination. The seeds with the impermeable seed coat were not easily digested and were excreted. Approximately 40% of the excreted seeds germinated. The impermeable seed coat may be a mechanism that aids in the long-term dispersal of *N. marina*, since seeds can remain in the ducks' digestive tracts for 10 to 12 h and the birds can fly between 100 and 200 km/day.

Some fish species also affect the germination of *N. marina* as well as that of *Ruppia maritima* (wigeon grass), another submerged brackish water plant (Agami and Waisel 1988). *R. maritima* has only one type of seed: hard with a dark-colored coat. The effects of ingestion by tilapia (*Oreochromis* sp.), grass carp (*Ctenopharyngodon idella*), and common carp (*Cyprinus carpio*) were studied. Common carp digested almost all of the seeds it ate, of all kinds. The two other fish digested the soft-coated seeds of *N. marina* but excreted 40 to 60% of the hard-coated seeds of both *N. marina* and *R. maritima*. Germination rates of *R. maritima* were 12.2 and 5.7% after passing through the tilapias and the grass carp, respectively (as compared to 0.5% with no treatment). Germination of *N. marina* was 16% after the tilapias and 10.5% with the grass carp. Seeds were retained in the digestive tracts of the fish for 30 to 65 h. The fishes' movements within the water body, and less frequently, to a different water body, influenced the distribution of these two submerged plant species.

C. Seedling Adaptations

1. Seedling Dispersal and Establishment

The period of seedling establishment is critical in the angiosperm life cycle. The seedling stage is generally one of great vulnerability to wetland plants because (Titus and Hoover 1991):

- Seedlings of most species cannot grow under anaerobic conditions. Low oxygen in the substrate can be stressful to young plants because seedlings are required to quickly develop a means of coping, such as aerenchyma. Even when seedlings manage to become established in aerobic substrates, there may not be an adequate level of oxygen to meet the nighttime respiratory demands of small seedlings with limited capacity for oxygen storage in aerenchyma.
- The substrates in many wetlands, particularly those influenced by tide, are unstable. This is at least partially countered by the early development of anchoring hairs in many young seedlings, but erosion and sediment disturbance by fish can lead to seedling loss.
- Low light can be a problem for seedlings, particularly when they are growing in the shade of mature plants or at water depths where light is limited.
- Seedlings must compete for nutrients with more mature clonally produced or already established neighboring plants. Nutrients may be in short supply or difficult to sequester due to low oxygen or salinity.
- Seedlings may be particularly susceptible to pathogens or herbivores.

The difficulties facing seedlings are illustrated by a number of wetland plant studies. In a New England salt marsh, seedlings of *Spartina alterniflora* (cordgrass) had high rates of survival when grown in bare areas. In the presence of a mature canopy of *S. alterniflora,* however, no seedlings survived the entire first growing season, probably because of competition with the adult plants (Metcalfe et al. 1986). Seedlings of bottomland hardwoods, *Acer rubrum* (red maple), *Betula nigra* (river birch), *Platanus occidentalis* (sycamore), *Quercus falcata* var. *pagodifolia* (cherrybark oak), and *Ulmus americana* (American elm), were grown under waterlogged conditions. The seedlings of all of these species required a water level at or below the soil surface in order to survive, probably to ensure adequate oxygen in the root zone (Jones et al. 1989). In New York lakes, Titus and Hoover (1991) observed that many seedlings of *Vallisneria americana* were uprooted due to physical disturbance, perhaps caused by fish.

In many forested wetlands, including mangrove swamps, scattered organic mounds accumulate. These mounds form from organic debris, root growth, and the enlargement of roots at the bases of large trees or after large trees fall. They are higher, and therefore more often above standing water, than the surrounding ground. The mounds are regeneration sites for tree seedlings and other plants (Lugo 1997). Such aerated microsites are vital in wetlands since bare areas with higher oxygen are required for seedling establishment in many species.

In cypress swamps (dominated by gymnosperms), seedling distribution depends on several factors, including the number of microsites available for successful seedling establishment. In two South Carolina *Taxodium distichum–Nyssa aquatica* (bald cypress–water tupelo) stands, Huenneke and Sharitz (1986) measured the abundance of microsites and the distribution of woody seedlings. They found that seedling densities were much higher in the swamp with more open areas and favorable microsites. Some seedlings germinated and grew on stumps or logs, probably because they were above the summer water level.

The stability of the substrate was of importance because seeds that ended up in muck were often moved or buried and not able to grow. Cypress seedlings require a prolonged dry period during the growing season. Any human activity that disturbs the hydrology or increases sediment loading decreases the cypress stand's ability to regenerate (Visser and Sasser 1995).

2. Vivipary

Vivipary is the germination of seeds within fruits that are still attached to the parent plant. Vivipary can occur after a period of dormancy or, as found in the tropics, as a continuation of seed development. The propagating unit is the seedling rather than the seed. Vivipary allows the seedling to grow in a protected site and to attain a large enough size to increase its chances for survival. Vivipary may be particularly beneficial for seedlings exposed to salinity, anaerobiosis, and tides.

The mangrove trees of the Rhizophoraceae produce viviparous seedlings that can float, some for prolonged periods. *Rhizophora mangle* (red mangrove) has a four-ovary flower that generally produces only one seed. Lacking a dormant phase, the seed germinates while still on the parent tree and grows into a cylindrical shoot. The seedling maintains a lower cellular salt concentration than the parent plant by the development of barrier cells. This mechanism protects seedlings from high salinity until they are mature. After seedlings reach 20 to 30 cm in length, they drop from the tree. Some remain in the sediment directly below, but most are carried away by the current. The seedlings that float away become increasingly waterlogged, tip, and then float upright. Within the first 5 to 15 days of floating, the seedlings develop a root system that helps them become anchored once they reach favorable conditions. When a seedling reaches a shallow area, the base becomes implanted in the sediments (Figure 5.17; Tomlinson 1986).

FIGURE 5.17
Rhizophora mangle (red mangrove) seedlings, seedling height = 20 to 30 cm, Florida Everglades. Red mangrove seedlings germinate on the parent plant and are dropped once they reach 20 to 30 cm. They either remain in the sediment below or are carried away by the current. When a seedling reaches a shallow area, the base becomes implanted in the sediments. (Photo by H. Crowell.)

The two other dominant mangrove species of the western hemisphere, *Laguncularia racemosa* (white mangrove) and *Avicennia germinans* (black mangrove), are also viviparous. The propagules of all three taxa float due to spongy tissue and remain viable for extended periods. Longevity of the propagules was estimated at 35 days for *L. racemosa* and 110 days for *A. germinans*, while *R. mangle* seedlings have remained viable after floating for more than 12 months (as reviewed by Odum and McIvor 1990). The ability to remain buoyant for extended periods enables mangroves to cross large bodies of water and thereby colonize islands (Tomlinson 1986).

The seagrass *Amphibolis antarctica* grows along the temperate coasts of Australia and Tasmania (Cook 1996) and produces a viviparous seedling. A comb-like structure with four spines develops at the apex of the fruit as an outgrowth from the pistillate flower. The outgrowth consists of vegetative tissue from the parent plant and it is firmly attached to the seedling. When the seedling is mature (at about 7 to 10 cm in length), it is released along with the outgrowth, which keeps the seedling upright as it floats and presumably assists the seedling in attaching to a suitable substrate (Sculthorpe 1967; Cox and Knox 1988; Philbrick and Les 1996).

III. Asexual Reproduction in Wetland Angiosperms

Asexual reproduction is defined as the "numerical increase in physiologically independent plant units by clonal means" (Grace 1993). The process is also called *vegetative regeneration* and *cloning*. The generation of new and independent plants arises through the lateral spread of creeping shoots or roots or from the separation and dispersal of plant fragments or modified leaf buds. The new plant that arises through vegetative reproduction is called a *ramet* or a *clone*, while *genet* refers to the entire set of ramets that results from a single seed (Grace and Wetzel 1982). Clones are usually genetically identical to the parent plant, although alterations are possible due to somatic mutations. Asexual reproduction is the dominant form of reproduction for wetland plants (Sculthorpe 1967; Hutchinson 1975; Philbrick and Les 1996). Indeed, virtually all wetland plants have the capacity to reproduce through vegetative propagules.

The timing of vegetative propagule production and the number and size of propagules produced are regulated by one or more of the following (Titus and Hoover 1991):

- *Environmental conditions:* For example, clonal structures that overwinter often develop at the end of the growing season as temperatures and photoperiod decrease.
- *The size of the parent plant:* Size can determine whether internal plant resources or *meristems* (undifferentiated cells at root and shoot tips in which mitosis occurs) capable of differentiating into propagules are available.
- *Competition between vegetative and sexual reproduction for meristems:* For example, in *Eichhornia crassipes*, vegetative reproduction is reduced during flowering and in *Lobelia dortmanna*, vegetative reproduction occurs only after the plant has flowered.

A few wetland plants are capable of reproducing asexually, not through vegetative propagules, but by producing embryos without sex. *Apomixis* (or *agamospermy*) occurs when fertilization is bypassed and the flower forms seeds that start from a maternal cell instead of a zygote (Rost et al. 1984). Apomixis is rare in wetland plants (Grace 1993), although some salt marsh plants do reproduce in this manner. In Australian salt marshes, at least two species, *Halosarica indica* and *H. pergranulata,* are apomictic. By reproducing in this manner, genetic traits that enable plants to germinate and grow in a saline environment are conserved. The resulting low level of genetic variation is characteristic of plants capable of exploiting a narrow ecological niche (Jefferies and Rudmik 1991).

A. Structures and Mechanisms of Cloning

A variety of structures are involved in asexual reproduction (Table 5.4). The number of offspring that are produced from vegetative structures varies widely. Submerged species that reproduce asexually through shoot fragmentation or winter buds tend to have a very high degree of regeneration, with thousands of plants produced from a single parent plant. New plants that root at the nodes of horizontal stems can give rise to a large number of clones closely spaced around the parent plant. Plants that reproduce vegetatively through tubers, bulbs, corms, or other belowground storage organs produce few clonal offspring. The dispersability of vegetative offspring depends upon the structure from which the plants develop. In general, clonal propagules from rhizomes, stolons, or tubers develop within several centimeters of the parent plant. The shoot fragments or winter buds of submerged plants have the potential to disperse farther from the parent plant since they are sometimes buoyant or attached to buoyant plant parts. Buoyant propagules are dispersed by water and wind as well as by biotic agents such as birds, amphibians, mammals, reptiles, and humans (Philbrick and Les 1996).

For all clonal structures the production of independent plants may be a secondary function, since all have functions other than reproduction. As outlined by Grace (1993), the adaptive functions of cloning structures include photosynthesis, resource acquisition from sediments, anchorage, carbohydrate storage, protection, and perennation (overwintering). Tap roots, which may become detached and generate new plants, are involved in carbohydrate storage and perennation. Rhizomes may be particularly important for anchorage in tidal zones and other areas of considerable water flow or high winds. They can also take up nutrients from the soil, store carbohydrates, and function as perennating structures. Shoots that detach and float in the water column are photosynthetic as are runners sent out along the soil surface. Subterranean storage structures like bulbs, tubers, corms, and rhizomes afford the plant protection from temperature extremes (freezing and fire), drying, mechanical damage (wind, flood, trampling), and grazing (Grace 1993). *Utricularia* produces clonal structures only in temperate climates where they serve as overwintering structures. In the tropics, species of *Utricularia* do not develop turions. Various clonal structures, their features, and adaptive functions are described below.

1. Shoot Fragments

During a storm, we often see leaves or whole branches blown from trees. These plant fragments, not capable of regeneration, simply decompose. In many submerged wetland plants, however, fragments broken from the parent plant can form a new individual. If a fragment contains meristematic tissue, it has the necessary cells with which to begin new growth (Leakey 1981). Even small, fragile shoots are able to survive. For example, root, stem, and leaf fragments of the partially submerged *Neobeckia aquatica* (lake cress; some include *Neobeckia* within the genus, *Rorippa;* Cook 1996) are capable of generating new individuals even when they measure less than 0.5 mm in length (Philbrick and Les 1996).

Shoot fragmentation is one of the most common means of vegetative reproduction in submerged and floating plants (Stevenson 1988; Grace 1993). Some examples of plants that are able to start new individuals from shoot fragments are *Myriophyllum, Elodea,* and the

TABLE 5.4

Vegetative Clonal Structures and Their Features

Clonal Structure	Dispersability	Degree of Reproduction	Adaptive Functions	Examples
Shoot fragments	High	High	Photosynthesis	*Elodea* spp. *Myriophyllum* spp.
Modified buds				
Turions	High	High	Perennation Photosynthesis Storage	*Myriophyllum* spp. *Potamogeton* spp. *Utricularia* spp.
Dormant apices	High	High	Photosynthesis Perennation	*Ceratophyllum demersum* *Elodea canadensis*
Pseudoviviparous buds	High	High	Photosynthesis	*Myriophyllum verticillatum*
Gemmiparous buds	High	High	Photosynthesis	*Armoracia aquatica* *Ceratopteris* spp.
Modified stems				
Layers	Low	Fairly high	Photosynthesis	*Ludwigia* spp. *Panicum* spp.
Runners	Low	Fairly high	Photosynthesis	*Decodon verticillatus* *Eleocharis rostellata*
Stolons	Low	High	Anchorage Photosynthesis	*Ranunculus* spp. *Eichhornia* spp. *Limnobium* spp.
Rhizomes	Low	High	Anchorage Protection Storage Perennation Resource acquisition	*Iris* spp. *Nymphaea odorata* *Papyrus* spp. *Phragmites australis* *Typha* spp.
Stem tubers	Low	Low	Perennation Protection Resource acquisition Storage	*Cyperus rotundus* *Eleocharis* spp. *Sagittaria latifolia* *Scirpus grossus*
Modified shoot bases				
Bulbs	Low	Low	Perennation Protection Storage	*Crinum americanum* *Cypella aquatica*
Corms	Low	Low	Perennation Protection Storage	*Hypoxis* spp.
Modified roots				
Creeping roots	Low	Low	Anchorage Resource acquisition	*Myrica heterophylla* *Rorippa sylvestris*
Tap roots	Low	Low	Perennation Protection Storage	*Cicuta* spp.
Root tubers	Low	Low	Perennation Protection Storage	*Eleocharis* spp. *Nymphoides aquatica*

Categories and some examples from Grace 1993, with additional information from Sculthorpe 1967; Leakey 1981; Wetzel 1983a; Bartley and Spence 1987; Philbrick and Les 1996.

Lemnaceae. Shoot fragmentation can occur as plants naturally senesce and plant parts separate, or by disturbance, say, by the stroke of an oar. Shoot fragments have high dispersability within a water body and they can be carried to different water bodies on animals' bodies or the hulls of ships. Submerged species can quickly colonize a new area by shoot fragmentation. For example, a 1000-ha South Carolina pond in which no *Myriophyllum spicatum* was present in 1970 was almost entirely taken over by this species only 4 years later (Grace 1993). This rapid rate of colonization was attributed to the plant's ability to regenerate from shoot fragments.

2. Modified Buds

a. Turions

Many submerged plants produce modified buds of tightly compressed leaves at the ends of stems (Figure 5.18a). The modified buds, or turions, abscise from the parent plant and either float or sink to the sediments. Many remain dormant during the winter, and then grow in the spring. Some turions are not involved in overwintering and never become dormant, but do bring about new individual plants. Other names for turions and their variations are *dormant apices, winter buds,* and *hibernacula*. Dormant apices are densely crowded leaves at the apices of lateral shoots seen in *Ceratophyllum demersum* and *Elodea canadensis*. Turion is sometimes used as a synonym for dormant apex; however, some authors make the distinction that turions are more compact apices with specialized leaves. Turions are very different from the normal foliage and are often protected by a layer of mucus, as seen in *Utricularia* and *Myriophyllum* (Sculthorpe 1967; Bartley and Spence 1987). *Winter bud* and *hibernaculum* are used as synonyms for turion. Contrary to what the names imply, they enable some plants, such as *Potamogeton crispus,* to survive high summer temperatures rather than persist in the winter (Philbrick and Les 1996). Like seeds, turions can be widely dispersed due to their own buoyancy, or they can sink near the parent plant.

Utricularia neglecta, a European bladderwort that flowers rarely and is perpetuated almost solely by turions, provides an example of turion development in a northern temperate climate. In a population from a pond outside Paris, turions form in September, probably triggered by shortening day length and decreasing water temperature. The turions consist of a central axis and about 20 leaves tightly compressed around each other. The leaves are wider and shorter than the plant's normal foliage, and covered with hairs. A relatively thick mucus produced by mucilaginous glands on the leaf exteriors covers the turion. The mucus may protect turions from decomposers that would otherwise consume them in the fall. As the parent plants decompose, the turions separate from the plant. Some sink while others float to the surface where they are trapped in the ice until spring. The germination of turions takes place slowly, beginning sometimes as early as January. The outermost leaves open and the stem begins to elongate, forming a new plant (Vintéjoux 1982).

Several factors can stimulate turion development in a laboratory setting: both low and high temperatures, short and long photoperiods, several days of long photoperiods followed by short ones, a lack of nutrients, heavy competition, overcrowding, the absence of a substrate for rooting, and the application of abscisic acid (Bartley and Spence 1987). Even plants that do not ordinarily form turions can be stimulated to do so when subjected to these conditions (Wetzel 1983a). Agami and others (1986) showed that *Najas marina,* considered an annual species in warm climates, develops turions when water temperatures drop below 13°C. Dormancy in turions is broken in much the same way as for many seeds: a period of chilling followed by high temperatures, increased light or photoperiod, and an increase in nutrient availability (Bartley and Spence 1987).

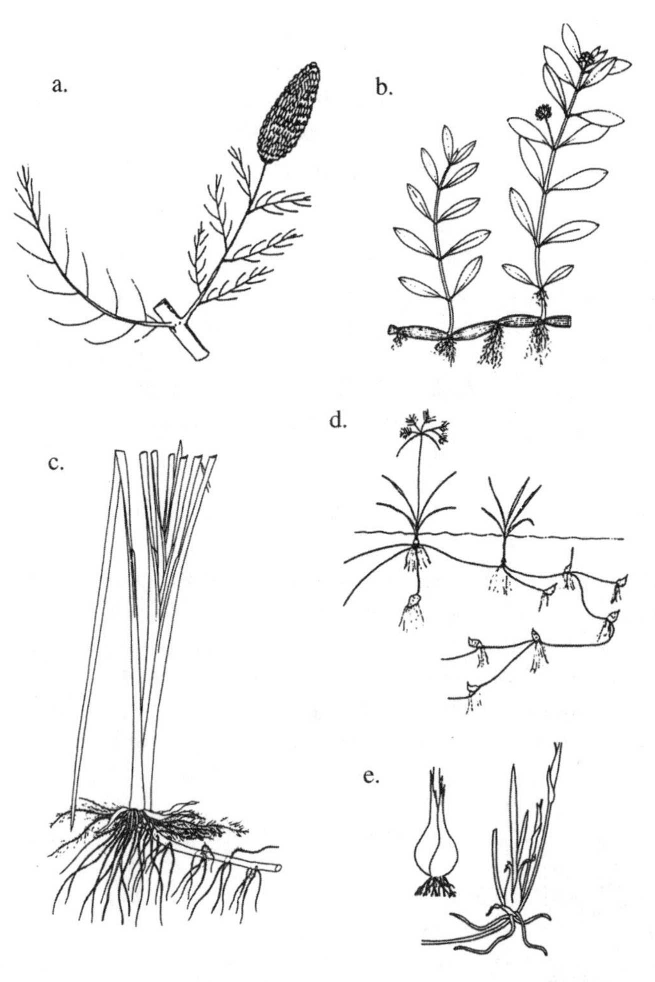

FIGURE 5.18
Examples of vegetative structures involved in asexual reproduction in wetland plants. (a) Turion of *Myriophyllum verticillatum*, length = 1 to 2.5 cm; (b) Layers of *Alternanthera philoxeroides*, diameter of layer = 3 to 5 mm; (c) Rhizome of *Typha latifolia*, diameter of rhizome = 0.5 to 5 cm; (d) Tubers, basal bulbs, rhizomes, and shoots of *Cyperus rotundus*, shoot height = 30 to 45 cm; (e) Bulb and vegetative proliferation from a bulb of *Cypella aquatilis*, bulb diameter at widest point = 1 to 1.5 cm. ([a and d] From Leakey, R.R.B. 1981. *Advances in Applied Biology* 6: 57–90. Reprinted with permission; [b, c, and e] from Cook, C.D.K. 1996. *Aquatic Plant Book.* The Hague. SPB Academic Publishing/Backhuys Publishers. Reprinted with permission.)

A different type of turion is formed from the modified leaves of the free-floating species, *Spirodela polyrhiza*. In *S. polyrhiza*, each plant is made of a group of leaflets that floats at the surface of the water. Roots are attached to the leaflets' undersides and extend a few millimeters into the water. Small, thick turions appear as separate, but attached leaflets that are brownish, smaller than the other leaflets, and without roots. The turions store carbohydrates and have few air spaces, so they sink when detached. The stored starch probably provides a food source for re-establishment. These turions are formed throughout the growing season. Flowering and seed set are rare in *S. polyrhiza* and turions are the main perennating and regenerating organs (Vintéjoux 1982; Bartley and Spence 1987).

In some species, subterranean turions develop as perennating structures. In northeastern and north-central U.S., *Vallisneria americana* may start the growing season as a turion buried 3 to 15 cm in the sediment. A short stem grows from the turion and a rosette of ribbon-like leaves develops. The rosette soon develops a system of stolons (horizontal stems) and fibrous roots (with as many as 24 interconnected rosettes per plant). The stolons play a role in both anchoring and spreading the plant. At the end of the growing season, underground turions develop at the ends of stolons. The parent plant decays, along with the roots and stolons, but the turions overwinter at a slowed rate of metabolism. In the spring, the turions germinate in response to rising sediment temperatures and increased light. Many do not survive until spring as they are subject to mortality from herbivory, fungi, pathogens, low oxygen levels, and inadequate energy stores. Over the course of the growing season, *V. americana* allocates far more biomass to turion production than to sexual reproduction. It must reach the water surface to flower and set seed, but not to produce turions; therefore turions may be particularly effective for generating new plants in areas of deep water. In general, the larger the turion, the greater the chance of germination (Titus and Hoover 1991).

In *Hydrilla verticillata*, an invasive species in the U.S. and elsewhere, both subterranean and axillary turions develop. Axillary turions form in the axils of leaves or branches while the subterranean turions form at the terminal nodes of rhizomes. Both kinds consist of overlapping leaf scales and leaves surrounding a dormant meristem. The subterranean turions are slightly larger and they store carbohydrates. Both types develop an abscission zone and detach from the parent plant. They are capable of remaining dormant in sediments for several years, thus forming a sort of meristem bank that allows for re-infestation after application of weed control measures (Netherland 1997).

b. Pseudoviviparous Buds

True vivipary occurs when a seedling germinates and begins growth on the parent plant, as in *Rhizophora mangle* (red mangrove). In pseudovivipary, vegetative propagules replace some or all of the normal sexual flowers in an inflorescence. Pseudoviviparous buds, for which the term turion is also used, develop as a flower replacement if flowers become submerged, or if the plant is growing at greater depths than usual. For example, when an inflorescence of *Myriophyllum verticillatum* is submerged, a turion develops at the tip of the flowering head. In *Nymphaea lotus*, tubers, rather than turions, can replace flowers and

FIGURE 5.19a
Tip rooting, a form of layering, in *Decodon verticillatus* (swamp loosestrife). (Photo by H. Crowell.)

FIGURE 5.19b
D. verticillatus tip rooting into the open water at the edge of a pond in western Michigan. (Photo by H. Crowell.)

then develop into independent plants. Many other aquatic plants are pseudoviviparous under the right conditions, including several in the Alismataceae such as *Baldellia ranunculoides, Caldesia parnassifolia,* and some species of *Echinodorus*. Pseudovivipary occurs in upland plants as well, so submergence is not a requirement (Sculthorpe 1967; Cook 1996).

c. Gemmiparous Buds

Buds of new plants are also produced from foliar tissue, particularly if the leaf tissue is injured or broken off from the rest of the plant. Small buds appear on leaves at sites of meristem tissue and then drop off or become independent when the leaf decays. Species within the tropical floating genus, *Ceratopteris,* can produce new plants at the ends of veins and at the base of marginal notches in the mature leaf. In the heterophyllous *Armoracia aquatica,* gemmipary is common on the submerged dissected leaves. Many buds can grow on each leaf and then remain floating after the parent plant has decayed (Sculthorpe 1967).

3. Modified Stems

Wetland plants can develop five types of horizontal, or creeping, stems that are not always easily identified or partitioned in the field. In the order shown here, they represent a continuum from aboveground to belowground, with each structure found deeper in the ground than the one before it (Leakey 1981; Grace 1993). All of the creeping stem types can effectively produce and disperse clonal offspring, though usually not far from the parent plant. Creeping stems that are above ground can photosynthesize and the new roots produced at the stem nodes can acquire resources from the substrate. Belowground rhizomes and stem tubers are involved in resource acquisition, carbohydrate storage, and perennation.

a. Layers

Layers are the simplest form of creeping stems. No specific control of growth keeps them horizontal. When layers come in contact with soil they develop adventitious roots, usually at the nodes (Leakey 1981). *Alternanthera philoxeroides* (alligator weed), a tropical emergent and an aggressive exotic in the southeastern U.S., develops buoyant layers that root at the nodes, permitting the plant to spread across the water surface and absorb nutrients from the water (Figure 5.18b; Cook 1996). *Decodon verticillatus* (swamp loosestrife) spreads vegetatively by tip rooting, a form of layering (Figures 5.19a and b). It usually lives at the water's edge in lakes and ponds and its stems grow downward into the water and root at the nodes. The new shoots expand the vegetated fringes of the lake into the previously open water.

b. Runners

Runners are normally upright shoots that become horizontal late in the season due to the weight of the plant. They can develop new plants wherever the tips of the shoots touch the ground (Leakey 1981). Runners are aboveground and they are largely photosynthetic. *Eleocharis rostellata* produces buds at the tips of jointed runners that root when the stem contacts the soil (Grace 1993).

c. Stolons

Stolons are horizontally growing stems, produced from basal nodes of parent plants, which spread along the soil surface. The main factor differentiating stolons from layers and runners is their controlled *plagiotropism* (prostrate growth habit). Light seems to be the controlling factor in their plagiotropism, which is presumably mediated by growth regulators (Leakey 1981). New plants arise at stolon nodes, occurring at almost every node in many species. *Phragmites australis* spreads extensively along stolons that have been observed to grow as long as 13 m (Voss 1972). Stolons, like roots and rhizomes, can be important in anchoring plants. Extensive networks of stolons anchor *Justicia americana* (water willow) to rocks and soil (Grace 1993).

d. Rhizomes

Rhizomes are highly modified subterranean stems that are often enlarged since they are a site of carbohydrate storage (Leakey 1981). Rhizomes are nearly universal among perennial grasses. For emergent aquatics, rhizomes are the most prevalent mode of vegetative reproduction (Grace 1993). *Typha latifolia* and *T. angustifolia* (cattail) colonize new sites at the water's edge with seeds, and they spread into deeper water by rhizomes (Figure 5.18c; Grace and Wetzel 1982). *Spartina alterniflora* (cordgrass) clones also spread through salt marshes by producing ramets along rhizomes which grow either in a fixed direction or non-directionally, filling all available space (Metcalfe et al. 1986). *Nypa* and *Acrostichum* are mangrove genera of the eastern hemisphere in which spread by rhizomes has been noted, making them unique among the mangroves (Tomlinson 1986).

Rhizomes can be of two types: *pachymorphic*, which are large with little branching (as in the storage rhizomes of *Nymphaea odorata* and *Nuphar lutea*; Figure 4.10), and *leptomorphic*, which are thin rhizomes that require a lower energy input (found in *Eleocharis acicularis;* Leakey 1981). Leptomorphic rhizomes are capable of greater dispersal of clones, while pachymorphic rhizomes provide carbohydrate storage (Grace 1993). Because rhizomes are underground, their growth may be advantageous to wetland plants that are subjected to periodic drought or fire (Philbrick and Les 1996). *Cladium jamaicense* (sawgrass), the dominant plant of much of the Florida Everglades, has rhizomes that help it persist and spread after fire destroys aboveground parts (Brewer 1996).

e. Stem Tubers

Stem tubers are swollen portions of underground stems, especially adapted as perennating organs (Leakey 1981). A sedge, *Cyperus rotundus*, has tubers that form in meristematic regions near the tips of rhizomes. With the accumulation of starch, the tubers swell and lateral meristems can develop new rhizome branches (Figure 5.18d). In the herbaceous emergent, *Scirpus fluviatilis* (river bulrush), small buds like potato "eyes" form on tubers at the end of the growing season. During the following growing season, the small buds are the site of new tuber, shoot, and rhizome growth (Klopatek and Stearns 1978).

4. Modified Shoot Bases

a. Bulbs

Bulbs are storage organs that consist of swollen leaf bases or scales at or slightly below ground level. They include the elements of a shoot system with very short internodes. Bulbs are normally perennating organs, but some axillary buds develop as daughter bulbs called *offsets* or *bulbils* that are capable of regeneration (Leakey 1981). The tropical emergent, *Crinum americanum*, spreads by both seeds and bulbs (Grace 1993; Cook 1996). *Cypella aquatica*, an emergent of Brazil and Argentina, produces bulbs as perennating structures that also serve as clonal structures (Figure 5.18e; Cook 1996).

b. Corms

Corms are swollen stems with dormant buds in the axils of scale-like leaf remains. Like bulbs, corms occur at the base of the shoot and are usually perennating organs. They regenerate by replacing the decomposed aboveground portions of the plant at the beginning of the growing season (Leakey 1981). The emergent sedge, *Cyperus strigosus* (straw-colored cyperus), produces corms.

5. Modified Roots

a. Creeping Roots

Creeping root systems are a series of vertical and horizontal roots. After a period in which roots grow horizontally, they begin to grow downward. At the points of downward growth, daughter plants can arise and when the horizontal connecting roots die, the daughter plants survive separate from the parent plant (Leakey 1981). Examples of wetland plants that spread vegetatively with creeping roots are *Rorippa sylvestris* (yellow cress) and *Myrica heterophylla* (bayberry; Grace 1993).

b. Tap Roots

Tap roots are downward-growing primary roots that develop a secondary thickening as they become storage areas for carbohydrates. Tap roots only serve as clonal structures when the top of the plant is defoliated or damaged. If the plant is fragmented, a new plant can arise from the tap root (Leakey 1981). *Cicuta maculata* (water hemlock) produces highly poisonous swollen roots called both tap roots and root tubers (Voss 1972; Grace 1993).

c. Root Tubers

Root tubers are swollen roots or root apices that store carbohydrates and function as perennating organs. Like stem tubers, bulbs, corms, and tap roots, root tubers are primarily involved in storage. Reproduction and dispersal are secondary functions (Leakey 1981). The floating-leaved *Nymphoides aquatica* (floating heart) develops banana-shaped root tubers (Cook 1996).

B. Occurrence and Success of Cloning among Wetland Plants

Asexual reproduction is dominant in wetland habitats and this may be for a number of reasons. One reason may simply be that there are more monocots than eudicots among wetland plants, and monocots have a high incidence of clonal growth, particularly through rhizomes. Depending on the definition of wetland plants, from one third to one half of monocot families contain wetland taxa compared to about 12 to 14% of eudicots (Les and Schneider 1995). Far fewer eudicots reproduce vegetatively. The eudicot families in which cloning occurs (about 10% of 300 eudicot families) are mostly semi-aquatic or aquatic. There are strong correlations between clonal reproduction and monocots and between clonal reproduction and the wetland habitat. Therefore, a significant factor leading to the high proportion of monocots among wetland plants may be the importance of vegetative reproduction in wetland habitats (Grace 1993).

The dominance of vegetative reproduction among wetland species may also be due to the importance of the perennial habit in wetlands. While terrestrial plants are fairly evenly divided among annuals (which reproduce only sexually) and perennials, most wetland plants are perennials. Perennials have specialized structures which "double" as overwintering and reproductive structures so that perennation and vegetative reproduction are inextricably linked (Leakey 1981; Grace 1993). Subterranean structures, such as tap roots, tubers, bulbs, corms, and rhizomes, persist even when aboveground parts have decayed. Plants also perennate through turions that overwinter either in the surface ice or in the sediments. All of these perennating structures have the potential to bring about new individuals.

In addition to the dominance of monocots and perennials among wetland plants, other factors may contribute to the importance of vegetative reproduction in wetlands. Asexual modes of reproduction may be favored because vegetative structures outcompete seedlings. Clonal structures enable plants to bypass the phase of seedling establishment.

TABLE 5.5

Introduced Plants That Reproduce Exclusively Asexually in Their New Range Due to Dioecy or Heterostyly

Species	New Range
Dioecy (populations are exclusively male or female)	
Elodea canadensis	Europe, Australia, New Zealand
Egeria densa	North America, New Zealand, Japan
Hydrilla verticillata	Southeastern United States, New Zealand
Lagarosiphon major	New Zealand
Myriophyllum aquaticum [a]	California
Stratiotes aloides	Europe
Heterostyly (only one flower type is present)	
Lythrum flagellare	Florida
Nymphoides indica	Lower Amazon
Pontederia rotundifolia	Costa Rica

[a]Formerly called *Myriophyllum sibiricum*.

From reviews by Barrett et al. 1993; Cook 1993; de Winton and Clayton 1996; Philbrick and Les 1996.

This may be beneficial since seedlings are vulnerable and may not compete well, particularly in cold, saline, or anaerobic conditions (Metcalfe et al. 1986; Jefferies and Rudmik 1991; Madsen 1991). When vegetative propagules compete for the same space as seeds and seedlings, the vegetative structures are often more successful. In a study of *Potamogeton crispus*, Rogers and Breen (1980) showed that the numbers of seeds and turions produced per unit sediment area in dense beds were similar, but germination was quite different: 0.001% for seeds and 60% for turions.

Several studies have illustrated the frequency of asexual reproduction in northern salt marshes, an environment that may be especially stressful to seedlings. *Puccinellia phryganodes*, a northern salt marsh grass, has been observed to flower only rarely in Alaska, failing to set seed, and to be sterile in the Hudson Bay area. *P. phryganodes* can develop new shoots from nearly any severed plant part, including leaves, that is torn from plants by ice movement and foraging birds (Vince and Snow 1984; Jefferies and Rudmik 1991). A study of a sedge in an Alaskan salt marsh, *Carex ramenskii*, showed that only 6% of the population had flowering shoots, so most reproduction of this species was attributed to cloning (Vince and Snow 1984). Only 5 to 10% of the population of *Spartina anglica* in a Suffolk, England salt marsh flowered and its spread was almost entirely rhizomatous (Jackson et al. 1986).

Asexual reproduction may appear to be favored in wetlands because its counterpart, sexual reproduction, is often inhibited, particularly in submerged species. Most submerged angiosperms have to produce aerial flowers in order to set seed. This becomes increasingly difficult with increased depth (Barrett et al. 1993). Even when many seeds are produced, some may be buried at unfavorable depths or seedlings may not be able to germinate or grow (Titus and Hoover 1991). Many wetland plants flower either infrequently or have a low seed set. This decreased reproductive ability may be caused, at least in part, by clonal reproduction. If somatic mutations that affect sexual function or early phases of life history occur within clones, then the species may be unable or unlikely to produce viable seeds. Wetland plants can be sterile due to hybridization, irregularities during meiosis, and mutations that impair reproductive function (Barrett et al. 1993).

Wetland plants show an extreme reliance on asexual reproduction where only a single mating type (i.e., either staminate or pistillate flowers) is present in a population (Table 5.5). For example, *Hydrilla verticillata* can be both dioecious and monoecious. Where it is dioecious, both sexes of the plant must be present for sexual reproduction to occur. Its pistillate form was introduced to the southeastern U.S. in about 1955, and despite the absence of staminate plants, it has become a nuisance species that clogs waterways and chokes out native submerged plants (Steward 1993). *Elodea canadensis* is a submerged plant with widespread distribution in Europe, where only the pistillate flowers are found (Philbrick and Les 1996). *Heterostylous* species (i.e., angiosperms with two or more flower forms in which the pollen is only compatible with stigmas of unlike flowers), such as *Lythrum salicaria* (purple loosestrife; Figure 8.9), also spread only vegetatively where only one flower type is introduced to a new range.

Within large clones or where a single clone inhabits a water body, nearly every seed is likely to be the product of self-fertilization, thus decreasing the chance for genetic variation within the population even when sexual reproduction does occur (Barrett et al. 1993). When sexual reproduction occurs within a large clone, or genet, the distance pollen must travel in order to reach a genetically different flower may decrease the likelihood of outcrossing. Self-pollination in species that are self-compatible does not necessarily bring about a decrease in seed set (e.g., *Decodon verticillatus*), but for self-incompatible plants (e.g., *Eichhornia paniculata*), seed production is diminished.

Asexual reproduction conserves existing genetic variation and in this way genotypes that are adapted to the local conditions are maintained (Jefferies and Rudmik 1991; Titus and Hoover 1991). However, whether asexual reproduction evolved and became prevalent in wetland habitats due to the selective advantage of genetic uniformity is questionable. It may be that clonal structures evolved for different reasons, with asexual reproduction and the conservation of genotypes as secondary consequences.

Summary

Wetland angiosperms reproduce both sexually and asexually. Sexual reproduction occurs in much the same way as for terrestrial angiosperms, with the pollination and fertilization of flowers, the formation and dispersal of seeds, and their eventual germination and growth. About two thirds of wetland plants are biotically pollinated and about one third are abiotically pollinated. Of the abiotically pollinated plants, about 98% are wind-pollinated (many of the dominant monocots of marshes are in this category). The few remaining plants are hydrophilous and use water as the vector of either pollen or male flowers. Hydrophily occurs either on the surface of the water (dry and wet epihydrophily) or underwater (hypohydrophily).

Asexual reproduction occurs through the formation of propagules, all of which have other functions such as the storage of carbohydrates, perennation, photosynthesis, anchorage, or resource acquisition. Clonal propagules are shoot fragments, and modified buds, stems, shoot bases, and roots.

Asexual reproduction is the predominant way in which new individuals are formed in wetlands. In part, this may be due to the fact that many wetlands are dominated by perennial monocots, which have a high incidence of clonal growth. In addition, wetland plants are vulnerable in the seedling stage and vegetative clones tend to outcompete them. Sexual reproduction is often unsuccessful in wetlands due to flower wetting and the loss of seeds in unsuitable areas for germination (i.e., in deep waters or anaerobic substrates).

Part III

Wetland Plant Communities: Function, Dynamics, Disturbance

6:

The Primary Productivity of Wetland Plants

I. Introduction

The primary productivity of many wetlands is quite high especially when compared to other natural communities or even to highly managed agricultural croplands (Table 6.1). A high value for the aboveground primary productivity of swamps and marshes in temperate zones is about 3500 grams dry weight per square meter per year (g m^{-2} yr^{-1}). In cold wetlands and peat bogs an upper limit of about 1000 g m^{-2} yr^{-1} is typical (Bradbury and Grace 1993). Wetlands with emergent herbaceous vegetation are often more productive than other wetland types, although high values are found in some mangrove swamps as well (Table 6.2). Wetland primary productivity depends upon the type of wetland and the vegetation found there as well as on hydrology, climate, and environmental variables such as soil type and nutrient availability. Wetlands that receive nutrient subsidies either naturally from flooding or from farm runoff tend to be more productive than those that receive nutrients only from rainwater, such as scrub cypress swamps or ombrotrophic bogs (Brown 1981). In a highly productive freshwater marsh in Wisconsin (from 2800 to 3800 g m^{-2} yr^{-1}), the soil nutrients were found in higher concentrations than in upland soils and in excess of what is needed for agricultural crops (Klopatek and Stearns 1978).

Still water wetlands such as bogs or scrub cypress swamps have low primary productivity (100 to 300 g m^{-2} yr^{-1}), but they may perform essential ecological functions by supporting wildlife or rare plant species or they may be sites of important storages of water or peat (Brown 1981). The salt marshes of the arctic and subarctic are among the least productive of coastal wetlands. Nonetheless, they are valuable as vital staging areas for large populations of migrating waterfowl (Roberts and Robertson 1986).

A. Definition of Terms

The terms used to report primary productivity results for wetland habitats are sometimes used interchangeably, making it difficult to directly compare the results from different studies. Some researchers have argued for the adoption of standard terms and methods; most studies use the terms as defined here (Wetzel 1964, 1966, 1983a; Wetzel and Hough 1973; Westlake 1975, 1982; Aloi 1990).

1. Standing Crop

Standing crop (synonymous with *standing stock*) is the dry weight of a plant population on any given date. The term *maximum standing crop* denotes the maximum dry weight of

TABLE 6.1

The Annual Aboveground Primary Productivity of Different Ecosystem Types (units are g dry weight m^{-2} yr^{-1})

Ecosystem Type	Mean Net Primary Productivity
Swamp and marsh	2500
Tropical rain forest	2000
Tropical seasonal forest	1500
Temperate evergreen forest	1300
Temperate deciduous forest	1200
Boreal forest	800
Savanna	700
Agricultural land	644
Woodland and shrubland	600
Temperate grassland	500
Lake and stream	500
Tundra and alpine	144
Desert scrub	71
Rock, ice, and sand	3
Weighted average land NPP	720
Algal bed and reef	2000
Estuaries	1800
Upwelling zones	500
Continental shelf	360
Open ocean	127
Weighted average ocean NPP	153
Average for biosphere	320

Data from Colinvaux 1993.

plants during the season. Strictly speaking, standing crop applies to only aboveground plant parts, so the term should not be used when the belowground portions of the plants are also sampled (Wetzel 1966; Wetzel 1983a).

2. Biomass

The term *biomass* is in wider use for ecological studies than standing crop. The biomass of a plant is its dry weight in grams. The biomass of a tree, for example, includes the weight of flowers + fruits + leaves + current twigs + branches + stems + roots (Brinson et al. 1981). If only the aboveground portions of the plant are measured, then this should be specified and called *aboveground biomass*. The biomass of a community is usually reported as grams of dry weight in an area (g m^{-2}). Dry weight is determined by drying plant matter in a drying oven usually for 24 to 72 h at temperatures from 60 to 105°C.

Sometimes biomass is reported as ash-free dry weight (AFDW; synonymous with organic dry weight). Ash-free dry weight is determined by combusting dried plant matter in a

TABLE 6.2
A Range of Net Aboveground Primary Productivity Values for Different Wetland Types

Wetland Type	Net Primary Productivity (g dry wt m^{-2} yr^{-1})
Salt marsh	130–3700
Tidal freshwater marsh	780–2300
Freshwater marsh[a]	900–5500
Mangrove	1270–5400
Southeastern bottomland hardwood	830–1610
Cypress swamp	200–1540
Forested northern peatland[a]	260–2000
Non-forested northern peatland[a]	100–2000

[a] Above- and belowground production.

Note: Most of the data are from North American wetlands (data from Mitsch and Gosselink 2000; some values were converted from g C to g dry weight assuming carbon is 45% of dry weight).

combustion oven at 550°C for 15 min. The organic carbon present in the plant tissues is released as a gas. The difference between the original dry weight of the material and its weight after combustion is roughly the weight of the organic matter, or the ash-free dry weight.

3. Peak Biomass

Peak biomass occurs when vegetation is at its highest biomass. After peak biomass, growth declines and the vegetation dies. Production studies often report peak biomass as the production for the growing season, even though much of the plants' production is unrecorded with this method (Wiegert and Evans 1964).

4. Primary Production

Primary production is the conversion of solar energy into chemical energy. The process of energy transformation, or photosynthesis, is a complex set of biochemical reactions that can be expressed in simple terms as a chemical equation:

$$6CO_2 + 12H_2O \xrightarrow[\text{chlorophyll}]{\text{light energy}} C_6H_{12}O_6 + 6O_2 + 6H_2O \quad (6.1)$$

Carbon dioxide and water are the raw materials necessary for the production of a simple carbohydrate (glucose), with the evolution of oxygen and the release of water as by-products.

In ecological studies, primary production is measured and reported as (Colinvaux 1993):

- Biomass, reported as the weight in grams of the dry matter produced by plants
- Mass of an element, such as the amount of oxygen evolved or the amount of

carbon fixed during photosynthesis, expressed in grams of oxygen or carbon or in terms of moles of the element
- Energy (calories produced or joules consumed in production)

In most wetland studies, primary production results are reported as amounts of biomass produced. Plant biomass reflects net primary production and does not include losses to respiration, excretion, secretion, injury, death, or herbivory.

To determine net primary production from biomass, it is necessary to measure plant biomass more than once. The change in biomass between two measurements is equal to the net production for that time period. Net production is calculated from biomass as follows (Newbould 1970):

B_1	Biomass of a plant community at a certain time t_1
B_2	Biomass of the same community at t_2
$\Delta B = B_2 - B_1$	Biomass change during the period $t_1 - t_2$
L	Plant losses by death and shedding during $t_1 - t_2$
G	Plant losses to consumer organisms such as herbivorous animals, parasitic plants, etc. during $t_1 - t_2$
P_n	Net production by the community during $t_1 - t_2$

If the amounts, DB, L, and G, are successfully estimated, we can calculate P_n as the sum

$$P_n = \Delta B + L + G \quad (6.2)$$

5. Respiration

Respiration is the process by which a plant cell oxidizes stored chemical energy in the form of sugars, lipids, and proteins and converts the energy released into a chemical form directly usable by cells (e.g., ATP). The equation for the respiration of glucose is essentially the reverse of Equation 6.1. During respiration the plant requires oxygen and releases carbon dioxide. Unlike photosynthesis, respiration takes place in both the light and the dark. In most ecological studies, respiration is measured in the dark as the evolution of carbon dioxide by the plant (usually enclosed in a gas chamber) or by the decrease in oxygen concentration surrounding the plant (Grace and Wetzel 1978). Respiration is usually expressed as an hourly rate and then multiplied by 24 for a daily rate under the assumption that daytime and nighttime respiration rates are equal. This assumption is probably false, since the daytime work of photosynthesis probably brings about a higher rate of respiration. Nonetheless, this assumption is often used in primary production studies and the underestimate of respiration that it represents is considered to be minimal.

Respiration can represent a high proportion of the gross productivity of a plant. Brinson and others (1981) reported the average respiration rate measured in nonforested wetlands to be 72% of gross primary productivity. Respiration rates change over time and are influenced by climatic variables. In a Florida riverine marsh, respiration was higher during the rainy season (77% of gross primary productivity) than during the cooler dry season (50% of gross primary productivity; Brinson et al. 1981). Respiration increases with higher temperatures or increasing rates of primary productivity, probably because of the increased availability of labile photosynthetic products.

6. Primary Productivity

Primary productivity is primary production over time, or the rate of primary production. If gaseous exchange methods are used to measure primary productivity, the time period is a day or an hour and the units are grams of oxygen evolved or carbon assimilated. In wetland macrophyte studies, results are usually given in units of dry plant matter production per unit area per year (g dry weight m^{-2} yr^{-1}). In the temperate zone, growth per year is actually growth during the growing season. It is important to specify the length of the growing season since it can be quite long in low latitudes and short in high latitudes.

a. Gross Primary Productivity

Gross primary productivity (GPP) is the measured change in plant biomass plus all of the predatory and nonpredatory losses (respiration) from the plant divided by the time interval. It includes all of the new organic matter produced by a plant plus all that is used or lost during the same time interval (Wetzel 1983a). It can be defined as the sum of daytime photosynthesis and day- and nighttime respiration (Brinson et al. 1981).

b. Net Primary Productivity

Net primary productivity (NPP) is the observed changes in plant organic matter over a time period. NPP is GPP minus all losses (such as respiration and herbivory). It is the value most often reported in wetland macrophyte production studies. Other terms and abbreviations for NPP are used in the literature that may be more precise because they include modifying terms such as annual (thereby giving the term a rate component), aboveground, aerial, or shoot (which indicate which portion of the biomass was measured). Some of these terms are:

- NAPP: net aerial primary production. Although rate is not implied in this term, reports are generally for 1 year of growth (Linthurst and Reimold 1978a, b; Groenendijk 1984; Cahoon and Stevenson 1986; Hik and Jefferies 1990; Dai and Wiegert 1996)
- ANPP: aboveground annual net primary production (Kistritz et al. 1983)
- NAAP: net annual aboveground production (Dickerman et al. 1986; Wetzel and Pickard 1996)
- ANPPs: annual net primary shoot production (Jackson et al. 1986)

7. Turnover

Turnover is the amount of biomass lost during the growing season (to leaf loss, herbivory, or other causes). The turnover rate is turnover (g m^{-2} yr^{-1}), divided by peak biomass (g m^{-2}). It is expressed in units of *year*$^{-1}$, which reflects the calculation involved (g m^{-2} yr^{-1} divided by g m^{-2}). Peak biomass (an underestimate of net primary productivity) can be corrected for leaf loss by multiplying by the turnover rate.

Leaf turnover is sometimes estimated for emergent plants so that peak biomass can be corrected for the weight of leaves that have been lost, dropped, or consumed, or that have died on the plant. Leaf turnover is determined by dividing the total number of leaves produced per shoot per year by the modal number of leaves per shoot per year (the mode is the value that occurs most frequently in a series of observations). Dickerman and others (1986) calculated leaf turnover in a Michigan *Typha latifolia* stand to be 1.38 leaves $leaf^{-1}$ yr^{-1}. Morris and Haskin (1990) showed that by adding leaf turnover to peak biomass of *Spartina alterniflora*, the result for NPP was 20 to 38% greater than peak biomass alone.

8. P/B Ratio

The *P/B ratio* is a measure of the amount of energy flow relative to biomass (Wetzel 1983a). The P/B ratio is unitless and it is estimated as the ratio of net primary productivity (P) to peak biomass (B). The P/B ratio is usually assumed to be equivalent to the turnover rate. In theory, however, the P/B ratio is greater than the turnover rate since the value for P includes turnover as well as the net production that occurs after peak biomass (Grace and Wetzel 1978). Typical values for P/B ratios in submerged plants are 1.0 to 2.0 (Kiorboe 1980; Wetzel 1983a). For emergents P/B ratios range from 0.3 to 7.0, with most values less than 1.0 (calculated from data in Wetzel 1983a). For large trees, P/B ratios are low (<0.5) since productivity of the wood and leaves in a single year is less than the total biomass of the tree (Brinson et al. 1981). If a P/B ratio for a species has been well established in other studies, researchers sometimes measure peak biomass and multiply it by the P/B ratio to estimate NPP (Kiorboe 1980).

B. Reasons for Measuring Wetland Primary Productivity

1. To Quantify an Ecosystem Function

An understanding of primary production is essential to the understanding of ecosystem functions (Reader and Stewart 1971; Gholz 1982; Kemp et al. 1986; Dame and Kenny 1986; Meyer and Edwards 1990). All of the heterotrophic organisms in a community depend on the energy supplied by plants. Primary productivity data are often the basis for quantitative studies of other ecosystem processes. Knowledge of oxygen and carbon fluxes can provide insight into the cycling and retention of other elements as well as the redox conditions in wetland water and soil. For example, primary producers retain nutrients during the growing season, and release them as they senesce. The mass of nutrients taken up by plants can be estimated from measurements of primary productivity and a knowledge of nutrient concentrations in plant tissue. Other pathways of nutrient removal (i.e., sedimentation or denitrification) can be estimated by the difference in plant uptake and the mass of the nutrients that remains. Information about productivity and nutrient fluxes is vital to ecosystem modelers who wish to describe an ecosystem's functions and make predictions about its future (Mitsch and Reeder 1991).

The primary productivity of wetlands is of particular interest because it provides a link between terrestrial communities and downstream aquatic ecosystems. Wetland plants use nutrients that flow into their habitat from upstream areas. The detritus generated by wetland plants may be carried out of the wetland into downstream waters where it is broken down and consumed. The plant matter produced in wetlands is vital both within the wetland and downstream. With wetland primary productivity measurements and information on hydrology, researchers are able to quantify the wetland's contribution to downstream ecosystems (Cahoon and Stevenson 1986).

2. To Make Comparisons within a Wetland

Results from primary productivity studies within a single wetland can reveal temporal and spatial patterns of growth and changes from year to year. If primary productivity studies are undertaken in wetlands that are subject to disturbance, such as water resource projects (Johnson and Bell 1976), fire (Ewel and Mitsch 1978), added nutrients (Mitsch and Ewel 1979; Brown 1981), impoundment (Conner et al. 1981; Conner and Day 1992), silvicultural activities (Gholz 1982), acid deposition (Rochefort et al. 1990), or the installation of roads or structures (Visser and Sasser 1995), the results can reveal the environmental impact of the disturbance. Such studies can also help managers and planners prepare for changes due to future disturbances (Bartsch and Moore 1985).

Primary productivity studies have been used to monitor the development of plant communities in constructed wetlands (Fennessy et al. 1994a; Cronk and Mitsch 1994a, 1994b; Niswander and Mitsch 1995). In the case of disturbed or constructed wetlands, primary productivity measurements can answer questions regarding the status of the ecosystem (i.e., whether the wetland is recovering from disturbance, responding to new inputs, or developing as a new wetland).

3. To Make Comparisons among Wetlands

Comparisons among wetlands allow us to generalize about primary productivity within categories of wetlands. For example, we usually expect the productivity of freshwater marshes in warm climates to exceed that of ombrotrophic bogs. When similar wetlands have different primary productivity results, the researcher is prompted to seek explanations for the differences.

The primary productivity of a restored or constructed wetland can be compared to that of nearby natural wetlands of the same type. The researcher can then begin to determine whether the new wetland is functioning like a natural one (Confer and Niering 1992; see Case Study 6.A, Salt Marsh Productivity: The Effect of Hydrological Alterations in Three Sites in San Diego County, California).

4. To Determine Forcing Functions and Limiting Factors of Primary Productivity

Variations in light, hydrology, nutrients, salinity, or substrate act to promote or limit plant growth in a particular site or type of wetland. Comparisons of the wetland type in different locations help distinguish the environmental and climatic variables that influence productivity. One reason that primary productivity is high in many wetlands is simply that the environment is favorable for growth. The plants are not under water stress, there are sufficient nutrients, and the wetland is located in an area with a suitable climate for plant growth.

Many wetland primary productivity studies and reviews have established that hydrology is a major forcing function in wetlands (Mitsch and Ewel 1979; Zedler et al. 1980; Brinson et al. 1981; Brown 1981; Grigal et al. 1985; Lugo et al. 1988; Brown 1990; Cronk and Mitsch 1994a). The source and amount of incoming water determine the structure and function of wetlands. Water can be nutrient-rich or nutrient-poor; it can arrive in the wetland as a gentle rain or a tidal surge. In general, the primary productivity of nontidal wetlands is higher with greater water throughflow (Brinson et al. 1981). For tidal wetlands, productivity is influenced by both tidal flushing and freshwater inputs from upland, with greater productivity in sites that receive substantial freshwater runoff (Lugo et al. 1988; see Chapter 3, Section II.B.1, Hydrology and Primary Productivity).

II. Methods for the Measurement of Primary Productivity in Wetlands

Many quantitative methods are available for wetland primary productivity studies, in part because wetlands encompass a variety of habitats. Methods have been adapted from both aquatic and terrestrial ecology. A combination of methods is sometimes used within a single study because several plant forms (e.g., algae, submerged, emergent, and floating-leaved plants) coexist within the wetland. Gas exchange techniques are usually used for the algal component of the community, harvest and biomass methods are used for macrophytes, and forestry techniques are used to determine the primary productivity of wetland trees.

Conclusions about a wetland's status or about its response to human-induced impacts are sometimes based on productivity studies, so it is critical to understand the methods used and to identify sources of error (Bradbury and Hofstra 1976). Difficulties in interpreting primary productivity measurements stem from the variety of methods used both within a single wetland for the different plant forms as well as among different studies (Talling 1975). It is difficult to compare primary productivity results generated with very different methods, sampling intervals, plot sizes, and conversion factors. The choice of methods depends on the type of plant community to be measured, as well as on the time, resources, and labor available. In wetland macrophyte primary productivity studies, most of the published results are from the harvest of vegetation that has an annual growth cycle. Sometimes plants are harvested once, at the end of a growth cycle. The growth cycle is usually based on the dominant species in the wetland. Alternatively, plants are measured several times throughout the growing season and their productivity is calculated from the findings at each sampling date. Some of the errors associated with these procedures include (Brinson et al. 1981; Dickerman et al. 1986; Bradbury and Grace 1993):

- *Statistical error based on the number of samples.* Statistically valid measurements of vegetation often require large sample sizes in many plots. Insufficient sample size can result in large statistical error.
- *Errors based on the frequency of sampling.* Estimates of production rates depend on the frequency of sampling throughout the growing season. More frequent sampling often results in higher estimates because less plant production is missed due to death, herbivory, and decay.
- *Errors due to plant mortality.* Some methods miss a large proportion of production because they fail to include biomass that dies between sampling periods or before peak biomass has occurred.
- *Errors due to seasonality of measurements.* Productivity is assumed to be zero between the end of one growing season and the beginning of the next. However, this assumption may not be valid, because growth continues throughout the year in the warmest areas of the temperate zone and in the subtropics and tropics (Hopkinson et al. 1978; Dai and Wiegert 1996).
- *Errors due to the omission (or unstated inclusion) of more than one of the plant components of the wetland.* Primary productivity estimates often reflect only the productivity of the dominant plant form in a community, such as the most common macrophytes or trees. If the productivity of other members of the plant community have been measured, such as phytoplankton, periphyton, epiphytes, vines, or the understory of forests, sometimes it is not clear whether these have been included in figures for total productivity. These values should be subtracted from total productivity when comparing to studies in which these measurements were not made.
- *The omission of belowground biomass.* Researchers frequently measure only aboveground biomass. Many wetland plants are long-lived perennials, with considerable belowground biomass, and failing to measure it can result in large underestimates of wetland productivity. Not all studies warrant this extra effort, especially if the goal is to determine effects of an aboveground phenomenon such as grazing or to compare to other studies of only aboveground results.

In this section we briefly describe several, although not all, of the primary productivity methods commonly used in wetland studies. The plant components of wetlands that we discuss are phytoplankton (floating algae), periphyton (attached algae), submerged plants, emergent plants, floating and floating-leaved plants, trees, shrubs, and moss. Although phytoplankton, periphyton, and moss are not the primary focus of our book, we include them here because of their importance in the primary productivity of the entire wetland plant community.

A. Phytoplankton

At any given time, algal standing stock is generally much smaller than that of macrophytes. Nonetheless, algae can constitute a high proportion of the annual primary productivity of an aquatic community. Algae have a high P/B ratio (>100; Wetzel 1983a) and can quickly take advantage of nutrient inputs. In addition, when higher plants are dormant during the winter, algal productivity may continue, thus increasing the relative contribution of algae to the total productivity of the system (Pomeroy and Wiegert 1981). Fontaine and Ewel (1981) showed that the plankton community of a shallow Florida lake contributed 44% of the total primary production for the system. Mitsch and Reeder (1991) found phytoplankton activity represented over 80% of primary production in a freshwater estuarine marsh adjacent to Lake Erie in Ohio. In four constructed emergent marshes in Illinois, phytoplankton contributed from 17 to 67% of the primary production of the wetlands (Cronk and Mitsch 1994a).

Several methods have been developed to measure phytoplankton primary productivity. We briefly describe two of them here. The first is the measurement of dissolved oxygen released during photosynthesis. The second is the measurement of carbon uptake during photosynthesis.

1. Dissolved Oxygen Concentration

The amount of dissolved oxygen present in water results from photosynthetic and respiratory activities of aquatic biota and from diffusion at the air–water interface (Odum 1956; Copeland and Duffer 1964; Lind 1985). Since dissolved oxygen concentrations fluctuate on a daily and seasonal basis, several measurements over time are necessary for an estimate of the system's productivity (Odum 1956; Penfound 1956; Jervis 1969). The method is based on the fact that oxygen is released into the water as a result of photosynthetic primary production during the day, and it is taken up throughout both the night and the day by autotrophic and heterotrophic organisms and by chemical oxidation.

a. Diurnal Dissolved Oxygen Method

Starting at dawn, oxygen production begins in response to daylight. On sunny days, oxygen production increases throughout the morning and early afternoon and then decreases before or at sunset. In this method, data are collected every 2 to 3 h during a 24-h period (from dawn on day 1 to dawn on day 2). Water samples are taken at pre-determined depths and poured into glass bottles designed for the measurement of biochemical oxygen demand (BOD; Figure 6.1). The dissolved oxygen concentration is determined with a dissolved oxygen meter, or with a chemical reaction known as the Winkler method (A.P.H.A. 1995). Alternatively, fully submersible dissolved oxygen meters with data loggers are left in place at the study site, and readings are taken as frequently as the researcher desires (although these data include oxygen production of submerged macrophytes and periphyton).

A plot of the results vs. time reveals the peak of oxygen production during the day as well as the nightly shutdown of oxygen production. The area under the curve represents the NPP of the phytoplankton sampled. The hourly rate of respiration (determined from

FIGURE 6.1
Biochemical oxygen demand bottles used for the measurement of oxygen production and consumption to estimate the primary productivity of phytoplankton. Shown are one 'light' bottle and one bottle darkened with aluminum foil and tape. (Photo by H. Crowell.)

the oxygen decrease during the night) is multiplied by 24 h and added to NPP for an estimate of GPP. Nighttime respiration is assumed to be equal to daytime respiration, although this may be a source of error since daytime respiration may exceed respiration in the dark by an unknown amount. Another source of error in the estimate for respiration comes from the inclusion of heterotrophs in the water sample. Their respiratory consumption is included in the measurement. The results of the diurnal method are expressed as mg O_2 l^{-1} d^{-1}. Results are multiplied by 1000 to convert to g O_2 m^{-3} d^{-1} and then multiplied by the depth at the sampling station for an areal result in g O_2 m^{-2} d^{-1}.

The diurnal method has been applied to many wetland and shallow aquatic systems, such as Narragansett Bay in Rhode Island (Nixon and Oviatt 1973), the Chesapeake Bay in Maryland (Kemp and Boynton 1980), the Illinois Fox Chain of Lakes (Mitsch and Kaltenborn 1980), a Florida lake (Fontaine and Ewel 1981), a Lake Erie coastal wetland in Ohio (Mitsch and Reeder 1991), a Georgia river (Meyer and Edwards 1990), and freshwater constructed wetlands in Illinois (Cronk and Mitsch 1994a).

b. Light Bottle/Dark Bottle Dissolved Oxygen Method

The light bottle/dark bottle technique provides estimates of GPP, NPP, and respiration based on incubated samples (Wetzel 1983a; Wetzel and Likens 1990). In this method, at least four water samples are taken at each depth under study. Two samples are kept in clear glass BOD bottles, and one in a BOD bottle darkened with aluminum foil or other opaque material. The fourth sample from each depth is analyzed for dissolved oxygen content immediately. This is the initial concentration (IB). The remaining light and dark samples are suspended in the water column at the depths from which they were taken, or they are kept in the laboratory under similar light and temperature conditions. The time of incubation must be sufficient for a change to occur (usually from 1 to 4 h). In highly productive waters typical of many wetlands, the incubation period can be shorter than in oligotrophic waters (more typical of deep lakes). During incubation, the dissolved oxygen in the light bottles should increase due to photosynthetic production of oxygen. The dissolved oxygen in the dark bottles should decrease from respiratory consumption of oxygen.

After incubation, the dissolved oxygen concentration within the bottles is determined. The average concentration of the two light bottles (LB) is greater than the original concentration (IB) and the difference is equal to the amount of oxygen produced:

$$(LB - IB) = NPP \tag{6.3}$$

The concentration of the dark bottle (DB) is subtracted from the original concentration (IB), to yield a rate of respiration:

$$(IB - DB) = \text{respiration} \quad (6.4)$$

The sum of the differences is the GPP:

$$(LB - IB) + (IB - DB) = GPP \quad (6.5)$$

To express the results as a daily rate of productivity (mg O_2 l^{-1} d^{-1}), samples are incubated for several 1- to 4-h periods from dawn to dusk. The results are plotted on a graph of time vs. productivity. The area under the curve for the day's measurements is divided by the area under the curve for a 1- to 4-h incubation period. This ratio provides a factor by which the shorter incubation time is expanded to a daily value (Wetzel and Likens 1990).

In the light bottle/dark bottle method, it is assumed that respiration is the same in the light and dark bottles. As for the diurnal oxygen method above, it is impossible to take a sample containing only phytoplankton, so the respiration rate includes that of bacteria and zooplankton. Isolating the samples in bottles also creates problems with "container effects" in which the environment is altered by excluding grazers, nutrients, and atmospheric exchange processes. The results from incubated samples may therefore be an underestimate of production (Hall and Moll 1975; Schindler and Fee 1975).

2. Carbon Assimilation: The ^{14}C Method

During photosynthesis, plants assimilate inorganic carbon and transform it into organic carbon compounds. In the ^{14}C method, the amount of inorganic carbon taken up by plants is measured and the results are expressed as a mass of carbon per unit volume per time interval (mg C m^{-3} $time^{-1}$).

Samples are taken at various depths or locations within the water column of the wetland. At the beginning of the sampling period, a portion of a sample is analyzed for alkalinity and pH (see A.P.H.A. 1995). The rest of the sample is poured into one dark and two light BOD bottles and a small amount of ^{14}C in the form of radioactive bicarbonate ($NaH^{14}CO_3$) is added to each using a syringe. The bottles are incubated for 3 to 4 h during the middle of the day at the depth from which they were taken. The dark bottle inhibits photosynthesis and the rate of carbon uptake should be close to zero. At the end of the incubation period, the amount of labeled carbon (^{14}C) that has been taken up in the phytoplankton is measured. The total amount of carbon assimilated is proportional to the amount of ^{14}C taken up. The average result for the light bottles minus the result for the dark bottle reflects the photosynthetic incorporation of carbon during the incubation period (C m^{-3} h^{-1}; Lind 1985; Wetzel and Likens 1990).

The result can be converted to daily rates by incubating samples for several 4-h periods throughout the day. As in the light bottle/dark bottle method, the ratio of the productivity for the day to the productivity for the shorter incubation period provides a factor by which to correct the hourly rate and estimate a daily rate (Wetzel and Likens 1990).

The ^{14}C method is more sensitive than the method of measuring oxygen change (Wetzel 1983a; Lind 1985). The smallest amount of photosynthesis that can be detected with dissolved oxygen readings is about 20 mg C m^{-3} h^{-1}, while the ^{14}C method is sensitive to changes as small as 0.1 to 1 mg C m^{-3} h^{-1} (Wetzel 1983a). In wetlands, this level of sensitivity may not be necessary since the water column is often highly productive. Instruments and supplies for the ^{14}C method are expensive, more training is necessary than for the oxygen method, analysis is not completed in the field, and some researchers have warned that results may underestimate GPP (Lind 1985; Stevenson 1988; Aloi 1990; Colinvaux 1993).

B. Periphyton

Periphyton are attached algae that may be found on nearly any submerged surface. More specific terms for periphyton are based on whether they are attached to inorganic or organic substrates (Aloi 1990):

- Epilithon: periphyton on rock substrates
- Epipelon: periphyton on mud or silt substrates
- Episammon: on sand substrates
- Epiphyton: on the submerged portions of aquatic macrophytes

In oceans, deep lakes, and downstream areas of rivers, phytoplankton dominates productivity, but when the ratio of sediment area to water volume increases, as is the case in wetlands, the macrophytes and periphyton become more significant contributors to the system's productivity (Sand-Jensen and Borum 1991). Periphyton are known to be significant producers in salt marshes, with productivity estimates ranging from 10 to 25% of macrophyte productivity in east coast U.S. salt marshes (Pomeroy 1959; Gallagher and Daiber 1974; Van Raalte et al. 1976; Pomeroy and Wiegert 1981) and 80 to 140% of macrophyte productivity in southern California salt marshes (Zedler, 1980). The higher ratios recorded in California are due to both higher algal production and lower macrophyte production. Twilley (1988) found that epiphytic algae took up as much as 16% of the total carbon fixed in mangrove wetlands of Florida and Puerto Rico. In four constructed freshwater marshes, from 1 to 37% of the primary productivity was attributed to periphyton, with the highest levels in wetlands with higher hydrologic throughflow (Cronk and Mitsch 1994b).

The literature concerning periphyton primary productivity is replete with variations in methodology (Wetzel 1983b; Robinson 1983; Aloi 1990). Periphyton primary productivity can be measured much like that of phytoplankton, i.e., as the evolution of oxygen or the uptake of carbon. Changes in periphyton biomass over time can also give an estimate of periphyton NPP. Biomass changes may be assessed using either natural or artificial substrates.

Natural substrates include rocks, macrophytes, and other underwater surfaces. To measure epilithon biomass in wetlands, rocks are collected and the attached algal growth is scraped off of the surface and then dried and weighed. To remove epiphyton from macrophytes, the macrophyte is placed in a jar with water and shaken. After shaking, the plants are gently scraped (Aloi 1990).

Artificial substrates are often used in periphyton studies to provide a uniform area and a means with which to control environmental variables. Artificial substrates provide a standard means of comparison between two sites with the benefit of decreased cost of sampling, decreased disruption of habitat, and decreased time required to obtain a quantitative sample (Aloi 1990). Suitable artificial substrates are uniform in size, shape, and material. Materials are chosen in part for their resistance to the effects of prolonged submersion. Many investigators have introduced artificial substrates such as microscope slides made of glass (many studies starting in 1916 as cited by Aloi 1990) or plastic (Figure 6.2; Cronk and Mitsch 1994b; Wu and Mitsch 1998), clay tiles (Barko et al. 1977), nutrient diffusing sand agar surfaces (Pringle 1987, 1990), or cylindrical rods of different textures (Goldsborough and Hickman 1991; Hann 1991). Substrates that can be suspended vertically are often

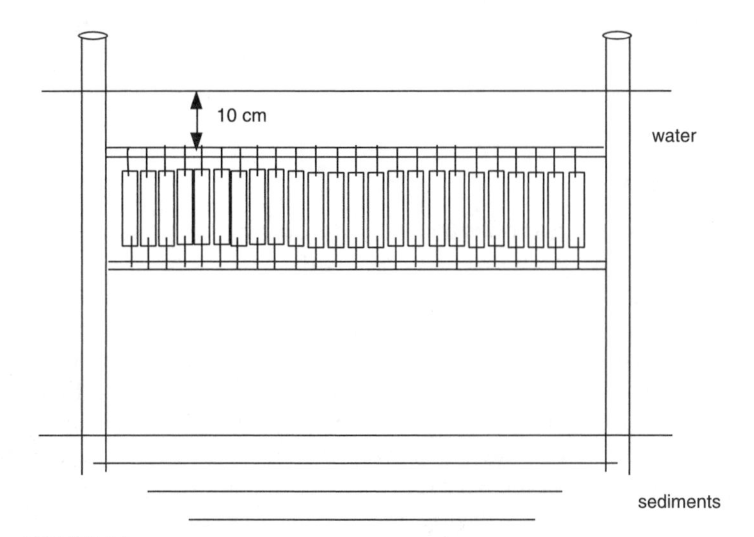

FIGURE 6.2
Periphyton sampler made of PVC pipe and plastic microscope slides from which slides were removed every 2 weeks during the growing season in order to assess the colonization rate, biomass accumulation, and community structure at four constructed wetlands near the Des Plaines River, Illinois. (From Cronk, J.K. and Mitsch, W.J. 1994b. *Aquatic Botany* 48: 325–342. Reprinted with permission.)

preferable in order to minimize the accumulation of inorganic solids (Aloi 1990). Artificial substrates are useful in studies of colonization, community development, herbivory, or in the comparison of the effects of an environmental variable or different habitats. They are less useful in primary productivity studies because they do not imitate naturally occurring growth and the algal species that grow on artificial substrates are not necessarily the same as those on natural substrates (Wetzel 1964, 1966, 1983a; Wetzel and Hough 1973).

Periphyton are harvested from a portion of the substrates at regular time intervals (for example, every 2 weeks) in order to detect initial colonization, as well as a peak and decline in growth (Carrick and Lowe 1988; APHA 1995). Colonization of introduced substrates generally occurs at an exponential rate for the first 2 weeks of exposure and then slows (Kevern et al. 1966; Lamberti and Resh 1985; Paul and Duthie 1988). Biomass measurements of periphyton are made by drying and weighing samples. Another common measure of periphyton biomass is ash-free dry weight because periphyton samples often include inorganic matter that could skew dry weight results upward. Biomass measurements provide an underestimate of NPP because they do not account for losses due to herbivory, sloughing, or dislodgement (Aloi 1990).

Periphyton NPP may be more accurately estimated by measuring gas exchange techniques. Rocks or other periphyton substrates are incubated in bottles and either the oxygen or ^{14}C method is used (as described for phytoplankton). Disturbing the community may skew productivity findings due to changes in flow regime and nutrient supply, so some researchers measure productivity *in situ* by enclosing the substrate in clear plastic chambers pushed into the substrate or in domes sealed to large rock surfaces. The chamber is left to incubate for a certain length of time and then changes in oxygen or carbon are measured. Measuring epiphytic productivity is more complicated because enclosing the substrate means that the productivity of the macrophyte substrate as well as the periphyton will be measured. Often, epiphytes are scraped or shaken from the plant and placed in bottles for incubation; however, the effects of this disturbance on productivity are not known (Twilley et al. 1985; Aloi 1990).

C. Submerged Macrophytes

The biomass of submerged macrophytes is measured by direct harvest or by incubating the plants and measuring oxygen production or carbon fixation.

1. Biomass

The biomass of submerged macrophytes is measured by harvesting and drying plants. Because they are underwater, the method is somewhat more involved than for emergent macrophytes. Plots of known area are placed randomly within an aquatic plant bed. Harvesting is usually carried out at the estimated peak biomass. If the depth of the site warrants, sampling is done by SCUBA divers. The plants are sorted, cleaned of sediments and periphyton, and dried in a drying oven at 70° to 105°C to a constant weight. Belowground biomass is estimated by collecting sediment cores, weighing the sorted, washed, and dried roots, and then extrapolating for the rest of the community.

Peak biomass is sometimes used as the estimate for that year's net production, or samples are taken through the growing season and net production is calculated as the sum of the positive biomass increments until peak biomass. Alternatively, peak biomass is measured and then multiplied by published values for P/B ratios. Examples of P/B ratios are 1.2 for *Potamogeton pectinatus,* 2.0 for *Ruppia cirrhosa* and *R. maritima,* 2.0 for *Myriophyllum spicatum,* and 1.16 for *Ranunculus baudauti* (from various sources cited by Kiorboe 1980). Peak biomass values may be difficult to determine for submerged species. For example, *M. spicatum* peaks earlier in shallow water than in deep water. Therefore, peak biomass for *M. spicatum* should be determined at different times depending upon water depth (Grace and Wetzel 1978).

2. Oxygen Production

To measure oxygen production by submerged plants, samples are harvested and cleaned of epiphytes and sediment. They are placed in light and dark BOD bottles and filled with water taken from the same site that has been filtered to remove algae. They are incubated at the approximate depth from which the sample was pulled and periodically shaken to reduce boundary-layer effects (an intact boundary layer can result in a decrease of nutrients near the plant surface). The incubation period is from 1 to 4 h. The dissolved oxygen in the bottles is measured using the same methods described for phytoplankton. Productivity is calculated as for the light bottle/dark bottle method for phytoplankton.

The disturbances inherent in this method can produce considerable error in the results. Plants are severed from their roots, which is the source of most of their nutrients. They are brought to the surface, exposed to intense surface light, and then returned to their original depth, thus imposing abnormal light and flow conditions (Wetzel 1964). NPP results are higher if only apical portions of the plant are used rather than lower parts of the plant. NPP may be underestimated because oxygen produced photosynthetically fills the plant's lacunae first, before it is released to the water (Wetzel 1966; Sondergaard 1979). For *Potamogeton perfoliatus* the lag time between initial light and the initial evolution of oxygen to the water has been measured at 5 to 25 min (Kemp et al. 1986). If oxygen is measured after the plant has been exposed to light for at least a few minutes, the error introduced into production estimates by lag time is reduced. Respiration in the light and dark bottles is usually assumed to be equal (Kemp et al. 1986).

3. Carbon Assimilation

Samples are collected in the same manner as for dissolved oxygen measurements. Radioactive bicarbonate is added to the bottles and the analysis and calculation of productivity are the same as described in the ^{14}C method for phytoplankton. Results are expressed as the weight of carbon fixed per dry biomass weight per unit time (g C g^{-1} h^{-1}) and incubation is generally 4 to 5 h (Wetzel 1966).

D. Emergent Macrophytes

The bulk of wetland primary productivity studies have been done in emergent marshes, so more methods exist for emergents than for other types of wetland plants. The primary productivity of emergent plants can be measured using gas exchange methods by enclosing the plants in clear chambers in which the changes in gas concentrations are monitored (Mathews and Westlake 1969; Blum et al. 1978; Pomeroy and Wiegert 1981; Brinson et al. 1981; Jones 1987; Bradbury and Grace 1993). However, most researchers use biomass methods and we describe several of these methods below.

1. Aboveground Biomass of Emergent Plants

Many researchers have compared aboveground emergent production methods in salt marshes (Kirby and Gosselink 1976; Turner 1976; Linthurst and Reimold 1978b; Gallagher et al. 1980; Hardisky 1980; Hopkinson et al. 1980; Shew et al. 1981; Groenendijk 1984; Houghton 1985; Dickerman et al. 1986; Jackson et al. 1986; Cranford et al. 1989; Kaswadji et al. 1990; Morris and Haskin 1990; Dai and Wiegert 1996) and freshwater marshes (Dickerman et al. 1986; Wetzel and Pickard 1996; Daoust and Childers 1998), yet a definitive method does not exist (Table 6.3). The comparison studies show that NPP results depend on the method chosen and they can vary up to sixfold (Table 6.4; Linthurst and Reimold 1978b). With such wide variation in the results, the choice of method can have important consequences.

The methods described here are based on measurements of plant biomass. The data are collected by harvesting, drying, and weighing plants. Alternatively, plant growth in study plots is monitored and biomass is estimated using regressions of height (or other parameters) to weight based on plants harvested outside of the plots. Production is expressed in grams dry weight per square meter and productivity is usually reported as a yearly rate (g dry weight m^{-2} yr^{-1}). Losses to respiration are not included so the results are a measure of NPP. The variation in the methods we describe arises from the inclusion of different components of the plant community (live, dead, decomposed matter) and from various ways of calculating NPP. The choice of method depends upon the plant community and on environmental and time constraints. We describe eight commonly used methods, but others are available. We use the names for these methods that are the most frequently used in the literature. Some have descriptive names while others use the originators' names:

Peak biomass	Milner and Hughes (1968)
Valiela et al. (1975)	Smalley (1959)
Wiegert and Evans (1964)	Lomnicki et al. (1968)
Allen curve method (1951)	Summed shoot maximum

TABLE 6.3

Some Methods for the Measurement of Net Aboveground Primary Productivity of Emergent Herbaceous Wetland Plants and Evaluations of the Methods Made in Comparisons or Reviews

Method	Calculation of Net Primary Productivity	Evaluation	Evaluated by
Peak biomass	Maximum biomass	Underestimate	Numerous studies; see text
Milner and Hughes (1968)	Sum of positive changes in live biomass	Underestimate	Linthurst and Reimold 1978b
			Shew et al. 1981
			Dickerman et al. 1986
			Morris and Haskin 1990
		Best estimate using tagged plants	Dai and Wiegert 1996
		Invalid method	Daoust and Childers 1998
Valiela et al. (1975)	Sum of changes in dead biomass plus an estimate for biomass decayed during sampling interval	Underestimate	Valiela et al. 1975
			Linthurst and Reimold 1978b
			Dickerman et al. 1986
			Dai and Wiegert 1996
Smalley (1959)	Sum of changes in live and dead biomass or, if sum is negative, equal to zero	Underestimate	Linthurst and Reimold 1978b
			Shew et al. 1981
			Dickerman et al. 1986
			Cranford et al. 1989
			Daoust and Childers 1998
Wiegert and Evans (1964)	Sum of changes in live and dead biomass plus an estimate for biomass decayed during sampling interval	Best estimate	Kirby and Gosselink 1976
			Groenendijk 1984
		Over- or underestimate	Linthurst and Reimold 1978b
		Overestimate	Hopkinson et al. 1980
			Shew et al. 1981
			Dickerman et al. 1986
			Dai and Wiegert 1996
			Daoust and Childers 1998

Lomnicki et al. (1968)	Sum of changes in live biomass plus the dead biomass measured at the end of each sampling interval	Overestimate, but slight modification was best estimate (see text)	Shew et al. 1981
Allen curve method (1951)	Area beneath curve of shoot density vs. average shoot biomass for estimates of plants growing as cohorts; requires new curve for each new cohort	Slight underestimate, not appropriate to all plant types	Dickerman et al. 1986
		Good estimate	Cranford et al. 1989
		Overestimate; for use with plants that develop in recognizable cohorts	Bradbury and Grace 1993
		Underestimate if plants have continuously emerging new shoots	Wetzel and Pickard 1996
Summed shoot maximum	Sum of the maximum biomass for each shoot in study plot	Best estimate	Dickerman et al. 1986

TABLE 6.4

Studies That Compared Results for the Estimation of Net Aboveground Primary Productivity of Wetland Emergents Using Various Measurement and Calculation Methods (units are g dry weight $m^{-2} yr^{-1}$)

Productivity Study	Plant Species	Location	Peak Biomass	Milner and Hughes	Valiela et al.	Smalley	Wiegert and Evans	Lomnicki et al.	Allen Curve	Summed Shoot Maximum
Kirby and Gosselink 1976	*Spartina alterniflora* (short)	Louisiana	788	748		1006	1323			
	S. alterniflora (tall)		1018	874		1410	2645			
Linthurst and Reimold 1978b[a]	*S. patens*	Maine	912	912	2523	3523	5833			
	S. patens	Delaware	807	522	1241	980	2753			
	S. patens	Georgia	946	705	1028	1674	3925			
Gallagher et al. 1980	*Juncus roemerianus*	Georgia				2500	2800			
	S. alterniflora (short)					700	1500			
	S. alterniflora (tall)					2300	3000			
Hardisky 1980	*S. alterniflora*	N. Carolina				931				
Hopkinson et al. 1980	*Distichlis spicata*	Louisiana	900		900	750	2800			
	J. roemerianus		1200		1800	1250	3200			
	Sagittaria falcata		550		1000	600	1300			
	S. alterniflora		700		1500	1000	2200			
	S. cynosuroides		750		300	1100	1600			
	S. patens		1200		2500	2000	5800			
Shew et al. 1981	*S. alterniflora*	N. Carolina	242	214		225	1029	1028		
Groenendijk 1984[b]	*Elytrigia pungens*	Netherlands				474–878	1416–1787			
	Spartina anglica					1162–1649	2139–2659			
Houghton 1985[c]	*S. alterniflora* (short)	New York	780	770	750	825				
	S. alterniflora (tall)		1220	1310	1460	1400				
Jackson et al. 1986[d]	*S. anglica*	England	160–180			590–760				
Dickerman et al. 1986[e]	*Typha latifolia* (harvested plots)	Michigan	768 –836	764–834	1557–1346	1604–1284	2762–2266			
	T. latifolia (study plots)		833–1318	833–1317		834–1301			927–1358	1005–1507

Cranford et al. 1989[f]	*S. alterniflora*	Nova Scotia	398			434			507	
Kaswadji et al. 1990	*S. alterniflora*		831	831		1231	1873	1437		
Morris and Haskin 1990[g]	*S. alterniflora* (short)	S. Carolina	209–694	295–820						402–1042
Dai and Wiegert 1996[h]	*S. alterniflora* (short)	Georgia		1105	993					1118
	S. alterniflora (tall)			1520	1315					1557
Wetzel and Pickard 1996[i]	*T. latifolia*	Lab	120–267						214–232	345–696
Daoust and Childers 1998[j]	*Cladium jamaicense*	Florida		944		2315	3593			
	Wet prairie: 8 dominant emergents			165		319	531			
Average ratio of method's results to peak biomass (±1 S.D.)				1.0 ± 0.2	2.0 ± 0.6	1.6 ± 1.0	3.3 ± 1.2	3.0 ± 1.8	1.2 ± 0.4	1.9 ± 0.7

Note: Methods are discussed in the text.

[a] Linthurst and Reimold 1978b: measured the NPP of other species, but only one is shown here to illustrate the different results among the methods.
[b] Groenendijk 1984: range is for different calculations used for each method.
[c] Houghton 1985: measured for 2 years, just 1 year's results are reported here.
[d] Jackson et al. 1986: range is for 2 years of data.
[e] Dickerman et al. 1986: data from the most frequent time interval used in their calculations; range is for 2 years; the first year's data are both > and < the second year's data, depending on the method.
[f] Cranford et al. 1989: modified the Smalley and Allen curve methods slightly.
[g] Morris and Haskin 1990: range is for 5 years of data.
[h] Dai and Wiegert 1996: used Milner and Hughes and Valiela et al. calculation methods with tagged plants in permanent quadrats rather than harvested plants.
[i] Wetzel and Pickard 1996: also used other methods not described here; range is for five different experimental treatments.
[j] Daoust and Childers 1998: used these computational methods in combination with phenometric techniques specific to the species in their study.

A general trend within the productivity literature has been to include more components of the plant or plant community in the measurements. The later methods tend to be more complex or more time-consuming as a result of these trends. The changes are an effort to more accurately estimate NPP. Most of the methods have been applied to stands of *Spartina alterniflora* (cordgrass) and *Typha latifolia* (broad-leaved cattail), so the methods here are probably most appropriate for monospecific stands of monocots. Areas with more diversity and those containing eudicots present problems to which researchers must adapt these methods.

a. The Peak Biomass Method

The primary productivity of emergent plants can be estimated by harvesting and weighing the aboveground portion of plants when they are at peak biomass. Peak biomass usually occurs in mid- to late summer in temperate areas. In subtropical and tropical zones, it can be difficult to detect peak biomass because growth continues throughout the year.

In this method, samples are selected either randomly, along transects within communities, within a specific moisture gradient, or according to other physical features of interest. At peak biomass, plant shoots within a plot, or *quadrat* (usually from 0.1 to 1 m^2) are clipped at the sediment surface. The plant matter is dried and weighed and results are expressed as g dry weight m^{-2}. NPP is usually expressed as yearly growth or growth per number of days of the growing season. In some studies, plants are harvested or monitored several times throughout the growing season and the greatest value is taken as the peak biomass. Frequent sampling enables the investigator to detect the time of peak biomass more accurately than with a single measurement (Boyd 1971; Boyd and Vickers 1971). The peak biomass method is simple to apply, requires minimal field and laboratory time, and the results are comparable from season to season within the same site.

The disadvantage of this method is that it does not provide a good estimate of NPP. The fact that peak biomass is almost always an underestimate of NPP, sometimes by severalfold, has been confirmed in many studies (Wiegert and Evans 1964; Wetzel 1966; Valiela et al. 1975; Bradbury and Hofstra 1976; Kirby and Gosselink 1976; Linthurst and Reimold 1978b; Whigham et al. 1978; Shew et al. 1981; Westlake 1982; Houghton 1985; Dickerman et al. 1986; Jackson et al. 1986; Dai and Wiegert 1996; Wetzel and Pickard 1996). This method does not include corrections for plant mortality before peak biomass, nor does it include any production that occurs after peak biomass. It also does not take into account any differences in the time at which different species attain peak biomass (Wiegert and Evans 1964). In most temperate wetlands, plants die during the winter and each year's plant growth is distinguishable from the previous year's crop (Nixon and Oviatt 1973). However, in warm areas, plant growth may occur throughout the year. The peak biomass method does not subtract carry-over live plant material that was present before the beginning of the current growing season (Linthurst and Reimold 1978b).

This method is not used as frequently now as it was in earlier ecological studies due to these problems. When it is used, the results should be clearly designated as peak biomass or maximum standing crop. Despite the method's drawbacks, use of the peak biomass method can be justified under some circumstances:

- *Cost*: It may be the only method that researchers can afford since it requires less effort than other methods.

- *Comparison among multiple sites:* When many sites are to be compared, it may be the only feasible method. For example, van der Valk and Bliss (1971) measured the peak biomass of emergent plants in 15 oxbow wetlands in Alberta, Canada. They sampled three times so as to detect peak biomass. Their goal was to compare the same parameter (peak biomass) among many wetlands rather than to report the primary productivity.
- *Difficult access to sites:* If the study site is difficult to access, sampling more than once during the growing season may be impractical. For example, Glooschenko (1978) studied a remote subarctic salt marsh in northern Ontario and reported results as aboveground biomass rather than productivity.
- *Subarctic sites:* The use of the peak biomass method in the preceding example (Glooschenko 1978) was also appropriate because peak biomass more accurately reflects true primary production in cold climates than it does in temperate zones (Hopkinson et al. 1980). During the short subarctic growing season, less turnover of plant tissues occurs, so there are fewer unaccounted losses. Thus, the accuracy of peak biomass measurements increases with increasing latitude.
- *Long-term study within a single marsh or area:* In a Netherlands salt marsh, the use of the peak biomass method by De Leeuw and others (1990) was justified because their goal was to compare results within the same marsh over a long time period. They measured peak biomass annually for 13 years. They acknowledged that their results are an underestimate of NPP, and they did not attempt to compare their results to those of other salt marshes.

b. The Milner and Hughes Method

In the Milner and Hughes (1968) method, NPP is calculated as the sum of the increases in live biomass between successive sampling dates. Dead biomass is not included in the sum. The results are for the growing season and expressed as g dry weight m^{-2} yr^{-1}. Usually plants are harvested each month during the growing season, and so this method is sometimes called "summation of the new monthly growth" (Morris and Haskin 1990; Dai and Wiegert 1996). The Milner and Hughes method yields results that are very similar to peak biomass results, particularly if all shoots die in the fall (Table 6.4; Dickerman et al. 1986). Results are less than peak biomass if plants continue to grow throughout the winter because live biomass remaining from the preceding growing season is subtracted from the sum for the current season (Kirby and Gosselink 1976; Linthurst and Reimold 1978b).

This method has been evaluated as an underestimate of NPP by several researchers (Table 6.3; Linthurst and Reimold 1978b; Shew et al. 1981; Dickerman et al. 1986; Morris and Haskin 1990). Daoust and Childers (1998) determined that the Milner and Hughes method was invalid for their study because their results with this method differed significantly from results obtained using other calculation methods. The results from the Milner and Hughes method are not corrected for mortality or loss of plant parts that occurs during the growing season (Dickerman et al. 1986). In monotypic communities, the method misses the peak of younger cohorts that may occur after the first peak in biomass. In diverse plant communities, the method misses the maximum growth of species that start and peak at different times, compared to the dominant species.

The Milner and Hughes method has been used in a number of studies (Smith et al. 1979; Cargill and Jefferies 1984; Dai and Wiegert 1996), particularly in salt marshes, where monotypic stands of *Spartina alterniflora* resemble the grasslands for which this method was designed. Results from a subarctic salt marsh probably come close to actual NPP, since the rates of leaf turnover and decomposition are low in cold climates with short growing seasons (Cargill and Jefferies 1984). Dai and Wiegert (1996) used a modified version of this method in which tagged plants of a single species were monitored closely throughout the growing season. They concluded that the results of this method were their best estimate for primary production because the method closely tracked actual growth.

c. The Valiela et al. Method

In the Valiela et al. (1975) method, NPP is calculated as the total of dead material measured over a growing season. This method was devised to measure the NPP of a *Spartina alterniflora* salt marsh in Massachusetts. Since standing crops varied little from year to year at their site, the researchers assumed that the dead matter that accumulated over a growing season was equal to net annual aboveground production. At each sampling period, they harvested and separated live and dead material. They also calculated losses that may have occurred due to decomposition during the sampling interval. The decomposed biomass was calculated as follows:

- If there is less dead material at the end of the sampling interval than at the beginning, then the change in dead material is a negative value. The absolute value of the loss in dead material is the amount of decomposed biomass. In the form of an equation, the decomposed biomass (e) is calculated as follows:

$$e = -\Delta d, \text{ if } \Delta l \geq 0 \text{ and } \Delta d < 0 \quad (6.6)$$

  where Δl is the change in live standing crop between any two sampling dates, and Δd is the change in standing dead matter for the same interval.

- If living plant material decreases during the sampling period, then the absolute value of the sum of the change in living material and the change in dead material is equal to the decomposed biomass (e):

$$e = -(\Delta l + \Delta d), \text{ if } \Delta l < 0 \quad (6.7)$$

Net primary production for each sampling interval is calculated as the sum of dead material plus the amount calculated for the decomposed biomass (e). NPP for the growing season is the sum of the values for each sampling interval.

Results from this method are usually considered an underestimate of NPP because when there is an increase in live material, concomitant mortality, and no apparent change in the standing dead material, then growth is unassessed for that period. In addition, if dead material is washed away by tides, it is not counted in the method (Valiela et al. 1975).

The Valiela et al. method is appropriate where the litter component is negligible and where the wetland is in a steady state with the rate of production balanced by the rate of decomposition (Dickerman et al. 1986). Dai and Wiegert (1996) modified the Valiela et al. method by summing the monthly dead biomass and correcting it for the net change of aerial living biomass between the beginning and the end of the growing season (their site was in Georgia, where growth continued during the winter). They monitored the same plants throughout the growing season. The height of the plants was measured, and biomass was determined from a regression of plant height on plant weight. The height of standing dead plants was used to estimate biomass. However, the height of the dead plants was less than that of their maximum live height, so biomass was underestimated.

d. The Smalley Method

To measure the NPP of *Spartina alterniflora* in a Georgia salt marsh, Smalley (1959) devised a method in which the biomass of both the live and dead material within quadrats is measured several times throughout the growing season. The changes in both live and dead biomass from one sampling period to the next determine how net production for that sampling interval is to be determined. The conditions for calculating net production are:

- If the change in living material is positive and the change in standing dead is positive, then net production is their sum.
- If both are negative, then net production is zero.
- If the change in living material is positive and the change in standing dead is negative, then net production is equal to the change in living material.
- If the change in standing dead is positive and the change in living material is negative, the two values are added for net production, but if the sum is negative, then net production is zero.

NPP for this method is the sum of net production results for each sampling period expressed per unit time (g m^{-2} yr^{-1}).

The Smalley method has been used in emergent wetland studies (Zedler et al. 1980; Kistritz et al. 1983) and tested against other methods by many researchers (Table 6.4; Kirby and Gosselink 1976; Turner 1976; Linthurst and Reimold 1978b; Gallagher et al. 1980; Hardisky 1980; Hopkinson et al. 1980; Shew et al. 1981; Groenendijk 1984; Houghton 1985; Jackson et al. 1986; Dickerman et al. 1986; Cranford et al. 1989; Kaswadji et al. 1990; Daoust and Childers 1998). The general conclusion from these studies is that the results are an underestimate of NPP (Table 6.3). Several reasons are given:

- The Smalley procedure does not correct for the instantaneous loss of plant litter (i.e., the plant material that is lost to decomposition during the sampling interval; Dickerman et al. 1986).
- Since negative values are reported as zero, the magnitude of change from one sampling period to the next is unknown. Negative net production is considered a contradiction in terms. Even though it is measured, it is disregarded for that sampling period. However, the errors in the estimate of the actual amounts of live and dead material present are carried forward and are not corrected (Turner 1976).
- In tidal marshes there is often export of dead leaf material, which is not measured in this method. The amount may be large (up to 35% of production) or small (where plants have a low rate of leaf loss; Kistritz et al. 1983).
- Tides can also import dead material into a study plot, but this added material does not represent growth within the plot (Linthurst and Reimold 1978b).

e. The Wiegert and Evans Method

The Wiegert and Evans method (1964) originally was devised for grassland studies. The method corrects for one of the sources of error in the Smalley method (namely, the instantaneous loss of litter). The Wiegert and Evans method sums the live and dead material produced during each sampling interval and it adds an estimate for the decomposed plant material. The decomposed biomass is estimated as a proportion of the dead biomass for each sampling period. Samples are usually taken once per month for a year or a growing season. NPP is the sum of the monthly values.

The changes in standing live vegetation and the changes in mortality are measured from one sampling period to the next. Mortality is the change in dead material plus the material that has decomposed. To determine NPP for any given time interval using this method, the following six parameters are needed (standing crop data are expressed in g m^{-2}; instantaneous rates in mg g^{-1} d^{-1}):

Let t_i = a time interval (in days)
a_{i-1} = standing crop dead material at start
a_i = standing crop dead material at end
b_{i-1} = standing crop live material at start
b_i = standing crop live material at end
r_i = instantaneous daily rate of disappearance of dead material during interval

These parameters are used to calculate the amount of dead material that disappears during an interval (x_i):

$$x_i = (a_i + a_{i-1}) / 2 \times r_i t_i \tag{6.8}$$

The change in standing crop of live material is

$$\Delta b_i = b_i - b_{i-1} \tag{6.9}$$

The change in standing crop of dead material is

$$\Delta a_i = a_i - a_{i-1} \tag{6.10}$$

Since Δa_i is the change in dead standing crop during the interval, then ($x_i + \Delta a_i$) is the amount of material added to the dead standing crop during the interval, i.e., the mortality of live material, symbolized here by d_i:

$$d_i = x_i + \Delta a_i \tag{6.11}$$

The last equation must be ≥ 0; negative values indicate an error in one or more of the measured parameters.

The growth during a given time interval (t_i) is then given by:

$$y_i = \Delta b_i + d_i \tag{6.12}$$

where y_i is expressed in grams per unit area.

These equations are used to calculate the mortality and growth of the vegetation for each sampling period. The amounts for each period are summed and expressed per unit time for NPP.

The estimate of the instantaneous loss of litter (r) needed for the Wiegert and Evans method is obtained using one of two procedures. The first is the *paired-plot method* in which each study plot is paired with a second plot that is identical in size and as similar to the first in vegetation size and type as possible. The dead material is removed from the first plot and weighed (W_{i-1}) at the initial sampling time (t_{i-1}). At the same time, the live material is removed from the second plot, leaving only dead material. At the second sampling time (t_1), the dead material from the second plot is removed and weighed (W_1). The instantaneous rate of disappearance of dead material from these plots is estimated as:

$$r = \ln (W_{i-1} / W_1) / t_1 - t_{i-1} \quad (6.13)$$

where r = disappearance rate in g g^{-1} d^{-1}, and $t_1 - t_{i-1}$ is in days.

The paired-plot method uses three key assumptions: (1) the rates of disappearance of dead material from the two plots are equal, (2) the biomass and the species composition of the two plots are identical, and (3) no additional material is added to the dead material of the second plot during the sampling interval.

The second way to estimate the rate of disappearance of dead plant material (r) is to use a *litter bag method*. Litter bags containing dead vegetation are staked to the ground at various locations (chosen at random) within the study area at the initial sampling time. The litter bags are made of mesh so that air, water, and decomposing organisms can be exchanged. Each bag contains the same mass of dead material. At the end of the sampling interval, the plant material remaining in the litter bags is weighed again. The decrease in weight is the amount of instantaneous loss of litter for that sampling period.

For Weigert and Evans (1964), the results for decomposition using the litter bag method were lower by an order of magnitude than those obtained using the paired plot method. They explained that the unweathered material in a litter bag is not likely to decompose as quickly as naturally stratified layers on the field. Litter bags also restrict the entry of larger decomposers and scavengers that could increase the rate of decomposition. They concluded that the paired-plot method provided a better estimate of the instantaneous loss rate than the litter bag method. However, Kirby and Gosselink (1976) evaluated this method for use in salt marshes and found that the litter bag method was preferable because tidal flushing moved dead material in and out of their paired plots, whereas the material within the litter bags remained the same.

The Wiegert and Evans method has been applied to emergent wetlands and compared to other methods by a number of researchers (Table 6.3; Kirby and Gosselink 1976; Linthurst and Reimold 1978b; Gallagher et al. 1980; Hopkinson et al. 1978, 1980; Shew et al. 1981; Groenendijk 1984; Dickerman et al. 1986; Kaswadji et al. 1990; Dai and Wiegert 1996; Daoust and Childers 1998). Estimates of NPP using this method are 1.7 to 6.4 times greater than estimates obtained using the peak biomass method (Table 6.4). The Wiegert and Evans method is considered to provide the best estimate by some investigators (Kirby and Gosselink 1976; Groenendijk 1984). However, others assert that the method overestimates NPP (Hopkinson et al. 1980; Shew et al. 1981; Dickerman et al. 1986; Dai and Wiegert 1996; Daoust and Childers 1998). The results may differ from actual NPP because any wind or water transport of litter out of the study area influences the results (Linthurst and Reimold 1978b; Carpenter 1980b; Shew et al. 1981). The actual value of NPP is often thought to lie between the overestimate from the Wiegert and Evans method and the underestimate from the Smalley method (Dickerman et al. 1986). The method is fairly complicated and requires far more work than some of the other methods. Hopkinson and others (1980) estimated that the Wiegert and Evans method requires at least 60 person hours per year per site. By comparison, they estimated that the peak biomass method requires about 10 person hours per year per site.

f. The Lomnicki et al. Method

The Lomnicki et al. (1968) method uses a modification of the paired-plots portion of the Wiegert and Evans method. The calculation for NPP includes the sum of the changes in both live and dead material. Plant mortality is measured directly, so the calculations for the instantaneous loss rate (r) in the Wiegert and Evans method are not necessary (Bradbury and Hofstra 1976; Shew et al. 1981). In this method, sampling is done more frequently in order to measure the dead material before it decomposes. In the Wiegert and Evans method, when paired plots are used to estimate instantaneous loss of dead material, all live material is removed from the second plot at the initial sampling time (t_{i-1}). Lomnicki and others (1968) suggested that the live material should remain on the second plot at time t_{i-1} to permit its normal growth and death. In their method, the live material in the first plot is harvested, dried, and weighed and the amount is represented as b_{i-1}. The standing dead material and litter are removed from the second plot at time t_{i-1} and the live material in the second plot is left standing. At the end of the sampling interval (t_1), the newly produced dead material is removed from the second plot (where the live vegetation was left standing) and measured (Shew et al. 1981).

The newly dead material from the second plot represents mortality (d_i) during the time interval. The live material present at the second sampling period (b_i) is used to determine the change in live material during the interval ($\Delta b_i = b_i - b_{i-1}$). The equation for NPP is

$$NPP = \Delta b_i + d_i \tag{6.14}$$

where net primary productivity is the amount of live material produced during a time interval plus the amount of live material dying during that interval (mortality). Negative monthly production values are set to zero.

The assumptions of this method are (1) the study site is homogeneous (so any two plots can be paired), (2) the removal of dead material has no effect on the mortality of live plants, and (3) there is no loss of material that died during the time interval. The first assumption is met by proper site selection. The second can be tested by comparing the weight of the live material to weights from plots where the dead material has not been removed. The third assumption is met by selecting a sampling interval that is short enough to prevent decomposition or other loss of dead material. If this method is used in a tidal marsh, screen cages around the study plots can be used to minimize the import or export of dead material by tidal flushing.

The Lomnicki et al. method has not been as extensively applied in wetlands as the Wiegert and Evans method. However, in at least one study, no significant differences were found between results from the Lomnicki et al. method and from the Wiegert and Evans method (Shew et al. 1981). The method may be an attractive alternative to the Wiegert and Evans method, particularly in salt marshes because it does not require estimates of decomposition. On the other hand, it is difficult to meet the requirement for short sampling intervals. Lomnicki and co-researchers were working in grasslands in Poland at 50°N where presumably less decomposition occurred during sampling intervals than would occur in a salt marsh at lower latitudes (Shew et al. 1981).

g. The Allen Curve Method

The Allen curve method (1951) was devised to measure fish productivity (Allen 1951) and was adapted for plants by Mathews and Westlake (1969). The Allen curve is based on the measurement of growth within cohorts of plants. A cohort is a group of plants of the same species that starts life and goes through the stages of the plant's life history at approximately the same time. The Allen curve method calculates cohort production graphically as the area beneath a curve where the mean biomass per shoot over a growing season is on the x-axis and the shoot density is on the y-axis (Figure 6.3). The points on the graph show the relationship between shoot density and mean shoot biomass on different sampling dates. Increases in cohort biomass from one sampling date to the next move the point upward or to the right and therefore increase the area beneath the curve. Declining numbers of shoots or dry weight losses move the point downward or to the left. The total area

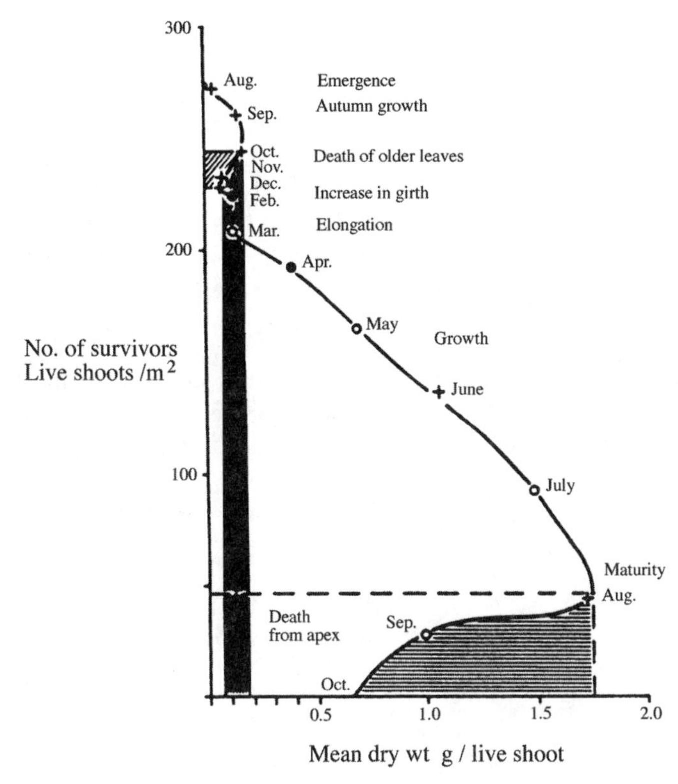

FIGURE 6.3
An Allen curve for an August cohort of aerial shoots of *Glyceria maxima*. +, plotted from complete observations; ●, plotted from observed numbers and interpolated weights; O, plotted from interpolations; diagonal shading, loss of biomass as dead shoots, October–December; horizontal shading, loss as dead stems and leaves, August–October; blackened area, loss as dead leaves, October–December. The horizontal dashed lines enclose seasonal biomass. To determine net production, it is necessary to add the weight lost as dead leaves. This is done by measuring the areas of negative production (shaded areas). Net production calculated from this graph is 293 g m^{-2}. (From Mathews, C.P. and Westlake, D.F. 1969. *Oikos* 20: 156–160. Reprinted with permission.)

beneath the curve is proportional to net aboveground primary productivity. The area under the curve can be quantified gravimetrically, by planimetry, or by computer analysis. To obtain the overall production of the stand this process is repeated for each cohort of each species.

Mathews and Westlake (1969) applied the Allen curve method to a population of *Glyceria maxima* (manna grass) from an English wetland (Figure 6.3). The grass grew in monotypic stands comprised of more or less monthly cohorts. With the Allen curve method, the plants within study plots are tagged and then checked at intervals. The average dry weight of stems is determined from plants harvested outside of the study plots on each sampling date. The number of shoots is plotted against the mean dry weight in grams per plant. The product of the x- and y-axes for any point on a curve gives the biomass present for that time and density. In other words, the mean shoot biomass (g) is multiplied by shoot density (number of shoots per m^2) to give an estimate of g dry weight per unit area. The area under the curve is added to the area in the shaded areas of the graph (areas of "negative production"). The shaded areas of Figure 6.3 represent the weights of the dead leaves and shoots lost through the growing season. These areas are included in the estimate for the total net production of the cohort. This process is repeated for each cohort to obtain the stand's overall estimate of NPP.

Carpenter (1980b) adapted the Allen curve method for use with the submerged species, *Myriophyllum spicatum* (Eurasian water milfoil), *Potamogeton pectinatus* (sago pondweed), and *Vallisneria americana* (wild celery). The modification entailed measuring plants and forming Allen curves for two parts of the plants: shoots and branches. This method accounted for the sloughing of branches that occurs in the species that he was measuring.

His conclusion was that the method was reliable, although the results were probably underestimates for these submerged species.

Dickerman and others (1986) applied the Allen curve method to a stand of *Typha latifolia* in a Michigan wetland (Figure 6.4). *T. latifolia* shoots emerged in three distinct cohorts in both years of their study for a total of six cohorts. Production of each of the six cohorts was calculated individually using separate Allen curves. The NPP for each year was determined by summing the production values for each of the cohorts for that year. The results were an underestimate by 8 to 10% of the "best estimate" for this study, which was calculated by the summed shoot maximum method (see next section). The Allen curve method did not correct for early shoot mortality, which accounted for 4 to 11% of the emerging shoots.

The Allen curve method is appropriate only where cohorts are easily distinguishable and where the losses of branches or parts of the shoots can be monitored so that their biomass is included in the total (Carpenter 1980b; Dickerman et al. 1986; Wetzel and Pickard 1996). Leaf loss throughout the growing season can be included by using a short time interval between sampling dates (Cranford et al. 1989). Although some researchers feel the Allen curve method underestimates (Carpenter 1980b; Dickerman et al. 1986) or overestimates NPP (Bradbury and Grace 1993), Cranford and others (1989) argue that the method has a number of advantages:

1. Field and lab measurements are straightforward and relatively simple.
2. Weekly sampling is practical.

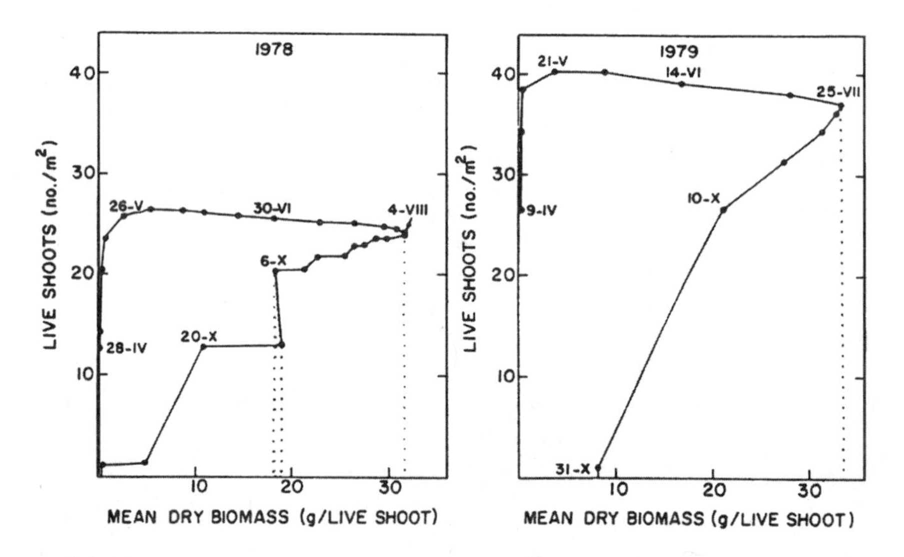

FIGURE 6.4
Representative Allen curves used for estimating production by *Typha latifolia* cohorts for 2 years in a marsh at Lawrence Lake, Michigan. In each graph, net annual aboveground production is proportional to the area bounded at the top by the solid line, on the right by the dashed vertical line, and on the bottom and on the left by the graph's axes. The row of double dots (left graph) indicates production occurring between October 6 (6-X) and October 20 (20-X), which must be added to calculate total production. Various sampling dates are indicated for reference purposes. The net production calculated by the Allen curve method in this study was 927 g m^{-2} yr^{-1} in 1978, and 1358 g m^{-2} yr^{-1} in 1979. (From Dickerman, J.A. et al. 1986. *Ecology* 67: 650–659. Reprinted with permission.)

3. A large number of shoots can be sampled and a different population is examined each time, which helps to account for heterogeneity within the marsh community.
4. Shoot loss during the growing season is accounted for and production of different cohorts can be separated if necessary.

In addition, the method provides useful information on growth and population dynamics that can be helpful in interpreting production estimates (Carpenter 1980b; Dickerman et al. 1986; Cranford et al. 1989; Wetzel and Pickard 1996). An Allen curve is a simple graphic method relating shoot numbers and the mean mass per shoot, and positive or negative changes in either of these two parameters are immediately apparent. In response to changes in the population, sampling effort can be optimized. More intensive sampling can be applied when rapid changes in the two parameters (shoot biomass and number of shoots) occur, and less sampling is needed when the population is less dynamic (Dickerman et al. 1986).

h. The Summed Shoot Maximum Method

In the summed shoot maximum method, plants are tagged and measured for height several times during the growing season. Their biomass is estimated using regressions of weight vs. height from plants harvested outside of the study plots. The maximum weight of each plant, regardless of when the maximum is achieved during the growing season, is used for the calculation of net primary production. The maximum masses of all shoots are summed and corrected for mean leaf turnover. The leaf turnover correction results in an increase in the net production result because it includes leaf death throughout the growing season.

The summed shoot maximum method and its modifications have been used in a number of saltwater (Bradbury and Hofstra 1976; Eilers 1979; Hardisky 1980; Cranford et al. 1989; Dai and Wiegert 1996) and freshwater (Dickerman et al. 1986; Fennessy et al. 1994a; Wetzel and Pickard 1996) marshes. According to some, the summed shoot maximum method provides a good estimate of NPP (Dickerman et al. 1986; Wetzel and Pickard 1996). However, it requires extensive field work and frequent monitoring. Daoust and Childers (1998) suggest that the intensive field work required can be decreased by measuring only a small sample of a total plot. They were able to estimate biomass to within 10% by measuring only 24 to 32% of the plants in 1-m^2 plots.

2. Belowground Biomass of Emergent Wetland Plants

Most wetland primary productivity studies provide results for only aboveground biomass. In reviews of several emergent macrophyte studies, Westlake (1975, 1982) found that root biomass is often two to five times the weight of the aerial parts.

A representative sample of root biomass is difficult to obtain because root biomass is extremely variable, both spatially and temporally (Gallagher and Plumley 1979). Root biomass may change seasonally in temperate zones, since translocation of aboveground material to the roots occurs in the fall and the reverse occurs in the spring. As a result, the same plant material may be measured as both above- and belowground biomass (Teal and Howes 1996). Changes in root biomass and in aboveground biomass occur at different times, making sampling for the maximum or minimum in root biomass a hit-or-miss process. In addition, taking soil cores to extract root samples probably leads to underestimates of biomass because of losses due to root exudation, sloughing of root hairs, rootcaps and cortical layers, and root grazers. Gas exchange methods, in which the uptake of carbon or the release of oxygen is measured, may be preferable since they do not rely on biomass data; however, they are complicated and difficult to compare to other studies (Blum et al. 1978; Drake and Read 1981; Houghton and Woodwell 1980; Singh et al. 1984).

Conversion factors, or *root-to-shoot ratios*, are sometimes used to estimate belowground biomass from aboveground biomass. No single conversion factor exists since the belowground/aboveground biomass ratio varies so widely depending upon species, moisture, and nutrient conditions (Bray 1963; Barko and Smart 1986). However, for broad comparisons among species, averaged over a number of sites, root-to-shoot ratios provide at least a rough idea of belowground biomass. Examples of some root-to-shoot ratios are 0.3 for annual species without rhizomes, 0.4 to 0.6 for *Typha latifolia*, 1.8 to 9.9 for *Phragmites australis*, and 2.3 to 3.9 for *Scirpus lacustris* (Westlake 1982).

We describe two methods for the measurement of belowground production: a harvest method and a decomposition method.

a. Harvest Method

To estimate belowground biomass in wetlands, sediment cores are taken, the material is washed and strained, and living matter is separated from dead. Belowground plant structures (roots, rhizomes, tubers, corms, and others) are dried and weighed and results are expressed as g dry weight per m^2. Complications in the harvest method arise for a number of reasons. Soil coring is difficult in wetlands because the soil is usually flooded and soft, the separation of organic material from soil is difficult, and the differentiation of live and dead tissues is extremely time-consuming (Schubauer and Hopkinson 1984). Coring devices need to be adapted for the specific conditions (Gallagher 1974). It is also difficult to sample often enough to keep track of quickly growing fine roots. In perennials, belowground biomass may include growth over several years, so it is difficult to give a rate of growth.

In some studies, belowground material is measured only once to give a rough estimate of the biomass (Reader and Stewart 1971). More frequent sampling can give an idea of the rate of growth (Valiela et al. 1976; Gallagher and Plumley 1979). NPP of roots can be estimated using calculation methods described for aboveground production, such as the Smalley method (Schubauer and Hopkinson 1984; Dame and Kenny 1986).

b. Decomposition Method

In the decomposition method, sediment cores are taken and aboveground organic matter is removed. The cores are sealed and incubated for 24 to 48 h. The production of carbon dioxide and methane is measured by drawing off gases from the core headspace and analyzing the samples with a gas chromatograph. Results are given in moles of carbon per square meter per time period (mol C m^{-2} d^{-1}). This measurement reflects the release of carbon through decomposition as well as the release of CO_2 by the respiration of living roots; however, the amount of carbon evolved due to respiration is assumed to be negligible. Results from this method showed annual belowground carbon mineralization to be between 60 and 67 mol C m^{-2} yr^{-1} in a Massachusetts *Spartina alterniflora* marsh (Howes et al. 1985). In terms of dry biomass production, this is roughly equal to 1560 to 1750 g m^{-2} yr^{-1}.

E. Floating and Floating-Leaved Plants

Methods for rooted floating-leaved plants are the same as for emergent plants since they can also be harvested, dried, and weighed. If plants are monitored in permanent plots, the leaf diameter of floating-leaved plants such as *Nymphaea odorata* can be measured and regressions of leaf diameter vs. dry weight can be established (Fennessy et al. 1994a). Because floating plants such as those in the Lemnaceae family may drift in and out of a permanent sampling site, collecting them and drying them may provide only a snapshot of productivity for the day on which the samples are taken, unless the quadrat has an enclosure.

F. Trees

Studies of tree primary productivity are based on a set of field measurements collectively known as *dimension analysis*. Study plots are located within the forest and a variety of nondestructive measurements such as height and diameter at breast height (dbh) are made. Biomass of trees is established by cutting down, measuring, and weighing the trunk, branches, and leaves of several representative sample trees. Dry weight measurements of the remaining trees are calculated from regressions of the biomass vs. one or more of the tree measurements (a detailed discussion of the development of tree biomass regression equations is given in Whittaker and Woodwell 1968). The object is to establish a statistically valid relationship between a comparatively small destructive sample and a larger nondestructive sample that is representative of the stand (Newbould 1970). When cutting down sample trees is not possible, regression equations from previous studies are used.

We describe the field measurements and calculations used for production estimates that are based on the biomass of trees and the rest of the forest community. While gas exchange methods have been used in forested wetlands (Golley et al. 1962; Lugo and Snedaker 1974; Brown 1981), we do not include them here.

1. Measures of Dimension Analysis

Foresters use dimension analysis to gauge the status of a forest with respect to wood products. The procedure normally includes more measurements than are given here. For primary productivity studies, the parameters of interest in dimension analysis are diameter at breast height and tree height.

a. Diameter at Breast Height

The diameter of the tree at breast height (dbh; breast height is defined as 1.3 m above the soil surface) is a basic measurement of forestry. The diameter of a tree is obtained by measuring the tree's circumference using a diameter tape or a tree caliper (Avery 1967; Husch et al. 1993). These instruments are calibrated in units of π so that the diameter can be read directly. Buttressed trees are often found in wetlands, and the diameter of these is measured 45 cm above the swell (Avery 1967; Conner and Day 1976; Ewel and Wickenheiser 1988). In mangroves, prop roots sometimes thrust the base of the main trunk far above the soil surface. In this case, the diameter is measured 1.3 m above the uppermost prop roots (Pool et al. 1977).

The dbh is the most frequently measured parameter in productivity studies. The biomass of unharvested trees is calculated by using a regression of dbh vs. biomass for harvested trees. To track the rate of growth, dbh is measured at an initial sampling time, and again after a time interval (usually 1 year). The difference in biomass between the two readings is reported as the wood production for that year. In some studies, aluminum vernier bands are installed on trees at breast height in order to track the changes and label the study trees (Mitsch and Ewel 1979; Conner et al. 1981; Conner and Day 1992).

b. Height

Instruments for measuring tree height are called *hypsometers, clinometers,* or *altimeters* (Avery 1967; Husch et al. 1993). The tree height measurement is based on the estimate of the angle from the measurer's eye level to the base and to the top of the tree and the lengths of the tangent of those angles (Figure 6.5). Height is measured in productivity studies because some regression equations relate both dbh and height to biomass (Mitsch et al. 1991). Height is also used with wood-specific gravity data to estimate productivity when harvesting is not possible (see Section II.F.3.a, Stem Production, Equation 6.18).

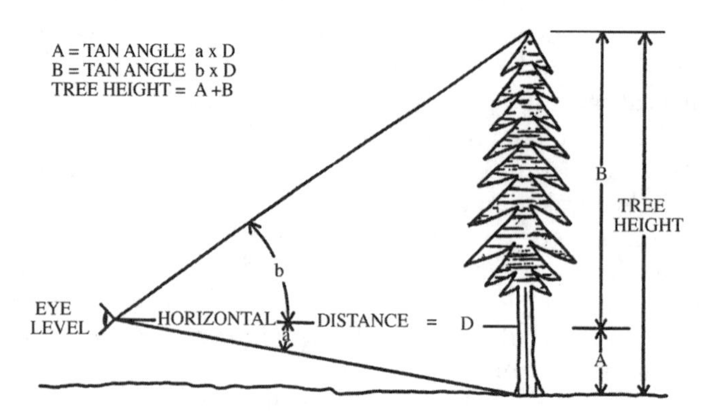

FIGURE 6.5
The principle of tree height measurement using a hypsometer. The observer's eye level intercepts the tree between stump height and tree top. The angular readings to the base and the top of the tree are added together to obtain the desired height value. (From Avery, T. 1967. *Forest Measurements*, p. 290. New York. McGraw-Hill. Reprinted with permission.)

2. Parameters Based on Dimension Analysis

The data collected in dimension analysis are used to calculate a number of forest community parameters such as basal area and basal area increment. These parameters, in turn, are related to productivity and are used in calculations of tree NPP (see Equation 6.18).

a. Basal Area

The basal area of a tree is the cross-sectional area of the tree at breast height. The basal area can be computed from the tree's diameter or circumference (Husch et al. 1993):

$$BA = \pi / 4 * d^2 \tag{6.15}$$

and since $d = c/\pi$,

$$BA = c^2 / 4 \pi \tag{6.16}$$

where

BA = tree cross-sectional area, or basal area in cm^2
d = diameter of cross-section in cm
c = circumference of cross-section in cm

The total basal area per unit area is the sum of the basal areas of all of the trees in the study plot (Husch et al. 1993). Basal area is usually used as an indicator of timber resources (Pool et al. 1977). It can be calculated for each species in a plot in order to determine dominance. Changes over time in the basal area of one species compared to the basal area of another species can reveal community changes.

b. Basal Area Increment

The basal area increment is the mean annual increase in wood area at breast height during the last 5 to 10 years (or other pertinent time period). It is determined from cores taken with an increment borer. An increment borer consists of a hollow cutting bit that is screwed into the tree. The core of wood that is forced into the hollow center of the bit is removed with an extractor (Husch et al. 1993). The ring widths for the last 5 to 10 years are measured and the average width is calculated. The basal area increment (A_i) is calculated as (Newbould 1970):

$$A_i = \pi \left[r^2 - (r - i)^2 \right] \tag{6.17}$$

where
- r = radius of tree at breast height
- i = mean radial increment per year

Basal area increment is used to estimate the past NPP of a tree (see Equation 6.18). In mangrove forests, tree rings are either not produced or are difficult to interpret since they may not be produced each year (Lugo 1997).

3. *Calculations of NPP of Trees*

The NPP of a tree is the portion of the biomass that is added during one growing season. It is calculated as the sum of production estimates for different portions of the tree: the stem (trunk), leaves, branches, and roots. The NPP of the trees in a community is the sum of the NPP values for the individual trees.

a. Stem Production

Stem production is generally the largest component of a tree's production in temperate wetlands. The relationship between a measured tree parameter (usually dbh) and wood biomass is established using harvested trees. The growth of wood is determined by the increase in the dbh of trees from one year to the next. The increase in diameter is converted to grams of wood by multiplying by the regression coefficient of wood biomass vs. the dbh (Golley et al. 1962; Newbould 1970; Mitsch and Ewel 1979; Conner et al. 1981; Brown 1981; Day et al. 1996).

The annual net stem production of the tree can also be calculated from tree height and the basal area increment if the specific gravity of the wood for that species is known, using the following equation (Brown and Peterson 1983; Mitsch et al. 1991):

$$P_n = 0.5 \rho A_i h \tag{6.18}$$

where
- P_n = annual net stem production (g dry weight yr^{-1})
- ρ = wood specific gravity (g dry weight cm^{-3})
- A_i = basal area increment (cm^2 yr^{-1})
- h = tree height (cm)

The specific gravity is the dry weight in grams of 1 cm^3 of fresh timber. Values of the specific gravity of wood for different species can be found in some foresters' manuals (e.g., U.S. Forest Products Laboratory 1974).

b. Leaf Production

Yearly leaf production is equal to the maximum dry weight of foliage present on the tree minus the minimum. In deciduous trees, the minimum is zero. Leaf biomass can be estimated from several representative samples taken throughout the year, or by litter fall collections every 1 to 4 weeks throughout the growing season and the autumn. Litter fall is collected in receptacles arranged throughout the community (usually randomly). Receptacles are cloth or mesh bags, trash cans, buckets, or containers that are specially designed for the community of interest (Brown and Peterson 1983; Day et al. 1996). The receptacles should drain freely to reduce moisture and losses to decomposition. One year of sample collection usually suffices for deciduous trees and 3 to 5 years are recommended for evergreen species. Either method provides an underestimate of NPP since some leaves or leaf parts are not measured due to herbivory or loss.

c. Branch Production

The branch biomass of harvested trees is measured by drying and weighing the branches. Regressions of the branch biomass vs. the diameter of the stem just below the joint of the lowest main branch or of the branch biomass vs. the basal diameter of each branch can be used to estimate the branch biomass of unharvested trees (Whittaker and Woodwell 1968; Newbould 1970). The production of new growth on each branch is the change in dry weight of the branch from one year to the next.

d. Root Production

Roots can be excavated and measured and weighed directly (Mitsch and Ewel 1979). The fine roots may be lost in this process so some root production as well as losses due to organic root secretions, death, and consumption are missed. The change in root biomass from one year to the next is the NPP. Alternatively, belowground production can be estimated from aboveground production using the following relationship (Newbould 1970):

$$AP / AB = k\,(BP / BB) \qquad (6.19)$$

where
AP = aboveground production
AB = aboveground biomass
BP = belowground production
BB = belowground biomass
k = a constant

The value of k is established by harvesting trees and roots and calculating root-to-shoot relationships (Whittaker 1966). Where harvesting is impossible, some researchers simplify further and make k equal to 1; however, this may not yield a valid estimate of root production (Whittaker and Woodwell 1968; Newbould 1970). Tree root biomass may vary with the hydrologic regime, with lower root-to-shoot ratios under continuously flooded conditions than under periodically flooded conditions. In studies of forested wetlands, estimates of root NPP provide information about the changes in biomass allocation (between the roots and the stem) that occur with changes in the hydrologic regime (Megonigal and Day 1992).

Many forested wetland studies do not include root production estimates, perhaps because of the following complications involved in sampling roots (Powell and Day 1991):

1. Production and mortality occur throughout the year, so periods of growth and decline are not as easy to distinguish as for aboveground biomass.

2. It is difficult to accurately distinguish live roots from dead.
3. Sampling dates may not coincide with the peaks and troughs in the seasonal pattern of growth, so the maximum and minimum are often missed. This can lead to both under- and overestimates of production.

It is particularly important to include root biomass in studies of mangrove primary productivity. The prop and drop roots of *Rhizophora* species can constitute 30 to 40% of a tree's biomass (Fromard et al. 1998).

4. Community Primary Productivity of Forested Wetlands

Forested wetland productivity studies sometimes include the NPP of the understory plants: shrubs, herbaceous vegetation, mosses, liverworts, vines, epiphytes, and floating or submerged vegetation (Reiners 1972; Schlesinger 1978; Conner and Day 1976; Conner et al. 1981; Grigal et al. 1985). The dry weight of clippings of herbaceous or shrub strata of the forest are determined once or several times throughout the growing season. In most forested wetland studies, the simple methods (usually peak biomass) used to estimate the NPP of understory components provide only a moderately reliable estimate of the NPP of the forest community (see Case Study 6.B, Mangrove Productivity: Laguna de Terminos, Mexico).

G. Shrubs

Primary productivity methods for shrubs are similar to methods for trees. Since shrubs are perennials, the challenge is to determine how much of the plant's biomass is from the current growing season. Methods for a detailed analysis of shrub NPP are given in Whittaker (1962) and Whittaker and Woodwell (1968). In published studies of wetland shrubs, less detailed methods have been used (Reader and Stewart 1971, 1972; Schlesinger 1978; Schwintzer 1983; Bartsch and Moore 1985).

Reader and Stewart (1971, 1972) studied the primary productivity of five ericaceous shrubs in a Manitoba wetland. They monitored permanent plots and determined the dry weight of the new growth of twigs, leaves, flowers, and fruit on shrubs weekly. Wood growth was determined by harvesting shrubs, determining weight and age from rings, and assuming an equal production of wood for every year of growth. The total of the weekly new growth plus the estimated annual production of woody tissue was multiplied by the number of shrubs of each species for an estimate of NPP. The average NPP of the five species ranged from 31 to 106 g m^{-2} yr^{-1}.

Using a similar method, Schwintzer (1983) determined the NPP of *Myrica gale* (sweet gale), a common shrub of peatlands. Stem production and NPP were estimated in the same manner as in the Reader and Stewart studies (1971, 1972). However, in addition to clipping and weighing leaf biomass, Schwintzer corrected for leaf loss before harvest based on leaves collected in litter buckets. Her results were 392 g m^{-2} yr^{-1} for aboveground NPP and 549 g m^{-2} yr^{-1} with belowground NPP included.

In a Georgia cypress swamp, Schlesinger (1978) determined the average NPP of small trees (<4 cm dbh) and several shrub species. He estimated the growth of the leaves, twigs, fruits, and flowers by clipping, drying, and weighing the various plant parts. The result was 53 g m^{-2} yr^{-1}. To account for stem and branch growth, he doubled the figure to yield 106 g m^{-2} yr^{-1}. This conversion was based on factors for converting clipping weight to NPP given in an article by Whittaker (1962; factors ranged from 1.5 to 2.3). In a study of shrubs in Quebec peatlands, Bartsch and Moore (1985) estimated a year's growth from foliar biomass alone. They assumed that woody production per year was negligible. Their result was 54 g m^{-2} yr^{-1}.

H. Moss

The measurement of moss NPP (mostly *Sphagnum* species) is challenging because there is seldom a clear division between live and dead plants. In order to calculate the productivity of *Sphagnum*, the following parameters are determined (Grigal 1985; Rochefort et al. 1990):

- *The surface area of the species.* The surface area of each moss species is measured directly. It is important to measure the depressions and hummocks formed by the peat since these add to the area of growth. Detailed methods for areal measurements are given in Grigal (1985).
- *The rate of linear growth.* Several methods for the measurement of linear growth have been devised (Clymo 1970). In a frequently used method, a wire with both horizontal and vertical portions (shaped like an old-fashioned car crank) is used. The bottom vertical portion is pushed into the moss and the horizontal portion is level with the moss surface. The top vertical portion of known length projects above the surface and into the air. After a time interval (usually of weeks or months), the portion of the top vertical wire that remains visible is measured. The difference between the total length and the remaining visible length is the linear growth (in cm; Figure 6.6).

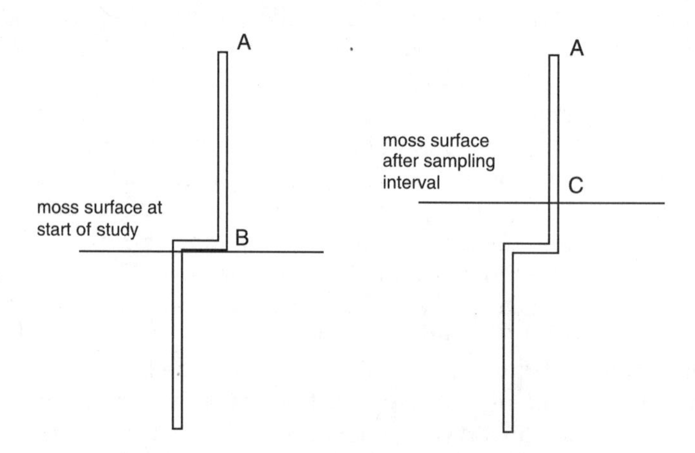

FIGURE 6.6
To estimate the rate of linear growth of moss, a wire with both horizontal and vertical portions is inserted into the moss mat. At the beginning of the sampling period, the bottom vertical portion is pushed into the moss and the horizontal portion is level with the moss surface. The top vertical portion of known length (the distance from point A to point B; usually 10 cm) projects above the surface into the air. After a time interval (usually of weeks or months), the portion of the top vertical wire that remains visible is measured (the distance from point A to point C). The difference between the total length and the remaining visible length is the linear growth in centimeters (the length of line AB minus the length of line AC). (Drawn according to descriptions given in Clymo 1970.)

- *The relationship of linear growth to mass increment.* Since results for moss growth are in centimeters of linear growth per unit of time, the weight per centimeter is determined by drying and weighing representative samples and using the average value.

From these parameters, production is calculated as length increment (cm) multiplied by the mean mass of moss per unit depth (g cm^{-1}). The product is divided by the surface area (m^2) for an estimate of g dry weight m^{-2}. For NPP, the production is expressed per unit time. For an example of the importance of moss primary productivity in peatlands, see Case Study 6.C, Peatland Productivity: Forested Bogs of Northern Minnesota).

Summary

Primary productivity is an important parameter in wetland studies. The level of primary productivity in some wetlands is as high or higher than for managed agricultural crops. Primary productivity measurements allow researchers to make comparisons among sites and to compare results before and after a disturbance. The net primary productivity (NPP) of a wetland includes the productivity of the phytoplankton, periphyton, submerged plants, floating plants, floating-leaved plants, emergent plants, woody plants, and moss.

Methods for measuring the NPP of phytoplankton include measuring the oxygen produced and the carbon fixed in a certain time period. Periphyton productivity may be estimated in the same way, or by collecting, drying, and weighing samples. Macrophytes of all growth habits may also be collected, dried, and weighed, and their dry biomass per unit area gives an estimate of the NPP for the site.

For the aboveground biomass of emergent wetland plants, methodology is as important to the results of NPP studies as the site-specific conditions that influence plant growth. Any attempt to compare NPP results should take into account the varying methods. In general, the best results are considered to come from methods that include live, dead, and decomposed biomass. The peak biomass method and the Milner and Hughes method measure live biomass only, and are considered to provide underestimates of NPP in most cases. Other methods account for plant losses throughout the growing season and may more closely approach "true" NPP. The Valiela et al. and Smalley methods usually yield underestimates of NPP. The Wiegert and Evans and Lomnicki et al. methods provide either good estimates or overestimates of NPP, depending upon wind and water transport of litter into or out of the study area. The Allen curve method has received mixed reviews. It is considered to provide a good estimate of NPP by some, but others think it over- or underestimates NPP. It is only applicable to plants that grow in easily distinguishable cohorts. The summed shoot maximum method is considered to provide a good estimate of NPP by those who have tested it against other methods.

The NPP of trees is measured using dimension analysis, which includes measuring dbh, tree height, and other parameters. Included in the total estimate for each tree are the stem, branches, leaves, and roots. In mangrove forests, it is important to include the biomass of aerial roots as they constitute a significant portion of the trees' biomass. The NPP of shrubs may be estimated by monitoring the new growth of twigs, leaves, flowers, and fruit. Wood growth may be determined by harvesting shrubs, determining weight and age from rings, and assuming an equal production of wood for every year of growth. The NPP of moss is measured using at least three parameters: the surface area of the moss, the linear growth of the moss, and the relationship of linear growth to mass increment.

Case Studies

6.A. Salt Marsh Productivity: The Effect of Hydrological Alterations in Three Sites in San Diego County, California

California has lost the vast majority of its original salt marshes and the few that remain often have altered hydrology. Incoming tidal waters from the sea and freshwater influxes from upland areas are restricted by obstructions and channelization. Reducing or eliminating a wetland's water inflow can bring about drastic changes in salinity and nutrient availability, which in turn leads to profound changes in plant and animal species composition. The primary productivity of the wetland may be either reduced or enhanced in response to the changed hydrology.

To find out what effect hydrological alterations had on southern California salt marshes, Zedler and others (1980) chose three sites in San Diego County in which to compare primary productivity (all about 32°N, 117°W). The first site is the Tijuana Estuary, the least disturbed coastal wetland in the county. The Tijuana Estuary is open to tides and fresh-water inputs and has three elevations: low, medium, and high, with the highest elevation the farthest from the ocean. The second site is Los Penasquitos Lagoon, where the hydrology was altered by a railroad in 1888 and further restricted by the construction of a road in 1932. Tidal waters flush the lagoon only at certain periods of the year. A sand bar blocks the tides during the summer, but it is usually washed away by rainfall or high tides during the winter and spring. The third site is the San Diego River Flood Control Channel. The Channel's opening to the ocean is indirect because a sand bar blocks the river's mouth and a dike separates the channel from an adjacent bay. The tides circulate there all year, but the dike slows both seawater and freshwater flow.

The plant community in western salt marshes is more diverse than the usually monotypic stands of *Spartina alterniflora* in the eastern U.S. Some of the species in these marshes were *Spartina foliosa, Salicornia virginica, Distichlis spicata*, and *Frankenia grandifolia*, with the greatest diversity at the Tijuana Estuary (Zedler 1977). The aboveground portions of live and dead plants and litter were harvested in 0.25-m^2 quadrats during three growing seasons. The NPP was calculated according to the Smalley (1959) method.

The results for the least disturbed site, Tijuana Estuary, showed a range of peak biomass between 800 and 1300 g m^{-2}, with no discernible pattern related to elevation or species composition (Table 6.A.1). In the Flood Control Channel, where the flows of seawater and freshwater are reduced from natural levels, the results for both peak biomass and NPP were similar to the medium elevation at the Tijuana Estuary. High salinity levels may limit primary productivity at this site. The highest biomass and primary productivity were in 1978 in the lower marsh of Los Penasquitos Lagoon. This was attributed to high rainfall (more than twice the average) for that year. In the upper marsh of the lagoon, the NPP was lower. Of course, the upper marsh also received the high rainfall, but evaporation and seepage lowered the standing water level rapidly, resulting in drier conditions.

Can conclusions be drawn concerning the influence of hydrology in these marshes? The authors point out that Los Penasquitos Lagoon, which had the least tidal circulation, had the highest live biomass, the most litter, and the highest NPP. Over 90% of the NPP was attributed to live biomass increases, so the accumulation of dead material that can occur with limited tidal flushing was not responsible for the high NPP. The emergent vegetation at this site seemed to benefit during the periods when the marsh was shut off from the tides. Freshwater runoff accumulated within the marsh when the sand bar was in place, resulting in lower salinity. A combination of low salinity and nutrient release from

TABLE 6.A.1

Peak Biomass (g m^{-2}) and Annual Aboveground Net Primary Productivity (g m^{-2} yr^{-1}) in Three Salt Marshes of San Diego County, California

Location and Elevation	Year	Peak Biomass	NPP
Tijuana Estuary (least disturbed)			
Low	1976	914	935
	1977	1099	916
	1978	1032	
Medium	1976	818	539
	1977	1035	724
	1978	925	
High	1976	869	412
	1977	1269	1046
	1978	920	
Los Penasquitos Lagoon (occasional tides)			
Lower	1977	2666	
	1978	4316	2858
Upper	1977	1879	
	1978	1458	1202
San Diego River Flood Control Channel (reduced flow)	1977	799	
	1978	970	599

Note: NPP was not calculated every year of the study. Peak biomass is greater than NPP because a portion of the peak biomass was from the standing stock of the previous year.

Data from Zedler et al. 1980.

decomposing litter probably contributed to the higher biomass and NPP at the Los Penasquitos Lagoon.

If reducing tidal circulation can bring about increased primary productivity, can we conclude that altering wetland hydrology is actually favorable to wetlands? No. In the first place, increased primary productivity is of no intrinsic value in terms of beneficial wetland functions or biodiversity. Further, the authors suggest that restricting the tide may bring about several negative consequences.

First, they suggest that restricting the tides is no guarantee that salinity will be lowered. In fact, hypersalinity could result in years of low rainfall. Fluctuations in salinity from year to year may lead to the elimination of some species from the marsh. *Spartina foliosa* was present in Los Penasquitos Lagoon in the 1940s, but has been replaced by *Salicornia virginica*, which can survive high salinities and quickly colonize newly opened habitat. *Spartina foliosa* is a desirable species because it provides habitat for the endangered light-footed clapper rail (*Rallus longirostris levipes*).

Second, algal mats are an important part of the productivity of natural southern California marshes. At Los Penasquitos Lagoon, the algal mats are restricted to open areas because they are shaded by the dense vegetation and thick litter accumulation. Many salt marsh invertebrates consume algae, so invertebrate numbers may be reduced when algae is limited.

Third, the productivity in the open channels of the salt marsh may increase with lowered tidal inputs. However, the dominance shifts from phytoplankton to floating algal mats. The algal mats shade the water column below and restrict water column productivity. Without tidal dilution, organic matter accumulates in the channels, leading to increased biochemical oxygen demand. The stagnant water in the channel aggravates the situation. As temperature increases, the solubility of oxygen decreases, evaporation increases, and salinity rises. Long periods of stagnation can lead to lower dissolved oxygen concentrations and eventually to invertebrate, fish, and shellfish death. Such effects have been observed at Los Penasquitos Lagoon.

This study of primary productivity served as a basis for further studies of these marshes and also helped to illuminate processes within the marshes that were influenced by hydrological alteration.

6.B. Mangrove Productivity: Laguna de Terminos, Mexico

Long-term primary productivity studies are relatively rare. They are of particular interest in forested wetlands, where tree growth is slow, because long-term trends may not be revealed by only 1 or 2 years of data. Inter-annual differences may be due to site-specific characteristics such as salinity, or hydrology, or to climatic factors such as rainfall or temperature (Day et al. 1996). Hoping to identify the causes of year-to-year variability of productivity, John Day, Jr. and seven other researchers (1996) spent 7 years examining the primary productivity of a mangrove forest bordering Laguna de Terminos, a large (1800 km^2), shallow (mean depth 3.5 m), coastal lagoon on the southern coast of the Gulf of Mexico in the state of Campeche, Mexico (18° 40'N, 91° 80'W). The area has a humid, tropical climate with 1680 mm of average annual rainfall and three seasons: dry season from mid-February to early June, wet season from June through October, and the 'norte' or cold front season from November to February. Two distinct forest types border a channel of the lagoon: a fringe forest 20 m wide with *Rhizophora mangle* (red mangrove), *Avicennia germinans* (black mangrove), and *Laguncularia racemosa* (white mangrove), and a basin forest inland from the fringe forest, 155 m wide and dominated by *A. germinans* (Table 6.B.1). Within the basin forest were three zones with large trees in the outer two zones (Zones I and III) and smaller trees in the middle zone (Zone II; Table 6.B.2).

Measurements were made in 1000-m^2 plots (two plots in each of the four zones) along a transect perpendicular to the channel. Net primary productivity was calculated as the sum of annual wood growth plus annual litterfall. The authors did not calculate NPP for the fringe forest, but we made a rough estimate from the average proportion of NPP that litterfall represented in the basin forest (72%). Wood growth was determined from changes

TABLE 6.B.1

The Distribution of Mangrove Species within Zones of Tidal Influence around Laguna de Terminos, Campeche, Mexico

Site	Location	Hydrology	Species
Fringe forest zone	Adjacent to channel about 20 m wide	Regularly inundated drains completely on each side	*Rhizophora mangle* *Avicennia germinans* *Laguncularia racemosa*
Basin forest	25–180 m inland from channel	Continuously flooded during wet season; often dry in dry season	*A. germinans*

From data in Day et al. 1996.

TABLE 6.B.2

Location of Study Zones within the Mangrove Forest of Laguna de Terminos

Zones of Growth	Distance from Channel	Average dbh	Average Height
Fringe forest	0–20 m	8.1	8–9 m
Zones within basin forest			
I	25–60 m	7.3 cm	6–7 m
II	60–120 m	4.8 cm	3–4 m
III	120–180 m	7.8 cm	6–7 m

Note: The average dbh and height in the fringe forest is for all three species found there. In the basin forest, *Avicennia germinans* dominated.

From data in Day et al. 1996.

in dbh and regressions of dbh vs. biomass. The dbh data were grouped by 2.5-cm-diameter classes and the number of trees per hectare in each diameter class was multiplied by the predicted biomass increase for that diameter class. The sum of the biomass increases for all diameter classes gave the total net wood production. Litterfall was collected monthly from five litter traps per plot and dried and weighed.

The annual average litterfall measurements were highly variable over the 7-year study period; however, there were discernible spatial and temporal trends (Figure 6.B.1). The highest amounts of litterfall were in the fringe forest, and the lowest in Zone II of the basin forest. Most of the litterfall occurred during the dry and rainy seasons with up to 50% less

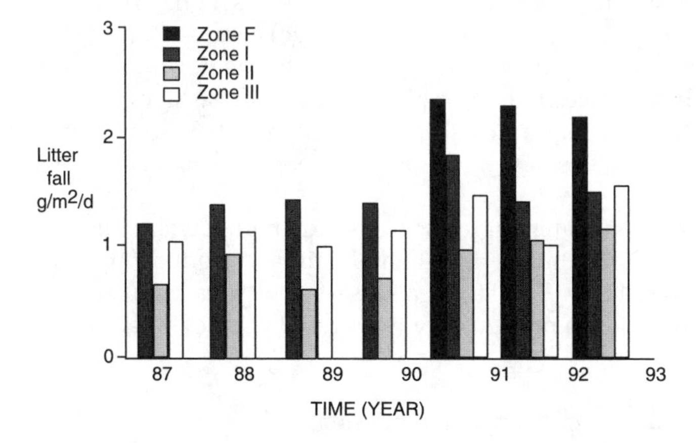

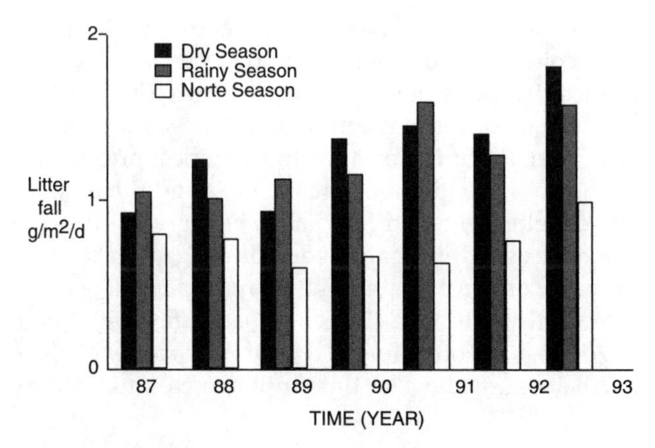

FIGURE 6.B.1
Inter-annual variation of total litter fall: (a) by zone and (b) the mean by season of both basin and fringe mangrove forests. Each bar is the mean for each year for the respective zone or season. (From Day, J.W. et al. 1996. *Aquatic Botany* 55: 39–60. Redrawn with permission.)

TABLE 6.B.3

Average Results of Litterfall and Aboveground NPP (g m^{-2} yr^{-1}) at Laguna de Terminos

Zones of Growth	Litterfall	NPP
Fringe forest	793	1100[a]
Zones within basin forest		
Zone I	496	695
Zone II	307	399
Zone III	410	612

[a]Gross estimate, not calculated by Day et al., but from the proportion of NPP that was litterfall (72%) in the basin forest.

From data in Day et al. 1996.

falling in the 'norte' season. Significantly lower rates of litterfall were measured during the first 3 years of the study than in subsequent years. NPP was highest nearest the channel in the fringe forest and within the basin forest it was highest in Zone I and lowest in Zone II (Table 6.B.3).

Interannual differences in litterfall were best explained by a combination of factors: average soil salinity, minimum air temperature, and minimum rainfall. Litterfall decreased with increasing average soil salinity, decreasing minimum air temperature, and decreasing minimum rainfall. Others have established that salinity stresses mangrove growth (Lugo et al. 1988; Twilley et al. 1986). Interstitial salinity within Zone II is higher than in the rest of the mangrove forest (60 to 75 ppt vs. 40 to 55 ppt in the fringe) because Zone II does not drain well. Salt water remains standing there for long periods during the rainy season. The productivity level of Zone II resembles that of scrub mangrove forests in other sites. The fringe area and Zones I and III of the Laguna de Terminos mangrove have greater flow and better drainage, which lead to lower salinity, and as a consequence, higher litterfall and wood growth.

This study supports the observation that mangrove forests with higher water turnover (riverine > fringe > basin; Lugo et al. 1988) also have higher primary productivity, even given pronounced interannual variability. The study also serves as a caution to those tempted to extrapolate long-term trends in productivity from 1 to 2 years of data.

6.C. Peatland Productivity: Forested Bogs of Northern Minnesota

The landscape of northern Minnesota is a patchwork of lakes and wetlands and the area's cold climate has created the appropriate conditions for the formation of peatlands. Grigal et al. (1985) carried out a primary productivity study in six bogs in northern Minnesota to determine the effects of different hydrologic and nutrient conditions. A bog is a peatland that is isolated from mineral-influenced water. Precipitation and atmospheric inputs are the primary sources of nutrients. The researchers studied three perched bogs and three raised bogs (Figure 6.C.1). *Raised bogs* usually develop on broad flat plains. Peat gradually accumulates there due to a rise in the water table as a result of impeded drainage. *Perched bogs* lie in small depressions in glacial moraines or outwash plains. They form as a result of gradual accretion of peat from the edge of open water toward the center (*paludification*). In the western Great Lakes region, both types of bogs are often forested with *Picea mariana* (black spruce) and *Larix laricina* (tamarack). All six of the bogs in this study were dominated by

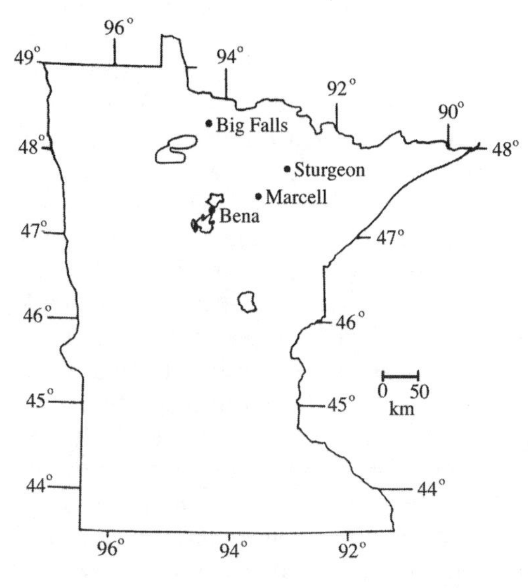

FIGURE 6.C.1
The location of three perched and three raised forested bogs in Minnesota. The three perched bogs are all within a few kilometers of each other at the site marked Marcell. The raised bogs are the Bena bog, the Sturgeon bog, and the Big Falls bog. (From Grigal et al. 1985. *Canadian Journal of Botany* 63: 2416–2424. Reprinted with permission.)

P. mariana with between 1 and 3% cover of *L. laricina*. The dominant overstory trees on the raised bogs were about 75 years old, and about 110 years old on the perched bogs.

The researchers measured the primary productivity of all of the plant components of the community, including trees, shrubs, herbaceous vegetation, and moss. To estimate the productivity of trees, they measured the dbh of all the trees in their study plots and took increment cores of a subsample of the trees. The dbh was related to biomass using regression equations established in a previous study (Grigal and Kernik 1984). Wood productivity was determined from the basal area increment and its relation to biomass. Litterfall was collected in five traps per bog, set 1 m above the bog surface, and the results were added to wood

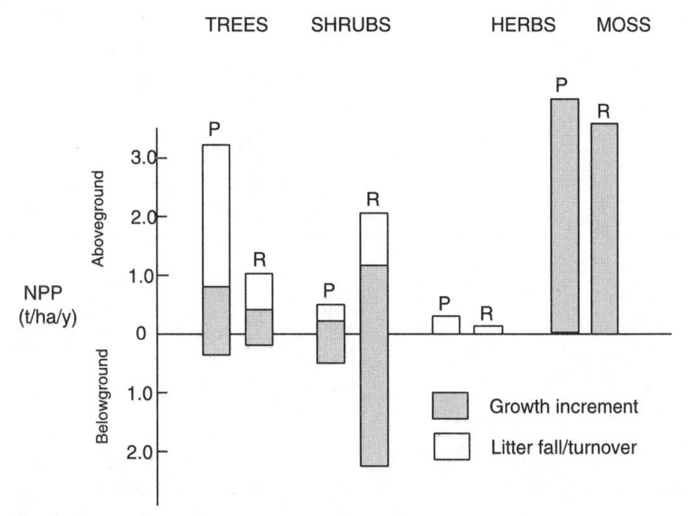

FIGURE 6.C.2
Net primary productivity for three perched (P) and three raised (R) bogs in Minnesota. Results in metric tons per hectare per year (t ha^{-1} yr^{-1}) can be multiplied by 100 to convert to g m^{-2} yr^{-1}. (From Grigal et al. 1985. *Canadian Journal of Botany* 63: 2416–2424. Redrawn with permission.)

productivity for total tree NPP. The understory consisted of *P. mariana* seedlings and two species of shrubs: *Ledum groenlandicum* (Labrador tea) and *Chamaedaphne calyculata* (leatherleaf). Their productivity was estimated by harvesting and drying seedlings and shrubs that were representative of the various size classes and determining growth per year using age rings. The underground productivity of the trees and shrubs was estimated from harvested samples. Root-to-shoot ratios were established by drying and weighing the samples and the ratios were multiplied by the aboveground production. The herbaceous plants were *Smilacina trifolia* (three-leaved Solomon's seal) and several species of *Carex* (sedge). They were clipped at the moss surface at peak biomass. The growth of the moss (all *Sphagnum* species) was determined as an increase in length of living tissue using wire cranks inserted into the *Sphagnum* mat (Clymo 1970). Grigal (1985) took samples of *Sphagnum* and determined dry weight per centimeter of length. These results were multiplied by the area of *Sphagnum* cover and divided by the time interval for an estimate of productivity.

The combined productivity of the various plant forms in the perched bogs was greater than in the raised bogs (Table 6.C.1). However, in the raised bogs the productivity of the understory seedlings and shrubs was higher than in the perched bogs (Figure 6.C.2), probabably because the canopy was more open. The results of the *Sphagnum* study showed higher productivity in the perched bogs, but productivity in both types of bogs depended on the position of the *Sphagnum* within the landscape. *Sphagnum* productivity in the low areas, or hollows, of both bog types was greater than on raised areas, or hummocks (520 g m^{-2} in hollows vs. 320 g m^{-2} on hummocks in perched bogs and 370 g m^{-2} on hollows and 300 g m^{-2} on hummocks in raised bogs). The hollows receive more water and greater nutrient inputs.

The productivity of the trees in these bogs is relatively low, but overall NPP is substantially increased by the relatively high productivity of the *Sphagnum*. Low productivity is typical in nutrient-poor bogs, but perched bogs occasionally receive runoff from surrounding mineral soils, and can therefore support greater plant productivity. This study illustrates that even small differences in hydrology among wetlands can significantly affect both the structure (more open canopy in the raised bogs) and function (higher productivity in the perched bogs) of wetlands.

TABLE 6.C.1

Mean Aboveground Biomass and Productivity of the Overstory and Understory Woody Strata, the Herbaceous Vegetation, and the *Sphagnum* Moss in Three Perched Bogs and Three Raised Bogs in Northern Minnesota

Vegetative Strata	Perched	Raised	Significance
Overstory woody stratum biomass			
Trees (*Picea mariana*)	10,073	3,098	**
Overstory woody stratum productivity			
Wood growth	83	45	**
Litterfall	231	54	**
NPP wood + litterfall	314	99	**
Understory woody stratum biomass			
Picea mariana seedlings	6	40	*
Ledum groenlandicum	80	288	n.s.
Chamaedaphne calyculata	17	167	n.s.
All species total	103	495	*
Understory woody stratum productivity			
Picea mariana seedlings	1	5	*
Ledum groenlandicum	20	70	n.s.
Chamaedaphne calyculata	6	52	n.s.
All species total	27	127	*
Litterfall	17	74	n.c.
NPP for understory litterfall + total	44	201	n.c.
Herbaceous vegetation productivity	22	14	n.s.
Sphagnum productivity	380	320	n.c.
Total productivity per year	760	634	n.c.

Note: Data for biomass are in g dry weight m^{-2} and data for productivity are in g dry weight m^{-2} yr^{-1}.

* Significant difference at $p < 0.05$;

** Significant difference at $p < 0.01$;

n.s. = no significant difference;

n.c. = not calculated.

Data from Grigal et al. 1985.

7

Community Dynamics in Wetlands

I. An Introduction to Community Dynamics

Plant communities change over time; the temporal scale for change ranges from a single growing season to many years. The study of community dynamics encompasses the many possible changes in the distribution and abundance of a species and the reasons for these changes in community structure. Changes in plant community composition, called *ecological succession*, are a result of both internal and external processes. Internal processes include competition between plants as well as the accumulation of peat. External processes include climatic or topographic changes, such as those due to glaciation. In wetlands, the most important external processes are usually associated with changes in water depth, flow rate, period of inundation, and water chemistry.

In this chapter, we discuss the definition of ecological succession and the development of successional theory during the 20th century. We discuss models of succession that have been applied to wetlands and we describe studies that have supported or refuted these models. We also describe the role of the seed bank in the formation of wetland communities. An important factor in community dynamics is competition among plants. We discuss theories of plant competition and their application to wetland plant communities. Natural disturbances, such as fire, flooding, and drought, as well as human-induced disturbances, also play an important role in community dynamics. Disturbances can remove species or inhibit their growth, or they can open areas where new species may become established.

II. Ecological Succession

Traditionally, succession has been defined as changes in the community structure of an ecosystem (discussions of ecological succession are often limited to plants) in which each new community has been thought of as a step, or *sere*. Often, the seres exhibit a predictable community structure. Over time, the seres eventually lead to a *climax community*, i.e., one that is stable and in which the species are long-lived and persist for many generations with no discernible changes in community structure. However, over long time periods, even so-called climax communities are mutable. Recognizing that both seral and climax communities are variable, and that many abiotic and biotic factors influence their structure, the definition of ecological succession has been somewhat broadened. Today, ecological succession may be defined as a change in the species present in a community (Morin 1999).

Changes in community structure come about for a number of reasons, including internal processes, such as competition among species, or herbivory, and external ones, such as natural or anthropogenic disturbances. In wetlands, a change in the plant community structure is often the result of a change in hydrology. Hydrologic changes are brought about by forces within the community (*autogenic*), such as the accumulation of peat, or by forces outside of the community (*allogenic*), such as the change of a river's course, an increased sediment load, or the breach of a sandspit that shelters a coastal wetland.

Two general types of succession are *primary* and *secondary* succession. Primary succession occurs where no plants have grown before, such as on newly exposed glacial till or on volcanic mudflows. In wetlands, primary succession occurs when a new area of wet soil is formed, such as a new deltaic lobe at the mouth of a river (see Case Study 7.A, Successional Processes in Deltaic Lobes of the Mississippi River). In unplanted constructed wetlands that do not occur on the site of previous wetlands, primary succession is also at work. Secondary succession occurs where a natural community has been disturbed, such as on abandoned agricultural fields. In wetlands, secondary succession occurs as a community recovers from a disturbance that opens large areas, such as a fire or a hurricane. Secondary succession also occurs when wetlands are restored. An example is the restoration and re-establishment of prairie potholes following the removal of tile drains from farm fields (Galatowitsch and van der Valk 1995, 1996).

Theories of ecological succession evolved throughout the 1900s, as ecologists added theory and data to the body of knowledge regarding plant communities and ecosystems. We briefly discuss the history of successional theories in the following three sections on holistic and individualistic approaches, the replacement of species in plant communities, and developing and mature communities.

A. Holistic and Individualistic Approaches to Ecological Succession

Much of the early theory regarding ecological succession was driven by the work of two ecologists, F.E. Clements and H.A. Gleason, near the beginning of the 1900s. Clements (1916) originated the *hypothesis of organization through succession* in which it is thought that whole communities or ecosystems are self-organizing entities of plants and animals. In studying prairie succession, Clements came to believe that communities operated cooperatively and that groups of plants functioned together. He likened the growth of a community to the ontogeny of an organism and put forth the concept of a *superorganism*, a group of organisms that migrated, reproduced, lived, and died together. The community moved through succession in a predictable series of steps, called *seres*, toward a climax community. The climax community, which was in large part a function of local environmental conditions, was thought to endure because the species replaced themselves and persisted without the invasion of new species. Under Clements' holistic hypothesis, succession is an autogenic process with each seral community preparing the environment for the next.

Gleason (1917) was one of the first public opponents to Clements' ideas. He proposed the *individualistic hypothesis of succession*, which states that each individual organism in a community is present due to its unique set of adaptations to the environment. From his perspective, changes in the community were brought about by allogenic forces and the response of each individual to the changes. In Gleason's view, any change in the relative abundance of a species or in the composition of the community was a successional change. Since environmental conditions change from year to year, even from season to season, the existing set of plants is variable and in flux, as species adapt to new conditions or are eliminated from a site (van der Valk 1981). In this view, the life history traits of an individual and the set of environmental conditions present at a site determine whether that species will become established. In order for a plant to become established, its propagules must reach a site where the conditions for germination and growth are suitable.

In the early part of the 20th century, Clements' ideas won widespread support, while Gleason's did not gain a foothold until the second half of the century. In 1953, Whittaker suggested that the theory of a single climax community for a specific region or set of environmental conditions was untenable. He thought that ecological succession was more complicated than Clements had believed it to be. He asserted that there is no absolute climax community for any area and that climax composition has meaning only relative to the specific set of topographical, edaphic, and biotic conditions at each site. He wrote that the species in both seral and so-called climax communities correspond to environmental gradients and that community diversity reflects the diversity of the environmental conditions.

In 1952, Egler proposed the *initial floristic composition hypothesis* in which secondary succession depends on the plant propagules present at the site before a disturbance. This hypothesis was counter to Clements' idea that plants and animals functioned together as a unit. Instead, the individual species' survival depended on the presence and longevity of its own propagules, and not on the propagules of a set of species.

B. The Replacement of Species

A part of the theory of ecological succession focuses on how one species replaces another in a community. Clements (1917) proposed that colonizing species in early successional communities have a net positive effect on later colonists; i.e., their presence facilitates the arrival of later species. For example, a newly opened area may be lacking in nutrients. Early colonizers are often nitrogen fixers, able to compensate for the lack of inorganic nitrogen. As these plants grow and decompose, they enhance the soil's fertility, thus facilitating the arrival of later species.

Connell and Slatyer (1977) added to Clements' idea and proposed that early colonists may have two other possible net effects on later ones: negative and null. They suggested that the three basic types of interaction between early species and later ones include: (1) *facilitation,* in which early species create a more favorable environment for the establishment of later species; (2) *tolerance,* in which there is no interaction between early and later species; and (3) *inhibition,* in which early species actively inhibit the establishment of later ones. Interactions with herbivores, predators, and pathogens were also of cricital importance in succession, but were outside of the scope of their model. They recognized that the three proposed mechanisms of interaction were extremes in a continuum of effects of earlier on later species (Connell et al. 1987).

It is interesting to note instances in wetland plant species replacement in which these mechanisms are at work. For example, plants with greater radial oxygen loss (see Chapter 4, Section II.A.5, Radial Oxygen Loss) facilitate the growth of other species with less oxygen loss. *Spartina maritima* aerates the surface sediments in salt marshes of southern Spain, making conditions favorable for the invasion of *Arthrocnemum perenne,* a rapidly spreading prostrate plant (Castellanos et al. 1994). In U.S. east coast salt marshes, *Juncus maritimus* increases the sediment redox potential and opens the way for the growth of *Iva frutescens,* a woody perennial (Hacker and Bertness 1995). In field trials with freshwater species, a non-aerenchymous wetland plant, *Salix exigua,* was planted with and without *Typha latifolia.* When planted with *T. latifolia,* which releases oxygen from its roots into the soil, *S. exigua* was able to survive flooded conditions. However, when planted alone, *S. exigua* was unable to tolerate the anoxic soil (Callaway and King 1996).

The second species replacement mechanism, tolerance of new species, is probably the most frequently seen mechanism. It occurs whenever a new species successfully colonizes a site without either the facilitation or inhibition of earlier species.

Some early colonists may inhibit the arrival of later species by shading the substrate or by the production of allelopathic phytochemicals (see Section IV.C, Allelopathy). In wetlands, some submerged and emergent plants can intercept light and keep it from reaching the substrate, thereby impeding the germination and growth of other species' seeds. Species that are able to quickly regenerate from stored reserves in the spring are able to establish a vegetative cover before the shoots of other species appear. *Glyceria maxima* (manna grass) has been observed to impede the growth of new *Phragmites australis* (common reed) shoots in this way. The rapid early growth of *Ruppia cirrhosa* (wigeon grass) has similar negative effects on *Potamogeton pectinatus* (sago pondweed; as reviewed in Breen et al. 1988).

C. Developing and Mature Ecosystems

An alternative view of ecological succession was put forth by Eugene Odum in 1969. He focused on the development of whole ecosystems, rather than on the replacement of species. Odum viewed succession as an orderly pattern of community development. Like Gleason, Odum did not believe that species grouped together to form recognizable superorganisms. However, like Clements, he suggested that the species in each community function together. His main emphasis was on ecosystem functions such as primary productivity and respiration as well as on other whole-system attributes such as the type of food chain, the amount of organic matter present, species diversity, mineral cycling, spatial heterogeneity, and the species' life cycles. Ecosystems were labeled as either developing or mature according to Odum's general schema. For example, he wrote that in a developing ecosystem, the food chain is linear, while in a mature ecosystem, the food chain is more complicated, better described as a food web, and detritus is an important component.

Odum's schema seems to adequately describe the succession of an open field to a forest community; however, he allowed that it did not fit wetlands, in which fluxes in hydrology, whether due to daily tides or seasonal changes, strongly affect community composition as well as ecosystem function. Mitsch and Gosselink (2000) provide a detailed analysis of Odum's description of ecological succession as it applies to wetlands. They conclude that wetlands display some features that are characteristic of developing ecosystems, while at the same time having features that are characteristic of mature systems. For example, the ratio of primary productivity to respiration is often greater than 1 in wetlands, a characteristic of developing ecosystems, while detrital-based food webs dominate, a characteristic of mature ecosystems.

Odum (1969) also described a concept called *pulse stability*, in which ecosystems are subject to more or less regular but acute physical disturbances imposed from outside the system. Pulse stability may describe ecosystem development in many wetlands better than the concept of developing and mature ecosystems. Regular disturbances maintain ecosystems at an intermediate point in the developmental sequence, resulting in a compromise between the developing and the mature ecosystem. Odum's examples of systems operating under pulse stability are wetlands with fluctuating water levels such as estuaries and intertidal zones, or systems adapted to periodic fires. Mangrove forests, subject to periodic hurricanes, and in the northern part of their range, to frost, seem to be maintained in a steady state by the pulsed nature of these disturbances (Lugo 1997).

III. Ecological Succession in Wetlands

Much of the study of ecological succession has focused on terrestrial ecosystems, namely, forests and old-field communities. Not all of these theories are applicable to wetlands, or they may only partially explain successional processes there. In many wetlands, abiotic factors, with hydrology chief among them, outweigh biotic factors (Mitsch and Gosselink 2000).

A. Models of Succession in Wetlands

Perhaps the most well-known model of succession in wetlands is the *hydrarch model,* in which wetlands are thought to be a seral community in the succession of an open freshwater lake to a terrestrial community. This model is concerned with ecosystem development and the accumulation of sediments that, in theory, lower the water table and open the area for the establishment of upland species. The hydrarch model has also been applied to both salt marshes and mangroves; however, as in freshwater wetlands, the theory has not been supported by research. Another model, called the *environmental sieve model,* is concerned with species replacement and the mechanisms that allow for species' arrival and establishment (van der Valk 1981).

1. Hydrarch Succession

Hydrarch succession is an autogenic process that begins with open water and purportedly ends, perhaps centuries later, with upland vegetation (Lindeman 1941; Gates 1942; Conway 1949; Dansereau and Segadas-Vianna 1952). In the final sere, an upland community fills a previous lake basin. It is the last step that has not been observed in nature. In the theory of hydrarch succession, sediments and peat accumulate on the lake bottom (Figure 7.1). Detritus accumulates slowly at first through the decomposition of algae, and then, as the lake becomes more shallow and suitable for the growth of submerged plants, detritus begins to accumulate more quickly. With more organic sediments and a shallower lake, emergent plants are able to grow. Their decomposition adds to the peat, and the lake becomes a marsh. Eventually woody plants along with *Sphagnum* moss are able to grow. They further lower the water table through higher evapotranspiration rates. A wet forest community can move in as the substrate becomes drier. The tenet of hydrarch succession, that upland communities form in former lake basins through autogenic changes, has not been upheld. Despite changes toward a drier community, the outcome is still a wetland, rather than an upland community.

Some lakes may have filled and become terrestrial habitats; however, it seems that the process was not caused by the internal accumulation of peat, but by allogenic changes in the water table. Allogenic processes that lower the water table, such as landslides, volcanoes, glaciation, and earth movement, may change flow patterns sufficiently to bring about the development of an upland community in a previous wetland or lake (Larsen 1982).

Autogenic changes do occur with the accumulation of plant matter and the gradual filling of lake basins. In some wetlands, it seems obvious that the edge community is closing in on the open water. We can stand at the edge of some lakes and feel the spongy peat below us, jump up and down and watch the trees around us shake, and know that we are on a *quaking bog,* in which the peat forms a cushion above the water (Figures 7.2a and b). Some plants, such as *Decodon verticillatus* (swamp loosestrife), seem particularly adapted to moving from the edge into open water, gradually increasing their area (Figure 5.19).

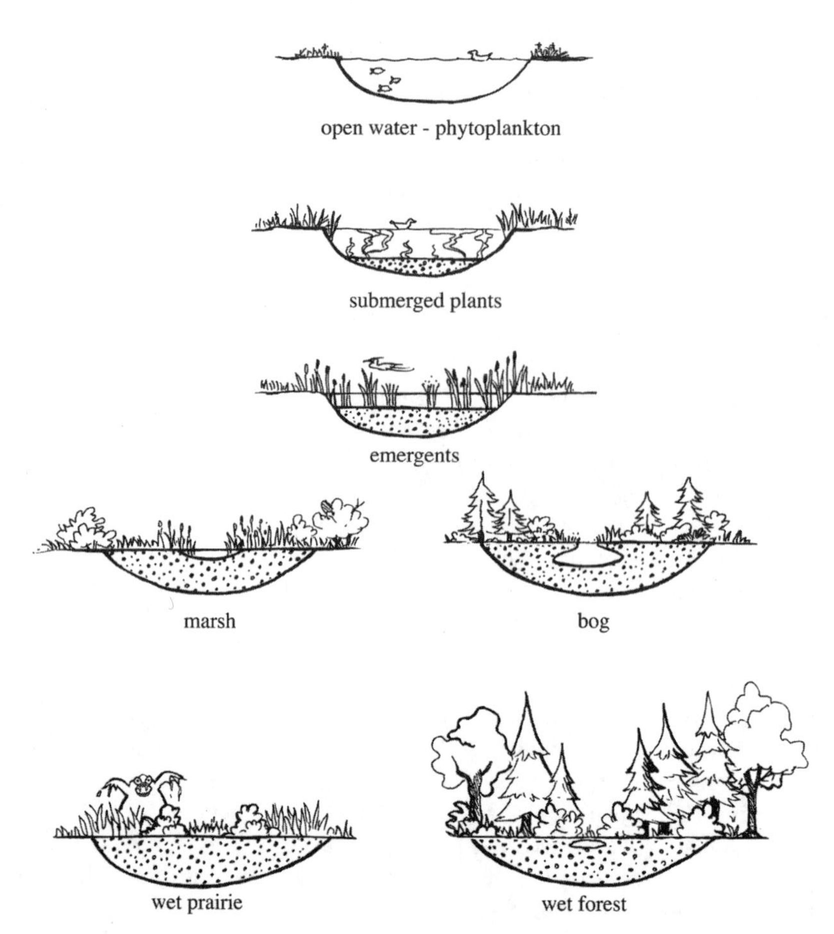

FIGURE 7.1
A classical view of hydrarch succession in which a lake slowly fills with detritus from the decomposition of algae, then from decomposed submerged plants, and later decomposed emergent plants and moss. The community eventually becomes drier. In theory, an upland forest is the climax state. However, this set of events rarely occurs in nature, and if filling does occur, the most likely ultimate stage is a wet prairie or wet forest rather than an upland community. (From Weller, M.W. 1994. *Freshwater Marshes Ecology and Wildlife Management*, p. 154. Minneapolis. University of Minnesota Press. Redrawn with permission by B. Zalokar.)

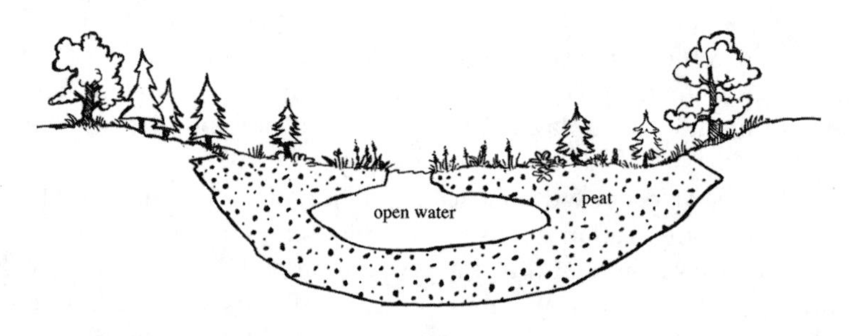

FIGURE 7.2a
A quaking bog seen in profile with peat closing in toward the center of the lake, still underlain by open water. (Drawn by B. Zalokar.)

FIGURE 7.2b
A peatland in northern Wisconsin in which the vegetated area seems to be engulfing the area of open water. As peat accumulates around the edges, larger plants such as emergents, shrubs, and trees are able to gain a foothold. However, barring any allogenic change in hydrology, a quaking bog such as this one remains a bog, rather than becoming an upland forest. (Photo by H. Crowell.)

The reason that upland communities do not result from autogenic changes is that the accumulation of peat only occurs under anoxic conditions. If oxygen is present, decomposition is enhanced and peat does not accumulate as rapidly as it does in wetlands. When organic peats are drained, they become oxidized and they subside. As peat accumulates and approaches the upper limit of the saturated zone, the rate of peat accretion becomes less than the rate of subsidence. The accumulation of peat ceases, and the peat layers do not continue to grow up out of the saturated zone. Without an outside force that lowers the water level, the peat will remain saturated, unable to support terrestrial vegetation (Mitsch and Gosselink 2000).

Heinselman (1963, 1975) studied peat accumulation in the Lake Agassiz region of northern Minnesota (the site of a former glacial lake that is currently characterized by lakes, bogs, and upland areas). He concluded that peat accumulation did not result in lake filling and the arrival of upland plants. Rather, peat grew upward and laterally, encroaching upon the forested land in a process called *paludification*. Wetland forests underlain with layers of peat indicate that entire watersheds in the region were subject to paludification. Heinselman found evidence that one lake in the region, Myrtle Lake, rose along with the surrounding peat, but remained an area of open water (Figure 7.3). Logs found in the peat indicated that trees had once inhabited the area, but were unable to persist in the nutrient- and oxygen-poor peat substrate.

Succession in the Lake Agassiz region was a complicated process, without a single model such as the autogenic accumulation of peat to adequately describe community development in each watershed. The processes involved included: (1) climatic changes, which led to increases or decreases in decay rates and changes in the regional flora and fauna, as well as the development or thawing of permafrost; (2) burning of peatlands and bog forests; (3) geologic factors such as erosion or uplift, which may eliminate peatlands by improving drainage; (4) flooding, often caused by beaver dams; (5) extensive plant

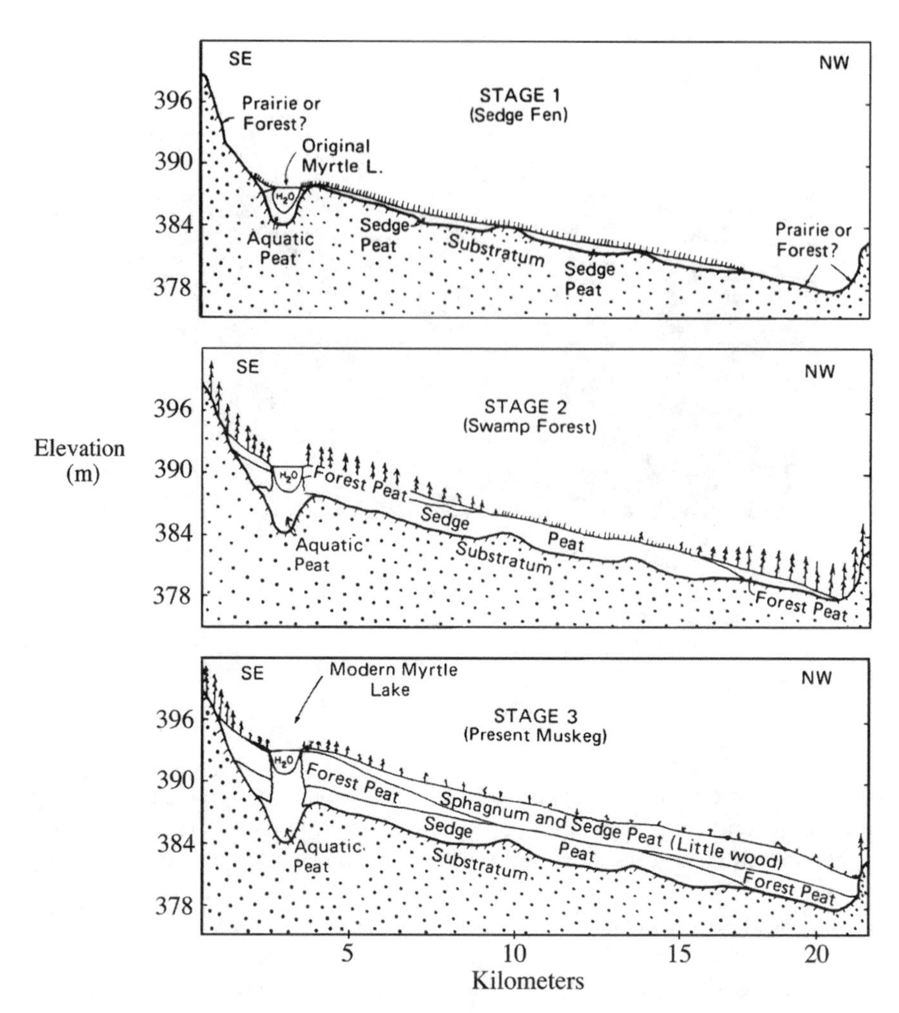

FIGURE 7.3
An idealized image of the stages of succession surrounding Myrtle Lake in Minnesota. In the first stage, the lake was surrounded by prairie or forest on the upslope side and by a sedge fen at the downslope side. In stage 2, detritus had accumulated in the lake bottom, keeping pace with paludification both up- and downslope from the lake. Parts of the peat blanket were inhabited by *Picea, Larix,* and *Thuja,* all trees of northern peatlands. In the current, or third, stage, the forest is overlain with *Sphagnum* peat, which extends approximately 10 miles from the lake. Heinselman (1963) calls the area a muskeg, which he defines as a large expanse of *Sphagnum* bearing stunted *Picea mariana* (black spruce) and *Larix laricina* (tamarack) as well as ericaceous shrubs. The lake is still open water, and the elevation of the lake bottom is over 20 ft higher than in stage 1. The development of the current peatlands, lakes, and forests in the Lake Agassiz region has taken from 9,200 to 11,000 years. (From Heinselman, M.L. 1963. *Ecological Monographs* 33: 327–374. Reprinted with permission.)

migration in the postglacial period; (6) human influences such as logging, agriculture, drainage, burning, and blocking drainage for the construction of roads. In general, most of the processes led to bog expansion in the Lake Agassiz region, with no consistent progress toward upland systems. As the bog surface has risen, so has the water table. In many areas where mesophytes formerly grew, bog and fen species have replaced them.

Similarly, in the northeastern U.S., the replacement of forested bogs by upland communities has not been observed (Damman and French 1987). In bogs of southern New

England, a wetland tree, *Chamaecyparis thyoides* (Atlantic white cedar), often replaces or surrounds *Osmunda–Vaccinium* (fern and shrub) communities. The bog mat surrounding the trees often consists of *Sphagnum* moss and *Chamaedaphne calyculata* (leatherleaf). Farther north, bogs are encircled by *Thuja occidentalis* (northern white cedar), but they are not inhabited by upland species.

In 15 oxbow lakes of different ages in the Pembina River valley of Alberta, Canada, newly formed lakes developed plant communities that progressed in the general sequence from submerged communities, to floating-leaved and emergent communities, to a sedge meadow, and eventually to a shrub and forest community (van der Valk and Bliss 1971). The trees and shrubs were wetland species, such as *Salix bebbiana, S. lutea,* and *Betula pumila* var. *glandulifera* with an understory of *Glyceria grandis, Urtica major,* and various species of *Aster*. Ultimately, *Populus balsamifera* (balsam poplar), a facultative wetland species, grew in some of the oldest oxbows (Figure 7.4). The mechanism at work appeared to be hydrarch succession, with the accumulation of peat and the arrival of longer-lived woody species. However, the climax community in the region is a wet forest and succession stopped short of an upland climax. The rate of succession from one community type to the next was variable and setbacks were frequent due to periodic flooding.

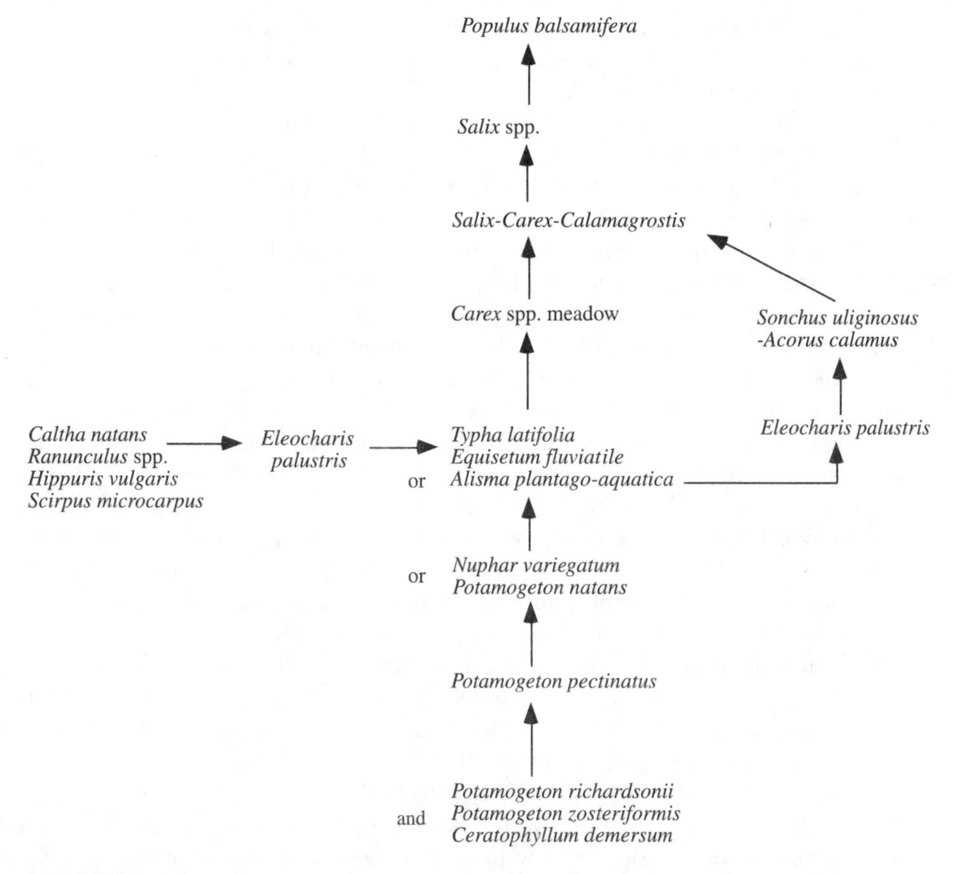

FIGURE 7.4
The successional pattern in oxbow lakes of the Pembina River valley in Alberta, Canada showing a change from submerged communities (at the left and bottom of the diagram) to emergent plants (in the center), to *Carex* (sedge) communities, ultimately leading to *Salix* (willow), and then *Populus balsamifera* (balsam poplar) forests. (From van der Valk, A.G. and Bliss, L.C. 1971. *Canadian Journal of Botany* 49: 1177–1199. Redrawn with permission.)

2. Succession in Coastal Wetlands

While the classic idea of hydrarch succession was developed for freshwater depressional ecosystems, the same model, i.e., that wetlands eventually become uplands, has been suggested for the succession of coastal systems as well. In the salt marshes of Louisiana, the following pattern of successional replacement was thought to be at work: open water → salt marsh → fresh marsh → swamp forest → wetland trees → upland dwelling live oaks and loblolly pine (Penfound and Hathaway 1938). However, studies since then have not supported the idea that coastal wetlands are replaced by upland ecosystems.

Both the accretion of land and its subsidence are important factors in the development of a salt marsh. The accretion of peat and sediments must equal sea level rise in order for salt marsh vegetation to become established. When accretion is less than subsidence, the wetland plant community remains in place and a move toward an upland sere is not possible. In New England salt marshes, accretion and subsidence have been at work for the last 3000 to 4000 years (Niering and Warren 1980). Before this time, the postglacial rise in sea level was approximately 2.3 mm/year. Sea level rise decreased to about 1 mm/year after that. This allowed sediment accretion to keep pace with the rise in sea level. Stands of *Spartina alterniflora* (cordgrass) were able to persist near the shore. As sediments accumulated on the landward side of the salt marshes, the elevation rose above mean high water and allowed less flood-tolerant species, such as *Spartina patens* (salt-meadow cordgrass), to colonize the area. This accretion and subsidence resulted in the salt marsh communities we see in New England today.

Redfield (1972) described the development of Barnstable Marsh on Cape Cod in Massachusetts (Figure 7.5). Barnstable Marsh developed as a result of several allogenic and autogenic factors. While tidal influences were the most significant environmental factor in the zonation of salt marsh vegetation, other factors were also important such as the physiology of the local vegetation, sedimentation, and changes in sea level relative to the land. In Barnstable Marsh, land increased in area from the landward side, by the erosion of cliffs, and from the seaward side, by the entrainment of sediments by tides that were subsequently trapped in the peat. The growth of land has been balanced by a rise in sea level.

In Barnstable Marsh, where sediment accumulated at a greater rate than the rise in sea level, *Spartina alterniflora* spread across the sediments (everywhere that the marsh's elevation exceeded the lower limit at which the plant can survive). In other locations, where the sea level rose in excess of sediment accretion, marshes drowned, were eroded away, or were buried by sediments. In Redfield's study site, a sandspit that protected the marsh from tides expanded during the last 4000 years. The marsh area grew in size in part due to the sandspit's increased size. Sand and silt accumulated behind the sandspit so that, in spite of the rising sea level, the water became shallow enough for *S. alterniflora* to extend its stands seaward, forming islands on the higher sand flats. The islands fused, forming peninsulas of intertidal marsh that later built up to become high marsh (which supports *S. alterniflora* and *S. patens*). The marsh's development was dependent on sedimentary processes that built up the sand flats to the level at which *S. alterniflora* could grow. Over time, the *S. alterniflora* community has proven to be very stable; the succession of this area from salt marsh to upland community has not occurred.

Like salt marshes, mangrove forests were once thought to be a sere in the development from coastal waters to upland systems with succession being driven primarily by autogenic forces. The seral stages started at sea level and advanced in the following order: sea (seagrass) → mangroves → strandline or freshwater swamp → terrestrial system (Davis 1940; Chapman 1976).

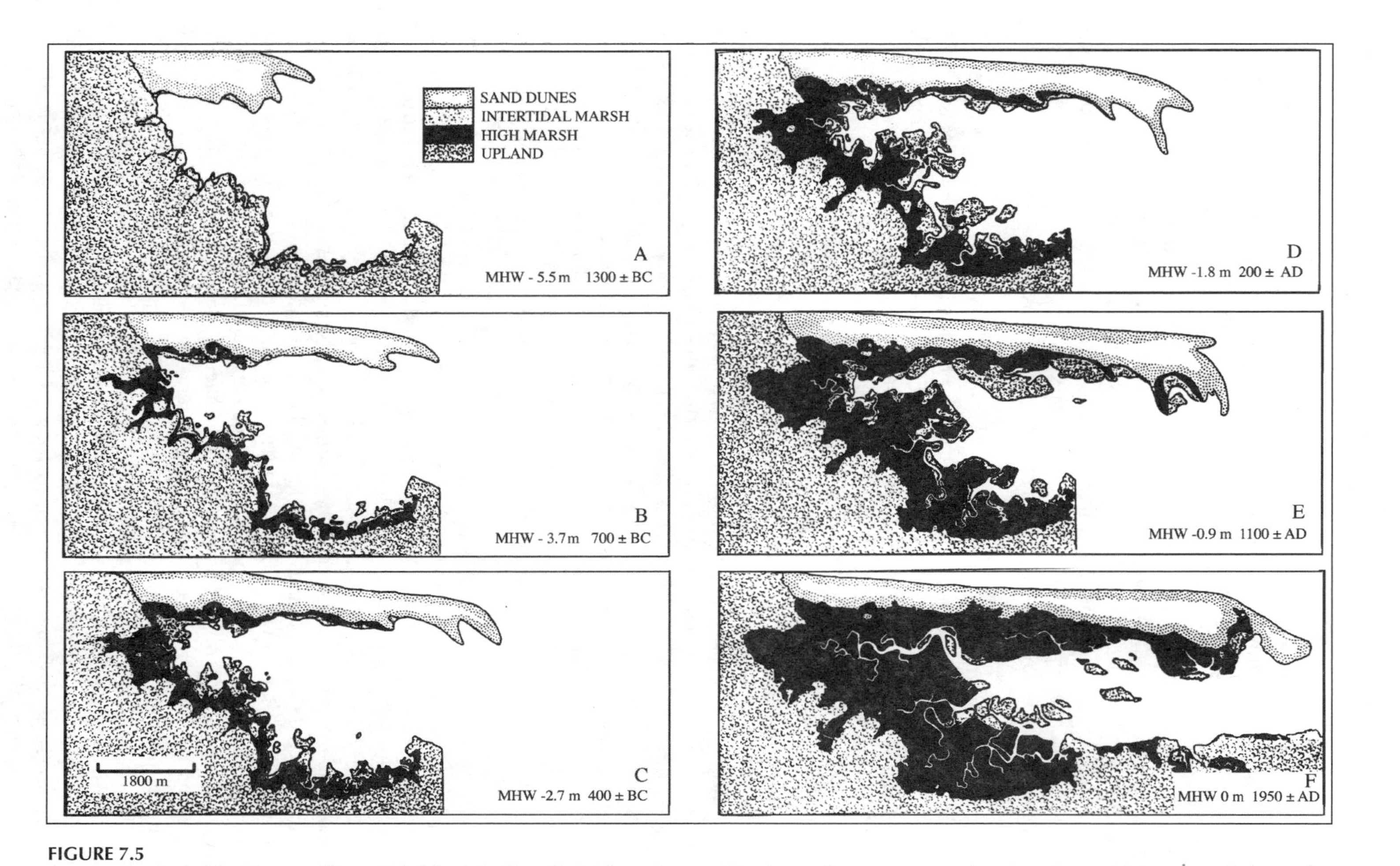

FIGURE 7.5
A reconstruction of the history of Barnstable Marsh in Cape Cod, Massachusetts. The date and contemporary elevation of mean high water, relative to the 1950s elevation, are indicated in the lower right-hand corner of each drawing. (From Redfield, A.C. 1972. *Ecological Monographs* 42: 201–237. Reprinted with permission.)

The replacement of mangrove forests by terrestrial communities has not been supported (Johnstone 1983). Distinct zonation does occur in many mangrove forests, with species best adapted to tidal fluxes nearest to the shore. These same species are often absent on the landward side of the mangrove forest where they are unable to compete with species growing there (Figure 2.12). It was thought that mangroves progress by growing seaward and that upland species typical of tropical or pine forests eventually replace mangroves on the same land. However, mangrove zones do not necessarily correlate to successional stages. The replacement of seaward species with those from farther inland has not been observed. As in salt marshes, the zones are maintained by environmental forces such as sea level rise and accretion of sediments, and succession is periodically reset to an earlier stage by hurricanes or other disturbances (Lugo 1980). The structural development, rate of change, and age expectancy of mangroves vary widely according to the environmental setting in which they are found (Cintrón et al. 1978; Lugo 1997).

3. The Environmental Sieve Model

van der Valk (1981) proposed a model of succession for freshwater wetlands that is rooted in Gleason's individualistic hypothesis (1917) and Egler's initial floristic composition hypothesis (1952). His model is based on life history features of the species involved. Three important life history traits are used to determine plant community composition under either flooded or drawndown conditions: life span, propagule longevity, and propagule establishment requirements. The model is a simple one, with the only possible allogenic change, or "environmental sieve," being the presence or absence of standing water. The degree of flooding permits the establishment of only certain species at any given time. In the model, succession occurs whenever one or more species become established, extirpated, or when both occur simultaneously. When the sieve changes in response to water level changes, different species become established. One of the model's assumptions is that competition and allelopathy cannot result in the extirpation of a species. In order to predict which species will become established under different hydrologic conditions, it is necessary to know about the life history of the plants that are likely to colonize the wetland.

van der Valk separates wetland species into three groups based on their life span: (1) annuals (A), which include mudflat species (ephemerals) that become established only during drawndown periods; (2) perennials (P), which may or may not reproduce vegetatively, but which have a limited life span; and (3) vegetatively reproducing perennials (V), the most prevalent type among wetland plants. van der Valk further divides wetland species into two groups according to propagule longevity and availability. One group has long-lived propagules that are in the wetland's seed bank and can become established whenever the conditions are right, called seed bank or S species. The second group, called dispersal dependent species or D species, has short-lived propagules or seeds that can only become established if the propagules reach the wetland during a period of suitable hydrologic conditions. The plants in all of these categories are further classified according to the seeds' germination requirements. Species with seeds and seedlings that can only become established when there is no standing water are called Type I species. Type II species become established only when there is standing water.

Combining the three classifications for wetland plants, there are 12 potential life history types (AS-I, AS-II, AD-I, AD-II, PS-I, PS-II, PD-I, PD-II, VS-I, VS-II, VD-I, VD-II). van der Valk gives examples of wetland species and how they fit into these categories. *Typha glauca* is a VS-I species, a perennial that reproduces vegetatively and becomes established from seeds in the seed bank only during drawdowns. *Phragmites australis* is classified as a VD-I plant, i.e., a vegetatively reproducing perennial whose seeds, which only germinate under drawndown conditions, are not present in the seed bank. *Bidens cernua* (AS-I) is a mudflat annual whose long-lived seeds are common in the seed bank and germinate during drawndown conditions. *Najas flexilis* (AS-II) is a submerged annual whose seeds are present in the seed bank and only germinate when flooded.

By changing the state of the hydrologic sieve, only Type I or Type II species can become established at any time (Figure 7.6). The S species are difficult to eliminate because they have seeds in the seed bank, while D species are eliminated once the hydrologic conditions are unsuitable for them. In order to apply the model to real wetlands, it is necessary to identify species in the seed bank. Information about each species' life history is also required in order to categorize species according to life span and propagule longevity. Once all of the species and potential species are categorized, one can predict the composition of the vegetation during future drawdowns or flooded periods.

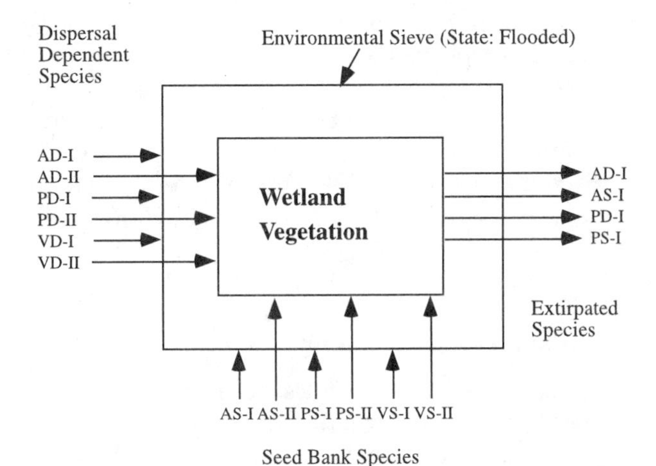

FIGURE 7.6
van der Valk's sieve model of wetland succession. The establishment and extirpation of species in this model are a function of the hydrologic regime which behaves as a sieve that alternates between two states: drawdown (without standing water) and flooded (with standing water). In this figure, the wetland is flooded. As a result, only the species with the proper life history features can become established in the wetland. Other species, because they are not adapted to flooded conditions, may be extirpated. When the wetland is drawndown, another set of species may become established while the set shown passing through the sieve will be extirpated. See the text for an explanation of the 12 life history types (i.e., AD-I, AS-I, etc.). (Redrawn from van der Valk, A.G. 1981. *Ecology* 62: 688–696. With permission.)

The model is qualitative because it cannot predict abundance. In addition, it ignores autogenic factors such as competition. Nonetheless, it can be used to predict the composition of the plant community under different hydrologic conditions. The model is based on hydrologic changes that are seen in depressional wetlands like prairie potholes and it does not apply to wetlands with tidal influences.

Additional filters may be added to the general model for some wetland types. For example, fire is a frequent and important disturbance in some wetlands and only fire-tolerant species persist after fires. Kirkman and others (2000) proposed a successional sequence using hydrology and fire as environmental sieves, or filters, for forested depressional wetlands of the southeastern U.S. Three different vegetation types were identified in these wetlands in southwestern Georgia. The first, called a cypress-gum swamp, is dominated by *Taxodium ascendens* (pond cypress) and *Nyssa biflora* (black gum). The second, a cypress savanna, is characterized by an open canopy of *T. ascendens* and a mixed grass-sedge ground cover. The third, a grass-sedge marsh, has no distinct overstory, but sometimes contains some *Pinus elliotii* (swamp pine) or *T. ascendens*. Their model of succession is an attempt to determine how each of these three vegetation types comes to dominate in any given depressional wetland in the area.

In their model, the potential frequency of fire increases as water depth and duration of flooding decrease (Figure 7.7). Conditions suitable for woody plant establishment depend on hydrologic conditions. For example, drawndown conditions are necessary for seed germination, and must be long enough for plants to grow sufficiently to survive subsequent fire and/or inundation. Seed dispersal must correspond with the prolonged drawdown. This combination of conditions may occur only infrequently at any given site. As a result, many depressional wetlands in the area are dominated by herbaceous plants. Therefore, the climate and its effects on water level provide the conditions for the establishment of either predominantly woody or predominantly herbaceous vegetation. Fire further determines the community composition. In areas with frequent fires that kill both cypress and hardwoods, a grass-sedge marsh results. With fires of intermediate frequency in which only hardwoods are killed, a cypress savanna results. Infrequent fires do not filter out hardwoods or cypress.

B. The Role of Seed Banks in Wetland Succession

In wetlands, as in other ecosystems, propagules may arrive from outside or they may already be present in the water or substrate. In secondary succession, the species composition of the seed bank can help determine the structure of the plant community. The seed bank can indicate which species will become established after a disturbance or when conditions are suitable for their germination. Seed banks are tested by observing the species that germinate from soil samples of a known volume taken from a study site.

The dispersal of seeds also determines which wetland species are found at any given site. Many wetland species have broad geographic ranges, while some are endemic to specific sites or areas. Dispersal agents such as water, air, birds, fish, and ocean currents can determine wetland flora (Leck 1989).

1. The Relationship of the Seed Bank to the Existing Plant Community

In general, the number of species represented in the seed bank reflects the diversity of the community. There are usually more seeds per square meter and a greater diversity of seeds in freshwater wetlands than in saline wetlands. Older wetlands tend to have a greater number of seeds than newly formed wetlands. With these generalizations in mind, it is important to note that there can be a great deal of variation among wetlands. It is possible to find low diversity and low seed numbers in freshwater wetlands, particularly in cold climates. In a review of 22 wetland seed bank studies, the results ranged from 0 to 59 species and from 0 to 377,041 seeds per square meter. The wetland with the poorest seed bank was an Alaskan floodplain, while a West Virginia bog had the largest seed bank with the greatest density of seeds. The greatest diversity was found in a seed bank from a South Carolina swamp (Leck 1989).

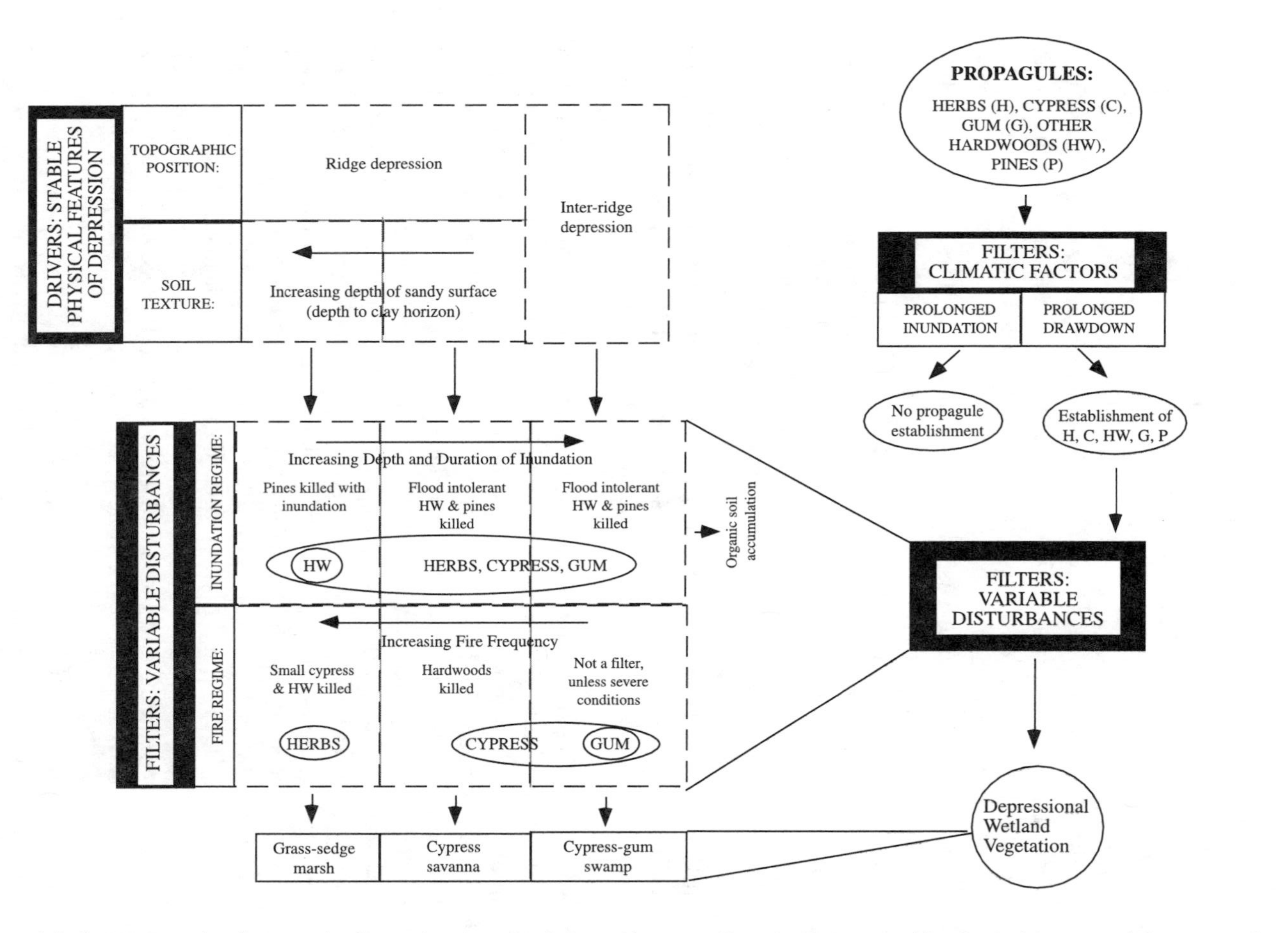

FIGURE 7.7
A conceptual model of ecosystem development in depressional wetlands in southeastern Georgia. Drivers (stable physical features of depression that control the establishment of species) and filters (climatic and disturbance factors that control the establishment of species) are identified in black boxes. Arrows from these boxes indicate the resulting environmental conditions or vegetation from each influencing factor. A generalization of the propagules and establishing vegetation that emerge through the filters are identified in ovals. Resulting depressional wetland vegetation is indicated in the boxes at the bottom of the figure (i.e., grass-sedge marsh, cypress savanna, and cypress-gum swamp). (From Kirkman et al. 2000. *Wetlands* 20: 373–385. Reprinted with permission.)

Seed bank tests indicate whether seeds are present and which species are likely to germinate under favorable conditions. In some wetlands, the composition of the seed bank is closely related to the composition of the plant community; however, there is a great deal of variation in the extent to which seed banks reflect the adult vegetation (Parker and Leck 1985; Leck and Simpson 1987). In a Canadian prairie marsh, in which much of the emergent vegetation had been destroyed due to high water levels, Welling and others (1988a, b) examined the recruitment of seven wetland species from the seed bank during a drawdown. The distribution of seedlings during the drawdown was similar to the pre-flooding distribution of the adult plants. *Phragmites australis* was the only exception; during the drawdown, its maximum density was at a lower elevation than the point where most *P. australis* adults had been prior to flooding. Tests of seed banks are particularly useful in restoration projects. The composition of the seed bank provides an indication of the species that will colonize the site once wetland hydrology is restored (see Chapter 9, Section I.A.3, Seed Banks in Restored Wetlands).

The vegetation often reflects the seed bank in coastal marshes, freshwater tidal marshes, lakeshores, and inland marshes. However, in forested wetlands, the seed bank often more closely resembles adjacent open areas (Leck 1989; Buckley et al. 1997). The seed bank can give an indication of the community that would replace trees should a natural disturbance result in an opening. The lack of woody species in the seed bank may be due to high predation and decomposition rates, delayed and variable reproduction rates, or a lack of dependency on long-lived seeds by these species. Disparities between the plant community and the seed bank may also occur when the dominant species reproduce asexually and contribute few seeds to the seed bank, such as *Acorus calamus, Phragmites australis,* and *Peltandra virginica*. Some taxa, such as *Eleocharis* and *Juncus* species, may have abundant seeds in the seed bank, but few adults in the standing vegetation (Leck 1989; Wilson et al. 1993).

In some cases, the seeds of only one or a few species constitute the majority of the seed bank. In many wetlands, the seed bank is dominated by graminoids (a notable exception is freshwater tidal wetlands where annuals often dominate; Leck 1989). Other propagules, such as turions, are also found in the soil and for many species, such as *Vallisneria americana* (Titus and Hoover 1991) and *Hydrilla verticillata* (Netherland 1997), they are more important than seeds in the species' recruitment (see Chapter 5, Section III.A.2.a, Turions).

In freshwater depressional wetlands such as prairie potholes, cyclic hydrology brings about dry and wet years on a fairly regular basis. This leads to the formation of a seed bank with at least two types of seeds: those produced by plants that thrive during flooded conditions and those produced by plants that grow during drawndown conditions. In this way, at least two community types may grow in the same location depending on the hydrology (van der Valk 1981). Similarly, the seed banks of two lacustrine wetlands of Long Island, New York were shown to contain seeds from two sets of species: those that grew under inundated conditions and those that grew during drawdowns. The phenomenon creates an increased diversity over time given fluctuations in the water level (Schneider 1994).

In tidal freshwater wetlands, where the tides create a daily period of inundation, the seed banks do not have two sets of species, adapted to different hydrologic regimes (Leck and Simpson 1987; Grelsson and Nilsson 1991). Rather, the species in freshwater tidal wetlands appear to have differing dependence on the seed bank. In general, the seed bank is depleted as new plants germinate at the beginning of the growing season. Some species' seeds, such as those of the annual, *Impatiens capensis,* are entirely depleted and have a complete turnover each year. Some perennials, such as *Peltandra virginica,* tend to have a high degree of turnover, with low numbers of seeds reserved in the seed bank. Other perennials, such as *Typha* species, have a persistent seed bank that lasts more than one growing season (Thompson and Grime 1979; Leck and Simpson 1987).

2. Factors Affecting Recruitment from the Seed Bank

Recruitment from the seed bank depends, to a large extent, on abiotic factors. The hydrologic regime is arguably the most important abiotic factor in wetland seed bank recruitment. The different growth forms vary in their response to flooding. For example, submerged species germinate under flooded conditions while emergents and mudflat annuals germinate under both flooded and drawndown conditions, with the number of seedlings higher under drawndown conditions (van der Valk and Davis 1978; Welling et al. 1988a; Leck 1989; Willis and Mitsch 1995).

Many species germinate only when oxygen levels in the soil are sufficient for respiration, however some, such as *Echinochloa crus-galli* (barnyard grass) and *Oryza sativa* (deep water rice) (Rumpho and Kennedy 1981; see Chapter 5, Section II.B.3, Seed Dormancy and Germination), germinate under anaerobic conditions. The depth of the overlying sediments is also of importance. When they are buried, large seeds produce seedlings that are generally better able to reach the soil surface than the seedlings of small seeds. Seed germination is also affected by competition from other seeds and seedlings, allelopathy, shading from adult plants, and herbivory. Humans can affect recruitment by disturbing hydrology, which can in turn affect the salinity level or other aspects of substrate chemistry, as well as sedimentation and the burial of seeds (Leck 1989). Because the environment is variable, it is often difficult to predict which species' seeds will germinate and successfully produce new seeds. Tests of seed banks alone do not enable us to predict succession or future communities and it is difficult to extrapolate between wetlands (Grelsson and Nilsson 1991; Wilson et al. 1993; ter Heerdt and Drost 1994; Leck and Simpson 1995).

IV. Competition and Community Dynamics

Competition, which has been defined as a "reciprocal negative interaction between two organisms" (Connell 1990), is one of the most important interactions in defining community structure (Gopal and Goel 1993; Keddy 2000). Resources must be limiting for competition to occur. When they are, a trade-off in the allocation of each individual's resources to growth, maintenance, and reproduction is necessary (Harper 1977; Grace and Tilman 1990; Wetzel and van der Valk 1998). Clements (1904, in Gopal and Goel 1993) was perhaps the first to recognize that competition is a major force in community succession. Since then, competition has been the subject of many investigative studies and it has been shown to be an important and "ubiquitous process in wetland plant communities" (Keddy 2000).

Three primary factors are thought to determine the distribution of species in a community: the relative competitive ability among the species present, the availability of resources in the system, and the type and frequency of disturbance (Chambers and Prespas 1988; Campbell and Grime 1992). Plants are distributed in response to environmental gradients, such as nutrient levels or water depth, and the degree of competition that they encounter along these gradients. Species that are weaker competitors tend to be restricted to marginal areas where competition is less (Barrat-Segretain 1996; Grace and Wetzel 1981, 1998).

Competition can take several forms, all of which influence plant community composition, and different forms of competitive interaction also influence successional processes. The most common form is *exploitative competition,* which occurs when individuals of the same or different species compete for a resource that is in short supply. Limiting resources might include nutrients, light, or space (Harper 1977). *Interference competition,* on the other hand, results when one competitor actively denies another access to a resource. Finally, *allelopathy* is a direct form of competition in which a competitor produces chemical substances that are released to the environment, reducing the growth of another. The production and release of phytotoxic compounds is seen as a means to gain a competitive advantage. Several species of wetland plants have been shown to produce allelopathic substances such as *Glyceria aquatica* (manna grass), *Hydrilla verticillata,* and *Nuphar lutea* (yellow water lily; Gopal and Goel 1993).

Most studies of competition in wetland ecosystems have focused on species of similar growth form, i.e., those that occupy similar positions in the water column (floating-leaved plants vs. floating-leaved plants, or emergents vs. other emergents). This is due to the assumption that species with different growth forms do not interact directly. When they do, competition is thought to be asymmetric, i.e., one species has a much stronger competitive effect than the other species. Because of this, species with the same growth form are expected to compete most directly with each other. However, there is evidence for competitive interactions between species of differing growth forms. For example, the growth of some submerged species has been shown to be suppressed in the presence of floating-leaved plants such as *Nelumbo nucifera* and *Trapa bispinosa* (as reviewed by Gopal and Goel 1993). In this case, light, one of the most important factors regulating wetland plant growth and distribution (Spence 1982), was captured more effectively by the floating-leaved plants. Interference competition for light has also been investigated between floating-leaved and emergent species. For example, *Potamogeton pectinatus* (sago pondweed) has been shown to be excluded by *Scirpus californicus* (giant bulrush) because *S. californicus* intercepts the available light (McLay 1974; in Gopal and Goel 1993).

In the following sections, we provide an overview of representative studies of competition between species of like growth form.

A. Intraspecific Competition

The competition between individuals of the same species, called *intraspecific competition,* occurs as a function of resource availability and population density. It is likely to be a factor in many wetland plant populations, particularly in those species that form dense, monospecific stands. For example, when conditions are favorable, free-floating species may reproduce vegetatively until the entire surface of the water is covered, making space a limiting factor. This represents density-dependent growth (Gopal and Goel 1993). Moen and Cohen (1989) studied *Potamogeton pectinatus* and *Myriophyllum sibiricum* (northern water milfoil; formerly *M. exalbescens*) in aquaria at high and low densities and found that growth rates decreased at higher densities for both species.

Phragmites australis (common reed) has also been the subject of intraspecific competition studies, particularly in Europe. Shoot density and shoot biomass have been shown to vary according to the *–3/2 power law* during the early stages of growth (Harper 1977). Under the –3/2 power law, as shoot density increases, shoot biomass declines and the line describing the relationship between the log of these two variables has a slope of approximately –3/2 (Figure 7.8; Mook and van der Toorn 1982). The –3/2 power law describes self-thinning. In *self-thinning,* as the number of individuals increases, the mean weight of individuals and the total biomass of the population decrease. Self-thinning occurs in even-aged, monotypic stands, and it is a function of the geometry of the space occupied by a plant.

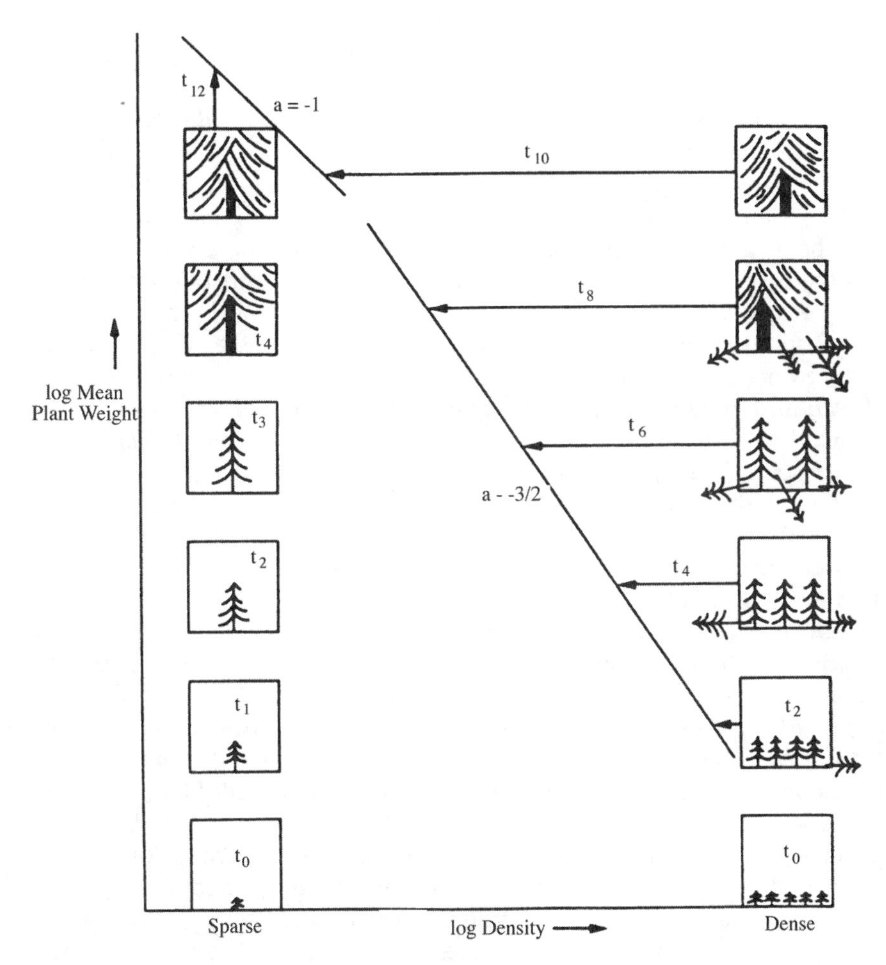

FIGURE 7.8
An illustration of the –3/2 power law of self-thinning that results from intraspecific competition. The slope of the line between the log mean plant weight and log density generally has a slope of –3/2. In dense populations (shown on the right) mortality will occur as density and biomass increase. In sparse populations (shown on the left) growth is not slowed by competition. Note in each case that growth starts at the bottom of the figure at t_0. (From Silvertown, J. 1987. *Introduction to Plant Population Ecology*, p. 229. Essex. Longman Scientific and Technical. Reprinted with permission.)

B. Interspecific Competition

Interspecific competition occurs between individuals of different species when both require the same limiting resource (such as nutrients, inorganic carbon, space, or light). The outcome of interspecific competition helps dictate species distribution and abundance within a community (Wilson and Keddy 1986; Gaudet and Keddy 1995). Plant characteristics that are correlated with competitive ability include biomass production, plant height (Gaudet and Keddy 1988), reproductive output (Weihe and Neely 1997; Rachich and Reader 1999), growth form, nutrient uptake efficiency (Tilman 1985), and the ability to oxygenate the root zone (Yamasaki 1984; Callaway and King 1996). The extent to which competition among wetland plants reduces both the fitness of the species involved and the available resources, is a function of the characteristics of the wetland and the species involved.

Many researchers have examined the competitive interaction between pairs of species. Weihe and Neely (1997), for example, investigated interspecific competition between *Lythrum salicaria* (purple loosestrife) and *Typha latifolia* (broad-leaved cattail). A total of five individuals were planted in each pot. The density ratio of *L. salicaria* to *T. latifolia* was 5:0, 4:1, 3:2, 2:3, 1:4, and 0:5. To investigate the effect of shade on competitive outcome, replicates were grown in unshaded and shaded conditions (40% of available sunlight). Shading decreased the growth of both species. In all cases, *L. salicaria* produced more above- and belowground biomass than did *T. latifolia*. The biomass of *T. latifolia* became smaller as the proportion of *L. salicaria* increased. Flower production was also measured as an index of competitive success. *L. salicaria* produced flowers in all treatments and showed no response to the presence of *T. latifolia* in this parameter. *T. latifolia,* by comparison, produced no flowering heads, although it was not clear if this was due to container effects. *L. salicaria* growth was also negatively related to *L. salicaria* density (i.e., growth was inhibited by self-thinning in this species). In fact, growth of *L. salicaria* was reduced more by intraspecific competition than by competition with *T. latifolia*. The authors concluded that under their experimental conditions, and regardless of the light regime, *L. salicaria* always outcompeted *T. latifolia*. In another study, competition between *L. salicaria* and the grass *Phleum pratense* (timothy) resulted in delayed flowering and reduced root dry weight in *L. salicaria* (Notzold et al. 1998).

In an unusually broad study, Gaudet and Keddy (1988) investigated the competitive performance of *L. salicaria,* using it as a "phytometer" to gauge the ability of 44 other species to reduce its growth. All species were monitored for biomass production, plant height, and canopy area. *L. salicaria* was found to be the most competitive species. The authors grouped species by their ability to produce biomass: those that produce similar biomass amounts were found to have similar competitive abilities (i.e., biomass production was a good predictor of competitive ability). Within each group, factors other than competition (e.g., nutrient availability) were important in influencing biomass production. The allocation of biomass (as measured, for example, by root/shoot ratios) has also been shown to change in response to competition. For instance, plants growing along hydrologic gradients have been found to adjust their biomass allocation as well as their distribution in response to interspecific competition (Carter and Grace 1990).

1. Competition and Physiological Adaptations

A number of physiological adaptations have been shown to convey competitive advantage. Grace and Wetzel (1981, 1998) investigated the dynamics of competing populations of *Typha latifolia* (broad-leaved cattail) and *T. angustifolia* (narrow-leaved cattail) over a 15-year period in a pond in Michigan. Their goal was to study the distribution of the two *Typha* species, which often co-occur but are segregated according to water depth (*T. latifolia* in shallow water and *T. angustifolia* in deeper waters). Their initial study (Grace and Wetzel 1981) demonstrated that both water depth and morphology determined the species' relative distribution. *T. latifolia* is able to displace *T. angustifolia* in shallow water. *T. angustifolia,* which is the competitively inferior species, appears to use deeper water as a refuge to escape competition. The authors explained the competitive advantage enjoyed by *T. latifolia* in terms of physiological differences between the species. In shallow areas, *T. latifolia* is a superior competitor for light because it has more leaf surface area. *T. angustifolia,* however, with thinner, taller leaves and smaller rhizomes, is better suited to deep water. Several years later, Grace and Wetzel (1998) returned to the pond to examine the long-term dynamics of the *Typha* populations. The density and distribution of the two species had not changed significantly. Using five experimental ponds that had both

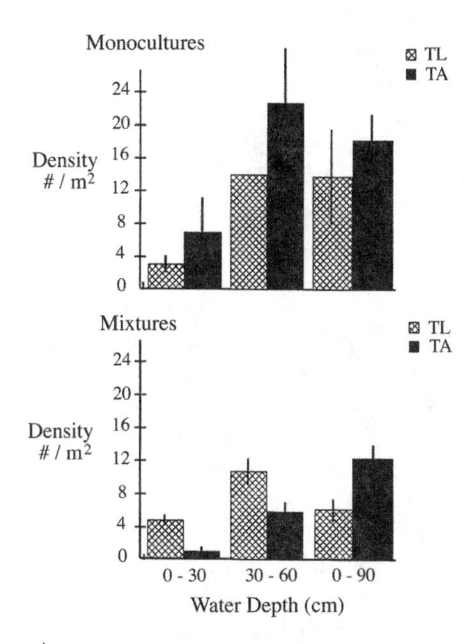

FIGURE 7.9
The density (numbers per 0.4 m^2) of *Typha latifolia* and *T. angustifolia* grown in monocultures (top) and mixed cultures (bottom) in five experimental ponds. The ponds were 23 years old at the time of measurement. Error bars represent ±one standard error. (From Grace, J.R. and Wetzel, R.G. 1998. *Aquatic Botany* 61: 137–146. Reprinted with permission.)

monocultures and mixtures of the two species, the authors confirmed the competitive displacement of *T. angustifolia* in shallow water. In monoculture, *T. angustifolia* densities were always higher than *T. latifolia* densities, regardless of water depth. However, in mixed cultures, *T. angustifolia* densities were greatly reduced (Figure 7.9).

Another physiological adaptation that appears to confer a competitive advantage is the ability to oxygenate the rhizosphere. For instance, when *Zizania latifolia* (wild rice) and *Phragmites australis* are grown together, *P. australis* is restricted to shallower water (10 to 35 cm) and *Z. latifolia* is found in deeper water (20 to 90 cm). Root aeration may determine this segregation, since *Z. latifolia* has been found to be efficient in transmitting oxygen from aerial shoots to roots (Yamasaki 1984). There is considerable variation in the degree to which different species are able to oxygenate the rhizosphere (Brix et al. 1992; Callaway and King 1996). In deeper water or where the sediments are saturated for prolonged periods, species with a high potential for gas throughflow (i.e., those with pressurized ventilation, underwater gas exchange, or Venturi-induced convection (see Chapter 4, Section II.A.4, Gas Transport Mechanisms in Wetland Plants) appear to have a competitive advantage over those that rely exclusively on diffusive gas flow.

Competition between submerged species may be affected by each species' relative ability to assimilate the inorganic carbon surrounding its leaves. If one submerged species is able to assimilate inorganic carbon more rapidly than others, it may negatively affect the growth and photosynthetic rates of adjacent species. In a study of three submerged species, *Elodea canadensis* (water weed), *E. nuttalli* (slender waterweed), and *Lagarosiphon major* (African elodea), *L. major* was found to photosynthesize at a more rapid rate than the two *Elodea* species. This ability may give it a competitive advantage since it depletes the supply of inorganic carbon, leaving less available for other submerged plants (James et al. 1999).

2. Competition and Life History Characteristics

One key to understanding the role of competition in structuring plant communities is a knowledge of the life history characteristics of the species involved (Grace 1991) and the implications of life histories in determining relative competitive ability. Although several models have been proposed, little consensus has been reached on the utility of using life histories to explain the distribution of species. Grime (1979) proposed a theory to explain the range of plant traits based on disturbance, stress, and competition, which he viewed as important forces in structuring communities (Grace 1991). This theory is based on the life-history traits of different species which are classified as either *ruderal, stress-tolerator,* or *competitor* (commonly referred to as R–C–S strategies; Figure 7.10). These are functional groups of species, grouped based on their response to stress (which reduces growth rates) and disturbance (which removes vegetation or biomass):

- *Ruderal species* are those with high reproductive abilities, fast growth rates, and short life spans that are generally found in disturbed, productive environments and are commonly annuals. These species are neither stress-tolerant nor competitive (Grime 1977, 1979). They tend to escape competition by dispersing. *Polygonum punctatum* (dotted smartweed) is an example of a ruderal species that grows and reproduces quickly and in doing so is able to complete its life cycle during short periods in which water levels are low (Middleton 1999).
- *Stress-tolerant species* have low reproductive effort and low growth rates. They generally occur in undisturbed, less productive areas (i.e., stressful habitats). They also tend to occur as sub-dominant species in late-successional, productive habitats where resource availability is low. Examples of stress-tolerant species are the shrubs of the Ericaceae that inhabit peatlands, such as *Vaccinium corymbosum* and *Chamaedaphne calyculata.*
- *Competitors* are species with low reproductive abilities and high growth rates. They are typical of undisturbed, productive habitats (unstressed conditions) and tend to flower late in the growing season (e.g., *Lilium michiganense*; Middleton 1999). *Elodea canadensis* and *Myriophyllum spicatum* have been classified as competitive strategists. Competitors have been described as 'resource capture specialists' (Grace 1991).

Kautsky (1988) elaborated on Grime's R–C–S model to include an additional category of stress-tolerant species which he dubbed, *'biomass storers'* (B-strategists; Figure 7.11). This group is characterized by the accumulation of biomass in storage organs such as rhizomes and turions. In this group, vegetative regeneration dominates over sexual reproduction. These species often thrive in conditions of low disturbance and high stress (low nutrient or light levels, or high salinity). An example is *Spartina alterniflora* (cordgrass), the dominant plant of many salt marshes. Kautsky's B-strategists are included in the S category in Grime's R–C–S model.

In many wetland restoration projects, desirable species often experience intense competition from aggressive (sometimes exotic) species such as *Phalaris arundinacea* (reed canary grass) and *Typha latifolia*. This phenomenon led Wetzel and van der Valk (1998) to test Grime's (1979) hypothesis that a superior competitive species will produce more biomass than an inferior competitor. This prediction should hold true under any environmental conditions. Both *P. arundinacea* and *T. latifolia* are able to maximize resource capture in fertile systems and are classified as C-strategists. In a greenhouse experiment, the authors measured biomass production by these two species as well as by *Carex stricta* (tussock sedge). They found that *P. arundinacea* significantly reduced the biomass production of both *T. latifolia* and *C. stricta,* regardless of water or nutrient levels. *P. arundinacea* initially grew taller more quickly, resulting in the shading of the other two species. The importance of plant architecture is implicated in these results: the horizontal leaf orientation of

P. arundinacea, compared to the vertical growth of *T. latifolia* and *C. stricta,* enabled individuals of this species to capture light more efficiently, thus increasing their competitive ability. *P. arundinacea* was determined to be a better competitor than the other two species.

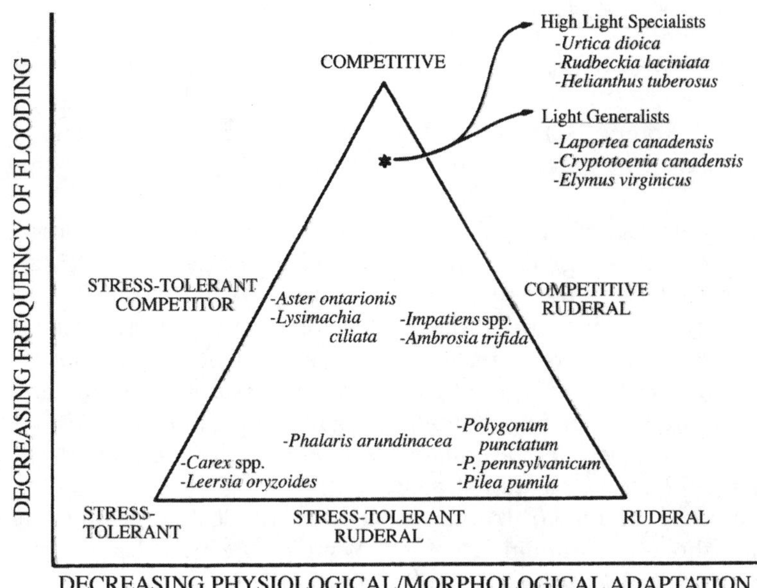

FIGURE 7.10
The R–C–S (Ruderal–Competitive–Stress-Tolerant) triangle developed by Grime adapted to show its relation to the frequency of flooding and the degree of physiological adaptations to flooding. (From Menges, E.S. and Waller, D.M. 1983. *American Naturalist* 122: 454–473. Reprinted with permission.)

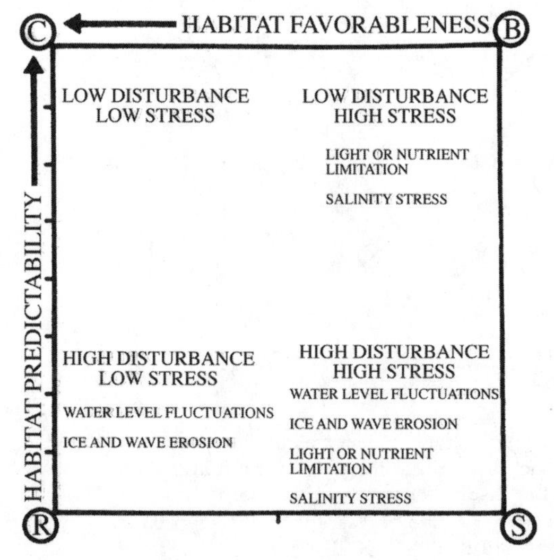

FIGURE 7.11
A square model showing life history characteristics based on Grime's triangle, with the addition of "B," the biomass storer. The corners represent extremes in environmental conditions where R is ruderal, C is competitive, and S is stunted guilds. In this model the B and S guilds are equivalent to the stress-tolerant (S) guild in Grime's R–C–S model. (From Kautsky, L. 1988. *Oikos* 53: 126–135. Reprinted with permission.)

Researchers have tested some aspects of the theory that life history traits can explain relative competitive abilities. Day and others (1988) investigated whether vegetation patterns in riverine wetlands were consistent with the R–C–S model of community organization proposed by Grime (1979). Two major gradients influenced community composition. The first was elevation, which is connected to many physical stressors such as the duration of flooding, wave and ice scour, mean water depth, rates of litter decomposition, and soil organic matter content. The second gradient was related to standing crop and litter, factors that are linked to species composition. This gradient was related to the rate of primary productivity and decomposition, rates of litter removal (e.g., by wave action), and standing crop. A conceptual model (Figure 7.12) summarized these interactions and led to an expansion of the life history strategies proposed by Grime. The authors classified the vegetation into five life-history types based on the range of fertility and disturbance found at their study site:

- *Clonal dominants* are large, rhizomatous species that form monotypic (or nearly monotypic) stands (e.g., *Typha latifolia, Sparganium eurycarpum*). These occur on undisturbed, fertile sites.
- *Gap colonizers* are large species lacking rhizomes but possessing high fecundity (i.e., high seed set) and the capacity for rapid germination (e.g., *Lythrum salicaria*).
- *Stress tolerators* have slow growth rates and evergreen tissues, and are typically found on infertile, flooded shoreline areas. Examples include *Eleocharis acicularis, Isoetes echinospora,* and *Ranunculus flammula*. This group is analogous to Grime's (1979) stress tolerator strategy.
- *Reeds* are deeply rooted (rhizomatous) species with leafless shoots that occur on infertile shores in running water. Examples include *Equisetum fluviatile, Eleocharis smallii, Scirpus acutus,* and *S. americanus*.
- *Ruderal species* are annuals that germinate and grow from the seed bank each year. They tend to be found in areas of high fertility and high disturbance.

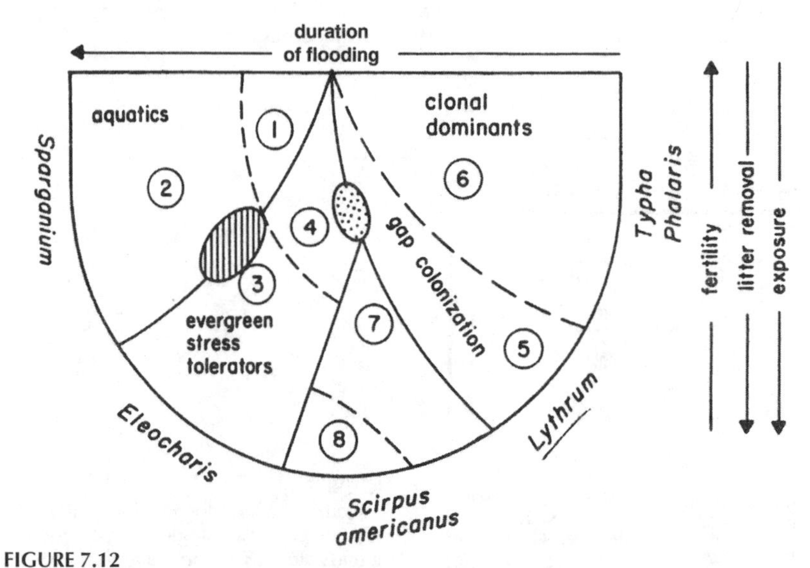

FIGURE 7.12
The conceptual model developed by Day et al. (1988) showing the Ottawa River vegetation. The primary environmental forces that act to structure the plant community are shown around the margin. Numbers refer to 8 TWINSPAN vegetation types: *Sporganium eurycarpum (1, 2), Eleocharis palustris (3, 4), Typha latifolia (5, 6),* and *Scirpus americanus* (7, 8). (From Day et al. 1988. *Ecology* 69: 1044–1054. Reprinted with permission.)

Biomass production peaked (>400 g m^{-2}) in areas of high fertility and low disturbance. When these conditions coincided, species richness was lowest due to competitive exclusion (with *Typha* sp. dominating). Conversely, species richness was highest in areas of low fertility and high disturbance.

Tilman (1982) developed a theory to predict competitive success among species based on resource levels, and the ability of different species to acquire those resources. Species composition in a given area can be explained as the result of competition for resources that are in short supply. The competitive 'victor' is the species that is most able to capture limiting nutrients. Thus, competitively superior species are those with the ability to gather resources when they are at low levels and to survive at those low levels (Grace 1991). In this model, a species' competitive ability changes as the availability of resources changes. For instance, a species that dominates when nutrients are low, because it can more effectively gather nutrients, may be outcompeted when nutrient availability increases. This explains why nutrient enrichment often leads to declines in species richness. The invasion of *Typha* sp. into the Florida Everglades is an example of this phenomenon (see Case Study 7.B, Eutrophication of the Florida Everglades: Changing the Balance of Competition).

Some species have evolved strategies to reduce or avoid competition. One example is related to niche theory (i.e., how different species subdivide the environment). For instance, community structure is influenced strongly by spatial and temporal environmental heterogeneity (Pickett and White 1985). Vivian-Smith (1997) studied the relationship between habitat heterogeneity, in this case due to microtopographic relief, and species richness using freshwater plants. In this mesocosm experiment, microtopography was manipulated to create flat soil surfaces (homogeneous environments) and 'hummock-hollow' surfaces (heterogeneous) where soil elevations varied by approximately 3 cm. The coexistence of species was facilitated by heterogeneity. Species richness, the number of individuals, and biomass were all significantly higher in the heterogeneous environments. The availability of different microsites allowed the interspecific differences in habitat preference to be exploited, reducing direct competition and resulting in more species with higher abundances. Most species showed a decided preference for one habitat over the other. The majority of species, including some woody perennials (e.g., *Cephalanthus occidentalis, Clethra alnifolia*), preferred hummocks.

3. Resource Availability and Competitive Outcome

A fundamental question in community ecology is how changes in resource availability, particularly nutrients, alter the intensity of competition (Twolan-Strutt and Keddy 1996). Changes in resource availability have been shown to change the outcome of competition between species (Weiher and Keddy 1995). As wetlands become enriched in nutrients, the relative competitive ability of species may shift, changing the competitive outcome. Increasing nutrients are often associated with the invasion of exotic species. In Australian heathlands, for example, the structure of the plant community changed and exotic species invaded when soil nutrient levels increased (Heddle and Specht 1975). Other environmental factors, including the availability of dissolved inorganic carbon (Svedang 1990; McCreary 1991), sediment texture, bulk density and organic matter content of the sediments (Barko and Smart 1986), and salinity (Callaway and Zedler 1997; Streever and Genders 1998), have also been shown to affect the outcome of competition. The number of factors that affect competition "increases the complexity of predicting competitive outcomes" (McCreary 1991).

McCreary (1991) points out that nitrogen and phosphorus limitations on macrophyte growth have been studied in two ways: through the use of plant tissue analyses (Gerloff and

Krombholz 1966) and through sediment analysis (Barko and Smart 1986). In a study of 13 species of freshwater plants, Gerloff and Krombholz (1966) established critical concentrations of nitrogen and phosphorus, below which sub-optimal growth occurred. The critical phosphorus concentration for the submerged species in their study was 0.13%. The critical nitrogen concentration was approximately 1.5%. None of the species had leaf tissue nitrogen concentrations below the critical concentrations, while 9 of 13 (69%) had leaf tissue phosphorus concentrations below the critical concentration for phosphorus. This is one line of evidence implicating phosphorus as the limiting nutrient in oligotrophic systems.

Barko and Smart (1986) tested the growth of *Myriophyllum spicatum* and *Hydrilla verticillata* on 40 different types of sediments collected from a variety of North American lakes. Plant growth was highly correlated with sediment nutrient concentrations. Of the seven sediment nutrients studied (N, P, K, Na, Mg, Fe, and Mn), all except N were significantly correlated with the growth of both species ($p < 0.001$). Growth on highly organic sediments was less than on mineral sediments. The authors concluded that sediment density was responsible for the observed differences in growth, rather than organic matter or an inhibitory organic compound.

Several competition studies have focused on invasive species (Barko et al. 1988, 1991; Hofstra et al. 1999; Van et al. 1999). *Hydrilla verticillata* has displaced native submerged vegetation in many areas of the world (see Chapter 8, Section V.B, *Hydrilla verticillata*; Sutton 1986; Chambers et al. 1993). Several studies provide evidence that sediment nitrogen availability might limit the growth of *H. verticillata* (Barko et al. 1988, 1991). For example, on nitrogen-poor soils, *Potamogeton americanus* was able to outcompete *H. verticillata* (McCreary 1991). Van and others (1999) investigated soil fertility in relation to competitive interactions between *H. verticillata* and *Vallisneria americana*. When soil nutrient levels were high, *H. verticillata* was the stronger competitor, but under low nutrient levels, *V. americana* was able to grow better.

In submerged communities, carbon dioxide is sometimes a limiting factor. Some species have physiological adaptations that aid in assimilating inorganic carbon, such as the uptake of bicarbonate (see Chapter 4, Section V.B, Submerged Plant Adaptations to Limited Carbon Dioxide). Other species, such as *Juncus bulbosus* (bulbous rush), are most productive at times when carbon dioxide levels are high. *J. bulbosus* can grow at low light levels and it is tolerant of low temperatures. It is most productive early in the growing season (occasionally even under ice) when CO_2 concentrations are high and before other species are active. This may be a strategy to avoid competition for inorganic carbon (Svedang 1990).

4. Light in Submerged Communities

Changes in underwater light levels are known to cause changes in the composition of submerged plant communities (Chambers 1987; Van den Berg et al. 1998). The response of species to light availability varies with photosynthetic efficiency and the ability of species to compete for the space where adequate light is available. For example, submerged species that form canopies just below the water surface (such as *Ceratophyllum demersum*, *Myriophyllum spicatum*, and *Potamogeton pectinatus*) may have a competitive advantage over those that grow in a pillar or bottom-covering growth form (Chambers 1987; Van den Berg et al. 1998). *P. pectinatus* was found to reduce the growth of *Chara aspera* by a maximum of 63% at high light levels, primarily due to its ability to create shade with its canopy. *Ranunculus* species have also been shown to shade *Chara*, reducing its growth (Grillas and Duncan 1986).

In a study of competition between *Myriophyllum sibiricum* and *Potamogeton pectinatus*, *P. pectinatus* was able to outcompete *M. sibiricum*. *P. pectinatus* initiates growth earlier in the growing season and it grows higher in the water column where it can shade *M. sibiricum* (Moen and Cohen 1989). A study using experimental tanks to investigate light and competition between *M. spicatum* (Eurasian watermilfoil) and *M. sibiricum* showed that both water clarity and depth affect competitive outcome. In shallow, clear water *M. spicatum* was the superior competitor in terms of biomass accumulation (Valley and Newman 1998).

A study of the competitive ability of *Elodea canadensis* (an exotic invasive in Europe where it co-occurs with *Myriophyllum spicatum*) and *M. spicatum* (an exotic invasive in North America where it co-occurs with *E. canadensis*) was conducted to investigate their relative response to stress (in the form of shade) and disturbance (in the form of cutting; Abernethy et al. 1996). When grown alone, *M. spicatum* grew longer (more stem length produced per plant) than *E. canadensis* in all light treatments (unshaded, low shade, and high shade). Neither species showed any response to shade in terms of biomass production, except for *M. spicatum* at high shade, which showed a 71% reduction in root biomass. In response to cutting, both species produced less biomass, but for *E. canadensis* the effect was less. In mixed stands, *E. canadensis* both grew longer and produced more biomass, indicating that it is the superior competitor.

5. Light in Emergent Communities

The morphological characteristics of each species influence its competitive ability. Plant architecture is correlated with a species' ability to capture light and, as such, is a determining factor in competitive outcome. Sculthorpe (1967) was an early observer of the ability of some emergent monocots (e.g., *Carex, Cyperus, Glyceria, Phalaris, Phragmites,* and *Typha*) to reproduce vegetatively and exclude competitors by reducing the amount of light that reaches the water's surface. This severely limits the ability of submerged or floating-leaved species to coexist with emergents.

Gaudet and Keddy (1988) found that when environmental conditions are kept uniform, wetland plants with tall shoots, leaf shapes with a moderate length/width ratio, and a large canopy diameter tended to be better competitors. For example, *Phalaris arundinacea* is a superior competitor because it is able to maximize light and nutrient capture by rapid vegetative growth, even when nutrients and soil moisture are limiting (Wetzel and van der Valk 1998). Generally, competition between emergent species and species with different life forms (floating-leaved or submerged) is asymmetric. That is, emergent species can shade the others, but the reverse cannot be true (Keddy 2000).

6. Competition and Salt Marsh Communities

The distribution of salt marsh plants is determined, to a large degree, by physical factors including salinity, nutrient availability, peat accumulation, and the degree of inundation (Valiela et al. 1978; Bertness 1991a; Pennings and Callaway 1992). Biotic interactions such as competition are also crucial in explaining species distribution. The interplay of physical factors and competition creates a filter (*sensu* van der Valk 1981) in determining species composition and zonation patterns (Figure 7.13). The significance of competition in maintaining zonation patterns is "suggested by the fact that boundaries between marsh vegetation zones are often abrupt, whereas edaphic (and other physical) factors change gradually across the marsh" (Pennings and Callaway 1992).

The typical zonation observed in perennial salt marsh species can be explained in terms of the plants' competitive abilities and physiological adaptations. For instance, in the high marsh ecosystems of New England, *Spartina patens* (salt marsh hay) dominates the seaward side and *Juncus gerardii* (salt marsh rush) dominates the upland side, while *Distichlis spicata* (spike grass) is found in disturbed habitats. *J. gerardii* is the competitive dominant

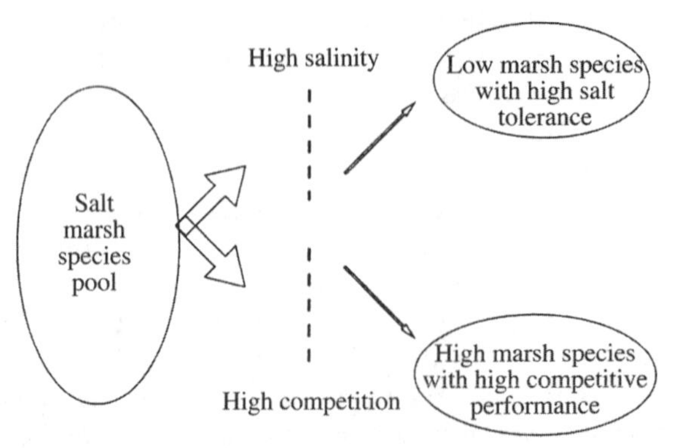

FIGURE 7.13
Model illustrating how salinity and competition can act as filters leading to different wetland communities. (From Keddy, P.A. 2000. *Cambridge Studies in Ecology*. H.J.B. Birks and J.A. Wiens, Eds. p. 614. Cambridge. Cambridge University Press. Reprinted with permission.)

and is able to restrict *S. patens* to lower marsh areas. *D. spicata,* on the other hand, is outcompeted by both of the other species and so is restricted to disturbed areas (Bertness 1991a). *Spartina alterniflora,* which dominates the low marsh habitats, is able to tolerate the harsh physical conditions found there, including daily inundation by tides. In the absence of competitors, *S. alterniflora* is capable of growing in the high marsh as well as the low marsh, but it is excluded from the high marsh by the other high marsh species (Bertness and Ellison 1987).

Snow and Vince (1984) performed reciprocal transplant experiments in Alaskan salt marshes where four vegetation zones extend from the lowest to the highest elevation, including: outer mudflat (dominated by *Puccinellia nutkaensis*, Alaska alkali grass), inner mudflat (dominated by *Triglochin maritimum,* arrow grass), outer sedge marsh (dominated by *Carex ramenskii,* Ramenski sedge), and inner sedge marsh (dominated by *Carex lyngbyei*). Their data showed that all species were capable of growth at all elevations when neighbors (i.e., potential competitors) were eliminated. Thus, zonation cannot be attributed to physical factors alone. Competition among species is an important force in determining species abundance. Snow and Vince (1984) concluded that these species were limited at lower elevations by their tolerance of salinity, and by competition at higher elevations.

Many other researchers have had similar findings (Valiela et al. 1978; Bertness 1991a; Pennings and Callaway 1992; Callaway and Zedler 1997). Pennings and Callaway (1992) generalized this pattern (derived from Connell's 1961 studies in rocky intertidal systems), saying that the upper limit of a species' distribution is set by competition, while the lower limit is set by harsh environmental conditions. Competition determines the limits of a species distribution in the zone where environmental conditions are relatively favorable (i.e., low in stress).

Pennings and Callaway (1992) investigated the validity of this model in mediterranean-climate salt marshes. Unlike many salt marshes, Mediterranean marshes do not exhibit a steady gradient of physical variables, from very stressful conditions in the lower marsh to mild conditions in the upper marsh. Rather, gradients are conflicting, making the middle marsh zones (and not the upper marsh) the most favorable for growth. Reciprocal transplant experiments were conducted using the perennials *Salicornia virginica* (glasswort), which grows in the low marsh, and *Arthrocnemum subterminale* (Parish's glasswort), which dominates at higher elevations. Both species grew in the middle marsh (the high *Salicornia* zone and the low *Arthrocnemum* zone). Therefore, the authors concluded that competition maintains the boundary between the two species in the middle marsh. In the two more stressful areas (the upper and lower marshes), physical stressors set the limit to species distribution. *S. virginica* is able to grow in the lower marsh because it has a high tolerance of flooding, while *A. subterminale* has greater salt tolerance and can grow under the high salinity conditions in the upper marsh.

Typically, zonation that results from competition can be altered by changes in soil fertility (Keddy 2000). Levine and others (1998) conducted a 2-year fertilization experiment using *Spartina alterniflora, S. patens, Juncus gerardii,* and *Distichilis spicata,* that resulted in a reversal in species patterns. *S. alterniflora,* normally restricted to the lowest elevations, invaded higher zones and was able to competitively exclude *S. patens* and *J. gerardii.*

Many U.K. salt marshes are dominated by *Spartina anglica,* a relatively new species that developed from a mutation in *S. townsendii* (which is the hybrid of *S. maritima* and *S. alterniflora*). From its place of origin (Hampshire, England), it spread northward along the coast, and south to the coast of France. Within a given marsh, its distribution is related to the duration and frequency of inundation and it tends to occur in the intertidal zone between mean high-water spring tide and mean high-water neap tides. It is also restricted by wave action. On a regional scale, the distribution of *S. anglica* is increasingly restricted as latitude increases. At northern sites, *S. anglica,* a C_4 species, is restricted on the landward side by competition with C_3 species, and the area it occupies tends to shrink with increasing latitude. The competitive ability of this species is strongest in southern marshes where temperatures are warmer (an environment that tends to favor C_4 species). In northern locations this advantage diminishes, and C_3 species encroach (Scholten and Rozema 1990 in Molles, Jr. 1999).

C. Allelopathy

Allelopathy is a specialized form of interference competition in which one species inhibits the growth of neighbors by releasing secondary metabolic compounds. These compounds may come from root exudates, or they may be leached from leaves or litter (Gopal and Goel 1993; Keddy 2000). Allelopathic effects are manifested in several ways including decreased germination rates of other species' seeds or of their own seeds, suppressed growth in neighboring species, or decreased chlorophyll content in another species (Barrat-Segretain 1996). Some commonly occurring wetland plant genera reported to release allelopathic compounds include *Cyperus, Eleocharis, Nelumbo, Nymphaea, Nuphar, Myriophyllum, Polygonum,* and *Vallisneria* (Gopal and Goel 1993). For example, *Nuphar lutea* has been shown to reduce the chlorophyll content of *Lemna minor* and inhibit the growth of lettuce seedlings (Elakovich and Wooten 1991). Perhaps the rarest form of allelopathy is autotoxic effects of leaf litter on seed germination. van der Valk and Davis (1978) found that *Typha latifolia* litter produced phytotoxic effects on its own seeds. *Phragmites australis* has also been reported to be autotoxic. More recently, Ervin and Wetzel (2000) found that *Juncus effusus* seedlings also had "autotoxic sensitivity" to the extracts of dead tissue of adult plants.

The chemical composition of allelopathic compounds varies by species. Some of the most common classes of alleochemicals include: alkaloids (found in many rooted floating-leaved species, particularly the Nymphaeaceae), terpenoids (found in some emergent species), and flavonoids (found in free-floating species). Some seagrasses have also been shown to produce sulphated phenolic compounds. In *Zostera marina* alone, six different phenolic acids have been identified (Zapata and McMillan 1979). Many submerged or floating-leaved species, including *Ceratophyllum demersum, Potamogeton crispus, P. pectinatus,* and *Vallisneria americana,* appear to use alkaloid compounds as a defense against herbivores, particularly aquatic invertebrates (Ostrofsky and Zettler 1986; see Chapter 4, Section VI, Adaptations to Herbivory). An explanation for the synthesis of allelopathic compounds in species with these life forms is that submerged plants have few structural defenses such as spines or thick cuticles. Since structural defenses are limited, chemical defenses prevail. Ostrofsky and Zettler (1986) reported the content of alkaloids in 15 species of plants and found concentrations ranging from 0.013% (in *Heteranthera dubia*) to 0.056% (*Potamogeton crispus*). These are levels high enough to be pharmacologically active. The authors report that sheep grazing on *Phalaris* sp. containing 0.017 to 0.059% alkaloids are often killed. The Nymphaeaceae have perhaps received the most study with respect to their production of alkaloids (Elakovich and Wooten 1991; Vance and Francko 1997).

One common technique to test for potential allelopathic effects is to wash leaves, leaf litter, or roots with distilled water and test the exudate for its ability to reduce seed germination (e.g., Elakovich and Yang 1996; Quayyum et al. 1999). While this bioassay technique might show strong phytotoxic effects in the lab, demonstrating strong allelopathic interactions in the field is much more difficult (Keddy 2000). Dilution of allelochemicals in flooded wetlands could result in concentrations that are too low to have an effect.

In a review of allelopathic interactions, Gopal and Goel (1993) proposed that the production of allelopathic compounds is inducible. They suggested that an individual plant will produce allelochemicals only when environmental conditions trigger their synthesis, such as crowding stress. Otherwise, an individual will compete using only its physiological adaptations. Little experimental work has been done to test this theory. However, given the biochemical "expense" of producing phytotoxic compounds, inducible allelopathy would be considered an advantage.

V. The Role of Disturbance in Community Dynamics

Disturbances, such as fires and storms, often maintain the current plant community or reset successional processes to an earlier stage, starting the process all over again (Connell and Slatyer 1977). The magnitude of the disturbance is important in determining its impact. Infrequent catastrophic events, such as glaciation, may eliminate an entire community and alter the habitat so that different species become established. Low intensity recurrent disturbances, such as storms or fires, cause a temporary disruption followed by recolonization. The patchiness of both recurrent disturbance and the recovery of the biota creates areas at different stages of succession within an ecosystem (White and Pickett 1985; Pickett et al. 1989). In wetlands, the disturbances that bring about community change or maintain communities in early successional stages include allogenic hydrologic alterations, severe weather events, and fire. Animals, particularly herbivores, can cause disturbance. Increasingly, humans are the most influential agents of disturbance in wetlands.

A. Hydrologic Disturbances

Hydrologic disturbances can occur on a large scale, such as the alteration of a river's course (see Case Study 7.A, Successional Processes in Deltaic Lobes of the Mississippi River), or on a smaller scale, such as when a stream is dammed, lake levels are stabilized, or drainage "improvements" are made. Alterations in hydrology can also result from climatic vagaries, which may result in wet and dry periods. Prairie potholes, which have cyclic hydrology, are an example of wetlands whose vegetative community is affected by climatic shifts. The plant community shifts from submerged, floating-leaved, and emergent plants in wet years, to emergents and mudflat annuals in dry years (van der Valk 1981).

Many wetlands are affected by the hydrology of an adjacent water body, so any change or hydrologic disturbance in the nearby lake, river, or stream can result in changes in the wetland's plant community. For example, in a depressional wetland adjacent to Lake Erie in Ohio, the plant community has changed many times in the last several thousand years, depending on whether the lake level was high or low. Pollen evidence suggests that under high-water conditions, floating-leaved plants and algae have dominated the wetland. When water levels were lower, emergent plants dominated (Reeder and Eisner 1994).

Human-induced hydrologic disturbances, including changes in water quantity, water level fluctuations, or water quality, have well-documented impacts to plant communities (Ehrenfeld 1983; Vitt and Bayley 1984; Ehrenfeld and Schneider 1991; Wilcox 1995). Zonation patterns may shift, species tolerant of disturbance may invade, or woody species may invade or die back as a result of drainage or flooding. Some of the responses in the plant community that have been shown to occur as a result of hydrologic change include (from Wilcox 1995; Ehrenfeld 2000):

- An increase in the numbers and dominance of invasive and exotic species, such as *Typha angustifolia* and *Lythrum salicaria*
- A decrease in species richness
- Mutualistic interactions, such as with pollinators or mycorrhizae, may decline
- An absence of species that are sensitive to disturbance
- Vegetation that is dominated by one species (monospecific) or by one structural type
- The presence of either very dense or sparse stands of vegetation (in response to water levels that were stabilized either at lower or higher than normal levels)

The normal hydrologic variations that are characteristic of many wetlands, on both a seasonal and inter-annual basis, contribute to the development of a diverse community, both in terms of species composition and structure. For example, periodic drydown of lower elevations prevents beds of dense submerged species from establishing, while flooding the upper elevations prevents shrubs and aggressive emergents from becoming dominant. Ice also plays a role by scouring lakeshores and opening up areas for the establishment of new species. Any hydrologic alteration that stabilizes water levels in this type of system represents a disturbance that can diminish the structural and taxonomic diversity of the community (Wilcox 1995). Ditching wetlands is done both to dry them down (e.g., drainage for agriculture) or to create channels for improved navigation (e.g., in the coastal wetlands of the U.S. Gulf Coast). In coastal zones, the channels created in salt marshes permit freshwater to drain away more quickly and saltwater to intrude farther inland than occur in unmodified wetlands. In the Gulf of Mexico marshes this has resulted in the die-off of salt-intolerant vegetation and the erosion of some marshes into shallow lakes (Mitsch and Gosselink 2000).

Hydrologic alterations, which cause changes in water quantity and quality, are increasingly common as a result of urban development. In a series of studies on the effects of anthropogenic hydrologic disturbance on wetlands in the New Jersey Pine Barrens, Joan Ehrenfeld has documented floristic changes that result from urbanization (Ehrenfeld 1983; Ehrenfeld and Schneider 1991, 1993; Ehrenfeld 2000). Of the total area of the Pine Barrens, approximately one quarter is wetland, including a diversity of community types such as marshes, bogs, ericaceous shrub and *Pinus rigida* (pitch pine) thickets, and *Chamaecyparis thyoides* (Atlantic white cedar) swamps. The wetlands contain a unique assemblage of species including carnivorous species (e.g., *Drosera rotundifolia, Sarracenia purpurea*), ericaceous shrubs (e.g., *Chamaedaphne calyculata*), and many species that are at the extremes of their geographic ranges.

In one study, Ehrenfeld (1983) compared 32 white cedar swamps, half of which were in watersheds dominated by agricultural or urban land use and half in undisturbed watersheds. Water in white cedar swamps typically has low pH and very low nutrient concentrations. Rapid urban expansion into the Barrens has resulted in dramatic hydrologic changes and associated floristic shifts (Table 7.1). Compared to undisturbed sites, wetlands in urbanizing areas had higher pH values and shorter periods of flooding annually. Conifers were less common in disturbed sites, presumably due to clearing of upland pine forests around the disturbed sites. Disturbed sites also had higher species richness due to the influx of cosmopolitan species such as *Leersia oryzoides* (rice cut grass) and *Parthenocissus quinquefolia* (Virginia creeper).

TABLE 7.1
Summary of Some Hydrological and Vegetation Changes Resulting from Urbanization in the New Jersey Pine Barrens

Parameter	Undisturbed Sites	Disturbed Sites
pH	3.82 ± 0.17	5.17 ± 0.35
% sites with standing water	69	38
Mean *Acer* sp. dbh (cm)	14.4	17.7
Dominant species at each site (%)		
Acer rubrum	81	100
Chamaecyparis thyoides	63	25
Pinus rigida	38	6
Mean species richness	27.8 ± 2.2	33.9 ± 2.2
# of species always rare	19	31
# of species always common	0	1

Data from Ehrenfeld 1983.

In a subsequent study, Ehrenfeld and Schneider (1991) further investigated the effects of hydrologic disturbance by selecting sites along a gradient of increasing urbanization. They found that water level fluctuations either increased or diminished compared to control sites due to human land use changes. In addition, concentrations of ammonia, orthophosphate, and chloride increase significantly in both surface and groundwater as urbanization increases. Using multivariate analysis, the authors concluded that it is the changes in water quality, and not the associated changes in water levels, that were most responsible for changes in the vegetation community, although this conclusion was based solely on the correlation between variables. Because waterborne nutrients move into and out of wetlands via water flow, separating the effects of water level changes and water quality on plants is difficult.

Floristic changes were assessed using the following index of species change:

$$\frac{\sum \text{indigenous species} - \sum \text{invader species}}{\sum \text{indigenous species} + \sum \text{invader species}} \quad (7.1)$$

The index shows changes in community composition relative to the undisturbed controls. Index values were 0.98 ± 0.2 for control sites, 0.72 ± 0.8 for sites near urban development, 0.28 ± 0.11 for wetlands within developed areas, and –0.16 ± 0.8 for sites receiving direct stormwater sewer outfalls. Thus the degree of hydrologic alteration is reflected in the degree of change in wetland flora.

B. Severe Weather

Strong weather conditions can result in vegetation changes. If they are severe, rainstorms, lightning, hail, snow, and ice all affect wetland plants. Drought can change the hydrology of a wetland and lead to the invasion of upland species (Hogenbirk and Wein 1991). In mangroves, drought can cause massive tree mortality (Lugo 1997). Two other types of weather-related disturbances that cause major disturbances in wetlands are floods and hurricanes.

1. Floods

Extreme floods occur irregularly, creating a high energy disturbance that effects quite different changes than those resulting from regular or seasonal floods (Breen et al. 1988). In flowing-water systems, floods scour the stream bed and the former channels. High water velocity breaks and uproots plants and removes fine sediment. After an extreme flood, recolonization occurs, and a new community develops. For example, two major floods along the Rhone River in 1990 and 1991 swept away all of the vegetation and imported coarser sediment than had previously been on the river's banks. Plants recolonized the banks gradually, with the first plants gaining a foothold at the least disturbed margins while later colonists moved into more disturbed areas. The first plants to become re-established, such as *Potamogeton natans, P. pusillus,* and *Myriophyllum verticillatum,* were those that produce turions or other vegetative organs that survived the flood (Henry et al. 1996).

In 1993, the midwestern U.S. was subject to extreme flooding which lasted from mid-June to early August. The floods were caused by record precipitation following heavy early spring rains. The floods had severe effects on forested bottomlands: trees were knocked over, the ground was scoured of herbaceous plants and tree seedlings, islands were eroded, and emergent and submerged vegetation beds were destroyed. After the floods receded, many of the disturbed areas were colonized by *Lythrum salicaria* (purple loosestrife), a fast-growing invasive species. The plant also moved into higher elevations because floodwaters carried its seeds into higher areas (Perry 2000).

In some flooding events, the plant community may be altered due to the import of sediments that can bury vegetation and prevent colonization. For example, in some South African rivers, the lack of submerged and floating-leaved plants is attributed to frequent flash floods that deposit silt and inhibit the plants' germination and growth (Edwards 1969). In oxbow wetlands of Alberta, Canada, floods have been shown to deposit large amounts of suspended material (van der Valk and Bliss 1971). During one 3-day flood, up to 4 cm of silt was deposited in newly formed oxbows, which was sufficient to bury the submerged plant community entirely. In older oxbows, flooding did not deposit large amounts of silt but did reduce light due to suspended solids in the water column. After floods the water transparency was drastically decreased: Secchi disk readings (a measure of water transparency) were measured at 1 to 4 cm, while normal readings were 2 to 2.5 m. It took 7 to 8 days for the water to clear. During that time only plants adapted to low light were able to persist. The standing crop decreased by about 50% after flooding in some oxbow wetlands, while in others the vegetation was destroyed. Floods did have beneficial effects on the oxbows. They replenished the water supply and provided nutrient inputs. Without floods, the oxbows became dry and evapotranspiration exceeded precipitation in some years.

2. Hurricanes

Hurricanes cause dramatic damage to coastal wetlands. Mangrove forests are often subject to hurricanes due to their location on tropical shores. Hurricanes uproot and destroy mangrove trees and open the area for renewed growth. Hurricanes are one of the factors influencing tree size in mangrove forests. In areas that are protected from hurricanes or only have them infrequently, mangroves grow to be much larger than in areas that experience more frequent hurricanes (see Chapter 2, Section III.B.1, Coastal Forested Wetlands: Mangrove Swamps; Odum and McIvor 1990; Lugo 1997). The effects of hurricanes can be patchy, destroying some stands while nearby stands escape damage. In this way, older mangrove forests are sometimes found within younger forests. Dwarf mangroves (trees that fail to increase in height due to salinity or anoxia stress) are often able to escape hurricance damage (Lugo 1997).

Mangroves of the Lesser Antilles suffered widespread tree mortality during Hurricane Hugo in 1989. Recovery began immediately after the hurricane with the growth of seedlings that survived the storm. However, recovery is still incomplete after 10 years, perhaps due to altered shore topography, the presence of fallen trees, and a lack of propagule production and dispersal. The young trees (mostly *Avicennia germinans*) suffered from caterpillar defoliation during the first months following the hurricane, a phenomenon that has been observed after other hurricanes as well (Imbert et al. 2000).

In 1992, Hurricane Andrew caused damage in the Florida Everglades, Biscayne Bay, and the Gulf of Mexico coast (U.S. Geological Survey 2000).Trees growing in about 70,000 acres of the Everglades and Biscayne Bay National Parks were broken or damaged. Many of the trees, including both mangroves and hardwoods, were defoliated, damaged, or knocked down. Within about 20 days, the surviving trees and shrubs had sprouted new growth. The hurricane had only minor effects on the vast *Cladium jamaicense* (sawgrass) stands in the Everglades. In Louisiana, Hurricane Andrew compressed floating marshes, resulting in a net decrease in marsh surface area. High sediment loads also damaged vegetation. Rooted plants were scoured from the bottom of marshes and many of the surviving plants were subjected to increased salt concentrations as the Gulf of Mexico water rose into normally freshwater areas. By clearing forested areas and killing off much of the existing vegetation, the hurricane acted as a reset mechanism to an earlier successional stage, altering the course of successional processes.

C. Fire

Fire can be an acute disturbance that maintains communities in their current successional stage. Fire removes the current season's aboveground standing stock and brings about a disruption of consumption, decomposition, and nutrient cycling processes. Fires in peatlands remove peat, thereby causing subsidence. New formations or changes of direction in water flow may also result (Breen et al. 1988).

Some wetland types are subject to frequent fires and some marsh plants are adapted to recurring fires. In this case, fire is not considered a disturbance, but part of the normal regime of the wetland. Many herbaceous plants in the Florida Everglades quickly grow back following a fire, including *Cladium jamaicense, Panicum hemitomon* (maidencane), *Pontederia cordata* (pickerelweed), and various species of *Sagittaria* (arrowhead). Their growth is enhanced due to the release of nutrients through oxidation and reduced competition for space. In the case of *C. jamaicense*, the buried rhizome is insulated from fire while the leaves are highly flammable and burn even when standing water is present. After fire, *C. jamaicense* responds quickly, growing as much as 40 cm in 2 weeks (Kushlan 1990).

In *Taxodium distichum* (bald cypress) swamps, fires tend to kill hardwoods and pines, leaving mature cypress trees (Ewel and Mitsch 1978). In this way, the cypress forest persists. Fire also maintains the pitcher plant community in Gulf Coast pitcher plant bogs (Folkerts 1982) and *Chamaecyparis thyoides* (Atlantic white cedar) stands in some U.S. east coast forested wetlands (Motzkin et al. 1993). In a *C. thyoides* wetland on Cape Cod in Massachusetts, cedar stands did not persist for more than 100 to 200 years in the past without regenerating fires. The current stand is about 150 years old and survival beyond this point would not be typical. Within the stand, new *C. thyoides* seedlings are few in number, suggesting that if fire is excluded there will eventually be a decline in the species.

Fire is sometimes sufficiently severe to reset a plant community to a previous successional stage. For example, Lake Powell in northern Wisconsin was previously a dense conifer bog forest, but repeated fires resulted in its reversion to open peatland with herbaceous species and large areas of open water (Larsen 1982).

D. Biotic Disturbance

Many animal activities cause disturbances in wetlands. Animals move through plant stands, and large ones, such as the hippopotamus (*Hippopotamus amphibius*) in Africa, maintain pathways of open water (Breen et al. 1988). Animals consume plants, both for food and for nesting materials. When animals consume emergent plant parts, the plants often die because their supply of oxygen is lost. Herbivores thus create areas of open water in many wetland types (Middleton 1999).

Herbivores sometimes have a profound impact on the plant species composition of salt marshes (Hik et al. 1992; Srivastava and Jefferies 1995, 1996; Olff et al. 1997). In salt marshes on the Hudson Bay coast in Manitoba, the seaward plant community is composed of *Puccinellia phryganodes* and *Carex subspathacea*. These species are also present in the upper marsh, but only where lesser snow geese (*Chen caerulescens caerulescens*) graze. When geese are excluded, the upper marsh becomes dominated by *Calamagrostis deschampsiodes* and *Festuca rubra*. This change in species composition occurs even in the presence of geese due to increases in elevation, but much more slowly (i.e., >10 years; Hik et al. 1992).

In a Netherlands salt marsh, Brent geese (*Branta bernicla*), cattle, and sedimentation are the most important determinants of plant community composition. The geese prefer early successional halophytes (e.g., *Festuca rubra*, *Puccinellia maritima*, and *Plantago maritima*) that grow at the lowest elevations. Their foraging activities and their nutrient-rich feces increase the productivity of their preferred plants. However, due to slow increases in elevation, these plants are eventually replaced by *Elymus athericus*. As this happens, the geese move to other forage areas. Grazing by cattle reverses the trend as the cattle remove *E. athericus*, opening the way for the re-establishment of the early successional halophytes preferred by the geese (Olff et al. 1997).

Some prairie potholes have a cyclic pattern of vegetation development due to precipitation and the activity of muskrats (*Ondatra zibethica*). The cycle may last from 5 to 20 years and the general stages are dry marsh → regenerating marsh → degenerating marsh → lake. A drought causes low water levels, exposing mudflats. Seeds of annuals (*Bidens*, *Cyperus*, *Rumex*, *Polygonum*) as well as emergent perennials (*Sagittaria*, *Scirpus*, *Sparganium*, *Typha*) are able to germinate. When rainfall increases, the mudflats are flooded and the annuals disappear, but the emergents persist. Submerged plants (*Potamogeton, Najas, Ceratophyllum, Myriophyllum, Chara*) are able to germinate and grow. The emergents increase in stand area and density during this regenerating stage. In response to their growth, muskrat populations increase. They build nests and dig up perennials to eat their roots; they also form trails throughout the marsh. With the muskrats' increased population, the density and vigor of the marsh plants decrease. With little emergent marsh remaining, much of the area reverts to an open shallow lake and then changes again during the following drought (van der Valk and Davis 1978; Weller 1994; Clark 1994).

Beavers (*Castor canadensis*) affect the distribution of wetlands across the landscape by impounding flowing waters. Following impoundment, the areas are converted from predominantly woody vegetation to open water and marshes (Figure 7.14). The North American population of beavers has largely recovered from overtrapping in the 19th and early 20th centuries. Their increased numbers have resulted in increased wetland area in some regions. For example, in a 298-km^2 region of northern Minnesota, beavers increased the area of wetlands and open ponds from about 2200 ha in 1940 to 3700 ha in 1988 (Johnston 1994).

FIGURE 7.14
A beaver impoundment in Virginia. Beaver-created wetlands often have standing dead trees, members of the pre-flooding plant community. (Photo by H. Crowell.)

E. Human-Induced Disturbance

Humans are increasingly the most important factor in wetland ecosystems, particularly in countries with intense population pressures. Eutrophication, acidification, river regulation and diversion, the construction of hydroelectric power plants, and other influences can change the plant community of wetlands (van Wirdum 1993; Müller 1995; see Section V.A, Hydrologic Disturbances). Other human impacts include industrial pollution, excessive sewage enrichment, agricultural runoff, and oil spills (Niering and Warren 1980).

Humans' impacts on wetland succession range in severity. Increased nutrient levels have resulted in plant community changes in some wetlands (see Case Study 7.B, Eutrophication of the Florida Everglades: Changing the Balance of Competition). For example, during the 1960s and 1970s, the growth and abundance of submerged plants in the Chesapeake Bay declined. The decline was attributed to shading from periphyton, whose growth was enhanced by high nutrient levels (Orth and Moore 1983). Although submerged aquatic vegetation has reappeared in the Bay, different species, such as the invasive *Hydrilla verticillata*, dominate many sites (Stevenson et al. 1993).

Other human impacts stem from construction near wetlands such as causeways near coastal wetlands that alter normal tidal flushing. In many U.S. east coast salt marshes, flooding tidal gates have been installed in an effort to prevent residential flooding. As a result, salinity of the marshes has decreased and *Spartina alterniflora* marshes have been converted to *Phragmites australis* marshes (Niering and Warren 1980; see Chapter 8, Section V.E, *Phragmites australis*).

Summary

The study of community dynamics includes changes in the distribution and abundance of a species and the reasons for these changes in community structure. Ecological succession is defined here as a change in the species present in a community. Changes in community structure arise due to internal processes (autogenic), such as competition among species or herbivory, and external ones (allogenic), such as natural or anthropogenic disturbances. Primary succession occurs where no plants have grown before, such as on newly exposed glacial till or on volcanic mudflows. Secondary succession occurs where a natural community has been disturbed, such as on abandoned agricultural fields.

The study of ecology is rooted in the theories of Clements (1916) and Gleason (1917). Clements (1916) proposed the hypothesis of organization through succession in which it is thought that whole communities or ecosystems are self-organizing entities of plants and animals. Under this hypothesis, succession is an autogenic process with each seral community preparing the environment for the next. Gleason (1917) proposed the individualistic hypothesis of succession, in which each individual organism in a community is present due to its unique set of adaptations to the environment. Connell and Slatyer (1977) proposed that early colonists have three basic types of interaction with later colonists: (1) *facilitation*, in which early species create a more favorable environment for the establishment of later species; (2) *tolerance*, in which there is no interaction between early and later species; and (3) *inhibition*, in which early species actively inhibit the establishment of later ones.

E.P. Odum (1969) focused on the development of whole ecosystems, and labeled ecosystems as either developing or mature. He emphasized ecosystem functions such as primary productivity and respiration as well as other whole-system attributes such as the type of food chain, the amount of organic matter present, species diversity, mineral cycling, spatial heterogeneity, and the species' life cycles. Odum (1969) also described pulse stability in which ecosystems are subject to more or less regular but acute physical disturbances imposed from outside the system. Pulse stability may describe ecosystem development in many wetlands with fluctuating water levels, such as estuaries and intertidal zones, better than the concept of developing and mature ecosystems.

In many wetlands, abiotic factors, with hydrology chief among them, outweigh biotic factors as the cause for changes in the plant community. Hydrarch succession has been proposed as a model of wetland succession. In hydrarch succession, open water lakes slowly fill with plant detritus and peat and they become increasingly shallow and dry. In theory, upland forests eventually replace open water habitats. This model of succession has not been upheld by research. Without an outside force that lowers the water level, the peat will remain saturated, unable to support terrestrial vegetation (Mitsch and Gosselink 2000). Hydrarch succession was also proposed as a theory of succession in saline wetlands, including mangroves; however, upland communities do not replace wetland communities there either. The environmental sieve model is based on life history features of the species involved and on the presence or absence of an environmental filter, standing water. The degree of flooding permits the establishment of only certain species at any given time. Additional filters may be added to the general model for some wetland types.

In wetlands, the seed bank can indicate which species will become established after a disturbance or when conditions are suitable for their germination. In general, the number of species represented in the seed bank reflects the diversity of the community. There is usually a greater density as well as a greater diversity of seeds in freshwater wetlands than in saline wetlands. Recruitment from the seed bank depends, to a large extent, on abiotic factors, such as the hydrologic regime.

Competition is defined as a "reciprocal negative interaction between two organisms" (Connell 1990), and it is one of the most important interactions in defining community structure (Gopal and Goel 1993; Keddy 2000). Intraspecific competition occurs between individuals of the same species as a function of resource availability and population density. Many species have been shown to have decreased biomass as the number of individuals increases (self-thinning). Interspecific competition occurs between individuals of different species when both require the same limiting resource (such as nutrients, inorganic carbon, space, or light). Plant characteristics that are correlated with competitive ability include biomass production, plant height, reproductive output, growth form, nutrient uptake efficiency, and the ability to oxygenate the root zone. The extent to which competition among wetland plants reduces both the fitness of the species involved and the available resources is a function of the characteristics of the wetland and the species involved. Allelopathy is a specialized form of competition in which one species inhibits the growth of neighbors by releasing secondary metabolic compounds. Some wetland plant genera reported to release allelopathic compounds include *Cyperus, Eleocharis, Nelumbo, Nymphaea, Nuphar, Myriophyllum, Polygonum,* and *Vallisneria* (Gopal and Goel 1993).

Disturbances, such as fires, hurricanes, floods, or herbivory, can maintain the current plant community or reset successional processes to an earlier stage, starting the process all over again. The magnitude of the disturbance is important in determining its impact. Infrequent catastrophic events, such as glaciation, may eliminate an entire community and alter the habitat so that different species become established. Low intensity recurrent disturbances, such as storms or fires, cause a temporary disruption followed by recolonization. In wetlands, the disturbances that bring about community change or maintain communities in early successional stages include allogenic hydrologic alterations, severe weather events, and fire. Animals, particularly herbivores, can cause disturbance. Increasingly, humans are the most influential agents of disturbance in wetlands.

Case Studies

7.A. Successional Processes in Deltaic Lobes of the Mississippi River

The Mississippi River has changed course upstream of its mouth several times in the last 6000 to 8000 years. With each change in course, a new deltaic lobe was formed and the abandoned delta began to degrade. The cycle of deltaic formation, abandonment, and degradation takes 1000 to 5000 years. Five lobes ranging in age from 30 to 4000 years can be distinguished in the Mississippi deltaic plain (Figure 7.A.1). Both abiotic and biotic features affect the plant communities of the lobes, including elevation, rate of sediment deposition and erosion, sediment grain size, salt inputs, and herbivore activity (Shaffer et al. 1992).

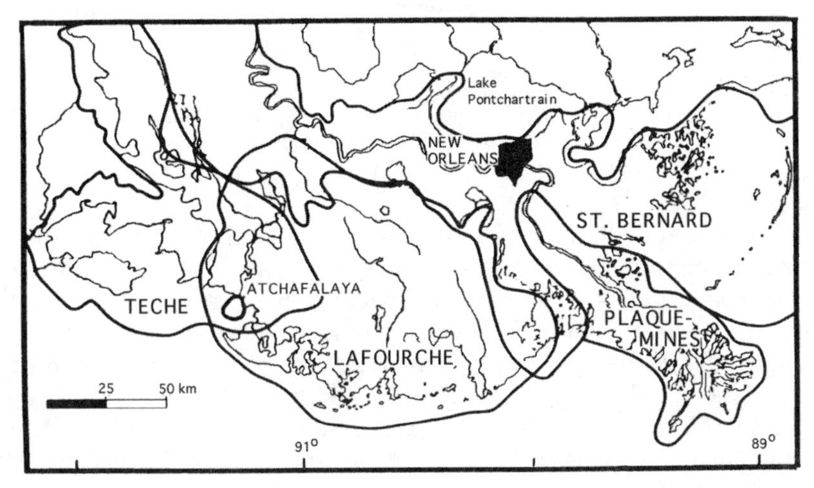

FIGURE 7.A.1
The Missisippi River deltaic plain region of South Louisiana showing historic delta lobes. The delta lobes from youngest to oldest are Atchafalaya, Plaquemines-Modern, Lafourche, St. Bernard, and Teche. (From Neill, C. and Deegan, L.A. 1986. *The American Midland Naturalist* 116: 296–305. Reprinted with permission.)

Three phases of lobe development have been described (Neill and Deegan 1986). During the initial phase, sediments are deposited underwater that decrease the depth of the stream. During the second phase, sustained land building continues until the river's hydraulic gradient becomes too flat to transport water and sediment to the Gulf of Mexico. Eventually, this serves as an impetus for an upstream diversion of the river to a steeper gradient. The salinity is lowest during the land-building phase of development. During the last stage, after which the river has been diverted from the lobe, there is a period of prolonged subsidence, land erosion, and lobe decay. Less and less water and soil are transported to the lobe and it begins to subside due to erosion and waves. The salinity increases during this phase.

Four wetland community types exist on all of the lobes: freshwater wetlands (including both marshes and swamps), and intermediate, brackish, and salt marshes. The dominance of one type over the others depends on the current phase of development. As the fresh water is diverted, the marsh becomes increasingly saline. This temporal pattern of

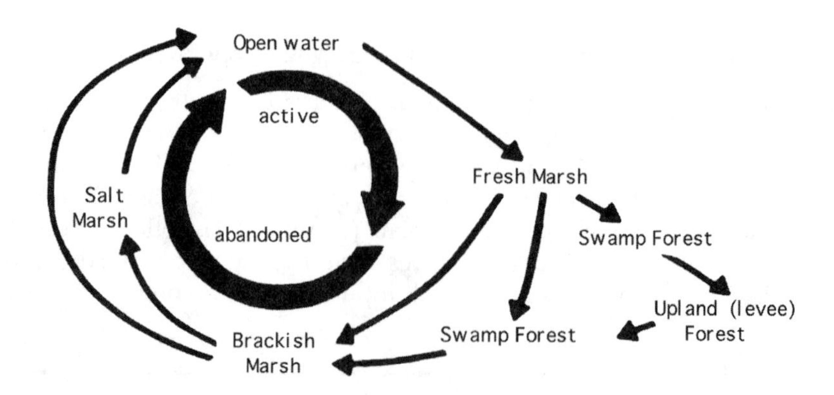

FIGURE 7.A.2
A model of cyclical vegetation change in the Mississippi River deltaic plain. (From Neill, C. and Deegan, L.A. 1986. *The American Midland Naturalist* 116: 296–305. Reprinted with permission.)

change is different from that predicted in the hydrarch model of salt marsh succession in which freshwater communities replace salt marshes. In the case of the delta lobes, the change in plant communities is cyclic, with allogenic forces creating much of the change (Figure 7.A.2). With a diversion in the river's course about once every 1000 years, fresh marshes colonize the newly created mudflats. The area of fresh marsh and freshwater swamp increases as the delta grows. The area of brackish marsh occurs farthest away from the active river mouth. As the lobe is abandoned by the river, the salinity increases and the brackish and salt marshes increase in size while the freshwater wetlands diminish.

The young deltaic lobes, with high inflows of fresh water and sediment, are dominated by freshwater marshes and mudflats (as seen in the Atchafalaya). Older lobes (Lafourche and Plaquemines-Modern) tend to support a higher number of different habitat types because elevation and salinity gradients create a variety of environments that support different vegetation types. Very old lobes (St. Bernard and Teche) support fewer habitat types because of erosion and decreased freshwater inputs.

7.B. Eutrophication of the Florida Everglades: Changing the Balance of Competition

The Everglades cover over 600,000 ha in southern Florida. *Cladium jamaicense* (sawgrass) covers an estimated 65 to 70% of the ecosystem either in monotypic stands or mixed communities that also include species of *Eleocharis* (spike rush) and *Rhynchospora* (beak rush). Other major habitat types include sloughs, wet prairies, and tree islands (classification by Loveless 1959). The sloughs are dominated by *Nymphaea odorata* (fragrant water lily), and species of *Utricularia* (bladderwort), *Eleocharis, Rhynchospora,* and *Panicum* (panic grass). The wet prairies and sloughs contain species such as *Hymenocallis palmeri* (alligator lily) and *Sagittaria lancifolia* (duck potato). The distribution and abundance of the different communities are a function of several environmental variables including soil type, nutrient availability, hydrologic regime, and fire frequency (Loveless 1959; Craft and Richardson 1993; Daoust and Childers 1999). Areas dominated by the nearly monotypic stands of *C. jamaicense* tend to be shallower areas (i.e., higher elevation) than the wet prairies and sloughs. They are inundated for shorter periods throughout the year and tend to burn more frequently (Jordan et al. 1997).

Historically, the Everglades were oligotrophic and highly deficient in phosphorus. During the last century an extensive system of canals and levees brought the hydrology of the region under human control. Agricultural runoff, which drains into the northern and eastern Everglades, has resulted in both phosphorus and nitrogen enrichment of the ecosystem (Craft et al. 1995; Daoust and Childers 1999). Most notably this has resulted in the formation of a nutrient enrichment gradient in the northern part of the Everglades where soil P concentrations range from more than 1000 mg P kg^{-1} to less than 500 mg P kg^{-1} (Richardson et al. 1999). Increases in soil bulk densities (over the range of 0.1 to 0.4 g cm^{-3}) as a result of drainage have also caused soil nutrients to concentrate (Newman et al. 1998). Coincident with increased nutrient loading there has been a shift in the composition of the Everglades plant communities. Native sawgrass and slough communities are being replaced by *Typha domingensis* (southern cattail) and *T. latifolia* (broad-leaved cattail). In areas where phosphorus enrichment has been greatest, cattails now dominate; areas still low in phosphorus remain dominated by sawgrass. Both nutrient enrichment and changes in hydroperiod are implicated in the successful invasion of cattails (Urban et al. 1993). Both *Typha* species are reportedly better able to assimilate nitrogen and phosphorus and produce biomass (a form of exploitative competition) than *C. jamaicense* (Davis 1994; Koch and Reddy 1992).

The response of Everglades plant communities to nutrient additions has been studied experimentally. For instance, in nutrient enrichment experiments, nitrogen (as NH_4^+) and phosphorus (as PO_4^{3-}) were added to sawgrass monocultures, mixed sawgrass-cattail communities, and slough communities dominated by *Utricularia, Eleocharis,* and *Panicum* (Craft et al. 1995; Chiang et al. 2000). Over a period of 4 years, phosphorus was added at a rate of 4.8 g m^{-2} yr^{-1}. The slough community responded immediately, showing a 90% decline in *Utricularia* biomass and its replacement by the macroalga, *Chara*. The composition of the sawgrass community was unchanged; however, biomass production increased in the sawgrass and mixed sawgrass–cattail communities.

In a field survey designed to investigate changes in macrophyte species distribution in the marshes and sloughs as a function of the surface water and soil phosphorus gradient that has developed in response to agricultural drainage, pronounced changes in species composition were observed. Unenriched areas were dominated by *C. jamaicense* while enriched areas were dominated by *Typha*. Species typical of slough areas also declined as phosphorus availability increased. Typical slough species such as *Utricularia purpurea, Eleocharis elongata* (slim spikerush), and *E. cellulosa* (gulf coast spikerush), which tend to be competitively displaced as eutrophication proceeds, have been suggested as indicators of background (unenriched) phosphorus levels (Vaithiyanathan and Richardson 1999).

Based on Gerloff and Krombholz's (1966) work on plant tissue nutrient concentrations, Daoust and Childers (1999) used plant tissue nutrient concentrations in *C. jamaicense* as an indicator of nutrient availability. They used nitrogen/phosphorus ratios to investigate nutrient limitations. Phosphorus was found to be strongly limiting in both the sawgrass and wet prairie communities. *C. jamaicense* had a significantly higher N/P ratio than two other species that coexist in the sawgrass community (N/P for *C. jamaicense* was 61.7; *Peltrandra virginica,* 33.2; and *Pontederia cordata,* 42.2), indicating that it has very low requirements for phosphorus. As a result, this species is able to capture and use phosphorus more efficiently when it is limiting and so is able to dominate large expanses of the ecosystem. This explanation of the structure of the Everglades communities, which are dominated by sawgrass under low nutrient conditions, has its basis in Tilman's resource-based theory of competition (Tilman 1982; Grace 1991).

8

Invasive Plants in Wetlands

I. Characterization of Invasive Plants

Wetlands and other water bodies around the world have been drastically altered by invasive species. Wetlands with strictly native vegetation are increasingly rare (Bazzaz 1986; Meffe and Carroll 1994; Cronk and Fuller 1995; Zedler and Rea 1998). Wetland invasives directly affect humans by obstructing water flow, reducing the recreational value of waters (lower accessibility, decreased fish production, clogged boat motors, increased habitat for hosts of parasitic diseases), and blocking hydroelectric and other installations (Van Zon 1977).

Before we can begin our discussion of invasive plants in wetlands, it is necessary to define some of the common terms used in this field. Terms used to describe plants that were historically absent from an area include *exotic, non-indigenous, alien, adventive, immigrant,* and *non-native* (Luken 1994), all of which are roughly synonymous. We have chosen to use the term *exotic* throughout this chapter. The 'opposite' category of plants, (i.e., those that originated in an area) are called *native* or *indigenous*. We use the term *native* in this chapter. Since species are naturally in flux and their distributions shift with time or disturbance, their status as *native* or *exotic* can be difficult to establish. Evidence from fossil and historical records and results from genetic studies are used to determine the origins of plants. Choosing a date or period after which newly arrived plants are considered exotic is problematic. Should we choose the last glaciation as a cutoff date? The introduction of agriculture? Post-colonial settlement? (Schwartz 1997). In this chapter, the plants we describe as exotic have been established as such by many others before us.

The focus of our chapter is *invasive* plants which may be either native or exotic. Invasive plants grow in profusion and produce a significant change in terms of community composition or ecosystem processes. They grow in agricultural or natural areas; we are mostly concerned with invasions of natural areas. Many use the term *weed* as a close synonym for invasive. An example of an exotic invasive is the widespread floating plant, *Eichhornia crassipes* (water hyacinth), which is native to South America, but exotic in waters throughout most of the tropics and subtropics. Another example is the purple-flowered emergent, *Lythrum salicaria* (purple loosestrife), which is native to Eurasia but exotic in the U.S. and Canada. Several species of *Typha* (cattail) are native to North America, but grow as invasives in areas that are disturbed, such as the Florida Everglades. Most of the invasives we describe in this chapter are exotics. While the majority of exotic plants become *naturalized* (integrated into the native flora without monopolizing space or resources or displacing native plants and animals), about 15% of them become invasive (Office of Technology Assessment 1993).

Often the same plants that are invasive in part of their range are desirable elsewhere. For example, *Heteranthera reniformis* (mud plantain) is on the list of endangered plants in Connecticut, but is among the worst invasives of northern Italian rice fields. *Trapa natans* (water chestnut) is extirpated or endangered in many parts of Europe and an important crop in India, yet it is a noxious invasive in eastern North America, and a serious threat to the sturgeon fisheries in the southern part of the Caspian Sea (Cook 1993). *Melaleuca quinquenervia* (melaleuca) has invaded and caused damage to wetland ecosystems in Florida, but in its native range in Australia, it has been nearly eliminated by habitat destruction (Bolton and Greenway 1997; Turner et al. 1998). *Phragmites australis* (common reed) is declining in parts of Europe and resource managers there are striving to understand its decline in order to restore its range. In North America, on the other hand, *P. australis* is considered an aggressive invasive and controlling it is vital to the restoration of many eastern salt marshes.

Wetland invasives are successful in new ranges for a number of reasons:

- Invasives usually spread rapidly by both sexual reproduction and vegetative regeneration. Some have prolific seed production, such as *Lythrum salicaria* (purple loosestrife), which produces up to 2.7 million seeds per plant (Mal et al. 1992). Many of the most noxious invasives, such as several members of the submerged Hydrocharitaceae (frogbit) family, spread entirely by vegetative regeneration in some habitats because only one sex of the plant is present. The vegetative spread of submerged or floating species is most rapid in the tropics and where water levels remain constant. In tropical waters, the floating plants *Salvinia minima* (water fern) and *Eichhornia crassipes* (water hyacinth) have been observed to double their areal extent in 3.5 and 13 days, respectively (McCann et al. 1996). *Salvinia molesta* (salvinia) doubles its area in 7 to 17 days. In Lake Kariba between Zimbabwe and Zambia, *S. molesta* was first reported in 1959. Its area had expanded to 39,000 ha 13 months later and by 1962 it occupied about 100,000 ha (Cook 1993).
- The aquatic environment is relatively uniform, and many species, particularly submerged and floating-leaved plants, are *cosmopolitan* (widely distributed throughout the world). Several species, such as *Ceratophyllum demersum* (hornwort), *Echinochloa crus-galli* (barnyard grass), *Eleocharis dulcis* (Chinese water chestnut), *Ipomoea aquatica* (water spinach), *Oryza rufipogon* (wild red rice), and *Pistia stratiotes* (water lettuce), grow in many parts of the world and are considered invasive in some habitats (Ashton and Mitchell 1989; Cook 1993).
- Many wetland plants have wide ecological tolerances. As generalists, they are capable of becoming dominant under the right circumstances (Cook 1985, 1993; Thompson et al. 1995; Daehler 1998; Pysek 1998). For example, an invasive loosestrife of Californian vernal pools, *Lythrum hyssopifolium*, is able to germinate in a variety of soil moisture and temperature conditions, making it a successful generalist among native vernal pool plants which require a specific set of conditions for germination (Bliss and Zedler 1998).
- Invasive exotics are usually not susceptible to pests or herbivores in the new habitat. The consumers or diseases that evolved in the same location as the exotic plant do not accompany the plant to its new range (Galatowitsch et al. 1999a).
- Invasive exotics encounter little competition from native plants in their new ranges (Lugo 1994). Native plants evolved to exploit separate niches, thereby minimizing competition with other plants of the same habitat. Since exotics'

competitor plants are usually not present in the new range, exotics are often without direct competitors. They displace native plants because they tend to grow quickly and monopolize resources and light. In New Zealand lakes, a viable shoot of the submerged genus *Lagarosiphon* (African elodea) may settle on a mixed native community of 15 to 150 cm height in shallow water (2 m). Long roots grow from the *Lagarosiphon* shoot to the sediment. Once the plant starts to grow side branches fall and produce more roots and small clumps of *Lagarosiphon*. The clumps may coalesce and eventually smother the native community (Howard-Williams 1993).

- Some invasives are resistant to flooding, fire, and drought (Flack and Benton 1998). The evergreen hardwood, *Melaleuca quinquenervia*, introduced to Florida in the 1880s, is highly flood-tolerant and fire-resistant and therefore capable of rapidly recolonizing burned wetlands (Ewel 1986). The invasive tree, *Tamarix ramosissima* (saltcedar) is more drought- and salt-tolerant than native inhabitants of many southwest riparian zones such as *Pluchea sreicea, Populus fremontii, Prosopis pubescens, Salix exigua,* and *S. gooddingii*, and it is able to dominate when periods of drought are prolonged (Figure 2.14; Busch and Smith 1995; Cleverly et al. 1997).

Plants have spread around the world by natural dispersal mechanisms throughout time. Recently, human transport and land use practices have increased the rate at which species are introduced to new habitats. People introduce wetland plants to new habitats in a number of ways:

- People introduce species to new habitats unintentionally. Such transport started centuries ago, and each new development in transportation has created new opportunities for the transport of exotic plants. Seeds travel along roads by hitching rides on vehicles. They are also carried by ships in food stores and ballast water. Three noxious invasives of Florida's waterways, *Pistia stratiotes, Salvinia minima,* and *Alternanthera philoxeroides* (alligatorweed), were probably accidentally released through the discharge of ship ballast (McCann et al. 1996).
- Some invasive wetland exotics are escapes from agriculture. Examples include *Trapa natans* (water chestnut) and *Eleocharis dulcis* (Chinese water chestnut). Both are Eurasian species grown as a food source; they are considered to be invasive in some North American waters. *Rorippa nasturtium-aquaticum* (=*Nasturtium officinale*; water cress), cultivated for its edible leaves, is an invasive in New Zealand (Howard-Williams 1993). The weeds of ricefields, such as *Cyperus squarrosus, Eleocharis olivacea, Lindernia anagallidea, L. dubia,* and *Najas gracillima*, spread to natural areas when their seeds are included in exported rice (Cook 1985). The North American *Acorus calumus* (sweet flag), grown for its oil that is used in medicine and perfume, is an invasive in Europe and South America (Cook 1996). *Arundo donax* (giant reed), used for canes and woodwind reeds and as an erosion control on shorelines, colonizes southwest riparian wetlands of the U.S. Several exotic grasses have been cultivated in the U.S. in the search for better cattle forage. *Brachiaria mutica* (paragrass), *Panicum repens* (torpedograss), and *Pennisetum purpureum* (napier grass) are adapted to wet soils and have become invasive in wetlands of the southeastern states (McCann et al. 1996).
- Some exotics, such as *Lythrum salicaria, Butomus umbellatus* (flowering rush), *Hydrocleys nymphoides* (water poppy), and *Aponogeton distachyos* (Cook 1996),

have escaped from horticultural uses. The tree *Schinus terebinthifolius* (Brazilian pepper) was intentionally planted throughout southern Florida for its dense masses of scarlet berries and evergreen foliage. It escaped to natural areas where it displaces native vegetation (Ewel 1986; McCann et al. 1996).

- People have transported several submerged and floating-leaved plants, such as species of *Cabomba, Egeria, Elodea, Hydrilla,* and *Vallisneria,* throughout the world because they have attractive foliage and are used in aquaria (Cook 1996). Most aquarium plants that have become invasive were deliberately stocked in natural waters to create wild populations to be harvested and sold at a later date (McCann et al. 1996).
- Sometimes people intentionally introduce an exotic species in the hope of solving a problem. The Australian tree *Melaleuca quinquenervia* was brought to the U.S. at the beginning of the 1900s because its high evapotranspiration rate lowers water levels. It was planted in the Everglades of Florida in an effort to make the area suitable for agriculture. Several species of *Casuarina (C. equisetifolia, C. glauca, and C. cunninghamiana;* Australian pine) were introduced to Florida before 1920 to form windbreaks along coastal areas and are now widespread in southern Florida (Ewel 1986; McCann et al. 1996; Turner et al. 1998).
- Some botanically interesting wetland plants have been transported to new habitats for study or teaching, such as species of *Azolla, Salvinia, Lagarosiphon,* and *Lilaeopsis* (Cook 1985). *Mimosa pellita* (commonly called both catclaw mimosa and giant sensitive plant; formerly *M. pigra*), an emergent South American plant of river banks, may have been introduced to North America as a botanical curiosity because its leaves fold on touch. Its presence in southern Florida is being closely watched as some believe it may displace native vegetation (McCann et al. 1996).

Once a species is introduced, its ability to become established and expand its territory depends on whether it has traits that are pre-adapted to the new habitat. If the new species' seeds or propagules are easily dispersed and dispersal agents such as waterfowl or humans are plentiful, then the likelihood it will spread throughout a region is enhanced (Chambers et al. 1993). Connections between regions such as ditches and canals, and activities such as increased nutrient loading, vegetation removal, altered hydrology, and changed salinity also increase the probability that invasive species will reach new habitats (Galatowitsch et al. 1999a).

II. The Extent of Exotic Invasions in Wetland Communities

It is estimated that at least 4000 foreign plant species (not including crop plants) and 2300 animal species have become established in the U.S., as well as hundreds of animal and plant pathogens. About 15% are nuisance species (Office of Technology Assessment 1993) and the effort to eradicate them costs U.S. taxpayers billions of dollars each year (the estimated annual cost in 1999 was $123 billion). This cost does not include the incalculable effects invasive plants and animals have on native ecosystems such as local extinction of species that are not of economic value (Simberloff 1996).

The success of an exotic species in a new range may reflect the conditions of the community being invaded rather than the aggressive traits of the exotic (Lugo 1994). On islands, for example, exotic invasions are especially dramatic (Vitousek 1994). Exotic species amount to as much as 20% of most continental nations' flora and fauna, but the proportion of exotics on islands is as much as 50% (Vitousek et al. 1996). Islands tend to import more plant species than they export. For example, the islands of New Zealand have received 42 wetland plant species and they have exported only one (Cook 1985). Exotics amount to about 20% of New Zealand's wetland flora, and many species, such as *Ceratophyllum demersum, Lagarosiphon major* (African elodea), *Elodea canadensis* (elodea), and *Egeria densa* (egeria), cause commercial losses to hydropower stations and threaten recreational waters. Species of the Hydrocharitaceae family dominate in almost every lake they have invaded in New Zealand, in part because New Zealand has no native canopy-forming submerged plants (Howard-Williams 1993).

In the U.S., the states most impacted by exotic invasives are Hawaii and Florida. Hawaii is the most remote island group in the world, separated from the continents by 4000 km of ocean. Few plant and animal species colonized the islands prior to human settlement and from them, thousands of endemic species evolved. Hawaii's tropical climate means it is subject to invasion by many species that would be eliminated in areas with frost. In addition, Hawaii is a transportation hub between Asia and North America. Heavy volumes of air and naval traffic increase the chances of exotics reaching the islands.

While Florida is not an island, it originally had relatively depauperate plant and animal communities since the waters surrounding most of the state excluded entry from tropical regions, and plants that thrived in temperate regions to the north were naturally excluded by the climate. Today its many routes of entry and rapidly growing human population have made controlling the entry of exotics nearly impossible. The subtropical climate makes it attractive for the year-round growth of ornamental and aquarium plants and many invasives have entered the state's natural areas as a result of these horticultural industries (Office of Technology Assessment 1993).

Disturbed sites are often susceptible to invasion. Disturbance can lead to opportunistic exploitation by invasive species, especially if they were present in small numbers before the disturbance. A disturbance may significantly alter environmental conditions (for example, making the habitat drier or more nutrient-rich) and invasives may be better suited to exploit them. Natural disturbances such as hurricanes and other storms (which affect whole geographic regions), fires (which affect a region or community), or fish nests and turtle trails (which create a disturbance within a community) can make a site susceptible to invasion. Humans cause disturbances to wetlands by altering wetland hydrology, developing wetlands or land adjacent to wetlands, and by releasing nutrients and pollutants into the air and water (Rejmanek 1989; Chambers et al. 1993; Vitousek 1994).

Some examples of human-caused disturbances that may lead to plant invasions are:

- Land use changes open formerly vegetated land and the most rapid colonizers (often with weedy tendencies) are the first to take over the open space. Such changes are seen in deforested watersheds, construction sites, abandoned farm land, drained or stressed wetlands, heavily grazed areas, roadsides, canals, and ditches (Rea and Storrs 1999).
- The damming and impoundment of nearly all of the major rivers in the U.S. have led to invasive problems by eliminating variations in the rivers' hydrology to which native species are adapted. In the southwest, the construction of dams along the Colorado River has lowered groundwater tables, and floods no longer scour river banks. Many western riparian communities have shifted from *Populus-Salix* (cottonwood–willow) forests to stands of *Tamarix* (Busch and Smith 1995; Vitousek et al. 1996).
- The fragmentation of natural habitats with agricultural and urban development has encouraged the spread of exotics. Weeds from farm fields and plants cultivated

in cities easily move from human-influenced habitats into natural ones (Vitousek et al. 1996). *Ambrosia trifida* (great ragweed) is a common weed in agricultural and urban landscapes that is also invasive in dry and wet natural areas.

- Freshwater inflows into salt marshes change the plant community structure. In California, plant invasions of tidal wetlands are often associated with storm drains, overflows of agricultural irrigation, and sewage spills, which bring about a decrease in salinity. The exotic grass *Polypogon monspeliensis* has colonized disturbed tidal marshes in southern California because it can outcompete the more salt-tolerant native plants (Kuhn and Zedler 1997).
- Climate change caused by increasing CO_2 levels may bring about shifts in the species composition of many communities. For example, California's vernal pool plant communities may be particularly susceptible to climate change. Increasing temperatures during the rainy season, or changes in the timing of the initial rains or in the occurrence of aseasonal rains that saturate the pools, may all bring about conditions to which native plants are not adapted. Vernal pools have already suffered enormous losses from human development and land use changes. Climate change may bring about a shift toward more widespread or exotic species and a further decrease in the species richness of vernal pools (Bliss and Zedler 1998).

III. Implications of Invasive Plant Infestations in Wetlands

Invasive wetland plants pose a serious threat to wetlands and waterways around the world. Invasive plants can replace desirable plants, displace animals, affect ecosystem functions by altering hydrology and nutrient cycling, and negatively affect humans by impeding waterways and harboring disease vectors (Simberloff 1996).

A. Changes in Community Structure

When exotics invade a new range, native plants, adapted to the environment, are sometimes displaced (Mills et al. 1993; Vitousek et al. 1996). Community changes arise through a variety of processes including interspecific competition and disturbance. A general trend is the loss of plant species diversity as communities shift from desirable plants to monospecific stands of the invasive species.The mechanisms by which invasives outcompete other species are not always known or quantified, but rapid growth and proliferation certainly play an important role.

Several introduced wetland tree species illustrate the capacity of invasives to alter the plant community structure and habitat. The Australian tree *Melaleuca quinquenervia* spreads rapidly. In the 1990s, it was increasing its range in south Florida by about 35 acres each day, replacing *Taxodium distichum* (bald cypress) and other native plants, particularly wherever cypress trees grow under stressful conditions (Figure 8.1; Myers 1984; Turner et al. 1998). An aggressive evergreen of Florida, *Schinus terebinthifolius,* typically moves into areas that have been at least partially drained by people. It forms dense stands that eliminate the herbaceous understory. *S. terebinthifolius* seedling survival is unusually high (66 to 100%) and their success impairs competition by native plants. In addition, *S. terebinthifolius* appears to be allelopathic, suppressing the growth of other plants (Ewel 1986; McCann et al. 1996). *Rhizophora mangle* (red mangrove) was planted on the Hawaiian island of Oahu where it has created dense forests up to 22 m high. Mangrove forests have affected native plants by creating shade, and the nearly impenetrable root system has

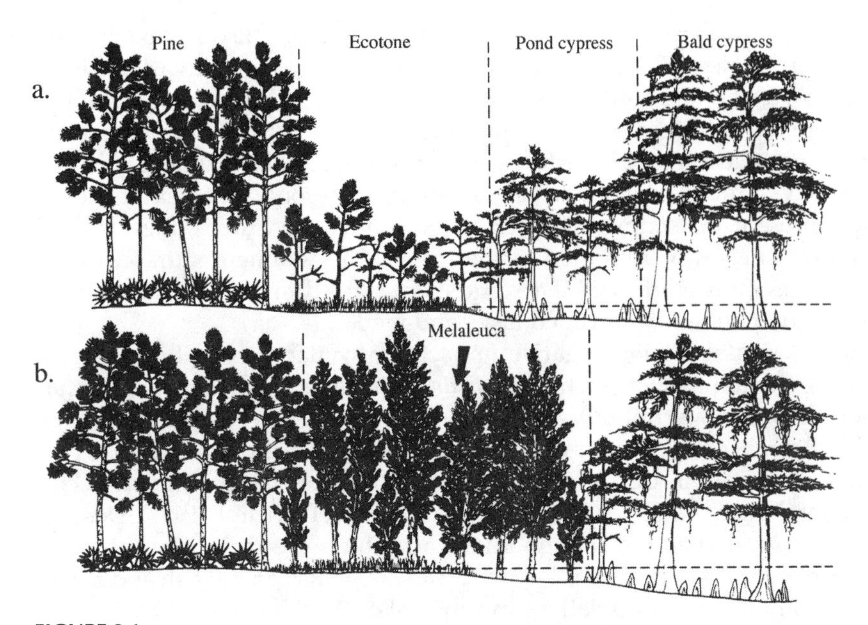

FIGURE 8.1
(a) A typical gradient in south Florida from dry pine forest to wet bald cypress forest with an intermediate zone that is not particularly favorable for either community. (b) Replacement of this intermediate zone by *Melaleuca quinquenervia* (melaleuca). (From Myers, R. 1984. *Cypress Swamps*. K.C. Ewel and H.T. Odum, Eds. Gainesville. University Presses of Florida. Reprinted with permission.)

altered the animal community and soil oxygen. In U.S. western riparian zones, *Tamarix ramosissima* (salt cedar) forms new forests or replaces native ones to the detriment of numerous native plant and bird species (Busch 1992; Busch and Smith 1995).

Many exotic emergents are able to outcompete native vegetation by rapidly filling in unvegetated areas and crowding out native plants. A native in tropical Asia, the emergent *Colocasia esculenta* (taro) grows in dense clumps along lake and river margins in Florida and crowds out native vegetation. *Brachiaria mutica* displaces native plants through rapid growth, and by producing allelopathic chemicals that inhibit other plants' growth. The Brazilian emergent *Alternanthera philoxeroides* has become an aggressive plant in many of Florida's waters. Its hollow stems, which grow up to 15 m in length, extend over the water's surface and enable these normally emergent plants to form dense floating mats. The mats reduce submerged native plants' habitat by shading the water column. Mats of the floating plants *Pistia stratiotes, Eichhornia crassipes,* and *Salvinia* species also eliminate submerged vegetation habitat by shading the water column (McCann et al. 1996; Rea and Storrs 1999).

Exotic species sometimes form hybrids with native plants, thus creating species that become new invasives and altering the genetic makeup of the community. For example, *Spartina alterniflora* (cordgrass), a native of eastern U.S. salt marshes, formed a hybrid with *S. maritima* when it was introduced to French and English marshes in the 1800s. The hybrid *S. townsendii* was sterile, but a mutation yielded a new species, *S. anglica,* that has proven to be an aggressive invasive along European coastlines (Beeftink 1977). In New Zealand, the exotic shrub *Viburnum opulus* (guelder rose) has become naturalized in bogs where it breeds with *V. americanum*. The resulting hybrid grows more rapidly than the original species (Flack and Benton 1998).

The seed banks of areas infested with invasive plants are also altered. In a study of the seed banks of 21 New Zealand lakes with varying degrees of invasion, deWinton and Clayton (1996) found that seed number and seed species richness were significantly lower at sites where the submerged community was dominated by exotics. The exotics, *Elodea canadensis, Egeria densa,* and *Hydrilla verticillata* (hydrilla) formed tall canopies with high biomass solely through vegetative regeneration, since only one sex of these dioecious plants was present. As sediments accumulated under the exotics, the seeds of formerly present native species were buried farther below the sediment surface. Even if control measures successfully eradicated the exotic species, the diminished seed bank would limit the revegetation potential of invaded lakes and wetlands.

Invasive plants also have negative impacts on wetland animal communities. Dense stands of submerged exotics, such as *Hydrilla verticillata, Egeria densa,* and *Myriophyllum spicatum* (Eurasian watermilfoil), provide refuge for young fish and allow high survival rates, which can lead to overpopulation and stunted fish growth. Because predator fish cannot forage as well in dense weed beds, their numbers and biomass decline as submerged plant density increases beyond an optimal level (Nichols 1991). Dense *Eichhornia crassipes* mats shade benthic communities and inhibit the diffusion of oxygen into the water. Low oxygen concentrations below *E. crassipes* mats can kill fish and dense mats can completely eliminate fish populations in small lakes (McCann et al. 1996).

Waterfowl and other bird habitats are also negatively impacted by the presence of wetland invasives. The Florida Everglades kite (*Rostrhamus sociabilis*) is endangered, in part because *E. crassipes* has invaded much of its habitat. *E. crassipes* outcompetes emergent vegetation which is the habitat of the kite's preferred food, the apple snail (*Pomacea paludosa*). The roots of *E. crassipes* can accumulate heavy metals and toxic organic compounds, which may pose a risk for the endangered West Indian manatee (*Trichechus manatus*) that consumes the plants. Also in Florida, *Casuarina* species have reduced the habitat area of cotton rats (*Sigmodon hispidus*), marsh rabbits (*Sylvilagus palustris*), gopher turtles (*Gopherus polyphemus*), loggerhead turtles (*Caretta caretta caretta*), green sea turtles (*Chelonia mydas mydas*), and American crocodiles (*Crocodylus acutus;* McCann et al. 1996). Dense stands of *Melaleuca quinquenervia* eliminate standing water habitats and create a shift in the local wildlife community from aquatic organisms to upland and arboreal species (O'Hare and Dalrymple 1997).

B. Changes in Ecosystem Functions

Invasive species can alter the abiotic components of their habitat. For example, floating mats of vegetation reduce dissolved oxygen levels in the water by shading the phytoplankton and submerged plants that produce oxygen. In addition, detritus accumulation can decrease the dissolved oxygen content of the water due to the oxygen demand created by its decomposition (Howard and Harley 1998). Floating plants can also alter the normal succession of a wetland. *Salvinia molesta* forms floating mats on which herbaceous plants grow. Eventually woody shrubs and small trees grow there as well. The larger plants have a higher water demand and thereby eliminate open water plants and animals (Cook 1993).

Hydrology can also be altered by plants with high evapotranspiration rates. *Melaleuca quinquenervia* lowers water tables through high evapotranspiration and now infests over 200,000 ha of South Florida, posing one of the greatest threats to the Everglades (U.S. Army Corps of Engineers 1999). *Tamarix ramosissima, T. chinensis,* and several other saltcedar species were originally introduced to the U.S. as a source of wood, shade, and erosion control, and are now considered nuisance species over nearly 400,000 ha of western riparian areas. *T. ramosissima* transpires water at a greater rate than native plants. It roots deeply and lowers water tables, thus eliminating surface water habitats that are vital in the arid southwest. When rain falls, the tree promotes flooding by blocking water channels with its dense growth (Busch 1992; Busch and Smith 1995; Flack and Benton 1998).

Fire regimes may also be altered when exotics take over a habitat. In the riparian areas of southwestern U.S., the Eurasian grass *Arundo donax* forms tall, dense monospecific stands. In the autumn, the dry leaves and stems can fuel intense fires. *A. donax* is fire-tolerant and quickly resprouts from rhizomes. The effect is to transform riparian swamps from flood- to fire-dominated systems where native species cannot survive. Because of the lack of trees and other natives, areas infested with *A. donax* suffer from increased erosion and reduced habitat value and biological diversity (Flack and Benton 1998). In the Australian tree *Melaleuca quinquenervia*, fire induces massive seed release which can create dense stands with up to 250,000 seedlings per hectare. In Florida, *M. quinquenervia* is able to replace less fire-tolerant native vegetation (Turner et al. 1998).

C. Effects on Human Endeavors

Wetland weeds are a nuisance to many human activities and their capacity to harbor disease vectors can seriously threaten human life, particularly in tropical countries. Vectors of human and animal diseases, such as malaria, schistosomiasis, and lymphatic filariasis of the brughian type (also called elephantiasis), have long been serious problems in tropical regions. Floating weeds such as *Eichhornia crassipes, Pistia stratiotes,* and *Salvinia auriculata* exacerbate the situation by expanding the disease vectors' habitat and inhibiting the movement of their fish predators (Hill et al. 1997).

Wetland weeds negatively impact human enterprise in a number of other ways as well (Bos 1997; Hoyer and Canfield 1997; Madsen 1997; Kay and Rice 1999):

- Due to their rapid growth, floating and submerged species fill water bodies and clog water intakes and distribution systems used for irrigation, public water supplies, and hydroelectric generating plants. If plants block water control gates during floods, there may be damage to crops, buildings, and equipment, and possibly loss of life.
- The roots of floating species bind suspended sediments and keep them within reservoirs. When they decompose, sedimentation in flood control reservoirs is increased, thus decreasing their holding capacity. A change in the sediment type (sand, clay, silt, and organic matter) affects plant establishment and growth, invertebrate populations, and fish spawning and feeding.
- Floating invasives interfere with aquaculture because they shade submerged plant refuges and the phytoplankton that fish eat. Oxygen levels are decreased beneath floating mats, resulting in fish kills.
- Both herbaceous and woody exotics can impede boating access and navigation by blocking boat ramps and boat trails. Floating and submerged plants hinder boat travel by covering or filling entire water bodies.
- Piles of live or dead vegetation along residential shorelines, on boat ramps, in swimming areas, and in commercial boating areas create odor problems and can provide a breeding location for mosquitoes and other nuisance organisms.
- Recreational activities such as swimming, boating, waterskiing, and sport fishing are difficult, if not impossible, in the presence of dense weed infestations.

IV. The Control of Invasive Plants in Wetlands

The control of wetland invasives entails eradicating or reducing the plant's growth and preventing its spread. Control also includes restoring native species and habitats to prevent further invasions (Clinton 1999). The most effective control for exotic species is to eliminate their introduction to new ranges entirely. Keeping exotics out of a new range requires legislation and its enforcement and such laws are not in practice worldwide.

In the U.S. the need for legislation regarding aquatic exotics became apparent in the late 1800s when *Eichhornia crassipes* began to impede river traffic in Florida and Louisiana. The U.S. Congress initiated the Removal of Aquatic Growths Project within the Rivers and Harbors Act of 1899. Since then the project has been renewed several times and today it is funded under the Water Resources Development Act of 1986. The state and federal governments generally share the cost of controls (U.S. Army Corps of Engineers 1999).

The acts passed by the U.S. Congress that have a bearing on wetland plant exotics are the Federal Noxious Weed Act of 1974 and the Non-Indigenous Aquatic Nuisance Prevention and Control Act of 1990. The Federal Noxious Weed Act of 1974 is administered by the Animal and Plant Health Inspection Service of the U.S. Department of Agriculture whose task is to identify actual and potential noxious weeds, prevent their entry into the U.S., and to detect and eradicate infestations in their early stages. The Non-Indigenous Aquatic Nuisance Prevention and Control Act of 1990 authorizes the U.S. Fish and Wildlife Service and the National Oceanic and Atmospheric Administration to regulate introductions of both plant and animal aquatic nuisance species such as the zebra mussel (*Dreissena polymorpha;* Hoyer and Canfield 1997).

Wetland invasives are usually controlled using a combination of methods which reduce the plant's growth rather than eliminate it entirely. Controlling invasives can bring about negative consequences if dead plants are left in place to decompose because decaying vegetation reduces oxygen levels, releases plant nutrients, and deposits large amounts of detritus (Nichols 1991). In addition, control of one plant can lead to the success of another unwanted species (Harris 1988). In some Florida waters, when *Eichhornia crassipes* was controlled, the exotic species *Pistia stratiotes* and *Alternanthera philoxeroides* moved in to exploit the newly opened habitat and caused similar problems (Schmitz et al. 1993; McCann et al. 1996). Wherever the decision is made to manage wetland invasives, planners must set specific goals and adapt them to the local situation (Luken 1997). Invasives are controlled using habitat alteration and mechanical, chemical, and biological controls.

A. Habitat Alterations

Habitat alterations such as shading the water or sediment surface, dredging the top layer of sediment, or changing the hydrologic regime of the water body can impede plant growth. Because none of these alterations is specific to nuisance species, they are used to control monospecific stands of a nuisance plant or all of the plant growth in an area.

1. Shading the Water's Surface

The growth of submerged plants can be inhibited by decreasing the amount of light in the water column. Water bodies may be shaded by planting trees, shrubs, or tall herbaceous plants along the shoreline. Tall plants at the edge of the water can provide an effective light barrier; however, their shade only reaches a narrow area near the shore, so this method is only effective in streams or small water bodies. Tall shade plants are an attractive solution because they are natural and do not alter the chemistry of the water or sediments. They also allow some light to penetrate to the water's surface so that some macrophyte production can be maintained (Barko et al. 1986; Nichols 1991).

Dyes can be added to the water that absorb light within the range needed for photosynthesis, thereby creating a chemical shade. Dyes last longest where the water turnover is slow and in clear water because suspended sediments can remove dyes from the water column (Nichols 1991).

Plastic sheeting over the water's surface can also provide shade, but it is not very practical in the long term. The sheets need to be removed periodically for cleaning and they cannot withstand high winds or waves. Their use is generally limited to small areas around boat docks or swimming areas (Nichols 1991).

2. Shading the Sediment Surface

Barriers placed on the sediments block light and inhibit the growth of rooted plants. Sediment barriers are usually black plastic sheets or layers of sand and gravel. Barriers are effective for only short periods because plants return as soon as sediments accumulate on the barriers or when there is a tear in the plastic sheet. Decomposing vegetation trapped underneath the barriers produces gases that can cause sheeting to lift and float to the surface. Some sediment barrier materials are gas-permeable, but they eventually become clogged by debris and microorganisms and then trap gases. Benthic organisms are unable to survive under the barriers (McCann et al. 1996; U.S. Army Corps of Engineers 1999).

3. Dredging Sediments

In the case of extensive aquatic infestations, dredging machines may be used to remove the vegetation and bottom sediments. Dredging is expensive and the plants must be removed to an upland site. Permits are often required for the discharge of dredged material, and dredging is considered an extreme measure that clears an area in the short term, but does not prevent re-infestation (Hoyer and Canfield 1997).

In a Wisconsin lake where managers were trying to eliminate all macrophyte growth, not just nuisance species, the sediments in one area were dredged to expose nutrient-poor soil. In three other vegetated areas sediment barriers of sand, gravel, and plastic were installed. Shortly after these changes were made, filamentous algae invaded the areas with barriers and *Chara* species invaded both the barrier-covered and dredged areas. By the third summer various species of *Potamogeton* dominated the dredged areas and *Najas flexilis* and *Elodea canadensis* grew on the barrier-covered areas. Within 3 to 7 years, all of the areas were densely covered with *Ceratophyllum demersum, Myriophyllum sibiricum* (watermilfoil), and *Potamogeton* and the plant biomass had recovered to pre-treatment levels. Neither dredging nor covering the sediments proved effective in the long term (Engel and Nichols 1984).

4. Altering Hydrology

In lakes, reservoirs, and wetlands with water control structures, water levels can be manipulated in order to control aquatic weeds. Raising the water level drowns emergents, while lowering the level exposes submerged plants to freezing, drying, or heat. Drawdown, which refers to the lowering of water level, is more commonly used than raising water levels. Drawdowns are usually conducted during the winter so that plants are exposed to both drying and freezing. Summer drawdowns negatively impact agricultural and recreational water use and stress fish populations (Hoyer and Canfield 1997).

Drawing down the water level of a water body to expose the sediments of the rooted plant zone can bring about short-term (1 to 2 years) control of some of the rooted species. The control is most effective if the sediments are nearly completely dewatered and subjected to more than a month of either freezing or heat. If there is groundwater seepage that maintains wet sediments, the drawdown may be ineffective. A thorough knowledge of the water budget is necessary before a drawdown is initiated. The advantage of using a drawdown as a control measure is that drawdowns do not entail the addition of chemicals or the cost of machinery for harvesting. Lakes with gradual basin slopes are ideal for drawdowns since small drops in water level can expose large areas (Cooke 1980).

The response to winter or summer drawdown depends on the plant species and on site specifics (Table 8.1). *Myriophyllum spicatum* has been controlled using a winter drawdown with a period of freezing temperatures longer than 3 weeks, although in some sites it has shown no response to drawdowns (Cooke 1980). Drawdowns have been used to successfully remove submerged invasive Hydrocharitaceae species in New Zealand lakes in the 1- to 4-m depth zone (Howard-Williams 1993).

Drawdowns have some disadvantages. Some plants increase growth under drawn-down conditions. Undesirable resistant plants, such as *Alternanthera philoxeroides* and *Hydrilla verticillata,* have extended their range during drawdowns (Table 8.1). If freezing is required, lakes in warm areas are not candidates for this control measure. Drawdowns are ineffective against emergents and can even encourage their spread, since many only germinate on mudflats (Cooke 1980). Drawndown wetlands can negatively impact aquatic furbearers, waterfowl, reptiles, amphibians, and fish (Nichols 1991).

TABLE 8.1

Responses of Some Common Nuisance Wetland Macrophytes to Drawdown

Species	Common Name	No. of Observations	Seasons
Decreased			
Chara vulgaris	muskgrass	1	Winter
Eichhornia crassipes	water hyacinth	2	Annual
Nuphar spp.	spatterdock	3	Winter
Increased			
Alternanthera philoxeroides	alligatorweed	3	All seasons
Najas flexilis	naiad	7	All seasons
Potamogeton spp.	pondweed		Most increase or do not change
Hydrilla verticillata	hydrilla	1	Winter
No Change or Clear Response			
Cabomba caroliniana	fanwort	3	Annual, winter
Elodea canadensis	elodea	2	Winter
Myriophyllum spp.	milfoil	5	Annual, winter
Utricularia macrorhiza	bladderwort	3	Winter

From Cooke, G.D. 1980. *Water Resources Bulletin* 16: 317–322. With permission from the American Water Resources Association.

B. Mechanical Controls

The mechanical control of nuisance plants entails harvesting plants or mechanically disturbing the sediments. Harvesting includes collecting plants and transporting them to shore for disposal. Harvesting usually does not completely remove a species, but by reducing the biomass and clearing an area, the hope is that desirable plants will be able to move in before the weed grows back. At the least, near-shore recreational areas are kept clear. The sediments may be disturbed by rolling over or tilling the soil. Several manual and mechanical harvesting methods are in use (Table 8.2).

Harvesting is usually practical only in small areas like marinas, swimming areas, and fishing trails or where other methods are undesirable or unfeasible. Harvesting with machinery can be expensive because of the cost and maintenance of equipment, but it is an efficient way to provide immediate, tangible results (McCann et al. 1996). In a New Zealand lake, mechanical control was used to harvest the submerged weed *Lagarosiphon major*. The regrowth of *L. major* after harvesting was patchy and slow and native *Nitella* (muskgrass) species were able to successfully recolonize the open areas (Howard-Williams 1993).

Harvesting can have undesired effects since it can increase the population of a submerged weed that regenerates from fragments. Fish and invertebrates may be affected since they are sometimes removed with the vegetation and their hiding, spawning, and

TABLE 8.2

Mechanical Methods Used to Control Wetland Weeds in the U.S. and New Zealand

Handpulling is used where labor is inexpensive or in small infestation areas. Handpulling is like weeding a garden and works best if the entire plant including the roots is removed. This method leaves beneficial species intact. It works best in soft sediments with shallow-rooted species. Handpulling usually needs to be repeated several times throughout the growing season. If plant removal might result in shoreline erosion, other plants are planted to replace the invasive species.

Manual cutting is done with scythes or specialized underwater weed cutters. Manual cutting reduces the plant's biomass, but does not remove the entire plant. The cut plants float to the surface and are removed to an upland location.

Floating booms are placed at an angle across the current to collect floating weed masses and concentrate them at a single site on the shore for removal.

Mechanical screen cleaners that rake the intake screens of hydropower stations pull off vegetation as it accumulates.

Mechanical harvesters are large machines that cut and collect submerged and emergent plants. They can cut up to 3 m below the water's surface in a swath 1.8 to 6 m wide. Mechanical harvesters are used to open boat lanes. As with manual cutting, the lower portion of the plant remains, so harvesting must be repeated. Mechanical harvesters may impact fish and invertebrate populations.

Weed rollers are used to compress plants and soil. The roller is anchored in place and is up to 30 ft long. It rolls over an area repeatedly and inhibits plant growth. The weed roller is left in place and requires minimal effort; however it can disturb benthic organisms and fish, and it can be dangerous if people swim into the area.

Rotovators, or underwater rototillers, dig into the sediments and dislodge roots. The uprooted plants are removed manually or with a rake attachment. The rotovator works best with short plants and in large waterbodies. It is an expensive method that creates a high degree of sediment disturbance. It is effective in rapidly clearing areas.

From Howard-Williams 1993; U.S. Army Corps of Engineers 1999.

TABLE 8.3

The Susceptibility of Selected Wetland Plants to Various Herbicides

	Complexed Copper	2,4-D Butoxyethyl Ester	2,4-D Dimethylamine (DMS)	Diquat	Diquat + Complexed Copper	Endothal Dipotassium Salt (K2)	Endothal K2 + Complexed Copper	Endothal Dimethylamine Salts	Fluridone	Glyphosate
Emergent and Floating-Leaved Plants										
Alternanthera philoxeroides (alligatorweed)			G						G	G
Brachiaria mutica (paragrass)									G	E
Brasenia schreberi (watershield)		E	E	F	F	G	G	F	G	
Cladium jamaicense (sawgrass)										G
Hydrocotyle spp. (water pennywort)		G	G	E	E					
Justicia americana (water willow)			E	F					G	
Leersia oryzoides (rice cutgrass)									G	
Ludwigia spp. (water primrose)		E	E	F		F	F	F	F	
Nelumbo lutea (American water lotus)		E	E	E		G	G	F		G
Nuphar spp. (spatterdock)		E	G			G	G	F	G	E
Nymphaea odorata (fragrant water lily)		E	G			G	G	F	G	E
Panicum hemitomon (maidencane)				F					F	E
P. repens (torpedograss)									G	E
Paspalum dilatatum (watergrass)									G	
P. paniculatum (water paspalum)				F					G	
Phragmites australis (common reed)										G
Polygonum spp. (smartweed)		G	G	F		G	G		F	F
Pontederia spp. (pickerelweed)		G	G							F
Scirpus spp. (bulrush)		E	E	G	F					E
Setaria magna (giant foxtail)										G
Trapa natans (water chestnut)		E								
Typha spp. (cattail)		G	G	G					G	E
Zizaniopsis milacea (giant cutgrass)									G	E
Floating Plants										
Eichhornia crassipes (water hyacinth)			E	E	E	F	F	F		F
Lemna spp. (duckweed)		G	G	E	E	F			E	
Pistia stratiotes (water lettuce)		F		E	E	G				G
Salvinia spp. (salvinia)				E	E	F			G	

Spirodela polyrhiza (giant duckweed)		G	G	E	E				G	
Wolffia spp. (watermeal)				G					F	
Submerged Plants										
Cabomba caroliniana (fanwort)		F	F	G	E	E	E	E	G	
Ceratophyllum demersum (hornwort)		F	F	E	E	E	E	E	G	
Egeria densa (egeria)				G	E	E	E	E	G	
Elodea canadensis (elodea)				E	E		F	G	G	
Hydrilla verticillata (hydrilla)	G			G	E	G	G	G	G	
Myriophyllum aquaticum (parrot feather)[a]		E	E	E	E	E	E	E		F
M. spicatum (Eurasian water milfoil)		E	E	E	E	E	E	E	G	
Najas spp. (naiad)		F		E	E	E	E	E	G	
Potamogeton spp. (pondweed)				G	G	E	E	E	G	
Ranunculus spp. (water buttercup)				E	E		F			
Ruppia maritima (wigeongrass)				G	E	F	F	F		
Utricularia spp. (bladderwort)		G		G					G	
Vallisneria americana (wild celery)				F	F			F		
Zannichellia spp. (horned pondweed)		F	F			E	E	E	F	

Note: Herbicide labels should be consulted for the most current information (F = fair, G = good, E = excellent).

[a] Formerly called *M. brasiliense.*

Adapted from Westerdahl and Getsinger 1988.

feeding areas in submerged plant beds are eliminated (Van Zon 1977; Nichols 1991). Mechanical harvesting of trees may prove ineffective since many trees resprout from the stump. In Florida an attempt to remove large stands of *Melaleuca quinquenervia* from islands was only partially successful; just 4 months after clear-cutting, 66% of the cut stumps had resprouted (Stocker 1999).

C. Chemical Controls

Chemical herbicides are used throughout the world to control nuisance wetland plants. In the U.S., the widespread use of relatively safe organic herbicides began in the 1940s, when researchers at the U.S. Department of Agriculture and the Everglades Experiment Station of the University of Florida experimented with the newly discovered herbicide 2,4-D as a control agent for *Eichhornia crassipes*. The herbicide was effective against the target plant and was not toxic to fish, cattle, or humans. In 1947, 2,4-D was widely applied in water bodies containing *E. crassipes*, and for the first time in decades, infested streams were open to navigation (McCann et al. 1996).

Today about 200 herbicides are registered in the U.S., but fewer than a dozen are labeled for use in aquatic sites. Two of these, xylene and acrolein, are highly toxic and used only in irrigation systems of the 17 western states under the jurisdiction of the U.S. Bureau of Reclamation. The remaining herbicides, sold under various trade names, contain combinations of seven active ingredients: copper, 2,4-D, dichlobenil, diquat, endothall, fluridone, and glyphosate (Table 8.3). Few herbicides are available for aquatic applications because the market for them is small compared to the agricultural market. In addition, the aquatic environment presents a challenge because herbicides are instantly diluted when they are applied to underwater plants or sediments. Herbicides should be quickly absorbed and application rates must be sufficient to harm the target plants without affecting other organisms or people (Hoyer and Canfield 1997).

Herbicides are either *selective*, meaning they affect only specific species, or they are *broad spectrum*, killing a variety of vascular plant species as well as algae. Glyphosate, diquat, endothall, and fluridone are used as broad-spectrum aquatic herbicides, but can also be used selectively on individual plants because they only kill the plants they contact. Since broad-spectrum herbicides kill all plants, the newly devegetated area is left open for opportunistic species unless desirable species are planted instead. Selective herbicides, such as the aquatic herbicide 2,4-D, control certain plants but not others. The amount of herbicide used controls its selectivity. For example, if 2,4-D is applied to *Eichhornia crassipes* at the recommended rate, it selectively kills that species. At a higher application rate it controls other species, such as *Nuphar* (spatterdock), as well (Hoyer and Canfield 1997). For all types of herbicides, it is beneficial to reduce herbicide use to the lowest effective application rate.

Herbicides work in two ways: through contact with exposed plant tissues (*contact herbicides*) or by moving through the plant from adsorption sites to critical areas (*systemic herbicides*). Contact herbicides (also called limited movement herbicides) harm the tissue to which they are applied by inhibiting photosynthesis almost immediately. The plant's oxygen is depleted by normal cellular respiration and by bacteria breakdown of the exposed tissue. The oxygen is not replenished by photosynthesis, and the plant releases nutrients soon after contact, so the tissue dies. Contact herbicides are not translocated to underground tissues, so perennial structures such as rhizomes or tubers are left intact and perennial plants are able to re-infest the area. Contact herbicides work quickly so that recreational uses of the water body can be restored. They are most effective on annual,

TABLE 8.4

The Advantages and Disadvantages of Herbicide Use to Control Wetland Invasives

Advantages

Herbicides

- Are usually easy to apply.
- Usually act rapidly to remove nuisance plants.
- Can be used in a variety of water depths and wetland types.
- Are often less expensive than other control methods.
- Can easily be applied around underwater obstructions and structures such as docks.
- Can be applied directly to problem areas of all sizes.

Disadvantages

Herbicides

- Can adversely influence non-target plants.
- Can be toxic to fish, birds, or other aquatic animals when not used according to the manufacturers' specifications. Fish kills are possible when a herbicide, such as copper or the amine salt of endothall, is applied in an enclosed water body. Fish kills can also occur as an indirect effect if the decaying plant matter causes an increase in the biochemical oxygen demand for prolonged periods, during which fish may die due to a lack of oxygen.
- Often have fishing, swimming, drinking, irrigation, and other water use restrictions. A waiting period of up to 30 days for some uses and for some herbicides is recommended. Users need to be aware of all restrictions.
- May take several days to weeks or several treatments during a growing season to control or kill target plants.
- Require special training and permitting.

From Nichols 1991; Hoyer and Canfield 1997; U.S. Army Corps of Engineers 1999.

slow-growing, or senescent plants. Contact herbicide treatments are usually repeated two or three times per year because parts of the plant survive. Endothall, diquat, and copper are contact herbicides.

Systemic herbicides are translocated from absorption sites to critical points in the plant. Death occurs more slowly, and increased oxygen demand does not occur as quickly as for contact herbicides. Nutrients are released from plant tissues over a longer time period. Systemic herbicides that are absorbed by plant roots are referred to as *soil-active* herbicides and those that are absorbed by leaves are called *foliar-active* herbicides. Systemic herbicides may cause fewer environmental problems than contact herbicides because the ecosystem has more time to assimilate the oxygen demand and nutrient release. However, systemic herbicides must be used with care. If the application rates are too high, systemic herbicides act like contact herbicides and stress the plants so much that translocation to critical plant growth areas does not occur. Systemic herbicides are generally more effective for controlling perennial and woody plants and they have more selectivity than contact herbicides. Dichlobenil, 2,4-D, fluridone, and glyphosate are systemic herbicides (Nichols 1991; Hoyer and Canfield 1997).

While herbicides are often effective and easy to use, concerns regarding the environmental safety and human health risks of herbicides and the other potential drawbacks of their use sometimes make planners and managers hesitant to use them (Table 8.4). In an effort to decrease the risks associated with herbicides, the U.S. Environmental Protection Agency requires that the effects and eventual fate of herbicides be thoroughly tested before they can be sold (Table 8.5). Herbicides are labeled with instructions for storage and

TABLE 8.5

Information Required by the U.S. Environmental Protection Agency Concerning the Safety of Herbicides before They Can Be Sold for Use in Aquatic Ecosystems

The potential residue in potable water, fish, shellfish, and crops that may be irrigated
The breakdown products of the herbicide
The environmental fate of the compound and its breakdown products
The entry route of the herbicide in animals (i.e., through the skin or by other means)
The short-term or acute toxicity of the compound to test animals
The potential for the compound to cause birth defects, tumors, or other abnormalities after long-term exposure
The toxicity of the compound to aquatic organisms such as waterfowl, fish, and invertebrates

From Hoyer and Canfield 1997.

disposal, uses of the product, restrictions, and precautions for the user and the environment, known as *safety and use guidelines* (Hoyer and Canfield 1997).

Other compounds, called *growth regulators,* may also be effective in controlling invasives. Growth regulators prevent plants from reaching normal stature by inhibiting protein synthesis and thereby preventing cell division and elongation. Two of the most commonly used are bensulfuron methyl and thiadiazuron. Both stunt the growth of *Hydrilla verticillata, Myriophyllum spicatum,* and *Potamogeton* (pondweed). In *H. verticillata,* both inhibit propagule formation. While growth regulators have the potential to keep some species in check, the means of delivery, mode of uptake by the plant, length of control, mode of action in the plant, differential plant responses to different products, and other efficiency and environmental questions have to be answered before they are widely used. In high dosages, they are as lethal as the aquatic herbicides (Nichols 1991).

Some chemical controls of wetland weeds do not involve herbicides. Salt is a cheap and easy chemical control in tidal wetlands where freshwater inputs have enabled exotics to outcompete more salt-tolerant natives. In salt marshes where tidal inputs have been restricted, the salt-tolerant natives, such as *Salicornia subterminalis* (glasswort) in southern California or *Spartina alterniflora* on the east coast of the U.S., are often replaced by exotics with lower salt tolerances. In California, one such nuisance species is the exotic grass *Polypogon monspeliensi.* On the east coast, many salt marshes are overrun with *Phragmites australis.* In a California marsh, a salt application of 850 g m^{-2} mo^{-1} for 3 months was sufficient to control the exotic *P. monspeliensis,* while not noticeably affecting the native *S. subterminalis* (Kuhn and Zedler 1997; Callaway and Zedler 1997). Restoring tidal influxes in some east coast salt marshes raised salinity and decreased *P. australis* stands (Roman et al. 1984).

D. Biological Controls

Biological control of weeds is the use of a plant's natural enemies (i.e., herbivores and pathogens) to decrease the weed's growth. Biological controls usually do not entirely eliminate a nuisance species; instead they maintain its population at a tolerable level (Malecki et al. 1993; Deloach 1997). Two general types of biological control have been used, *selective agents* and *polyphagous organisms.* Selective agents, such as some insects, birds, crustaceans, fungi, pathogens, and allelopathic plants, consume or harm only the target species and have no effect on other species. Polyphagous organisms, such as some herbivorous fish species, snails, turtles, and manatees, consume both the target species and others.

Polyphagous control agents are used to clear or decrease plant growth in water bodies rather than to limit a specific plant.

When an exotic species has no enemies in its new range, biological control agents may be imported from the plant's native range. Before an exotic biological control agent is released, expensive and lengthy testing in quarantine is required in order to minimize the risks involved in introducing a second exotic species to fight the first. The exotic control agent must be proven to be host-specific so that it will not affect native plants (Hoyer and Canfield 1997). When the weed species is related to native plants, the enemies of the native plants may control the exotic. In such cases, fewer tests are required since the introduction of an exotic species is not involved (Sheldon 1997).

Biological control methods have a number of advantages in the fight against exotic plant species in wetlands. Biological control agents cause less ecological disruption than herbicides and mechanical control methods, so in most cases, biodiversity is maintained. Once the biocontrol agent is established, the method is usually long-lasting and less expensive than other methods. Biological control is also very effective against some plants (e.g., *Alternanthera philoxeroides* and *Eichhornia crassipes*). When fish are used as the biocontrol agent, there is the added benefit that the weeds are converted to a useful protein product (fish flesh) for human consumption (McCann et al. 1996; Weeden et al. 1996b).

Authorities and managers are sometimes unwilling to introduce biological control agents, largely because of a fear of creating additional ecological problems. Often it is difficult for the control organism to become established because so many features of the organism's home range are impossible to replicate (Malecki et al. 1993). The biggest disadvantage to biological control is that in about 75 to 80% of the attempts, it simply has not worked to control the invasive species (Simberloff 1996; Rea 1998).

The most widely used biological control agents in wetlands are insects and fish. Trials with pathogens, fungi, and other organisms have been less successful. Biological control is usually used in conjunction with chemical or mechanical controls.

1. Insects

Insects control invasives by feeding on seeds, flowers, leaves, stems, roots, or combinations of these, or by transmitting plant pathogens, which infect plants (Weeden et al. 1996b). Before insect controls are used, the selectivity of the insect for the plant is determined and the ecological consequences of both using and not using the insect are considered (Harris 1988).

In the U.S., the first introduction of an insect to control a wetland species occurred in 1964 when the South American alligatorweed flea beetle (*Agasicles hygrophila*) was introduced to control *Alternanthera philoxeroides*. Two other insects, the alligatorweed thrips (*Amynothrips andersoni*) and the alligatorweed stem borer (*Vogtia malloi*), were released in 1967 and 1971. In combination, the three successfully control *A. philoxeroides* in the southeastern U.S. (Hoyer and Canfield 1997).

Weevils have also been imported and tested to determine whether they will be able to control *Melaleuca quinquenervia* in the Everglades. The melaleuca weevil (*Oxyops vitiosa*), like the tree, is native to Australia. The adults feed on the leaves and stems of seedlings and on the new growth of older trees, causing foliar and stem damage. The weevils cause stems to droop by digging small trenches in the stems. Adults lay eggs near areas of leaf damage and the larvae consume about ten times more leaf tissue than the adults. The weevil is host-specific and its effectiveness is still being evaluated. It is likely that the weevil will slow the spread of the tree and make it more susceptible to other control measures (Stocker 1999; U.S. Army Corps of Engineers 1999). Along with the leaf weevil, seven other insects are under study for the control of *M. quinquenervia* (Turner et al. 1998). Several other insect control agents are covered in our case studies.

2. Fish

Some fish species have been used in the control of submerged species. They are all polyphagous, and for that reason are used to control overgrowth rather than a specific nuisance species. Fish provide a safe alternative to herbicides. Stocking with fish requires less labor, fewer treatments, and less expense than other shorter-term strategies (Kay and Rice 1999).

Several types of fish have been used to control vascular submerged plants, including redbelly tilapia (*Tilapia zillii*), common carp (*Cyprinus carpio*), and triploid sterile grass carp (*Ctenopharyngodon idella*). Blue tilapia (*T. aurea*) are also used to control algae. Both blue and redbelly tilapia are tropical and do not survive in waters below 10° to 18°C. They control soft submerged species, particularly *Utricularia*, but they reproduce rapidly and consume vegetation and small animals that are important food sources for other more desirable fish species. Tilapia are not recommended for plant control in U.S. water bodies (Hoyer and Canfield 1997). Common carp are omnivorous and usually not very effective in controlling nuisance species.

The most frequently used fish in the U.S. is the triploid sterile grass carp, from China and Siberia. The eggs of normal grass carp are treated to form an extra set of chromosomes and the resulting fish is normal, but sterile. The sterile grass carp consumes a large quantity of vegetation, grows quickly to an adult weight of 20 to 25 lb, and lives about 10 years. In many states, the sterile grass carp is the only fish species permitted for the control of exotic plants. They are effective in the control of soft submerged plants, such as *Hydrilla verticillata, Najas* (naiad), *Cabomba caroliniana* (fanwort), *Ceratophyllum demersum, Potamogeton, Utricularia* (bladderwort), *Myriophyllum spicatum* (though they consume other soft plants first), *M. aquaticum* (parrot feather; formerly called *M. brasiliense*), *Ruppia maritima* (wigeongrass), *Elodea canadensis*, and *Chara* (muskgrass). Triploid sterile grass carp usually do not consume plants with a coarse or woody texture. They are not stocked in rivers or large lakes, but only in ponds and other enclosed water bodies to prevent their escape (Kay and Rice 1999).

Results using the sterile grass carp have been varied due to problems in calculating the correct stocking density. In many cases grass carp have either consumed all of the edible vegetation in a pond (when stocked at high densities), or none (due to inadequate stocking; Kay and Rice 1999). In some cases, the carp's preferred food plants are desirable species and control of the target plant does not occur until after the more valuable plants are consumed (McCann et al. 1996). Usually a low stocking rate of grass carp is integrated with other plant control methods (Nichols 1991).

3. Pathogens

The use of pathogens to control weeds is somewhat limited by quarantine regulations that restrict the introduction of exotic pathogens. Extensive testing under natural conditions is necessary to determine the effectiveness of pathogens and whether they will spread to desirable plants (Chambers et al. 1993). Pathogen populations usually do not remain high enough for the sustained suppression of weeds. Pathogens may be most suitable for weakening a population and making it more susceptible to other kinds of control (Hoyer and Canfield 1997).

4. Fungi

Mycoherbicides are fungal pathogens that cause root rot or decay on leaves and other plant parts. In the 1970s, the decline of *Eichhornia crassipes* in a Florida reservoir was linked to a naturally occurring fungus (*Cercosporta rodmanni*). The fungus was found to be host-specific and after a period of infestation, it caused the plant to die. The fungus has been formulated as a mycoherbide, but it has not been effective (Hoyer and Canfield 1997). Since it is not cost-effective for companies to research and market a product that attacks only one weed, the development of mycoherbicides has been limited (Forno and Cofrancesco 1993).

5. Other Organisms

Other organisms that have been suggested or tested as biological control agents for nuisance wetland plants include ducks, geese, crayfish, nematodes, manatees, and water buffalo. So far, none of these has proven practical and they may actually do more harm than good. For example, rusty crayfish (*Orconectes rusticus*) indiscriminantly reduce the biomass of all submerged macrophyte species in some northern U.S. lakes, as well as the abundance of macrophyte-associated snails (Lodge and Lorman 1987). Trials using manatees (*Trichechus manatus*) to remove *Hydrilla verticillata* from some of Florida's canals have met with little success because the manatees do not keep up with the plant's rapid growth (Hoyer and Canfield 1997).

V. Case Studies of Invasive Plants in Wetland Communities

We describe here the biology, range, effects, and control of five plants that are particularly noxious in North America. Two submerged species, *Myriophyllum spicatum* and *Hydrilla verticillata,* have spread throughout most of the U.S. in different, but overlapping, ranges. The floating species *Eichhornia crassipes* grows in warm water bodies throughout the world and is a threat to natural ecosystems and human endeavors in subtropical and tropical regions. *Lythrum salicaria* is a Eurasian emergent with bright purple flowers that has formed dense monocultures in many freshwater wetlands of the eastern and midwestern states and the southern provinces of Canada. *Phragmites australis* is considered an invasive emergent in the U.S., but in some areas of Europe, where it is considered a desirable species, it is in decline and researchers are trying to restore its growth.

A. *Myriophyllum spicatum* (Eurasian Watermilfoil)

1. Biology

Myriophyllum spicatum is a rooted submerged eudicotyledon of the Haloragaceae, with long, flexible stems and finely dissected leaves (Figure 8.2). The leaves are arranged in whorls of four with 10 to 26 pairs of leaf divisions. *M. spicatum* grows best in water depths between 1.5 and 4.0 m. Plants at shallow depths reach peak biomass earlier in the growing season than those in deeper water. The lacunae, or air spaces, in *M. spicatum* are extensive; they aid in gas movement and help keep the plant buoyant. The shoots of *M. spicatum* branch profusely and form a canopy near the water's surface. The canopy allows *M. spicatum* to take advantage of near-surface light levels.

M. spicatum reproduces sexually as well as through vegetative regeneration. The plants are monoecious and the flowers are predominantly wind-pollinated (though some insect pollination may occur), so the flowers must emerge above the water's surface in order for fruits to develop. The seeds are dispersed by waterfowl and along the surface of the water

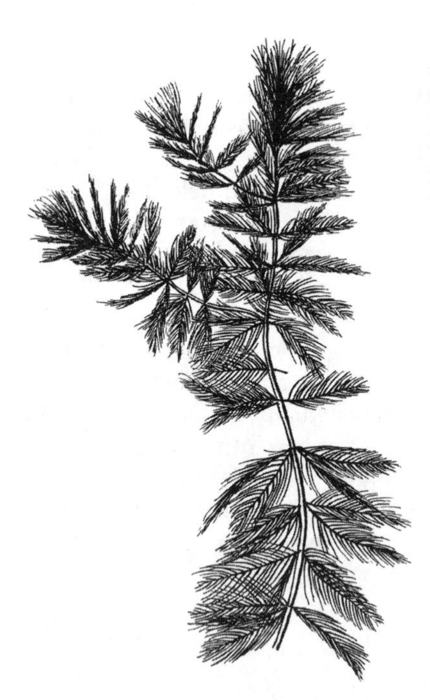

FIGURE 8.2
Myriophyllum spicatum (Eurasian water milfoil) is a rooted, submerged eudicotyledon that is invasive in fresh waters of most of the U.S. and parts of Canada (leaves are from 1 to 2.5 cm). (From Hotchkiss, N. 1972. *Common Marsh, Underwater and Floating-Leaved Plants of the United States and Canada.* New York. Dover Publications, Inc. Reprinted with permission.)

within the floating inflorescence. Seeds may be important for long-distance dispersal and, since many germinate after 2 years or more, they become a part of the seed bank and provide insurance against local extinction. *M. spicatum* is a perennial and can successfully overwinter as an evergreen or as a new, unexpanded shoot attached to rootstocks. New plants can arise from fragments that are separated accidentally or abscise. It is common for abscissing fragments to develop roots before they are released, which speeds their establishment after sinking to the lake bottom. Abscission often occurs following flowering. Fragmentation is probably the most important means of dispersal within a water body or from one water body to another. The plants also spread by growing new shoots at the nodes of stolons (Grace and Wetzel 1978).

M. spicatum can withstand a broad range of abiotic conditions, from oligotrophic to eutrophic waters, depths from 0.5 to 8 m, substrates that are sandy to organic, and waters with a pH from 5.4 to 10.0. *M. spicatum* usually grows in fresh water, but it can survive brackish water as well. It grows in both northern temperate and subtropical water bodies (Sheldon 1997).

2. Origin and Extent

Myriophyllum spicatum originated in Europe, Asia, and Northern Africa. It probably arrived in North America between the late 1700s and the 1880s. The date of the earliest confirmed record is in dispute, with some authors reporting that *M. spicatum* was found in 1902 in the Chesapeake Bay, while others report the species' migration to North America was not confirmed until 1942, near Washington, D.C. (Grace and Wetzel 1978; McCann et al. 1996). *M. spicatum* is now found in 44 states and 3 Canadian provinces (Quebec, Ontario, and British Columbia; Creed 1998). *M. spicatum* tends to be most abundant in mesotrophic to moderately eutrophic lakes (based on total phosphorus concentrations) of temperate areas (Madsen 1998). The plant's spread, at least in some cases, is attributed to intentional releases by people who grew the plant in natural waters for use in aquaria. Once it is established in a water body its fragments are easily transported to other water bodies on boat hulls and its spread can be quite rapid. In Minnesota it was first seen in 1987 in just a few lakes. After only 3 years it had spread to 50 lakes (Sheldon 1997).

3. Effects in New Range

When *Myriophyllum spicatum* arrives in a new water body, it does not always become the dominant plant. In some water bodies, *M. spicatum* coexists with native submerged species while in others it displaces them (Sheldon 1997). *M. spicatum* forms a dense canopy near the water's surface, and where it is dominant, its primary effect on water bodies is to shade the water column and thereby inhibit the growth of native submerged plants (Grace and Wetzel 1978; Madsen et al. 1991). The complex species-rich assemblage of native macrophytes are of variable height, growth form, and leaf shape, and provide a habitat for a diverse invertebrate and fish community. *M. spicatum* beds, on the other hand, are thick walls of shoots with uniform height and leaf shape. Their stem density can exceed 300 stems m^{-2} by midsummer, a density that inhibits the use of the bed as a fish refuge. Fish populations are generally lower in dense *M. spicatum* beds than in native macrophyte communities (Keast 1984; Sheldon 1997).

The decomposition of the plentiful fragments produced by *M. spicatum* can lead to an increase in nutrients and a decrease in dissolved oxygen that may have detrimental effects on other aquatic life. Floating detached plant material may interfere with water intake structures and other industrial uses of the water (Grace and Wetzel 1978; Mills et al. 1993). Dense *M. spicatum* beds can cause a decline in lakeshore property values and a decrease in sportfishing and tourism (Sheldon 1997).

4. Control

The best control of *Myriophyllum spicatum* is to minimize its introduction to new ranges. Public education campaigns and warnings at boat launchings are an attempt to prevent the plant's spread (Figure 8.3; Sheldon 1997). In established populations, mechanical, chemical, and biological control methods as well as habitat alterations have been used.

Habitat alterations such as drawdowns do not always produce the desired decrease in *M. spicatum* populations (Table 8.1). Winter drawdown followed by rototilling has been used to cut and remove stoloniferous rhizomes, and thereby reduce vegetative regeneration (Aiken et al. 1979). Bottom barriers have been used to shade new *M. spicatum* growth, but they are expensive (from $5,000 to $12,000 per hectare) and also exclude desirable plants (Sheldon 1997).

Handpulling is effective with small populations of *M. spicatum*; however, mechanical harvesting is usually necessary to control dense growth. Mechanical harvesting fragments the plants and can increase the number of growing shoots. To inhibit regrowth from fragments, barriers may be placed around harvesting operations. In Lake George, New York, suction harvesting (using a hydraulic vacuum system powered by a pump on a boat) resulted in a substantial reduction of *M. spicatum* biomass. A year after harvest, the submerged community included a greater number of native species and reduced *M. spicatum* biomass (Eichler et al. 1993).

A number of herbicides provide excellent control of *M. spicatum* (Table 8.3) and they are widely used (Christopher and Bird 1992; Bird 1993; Netherland et al. 1993; U.S. Army Corps of Engineers 1999). The growth regulator, bensulfuron methyl, provokes a number of symptoms in *M. spicatum* such as leaf chlorosis, deformed leaves on shoot tips, downward bending of leaves at upper nodes, stem necrosis, and the formation of axillary buds. The results are a reduction in biomass of 10 to 90%. The roots survive, however, and

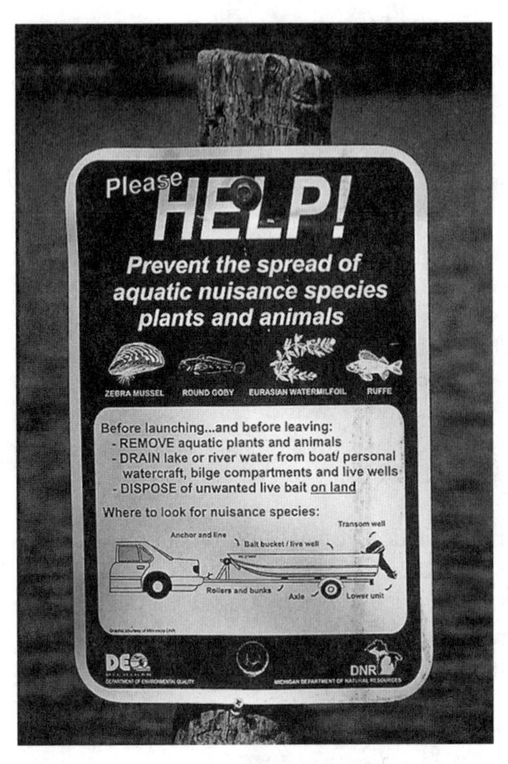

FIGURE 8.3
Public education efforts to prevent the spread of invasive species include signs near boat launches like this one in southwestern Michigan. (Photo by H. Crowell.)

regrowth from root crowns occurs soon after bensulfuron methyl concentrations are diluted (Nelson et al. 1993).

A North American weevil (*Euhrychiopsis lecontei*) has been found on *M. spicatum* plants in many infested water bodies of the upper midwest and eastern states. In enclosures it has been shown to have a significant negative effect on only *M. spicatum*. The weevil has been tested as a control agent in Vermont lakes with mixed results. *M. spicatum* biomass was reduced in its presence, however, in one lake. *M. spicatum* was also reduced in a non-weevil control area (Sheldon and Creed 1995; Creed 1998).

Attempts to use the triploid sterile grass carp to control *M. spicatum* have been unsuccessful, since the carp only consume *M. spicatum* once softer-textured submerged macrophytes have been depleted (McCann et al. 1996). Bacterial and fungal pathogens are under study, but not in use (Sheldon 1997).

5. The Natural Decline of Some Myriophyllum spicatum Populations

Myriophyllum spicatum infestations are often characterized by rapid colonization and dominance in a water body, followed by a decline that usually has nothing to do with control measures. In the 1960s *M. spicatum* declined in the Chesapeake Bay; 11 more declines were reported in the northeastern and midwestern U.S. in the 1970s. This trend has continued at an increasing rate, with 14 declines reported in the 1980s and 28 in the 1990s. The persistence of the declines varied from one season to several years (Creed 1998). In some southeastern U.S. water bodies, *M. spicatum* declined due to competition with another invasive exotic, *Hydrilla verticillata*, but elsewhere the reasons for the decline are unexplained. After a decline, *M. spicatum* usually persists, but native submerged macrophytes return. This "boom and bust" growth pattern is exhibited by other invasive submerged plants around the world (Barko et al. 1994). In Europe, the North American *Elodea canadensis* is being replaced by another North American invasive, *Elodea nuttalli*, and by

Lagarosiphon major (from South Africa). *E. canadensis* has also declined in New Zealand where it has been largely replaced by *L. major* and *Egeria densa* (Cook 1993).

The infestation and subsequent decline of *M. spicatum* were chronicled in detail in studies of Lake Wingra, Wisconsin (Carpenter 1980a; Trebitz et al. 1993). Lake Wingra is approximately 140 ha with a 43-ha littoral zone (Adams and McCracken 1974). In 1962, there was no record of *M. spicatum*, but by 1969, the plant covered 92.7% of the lake's littoral zone. It flowered twice every growing season. Other species, such as *M. sibiricum* (water milfoil; formerly called *M. exalbescens*), *Vallisneria americana* (wild celery), *Potamogeton amplifolius, P. illinoensis, P. freisii,* and *P. praelongus* (pondweeds) all disappeared by 1969. By 1977, *M. spicatum* still occupied the same area, but with decreases in shoot density and biomass. The next growing season, the area of the beds decreased and flowering occurred in only a few scattered locations.

By the 1990s, *M. spicatum* shared dominance with *Ceratophyllum demersum* and both species occupied about one quarter of the littoral zone. Sixteen native species that had been absent or rare in the 1960s returned to the lake and were growing well. The biomass of the new community was equal to that of the *M. spicatum* stands when they were at their peak (about 300 to 400 g dry weight m^{-2}). These community changes occurred in the absence of major changes in the trophic status, management, or use of the lake. A number of causes for the decline were suggested, including nutrient depletion, decreased light levels, competition, herbivory, parasites or pathogens, altered sediment characteristics, and management effects; however, the data from Lake Wingra do not clearly support any of these as the cause.

In some lakes, herbivorous insects may be the explanation for the decline. In Cayuga Lake, New York, *M. spicatum* declined markedly in the early 1990s while native macrophytes increased in abundance. The decline was attributed to increased density of the moth larva, *Accentria ephemerella,* which consumes the apical meristem of *M. spicatum.* Also present in the lake was another milfoil herbivore, the weevil, *Euhrychiopsis lecontei* (Johnson et al. 1997). A connection between the weevil's presence and a decline in *M. spicatum* has been drawn because the the weevil's range coincides with the area of *M. spicatum* declines across North America. Significantly more declines have occurred within the range of the weevil than would be expected by chance. The declines began occurring after *M. spicatum* was well established in many lakes. It may have taken 5 to 10 years for the weevil to shift from its native host, *Myriophyllum sibiricum,* to *M. spicatum*. Still more time was needed for the weevil population to grow to a level where it might noticeably affect a *M. spicatum* population. The lag in the weevil's effect would be even more pronounced in rapidly expanding *M. spicatum* populations (Painter and McCabe 1988; Creed and Sheldon 1993, 1994; Sheldon and Creed 1995; Creed 1998).

B. *Hydrilla verticillata* (Hydrilla)

1. Biology

Hydrilla verticillata, a native of southeast Asia, is a monocotyledon in the Hydrocharitaceae. It is a rooted, submerged perennial with leaves from 5 to 15 mm long and 2 to 4 mm wide, arranged in pairs on the lower nodes and in whorls of 3 to 10 leaves on the upper nodes (Figure 8.4). Its stems are varied in length from a few centimeters to several meters and are either erect, horizontal, or subterranean. Erect stems support the branches, leaves, flowers, and turions. Several erect stems form at a single node of a horizontal stem and together the branches form a dense canopy with 70% of the biomass

FIGURE 8.4
Hydrilla verticillata (hydrilla) is a native of southeast Asia. It is a rooted submerged perennial with leaves from 5 to 15 mm long and 2 to 4 mm wide, arranged in pairs on the lower nodes and in whorls of 3 to 10 leaves on the upper nodes. (From Cook, C.D.K. 1996. *Aquatic Plant Book.* The Hague. SPB Academic Publishing/Backhuys Publishers. Reprinted with permission.)

concentrated near the water's surface. *H. verticillata* typically grows in water 0.7 to 2.4 m deep, but it can grow in depths up to 15 m (Yeo et al. 1984).

Different strains of *H. verticillata* exist that are either monoecious (male and female flowers are separate, but on the same plant) or dioecious (each plant has either male or female flowers). Pollination is on the water's surface and the pollen is released explosively from free-floating male flowers and caught by floating female flowers (see Chapter 5, Section II.A.3, Water Pollination; Haynes 1988; Cook 1996). In the southeastern U.S., *H. verticillata* populations are dioecious and entirely female, so all of its spread is through vegetative regeneration. Sexual reproduction may occur in the populations of the Potomac River near Washington, D.C., where the monoecious strain is found (Steward et al. 1984).

H. verticillata reproduces vegetatively from subterranean tubers, axillary turions, and fragments. Turions form at leaf axils or at the ends of stems. They abscise and fall to the sediment, where they overwinter and sprout new growth in the spring. Subterranean stems form when *H. verticillata* is subjected to periods of decreased light and temperature. They grow downward from nodes at the base of an erect stem and their lengths vary from 1 to 15 cm. At the terminal end, the stems develop subterranean turions (often called tubers to differentiate them from the stem turions). Both tubers and turions are important to the plant's capacity to overwinter and to regenerate vegetatively (Yeo et al. 1984; Haynes 1988; Schmitz et al. 1993). The tubers are sometimes buried several centimeters below the surface, and they can sprout shoots even after a period of up to 4 years dormancy. Millions of turions and tubers are found in the sediments of infested water bodies (Schmitz et al. 1993; McCann et al. 1996; Netherland 1997). For example, in a Florida canal, an average of 918 turions m^{-2} (both subterranean and axillary) were found (Sutton 1996).

2. Origin and Extent

Hydrilla verticillata is a native of southeast Asia. It is widely distributed and found on all continents except Antarctica. *H. verticillata* is a noxious invasive in many states of the U.S. including most of the Atlantic and Gulf Coast states and the western states, Arizona, California, and Washington. *H. verticillata* is also considered an invasive in Europe, Central America, and elsewhere (Steward 1993). It usually grows in warm waters, but it has been found as far north as Poland and Ireland (53° N). The dioecious pistillate (female) form of *H. verticillata* was first introduced in the southeastern U.S. in about 1955 as an aquarium plant. Its current northernmost locality in the U.S. is in Connecticut (about 41° N; Les et al. 1997). By the mid-1960s, *H. verticillata* had become a nuisance species and it has been the most abundant submerged plant in Florida since 1983 (McCann et al. 1996). The first report of the monoecious strain in the U.S. was in the Potomac River near Washington, D.C. in 1982. Since the Potomac is subject to tides from the Chesapeake Bay, the plant has moved both upstream and downstream from its original location. The presence of both sexes increases the potential for sexual reproduction, genetic diversification, and adaptation to different habitats (Steward et al. 1984).

Once established in a water body, *H. verticillata* spreads very quickly. For example, in a south central Florida lake (11,332 ha) *H. verticillata* covered approximately 14% of the lake's total area in 1987. A year later, it covered 70% of the lake (Schmitz et al. 1993).

3. Effects in New Range

Hydrilla verticillata forms a dense canopy at the water's surface and shades growth in the water column below. Its tolerance for low light levels gives it a longer growing season than many other submerged species and makes it capable of outcompeting other submerged plants, such as *Vallisneria americana, Ceratophyllum demersum*, and even the invasive *Myriophyllum spicatum* (Grace and Wetzel 1978; Schmitz et al. 1993). Plant and animal diversity are decreased in areas of *H. verticillata* (McCann et al. 1996). *H. verticillata* produces allelopathic compounds that have been shown to inhibit the growth of *C. demersum* (Schmitz et al. 1993).

H. verticillata may alter the trophic status of a water body. In a hydrilla-infested lake near Orlando, Florida, the chlorophyll content of the water decreased from 82 mg m^{-3} to less than 6 mg m^{-3}, signifying a sharp decrease in the phytoplankton population. Such a decrease could alter the lake's food web and dissolved oxygen concentrations (Schmitz et al. 1993).

Sport fish are stunted in infested water bodies because *H. verticillata* eliminates open-water feeding and spawning areas. Dense coverage by hydrilla significantly reduces bluegill (*Lepomis macrochirus*), redear (*L. microlophus*), and black crappie (*Pomoxis nigromaculatus*) fisheries. Populations of these fish become skewed to smaller individuals because the *H. verticillata* beds provide refuge against larger predators. The smaller fish are of little value for sportfishing. Recreational use and tourism income decrease as a result of lower fishing success and because *H. verticillata* clogs waterways and impedes boat passage (Office of Technology Assessment 1993; McCann et al. 1996). Extensive growth can reduce waterflow by clogging the filters of irrigation pumps and it can reduce the holding capacity of storage ponds (Yeo et al. 1984).

4. Control

Hydrilla verticillata is an aggressive weed, and the cost of its control is quite high. In the 1980s, the state of Florida spent $50 million to control it (Office of Technology Assessment 1993). Mechanical removal methods can actually increase *H. verticillata* populations by spreading fragments, so they are considered only a short-term solution. They are sometimes used in small areas or in order to quickly clear boat trails (McCann et al. 1996).

H. verticillata responds well to several herbicides (Table 8.3). Fluridone may be used in varying concentrations to either kill the plant or prevent the production of turions (MacDonald et al. 1993; Netherland et al. 1993). In Lake Seminole, which lies between Florida and Georgia, fluridone use led to the re-establishment of some native species (Chambers et al. 1993). The growth regulator, bensulfuron methyl, is effective in reducing the growth and production of *H. verticillata* tubers (Langeland 1993; Rattray et al. 1993; Langeland and Laroche 1994; Van and Vandiver 1994).

A number of insects that may control *H. verticillata* are under study. Some have already been introduced in small numbers and while they reduce *H. verticillata* growth, results have not been dramatic (Table 8.6; U.S. Army Corps of Engineers 1999). In laboratory studies, an endemic fungal pathogen (*Mycoleptodiscus terrestris*) colonized host cells in *H. verticillata* resulting in a disruption of cell walls and subsequent cell death. The fungus brought about chlorosis of the shoot tissue within 2 weeks of exposure. The leaves and stems became flaccid, dropped from the plant, and floated to the water's surface. Since the fungus is native to the U.S., it does not have to be thoroughly tested before it is introduced to U.S. waters; however, it is not currently in use (Shearer 1998). Successful biological control has been observed with the triploid sterile grass carp; however, the carp also consumes desirable species, so it is only useful where *H. verticillata* grows in monospecific stands (Hoyer and Canfield 1997).

TABLE 8.6

Insects Used in the Biological Control of *Hydrilla verticillata* (hydrilla)

Hydrilla tuber weevil (*Bagous affinis*)

Adults and larvae damage *H. verticillata* tubers. The tuber weevil cannot withstand submergence for more than a few days, so they function best in waterbodies with periodic drawdowns. The tuber weevil has been introduced in Florida but it has not established a long-standing population.

Hydrilla stem borer (*Bagous hydrillae*)

Adults feed on *H. verticillata* stems and leaves and lay eggs within stems. When the larvae hatch, they chew through the stem and float to shore within the stem. The stem borer removes the top 100 cm of the plant. The stem borer has been introduced in Florida, but does not have a well-established population.

Australian hydrilla mining fly (*Hydrellia balciunasi*)

The larvae are aquatic and consume various parts of the plant; the adults are not aquatic. It was released in the U.S. in 1989 and there is an established population near Houston, Texas.

Hydrilla mining fly (*Hydrellia pakistanae*)

The larvae are aquatic and consume various parts of the plant; the adults are not aquatic. The mining fly was released in 1987, and it is often found in association with *H. verticillata* in the southeastern U.S.

Aquatic moth (*Parapoynx diminutalis*)

It is widespread in Florida and the larvae feed on *H. verticillata*, but it is slow to control the plant. The moth was introduced to the U.S. from Asia.

Leaf mining fly (*Hydrellia bilobifera*)

The leaf mining fly is a native species whose larvae feed on *H. verticillata* leaves and each individual may damage up to 20 leaves during its development.

Weeden et al. 1997; U.S. Army Corps of Engineers 1999.

C. *Eichhornia crassipes* (Water Hyacinth)

1. Biology

Eichhornia crassipes, of the Pontederiaceae (pickerelweed family), is a native of tropical South America and an exotic weed throughout warm waters of the world. Its rosettes, stolons, and showy blue flowers float on the water's surface and its roots and rhizomes extend into the water column (Figure 8.5). At the base of each leaf is a bulb-shaped petiole filled with air spaces that keep the plant buoyant. *E. crassipes* can grow up to 1 m tall, but

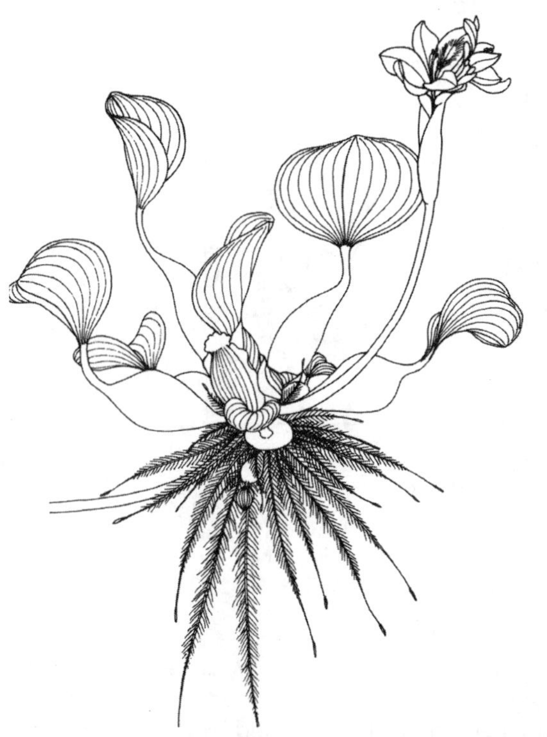

FIGURE 8.5
Eichhornia crassipes (water hyacinth) is a native of tropical South America and an exotic weed throughout warm waters of the world. Its rosettes, stolons, and showy blue flowers float on the water's surface and its roots and rhizomes extend into the water column. The leaves are approximately 10 cm in width at their widest point. (From Cook, C.D.K. 1996. *Aquatic Plant Book.* The Hague. SPB Academic Publishing/Backhuys Publishers. Reprinted with permission.)

it is usually less than 0.5 m. Its flowers are insect-pollinated and self-compatible. Pollination appears to be more successful in tropical areas where insect pollinators are more numerous. Flowering begins 10 to 15 weeks after germination and one inflorescence with 20 flowers produces up to 3000 seeds. Up to four inflorescences can be produced by a single rosette in a 21-day period. In the tropics and subtropics, *E. crassipes* flowers for 5 to 9 months. The seeds are released in capsules of 40 to 300 seeds each that either sink or accumulate in the floating mat. Germination of seeds in the sediments is prevented if the sediments are shaded or light levels and temperatures are low. Most new individuals arise through asexual reproduction when daughter rosettes formed on nodes along stolons detach from the parent plant (Barrett 1980a, b; Haynes 1988; Harley et al. 1996).

E. crassipes grows and generates new rosettes rapidly and forms dense floating mats. Leaves make up 60 to 70% of the plant's biomass and the leaf turnover rate is high, with about 60 to 70% of leaves being replaced each month. Its rate of biomass accumulation can be as high as 60 g dry weight m^{-2} day^{-1} and its average doubling time is 13 days (Cook 1993). The plant ceases to grow at about 10°C and its inability to tolerate freezing temperatures limits its range to warm waters (Schmitz et al. 1993).

2. Origin and Extent

Eichhornia crassipes spread from South America throughout the world in the late 1800s (Cook 1993). It is considered a serious pest in many countries, including Australia (Finlayson and Von Oertzen 1993), Papua New Guinea (Osborne 1993), India (Gopal and Krishnamurthy 1993), Java, Japan (Cook 1993), Thailand, Malaysia, Cuba, Mexico, and Bolivia (Harley et al. 1996). It has spread over the entire continent of Africa and has infested all of Africa's major rivers and lakes (Hill et al. 1997). Even in its native range in South America it can become a nuisance species when the hydrology is modified or nutrient levels are increased (Harley et al. 1996). Due to its rapid spread and the many problems it causes, *E. crassipes* is widely considered the worst aquatic weed in the world (Figure 8.6; Cook 1993).

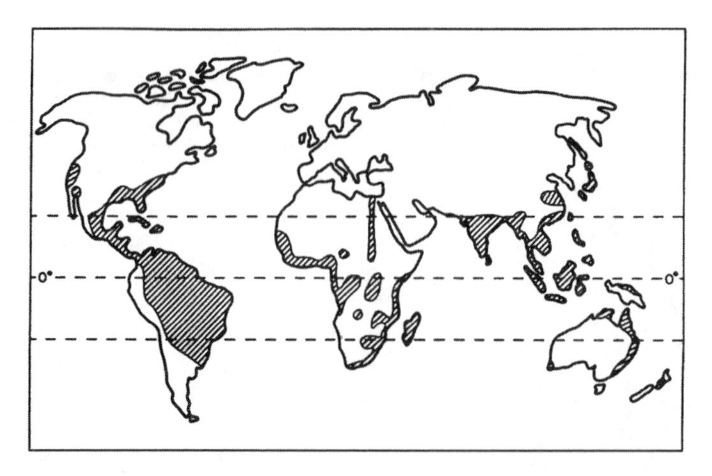

FIGURE 8.6
The world distribution of *Eichhornia crassipes* (water hyacinth). It is invasive outside of its native range in South America. (From Ashton, P.J. and Mitchell, D.S. 1989. *Biological Invasions — A Global Perspective*. J.A. Drake, H.A. Mooney, F. di Castri, R.H. Groves, F.J. Kruger, M. Rejmanek, and M. Williamson, Eds. Chichester. John Wiley & Sons. Reprinted with permission.)

E. crassipes was introduced to the U. S. at an exposition in New Orleans, Louisiana in 1884. A visitor to the exposition carried it to Florida where it was released into the St. Johns River. By 1893, *E. crassipes* was hindering navigation on the river and by 1896, the plants had spread throughout the St. Johns River watershed, mostly by farmers who used the plant as cattle fodder. By the turn of the century, *E. crassipes* had attained widespread distribution throughout the southeastern states (McCann et al. 1996).

3. Effects in New Range

When *Eichhornia crassipes* grows over a large area, the sheer mass and density of the floating mat can bring about changes in its habitat. Dissolved oxygen below the mat decreases because oxygen-producing plants beneath are shaded and the diffusion of oxygen from the air is blocked. The reduced dissolved oxygen concentrations associated with *E. crassipes* can decrease or eliminate fish populations. One hectare of *E. crassipes* can contain more than 1.6 million plants, which can deposit between 400 and 1900 metric tons (wet weight) of detritus per year. The decomposition of the plants results in anoxia, which also leads to fishkills. Mats of *E. crassipes* smother submerged vegetation and littoral emergent plants, many of which are important to waterfowl (Ramaprabhu et al. 1987; Schmitz et al. 1993; Hill et al. 1997; Howard and Harley 1998).

The hydrology of an infested wetland may also be altered. Losses of water to evapotranspiration have been measured as 3.2 times greater than losses from adjacent open water areas in Louisiana and 3.7 to 6.0 times greater in a Florida study (Schmitz et al. 1993). The rate of water flow is reduced by 40 to 95% in the presence of dense *E. crassipes* mats (McCann et al. 1996).

E. crassipes is a threat to human health in many countries where the increased evapotranspiration and blockage of water flow decrease the availability of water for human, agricultural, and industrial uses. *E. crassipes* increases the area available to vectors of human and animal diseases. Reduced fish biomass can lead to reduced nutrition and health. The worst effects of *E. crassipes* may be felt by people in developing countries who depend on fisheries for their diet and livelihood (Hill et al. 1997).

E. crassipes infestations cause economic problems because the plants impede boat traffic and hinder access to docks, bridges, or any narrow point in a river (McCann et al. 1996). In addition, *E. crassipes* mats block the pumping structures of hydroelectric plants and other water intake structures (Harley et al. 1996).

4. Control

The control of *Eichhornia crassipes* is a worldwide problem and many nations spend millions of dollars each year in an effort to control it. Many control methods are in use to reduce or eliminate *E. crassipes* populations.

It is difficult to harvest *E. crassipes* by hand, but manual removal is used for small infestations or where labor is inexpensive. In many developing countries, manual removal is the only control practiced because mechanical control is expensive and the environmental safety and operational problems associated with chemical controls are considered questionable. Manual harvesting is a slow process because the interwoven roots must be cut apart with knives. The heavy plants (>90% water content) are either loaded onto boats or dragged to the shore where they are burned, consumed by cattle, or left to decompose (Ramaprabhu et al. 1987).

Mechanical controls require sufficient funds to purchase and maintain equipment and pay for fuel. Land-based scoops lift *E. crassipes* out of the water and deposit it on shore. Draglines are used to clear water intake structures and floating barges are loaded with *E. crassipes* and unloaded on the shore. The plants are carried away in trucks for disposal. The upkeep of the machinery required for mechanical control of *E. crassipes* is expensive, and the removal process must keep up with the plant's rapid growth rate; however, it is effective for clearing channels or restoring access to the shore (U.S. Army Corps of Engineers 1999).

E. crassipes is susceptible to several chemical herbicides with excellent results from 2,4-D dimethylamine (DMS), Diquat, and Diquat + complexed copper (Table 8.3); however, their long-term success is limited. Herbicides may successfully eradicate small populations, but large infestations are much more difficult to control and herbicide use must be repeated (Schmitz et al. 1993).

Three insects are used to control *E. crassipes* in the U.S. The mottled water hyacinth weevil (*Neochetina eichhorniae*) was introduced in Florida from South America in 1972 and it is the most effective control agent of the three. In 1974, the chevroned water hyacinth weevil (*Neochetina bruchi*), also from South America, was released. Both feed on *E. crassipes* leaves and their larvae inhabit the plant's petiole bases or stems. *N. eichhorniae* lays more than 400 eggs in a lifetime, and *N. bruchi* lays about 300. The third insect, the water hyacinth borer (*Sameodes albiguttalis*), was released in 1977. The borer's larvae create tunnels and form cocoons in the plant's spongy petiole. Together the three insect species have brought about large-scale reductions and have even eliminated *E. crassipes* at some sites. More typically they cause a reduction in plant height, and a decrease in flowering and seasonal growth (Hoyer and Canfield 1997; Center et al. 1998a, b; Grodowitz 1998). In combination with herbicides, the insects brought about a reduction in the total *E. crassipes* mat area in Florida from 80,940 ha in 1972 to 1,050 ha in 1989. Once the insect populations are established, few human inputs are required to attain ongoing control. It usually takes 3 to 5 years for insect populations to reach a density that can cause weed declines (U.S. Army Corps of Engineers 1999).

A fungus, *Cercospora piaropi*, causes *E. crassipes* leaves to yellow and develop small sunken brown lesions. It may reduce plant populations (Martyn 1985). The effectiveness and feasibility of other fungi that cause disease in *E. crassipes* are still under study (Charudattan 1997).

D. *Lythrum salicaria* (Purple Loosestrife)

1. Biology

Lythrum salicaria, of the family Lythraceae, is an emergent herbaceous perennial that grows in freshwater wetlands in temperate areas (Figure 8.7). It grows from 0.5 to 2.7 m tall. The stems are squarish with evenly spaced nodes and short slender branches. The base of the plant is up to 0.5 m in diameter and the branches spread to form a crown that is up to 1.5 m in diameter (Stuckey 1980; Mal et al. 1992).

The purple flowers exhibit a tristylous breeding system, with the styles and anthers arranged in three different ways (Figure 8.8). The breeding system promotes outcrossing. Plants produce more seeds in ranges where all three flower types are present. The flowers are both insect- and self-pollinated. Seed production depends on age, size, and vigor of the plant. A single stem produces 900 to 1000 capsules and the number of seeds in each capsule varies from 80 to 130. The average number of seeds produced per plant is 2.7 million. The seeds are dispersed by wind and water and they adhere to aquatic wildlife as well as to people and their vehicles. The seeds have a high viability if they are fresh (up to 100%) and seed density in a *L. salicaria* bed has been measured at 410,000 seeds m^{-2} in the top 5 cm of soil (Mal et al. 1992). Seedling densities approach 10,000 to 20,000 plants m^{-2} under natural conditions (Malecki et al. 1993).

Many believe that *L. salicaria* generates new plants from rhizomes or rootstock; however, a study of the plant's clonal spread showed that stem meristem tissue must be present for new plants to arise. Vegetative regeneration occurs along lateral branches that bend downward as they compete with the upper branches for light. Eventually the lateral branches come into contact with the substrate. Adventitious roots form along such stems and new shoots arise at the stems' nodes (Stevens et al. 1997a).

L. salicaria tolerates a wide range of ecological conditions and it is able to adapt to changing conditions. Under flooded conditions, its stems produce aerenchyma. In response to changes in illumination, its leaves are able to increase in area, thereby capturing more sunlight. *L. salicaria* grows in low-lying coastal areas, emergent marshes, stream banks, floodplains, and temporarily flooded habitats such as roadside ditches. *L. salicaria* is often found in open sunny emergent marshes, but it also grows in wet woods, as it can survive in 50% of full sun. It thrives in both calcareous and slightly acidic soils and in a range of soil textures including sand, clay, gravel, organic soils, and even crushed rock ballast. It grows best in somewhat hydric soils, but it can grow in dry soil as well (Mal et al. 1992; Stevens et al. 1997b).

2. Origin and Extent

Lythrum salicaria originated in Europe and Asia. Its current range extends around the globe in the northern hemisphere, and it is also found in temperate areas of the southern hemisphere. In Europe, it grows west to east from Great Britain to central Russia, and as far north as 65° N in southern Scandinavia, and south to northern Africa. In Asia it is native in Japan, and also grows in China, Korea, Southeast Asia, North India, and Pakistan. It has been introduced to Australia, New Zealand, Tasmania, and North America.

L. salicaria seeds probably arrived in North America in ship ballast water in the early 1800s. Early settlers probably also intentionally introduced *L. salicaria* for its showy flowers and for the treatment of diarrhea, dysentery, bleeding, wounds, ulcers, and sores. Within 30 years, it was well established in freshwater bodies along the New England seashore. The plant migrated westward along the St. Lawrence River Valley and the Great Lakes region and north into the Hudson River valley. The spread of *L. salicaria* has been

FIGURE 8.7
Lythrum salicaria (purple loosestrife) is an invasive plant in the U.S. and Canada. It has replaced native species in freshwater marshes throughout the Great Lakes region. (Photo by H. Crowell.)

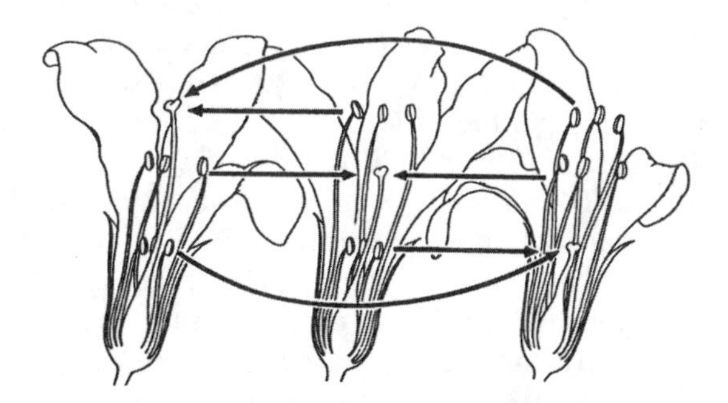

FIGURE 8.8
The flowers of *Lythrum salicaria* are trimorphic, i.e., they exhibit three different levels of anthers (which bear pollen) and stigmas (which receive pollen). The flowers are categorized according to stylar morphs as short-, medium- and long-styled (the style connects the stigma to the ovary). From left to right, the flowers are long-styled, medium-styled, and short-styled. Long-styled flowers have short and medium filaments (which bear the anthers), medium-styled flowers have long and short filaments, and short-styled flowers have medium and long filaments. The filaments are not at the same height as the stigma of the same flower and self-fertilization is thereby inhibited. The dark arrows show the pathways of 'legitimate' pollen flow among the three style types. Pollen from short filaments pollinates the short-styled flowers, pollen from medium filaments pollinates the medium-styled flowers, and pollen from long filaments pollinates the long-styled filaments. (From Mal et al. 1992. *Canadian Journal of Plant Science* 72: 1305–1330. Reprinted with permission.)

enhanced by gardeners who cultivate it for its flowers; it is still widely used in landscaping and private gardens (Figure 8.9). It appears to have escaped from cultivation several times west of the Mississippi because its distribution there is far more scattered than in the east and midwest (Stuckey 1980). Today the worst *L. salicaria* infestations are in the midwest and east, with the St. Lawrence River watershed and the Great Lakes region particulary affected. It is now found in all of the contiguous U.S. except Florida and in all of the Canadian provinces (Weeden et al. 1996a).

FIGURE 8.9
Lythrum salicaria (purple loosestrife) is widely used as an ornamental in gardens and landscaping. (Photo by H. Crowell.)

In most cases, there is a period of latency between the time when *L. salicaria* is first introduced to a new site and the time when it becomes a troublesome weed. *L. salicaria* infestations often grow scattered among other vegetation for 20 to 40 years and then proliferate and spread so rapidly that they outcompete native species. This pattern has been repeated in eight major geographic regions of the Laurentian Great Lakes area (i.e., (1) New England, (2) Delaware River Valley and New Jersey, (3) Hudson River Valley, (4) Western New York and Finger Lakes Region, (5) Michigan, (6) Ohio, (7) Indiana and Illinois, and (8) Wisconsin and Minnesota; Stuckey 1980). It is thought that *L. salicaria* cannot spread rapidly until all three of the plant's flower types are present. Once the three flower types become established, prolific seed production ensues and the seed bank is formed. If a disturbance such as drought or drawdown occurs, enabling more seeds to germinate, *L. salicaria* quickly colonizes the newly opened area and outcompetes other species (Mal et al. 1992).

3. Effects in New Range

Where *Lythrum salicaria* is native, it grows in mixed species stands, but where it is an exotic, it eventually forms dense monospecific stands that exclude native plants (Mal et al. 1992). It outcompetes common hardy plants like *Typha* as well as rare and endangered species. *L. salicaria* infestations extirpated *Scirpus longii* (Long's bulrush) in Massachusetts and the rare *Eleocharis parvula* (dwarf spike rush) in New York (Harris 1988). *L. salicaria* eliminates the animal habitat and food that other wetland plants provide and thereby endangers wetland animals such as the bog turtle (*Clemmys muhlenbergii;* Malecki and Rawinski 1985). Its seeds are rarely consumed by birds, although the indigo bunting (*Passerina cyanea*) and the red-winged blackbird (*Agelaius phoeniceus*) sometimes use it for nesting (Eastman 1995). The plant's dense stands restrict waterfowl access to open water. *L. salicaria* renders habitats unsuitable for the black tern (*Chlidonias niger*) which inhabits freshwater marshes throughout the St. Lawrence River watershed. Muskrats (*Ondatra zibethicus*) do not use *L. salicaria* for food or hut-building (Harris 1988; Mal et al. 1992).

4. Control

As with other invasive plants, prevention is the best control method. When *Lythrum salicaria* arrives in a new location, a rapid reaction may be the best way to prevent its spread and eventual dominance (Stuckey 1980). All types of controls have been attempted with *L. salicaria,* with varied success. Handpulling may be the best means of control in small stands. Cutting the plants can bring about a short-term solution, but the plants recover by the following growing season. Late summer cutting results in better control than mid-summer cutting, perhaps because the plants are less able to replenish the carbohydrate reserves necessary for growth the next spring. In one trial, when late summer cutting was followed by flooding, *L. salicaria* was significantly stressed. When flooded for two or more growing seasons the number of surviving plants was significantly reduced (Malecki and Rawinski 1985). A study of wetland plant germination indicated that consistent spring and early summer flooding can inhibit *L. salicaria* establishment (Weiher et al. 1996). Flooding is not always effective, however, and in one study, shallow flooding (30 cm) did not reduce *L. salicaria* seedling abundance (Haworth-Brockman et al. 1993).

The herbicide glyphosate is effective against *L. salicaria;* however, it is a broad-spectrum non-selective herbicide that may affect desirable species (Mal et al. 1992). Triclopyr amine, a systemic herbicide used for upland broad-leaved plants, is also effective on *L. salicaria.* It is absorbed by the upper portions of the plant and translocated to the roots. It breaks down in soil and water. Seedlings reappear the following season, so re-application of the herbicide may be necessary for long-term control (Gabor et al. 1995). Mycoherbicides may eventually prove to be effective against *L. salicaria* and several native fungal taxa found in association with the plant are under study for this purpose (Nyvall 1995; Nyvall and Hu 1997).

The European weevil, *Hylobius transversovittatus,* was introduced to the U.S. and Canada in 1992 and is part of a long-term plan to control *L. salicaria.* The weevils feed on the new leaves of *L. salicaria* and their effects can be observed along leaf edges (Figure 8.10). The weevil's larvae develop in *L. salicaria* roots and may severely damage plants after several years. The European beetles, *Galerucella calmariensis* and *G. pusilla,* feed on new leaves and shoots and can completely defoliate plants. The plants compensate for the loss of photosynthetic tissue by replacing foliage at the expense of belowground carbohydrate storage. The adult beetles are mobile and able to move to other *L. salicaria* stands (Figure 8.11). Two flower-feeding weevils, *Nanophyes marmoratus* and *N. brevis,* and a gall midge, *Bayeriola salicariae,* may eventually be released as well. They reduce seed production by attacking the flower buds. It may take 7 to 10 years to establish large and effective colonies of these insects (Hight 1993; Malecki et al. 1993; Blossey and Schat 1997).

E. *Phragmites australis* (Common Reed)

1. Phragmites australis as an Invasive Species in North America

Phragmites australis (formerly also called *Phragmites communis*) is distributed around the world. It is classified as a native species in North America and in Europe. It has been considered a nuisance species in the U.S. since the 1940s because its presence is often indicative of human disturbance. It quickly takes advantage of newly opened terrain and displaces more beneficial plants. It has little wildlife value and many wildlife and wetland managers in the U.S. strive to control it (Wijte and Gallagher 1996a, b).

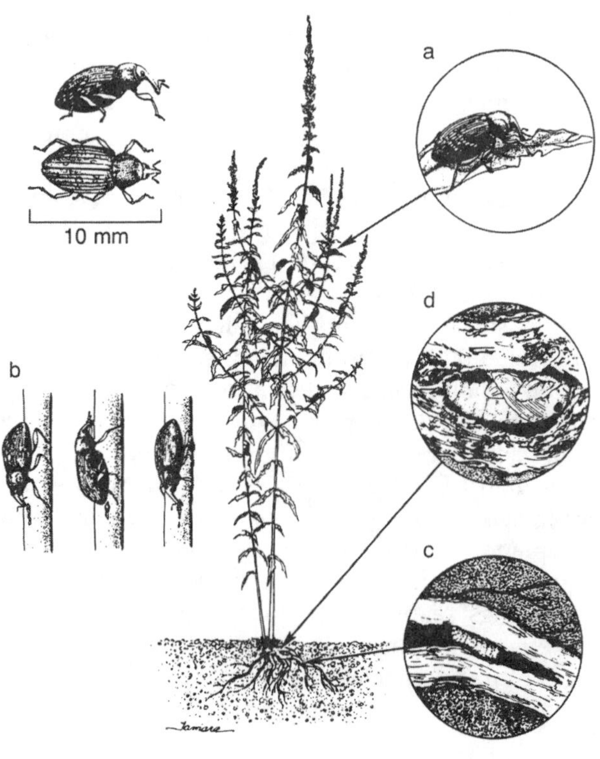

FIGURE 8.10
The European weevil, *Hylobius transversovittatus,* has been released in the U.S. and Canada as part of a long-term plan to control *Lythrum salicaria*. The life cycle of the weevil: (a) adults emerge in spring and feed nocturnally on newly formed leaves of *L. salicaria*; (b) oviposition lasts 2 to 3 months and consists of one to three eggs individually deposited each day into the stem; (c) developing larvae mine to the roots, where they feed extensively on root tissue; (d) mature larvae form a pupation chamber in the upper part of the root, emerging as adults in late summer or the following spring. Adults can live for several years. (From Malecki et al. 1993. *BioScience* 43: 680–686. Reprinted with permission © 1993 American Institute of Biological Sciences.)

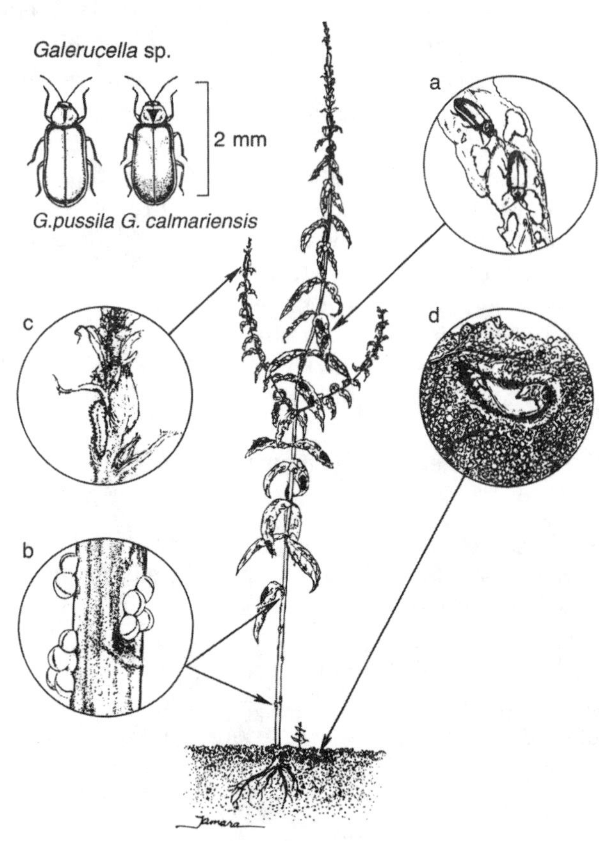

FIGURE 8.11
The European beetles, *Galerucella calmariensis* and *G. pusilla,* have also been released to control *Lythrum salicaria* in the U.S. and Canada. The life cycle of the beetles: (a) adults emerge in spring and feed on newly formed leaf tissue of *L. salicaria*; (b) spring oviposition lasts approximately 2 months; batches of two to ten eggs are laid daily on the plant stem or in leaf axils; (c) developing larvae feed extensively on bud, leaf, and stem tissue; (d) pupation to adult occurs in the soil or litter near the host plant. Adults are short-lived, dying soon after the spring oviposition period. (From Malecki et al. 1993. *BioScience* 43: 680–686. Reprinted with permission © 1993 American Institute of Biological Sciences.)

a. Biology

Phragmites australis is a tall perennial grass (in the family Poaceae) with large, pennant-like leaves that grows in both salt and freshwater marshes, in swamps and ditches, and along shorelines (Figure 8.12). *P. australis* grows up to 4 m or more in height and it towers above most other emergent wetland vegetation. Its stalks bear a conspicuous inflorescence with bisexual wind-pollinated florets. Each flowering stalk typically produces 500 to 2000 wind-dispersed seeds. The seeds germinate in bare areas that arise naturally in marshes due to fires or storms or on tidal wrack deposits. *P. australis* rapidly colonizes areas opened by human development and marsh construction or restoration activities.

FIGURE 8.12
Phragmites australis (common reed) grows in roadside ditches and in both fresh- and saltwater marshes. It is considered to be an invasive in many North American wetlands, while in Europe it is considered desirable and its decline has spurred restoration efforts. (Photo by H. Crowell.)

P. australis spreads vegetatively along stolons and rhizomes that grow out from the main stem, usually from the upland boundary toward the wetter parts of a marsh. The stolons have been observed to grow at a rate of 10.8 cm day^{-1} and a new plant can arise at each stem node. *P. australis* tolerates a wide range of inundation and salinity levels and can grow in salinities as high as 30 ppm (Voss 1972; Shay and Shay 1986; Cook 1996; Wijte and Gallagher 1996a, b).

b. Origin and Extent

In North America, *Phragmites australis* is found throughout the U.S. and north to Nova Scotia, Manitoba, Saskatchewan, and British Columbia. The categorization of *P. australis* as an invasive species is relatively recent. Remnants of *P. australis* have been identified in peat cores dating 3000 years or older; however, the individual fragments were not dated and could have grown down through the peat. The plant was first positively identified in 1843 on the Atlantic coast. Genetic evidence suggests that a more aggressive biotype was introduced to the Atlantic coast sometime before the 1900s. The expansion of *P. australis* along the Atlantic coast may be associated with this exotic biotype (as reviewed by Rice et al. 2000).

FIGURE 8.13
A salt marsh in Sandwich, Massachusetts with *Spartina alterniflora* in the foreground and *Phragmites australis* in the background. (Photo by H. Crowell.)

In salt marshes where tides are restricted, the interstitial salinity decreases, allowing less salt-tolerant plants to compete with halophytes (Figure 8.13). In such situations, such as the Hackensack Meadowlands of New Jersey, many of Connecticut's salt marshes, and the salt marshes of Delaware Bay, *P. australis* dominates where *Spartina alterniflora* previously thrived. Sites in Connecticut that have been tidally restricted for 20 years or more have dense and nearly pure stands of *P. australis* (Roman et al. 1984).

c. Effects on the Habitat

Wherever *Phragmites australis* grows, it is usually the dominant or co-dominant plant, sometimes coexisting with *Typha* (cattail), *Scirpus* (bulrush), or *Spartina alterniflora*. *P. australis* displaces other plants because it grows and spreads rapidly, shades other plants, and accumulates a large amount of litter that covers and shades the substrate. Dense monotypic stands of *P. australis* provide unsuitable or less-preferred food and habitat for waterfowl and other wildlife (Roman et al. 1984; Thompson and Shay 1989; Chambers et al. 1999).

Once *P. australis* dominates a marsh, there are notable differences in the physical environment. *P. australis* has a high evapotranspiration rate, and it can have a profound effect on the water table. In a Manitoba marsh, the annual evapotranspiration rate in a *P. australis* stand was about 300 mm yr^{-1}, almost double the precipitation during the growing season (as reviewed in Shay and Shay 1986). In many marshes where *P. australis* is dominant, the water table lowers and the accumulated peat compacts so that the elevation of the marsh is lowered. Aerobic bacterial populations increase once the peat is dry and the oxygen content increases, thereby enhancing decomposition of the peat and further lowering the marsh's elevation. In tidally restricted marshes, there is less exchange with the shoreline systems and the contribution of the salt marshes to estuarine trophic dynamics decreases. Because *P. australis* dries out the substrate, it removes habitat for organisms such as aquatic benthic invertebrates and crabs (Roman et al. 1984; Wijte and Gallagher 1996a).

d. Control

The controls for *Phragmites australis* consist mostly of restoring the pre-disturbance conditions since the plant's presence can usually be traced to human disturbance. *P. australis* can be controlled in tidally restricted salt marshes by restoring the natural hydrology (Sinicrope et al. 1990). In some east coast salt marshes, where *P. australis* had become dominant, gates were installed to allow tidal exchange and thereby restore higher salinity levels. A significant reduction in *P. australis* height was observed after one growing season, particularly in the areas of the highest salinity (along creeks and ditches). At one site, the reintroduced tidal flow reduced the height of *P. australis* from 3 m to 1 m and the plant's population density decreased by 50%. *Spartina alterniflora, S. patens* (salt marsh hay), and *Distichlis spicata* (spike grass) re-colonized the areas along the well-flushed creeks (Roman et al. 1984). Higher salinity inhibits *P. australis* expansion, since germination does not occur at salinities greater than 25 ppm (Wijte and Gallagher 1996a). Both adults and seedlings die when subjected to salinities of 35 ppm and greater (Lissner and Schierup 1997).

Hydrologic manipulation in freshwater marshes can also provide control of *P. australis*. *P. australis* thrives in a variety of water depths, from 50 cm during the spring to 100 cm below the soil surface during the summer. It does not survive where the water table is below 100 cm, nor does it spread where the water depth is more than 50 cm. *P. australis* has been eliminated from freshwater marshes by raising the water levels to more than 1 m depth for 3 years; however, it can recover once flooding has ceased (Shay and Shay 1986). In a Manitoba lake, the water level was controlled and natural fluctuations ceased. The stable shallow water levels allowed long-lived emergents, such as *P. australis*, to spread and eliminate other species from the lake edge. Restoring the lake's natural variations in hydrology brought the *P. australis* population under control (Thompson and Shay 1989).

Where the hydrology cannot be altered, burning offers a means to control *P. australis*. In Manitoba, Thompson and Shay (1989) burned *P. australis* stands at three times of the year: summer, fall, and spring. The following growing season, no differences were noted in the stands of the spring and fall burns, but when the plants were burned in summer, the plant's dominance decreased and species diversity increased.

2. Phragmites australis as a Declining Species in Europe

a. Extent of the Problem

Phragmites australis is the dominant plant of many European marshes. It is distributed along the banks of rivers and lakes, and in wet meadows. As in the U.S., *P. australis* is frequently found along roadsides and other disturbed areas. Extensive stands are found in areas of shallow spring flooding followed by sub-surface water levels during the summer, such as the floodplain of the Danube River and on lake fringes throughout much of Europe (Rea 1996).

In the last 40 years, many of the reed stands of central and eastern Europe have declined. Their decline, or dieback, is defined as "a visible abnormal and non-reversible spontaneous retreat, disintegration or disappearance of a mature stand of common reed (*P. australis*) within a period not longer than a decade (±2.5 years)" (Ostendorp 1989). Plants in dieback stands usually share a number of traits (Table 8.7). A dieback usually begins with a retreat from deep water, a gradual thinning, or a clumped growth pattern. In 1951, the first report that *P. australis* was retreating from lake edges came from Switzerland. In the following decades, many more retreats were reported in eastern and central Europe. In the Netherlands, the *P. australis* belt around former estuaries that are tidally restricted have reduced in size and many of the Netherlands' freshwater stands of *P. australis* have

TABLE 8.7

Characteristics of *Phragmites australis* in Dieback Stands in Europe

Relatively low primary productivity
Low shoot density (fewer than 100 stems m^{-2})
Clumped stands (because rhizome apices die and the plant is inhibited from spreading outward)
Delayed flowering or none at all
Fewer living rhizomes than in vigorous stands under comparable conditions
Abnormal formation of structural parts of the plants (lignin, suberin, callus, and tylose) that results in blocking gas exchange
Lower seasonal carbohydrate accumulation
Fungal and insect damage (in some cases)

From Armstrong et al. 1996; van der Putten 1997.

either disappeared or diminished in size (Figure 8.14). In the warmer and drier Mediterranean countries, in parts of Scandinavia, and in some fen meadows of the Swiss plateau, *P. australis* continues to grow vigorously and even to expand (Ostendorp 1989; van der Putten 1997; Gusewell and Klotzli 1998).

Because *P. australis* stands protect lake shorelines from wave action and provide refuge for wildlife, its decline may have severe ramifications for littoral communities, including (Ostendorp et al. 1995):

- A lack of bank protection creating opportunities for erosion and drift of littoral sediments offshore
- The uprooting of trees and bushes

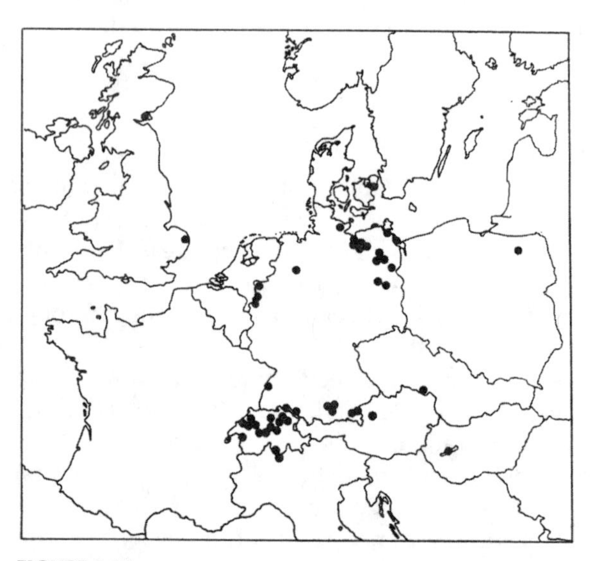

FIGURE 8.14
Phragmites australis declines have been noted at over 45 European lakes, most in central and eastern Europe. The dots denote lakes where reed declines have been documented. (From Ostendorp et al. 1995. *Ecological Engineering* 5: 51–75. Reprinted with permission from Elsevier Science.)

- The disappearance of bird species like great reed warbler (*Acrocephalus arundinaceus*), little bittern (*Ixobrychus minutus*), little grebe (*Podiceps ruficollis*) and purple heron (*Ardea purpurea*)
- Reduction in size of some fish species such as pike (*Exos lucius*), tench (*Tinca tinca*), carp (*Cyprinus carpio*), and rudd (*Scardinius erythrophthalmus*)

b. Causes of the Decline

The alteration of hydrology is one of the major causes of the decline of reed stands throughout Europe. Many of the documented cases of European reed decline have occurred where water levels are controlled by locks, weirs, and dams (Rea 1996; van der Putten 1997). Alterations in hydrology that restrict extreme fluctuations and thereby avoid both flooding and exposed sediments can result in the decline of reed beds. In Sweden, Switzerland, and other European countries, *Phragmites australis* has declined because lake water levels are regulated and the natural summertime dry periods no longer occur. *P. australis* is under stress when constantly submerged because less oxygen and carbon dioxide are available. Growth and germination are inhibited and the plants deplete carbohydrate reserves.

Other factors may further influence the decline of *P. australis* stands, such as mechanical damage, organic matter accumulation, development of intensely reducing soil conditions, eutrophication, phytotoxin accumulation, raised temperatures, and insect or fungal damage. These factors interfere with the reed's internal aeration as well as its water and nutrient uptake, leading to stunted growth and the death of underground plant parts. The accumulation of decaying organic matter exacerbates the problem because levels of phytotoxins such as hydrogen sulfide and volatile organic acids (formic, acetic, butyric, propionic, caproic) increase, further contributing to root, rhizome, bud, and shoot death, premature shoot senescence, and ultimately leading to a full dieback (Figure 8.15; Ostendorp et al. 1995; Armstrong et al. 1996; Clevering 1998). Allochthonous sources of organic matter such as wastewater effluent can also bring about the accumulation of phytotoxins and a high oxygen demand, and lead to reed decline (Cizkova et al. 1996; Kubin and Melzer 1997).

c. Solutions to the Phragmites australis Decline

Phragmites australis does exist in healthy and expanding stands in Europe: in oligotrophic waters, sand pits, temporarily wet roadside ditches, and abandoned meadows. In all of these sites, dry summertime conditions and low nutrient inputs seem to be factors in the plant's success. The solution to the decline may be to restore low water levels, at least during part of the growing season, and to reduce nutrient inputs (Rea 1996; van der Putten 1997).

Lake managers and ecologists are attempting to restore *P. australis* stands in a number of Swiss lakes. Restoration measures include reed protection against mechanical damage (fences), wave-dissipating structures (brushwood piles and refilling of substrate), nutrient export from the reed stands (winter mowing), and other measures such as reed plantings and the prohibition of public access (Ostendorp et al. 1995).

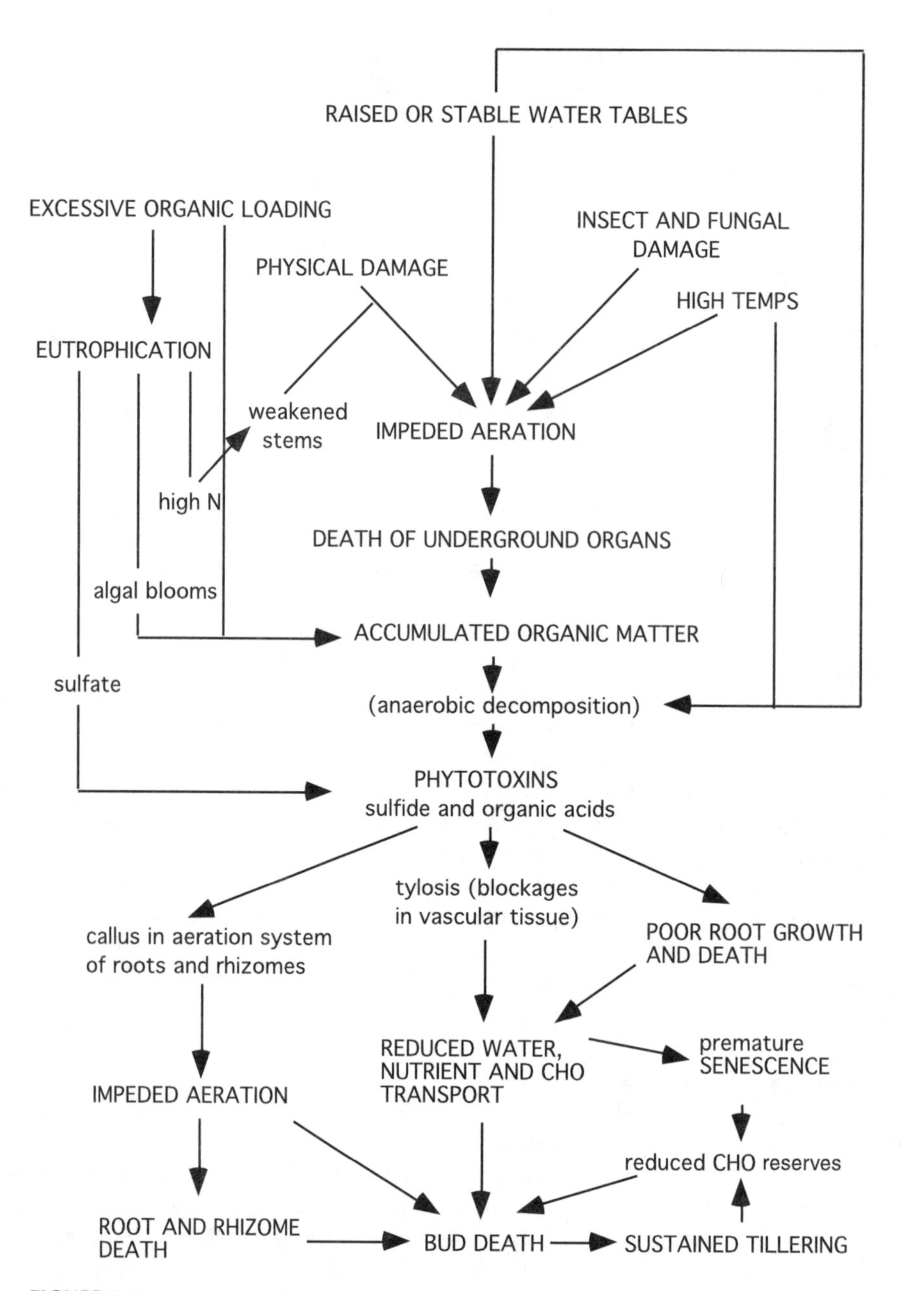

FIGURE 8.15
Scheme showing possible reasons for the dieback of *Phragmites australis*. Causes of the diebacks may include excessive organic loading, physical damage to the plants, raised or stable water tables, and insect and fungal damage that lead to impeded aeration, death of underground organs, and the accumulation of organic matter and phytotoxins. Within the plant (in the bottom half of the figure), phytotoxins further exacerbate low oxygen levels and reduce water, nutrient, and carbohydrate transport, leading to plant death (CHO = carbohydrate). (From Armstrong et al. 1996. *Folia Geobotanica Phytotaxonomica* 31: 127–142. Reprinted with permission.)

Summary

Invasive species grow in profusion and produce a significant change in terms of composition, structure, or ecosystem processes. Exotic species may become naturalized or invasive in a habitat. Native species may also be invasive. Invasive species tend to spread rapidly through both sexual and asexual reproduction. They often have wide ecological tolerances and can be successful in a range of habitats. Invasives are usually not susceptible to pests or herbivores in the new habitat. They are able to outcompete native plants for space, light, and resources.

Many invasives have been introduced to new habitats by humans, both intentionally and unintentionally. Disturbed habitats and islands tend to be more susceptible to plant invasion. Plant infestations in wetlands may bring about changes in both plant and animal communities. Floating invasives such as *Eichhornia crassipes* and *Salvinia molesta* can shade the water column, reducing light and the phytoplanktonic production of oxygen. Invasive plants such as *Melaleuca quinquenervia* and *Tamarix* species may change the local hydrology through increased transpiration. Invasives affect humans by expanding the habitat of vectors of waterborne or water-related diseases. They also clog water intakes, interfere with aquaculture, and block access to boat docks or swimming areas. Invasives are controlled by altering their habitat or by using mechanical, chemical, or biological controls. A combination of control methods is usually the most effective.

Five of the most noxious of wetland invasives in the U.S. and elsewhere are *Myriophyllum spicatum, Hydrilla verticillata, Eichhornia crassipes, Lythrum salicaria,* and *Phragmites australis*. Sometimes invasives are undesirable in one area and desirable in another. Such is the case with *P. australis,* which is invasive in many North American wetlands, but in decline in some parts of Europe where wetland managers are striving to restore it.

Part IV

Applications of Wetland Plant Studies

9

Wetland Plants in Restored and Constructed Wetlands

Around the world, wetland area has diminished due to ever-increasing human pressures. Our increased understanding and appreciation of wetland functions and values have spurred legislation to protect wetlands as well as popular interest in wetland preservation. Today, in an effort to stem the rate of wetland loss, wetlands are being restored or new wetlands are being created in many parts of the world. In the U.S., although wetlands continue to be lost to development, agriculture, and other landscape alterations, many of these losses are compensated by the construction of new wetlands. In addition, hundreds of wetlands have been built to treat wastewater of a variety of types. These treatment wetlands are an application of the natural water-cleansing functions of wetlands.

A number of terms concerning wetland restoration and creation are in use (Table 9.1). In this chapter, we use the term *restored* wetlands to refer to wetlands that are reinstated where they once were. Within our definition of *restored* wetlands, we include those that are *enhanced* by, for example, the removal of an invasive species or the introduction of a desirable plant or animal species. Entirely new wetlands, built where there were previously

TABLE 9.1

Definitions of Some of the Terms Related to Restored and Constructed Wetlands

Constructed	Any wetland that is made by humans rather than naturally occurring; refers to new wetlands built on a site where there were previously no wetlands; it can also refer to treatment wetlands
Restored	Includes enhancing an existing wetland by removing an invasive species, restoring some aspect such as the hydrology or topography of an existing wetland, building a wetland where one existed previously, and building a wetland in an area where wetlands probably were, such as in a riparian zone
Enhanced	The enhancement of an existing wetland by removing invasive species or restoring past animal or plant species or other aspects of the wetland (we include enhanced wetlands in restored wetlands)
Created	A new wetland, made on a site where there were not wetlands in the past
Mitigation	Wetlands constructed to replace wetlands that have been destroyed; may be created, preserved, or restored wetlands
Replacement	The same as mitigation wetlands
Treatment	Built to treat a specific wastewater problem such as domestic sewage, nonpoint source pollution, mine drainage, or animal farm wastewater
Artificial	Can refer to a created or treatment wetland, not widely used

none, are called *created* or *constructed* wetlands. Wetlands created, restored, or preserved to compensate for the loss of natural wetlands due to agriculture and development are called *mitigation* or *replacement* wetlands. We use the term *treatment* wetlands to refer to wetlands built to improve water quality.

While we discuss some aspects of these wetlands in general terms, we concentrate on the plants and plant communities. We discuss the development of wetland plant communities in newly created and restored wetlands and the role of plants in treatment wetlands.

I. Wetland Restoration and Creation

The restoration and creation of wetlands challenge our knowledge of ecosystem ecology. Can humans restore or create peatlands, swamps, marshes, and other wetland types? Can we duplicate the many complex functions of natural wetlands? Is it possible to re-create in a short period of time ecosystems that have taken centuries or longer to develop? Some types of wetlands, such as freshwater marshes, are easier to restore than rare wetland types with specialized plant species, such as peatlands, sedge meadows, and wetlands fringing oligotrophic rivers and lakes (Galatowitsch and van der Valk 1996; Weiher et al. 1996). Because natural wetlands are in constant flux, due to periodic disturbance or climatic variability, the goal of wetland restoration or creation can be a shifting target (Clewell and Lea 1990).

The most important aspect of restoring or creating wetlands is restoring or providing for the natural hydrology. There must be sufficient water flow to maintain hydric soils and hydrophytic vegetation. A key challenge is to reinstate the correct hydroperiod and allow for the hydrologic variability that occurs in natural wetlands. Restoring hydrology may involve providing or removing control structures in order to re-establish water flow or flooding regimes. In agricultural land, tile drains may need to be removed or broken. In some cases, fill material has to be removed. In tidal marsh restoration, the tidal regime and elevation are vital parameters because they determine the extent, duration, and timing of submergence (U.S. National Research Council 1992). Beyond hydrological remediation, steps to ensure sediment restoration may also be necessary. For example, the input of sediments from upland may need to be controlled, sediment dams in streams may need to be removed, and protective beaches or sand spits may need to be restored. Water quality is also important; controlling contaminant loadings is a vital step in many restoration efforts (Wilcox and Whillans 1999).

Wetland restoration includes a variety of activities. The restoration could involve diverting or eliminating a source of pollution, repairing damage caused by nearby development, reintroducing desirable species, reducing the population of exotics, or restoring wetlands where they existed previously (Wheeler 1995). Clewell and Lea (1990) described three levels of restoration for forested wetlands that apply to all wetland types:

- Enhancing an existing wetland to accelerate succession (or slow it down), or to provide suitable habitat for an endangered species
- Restoring a wetland so that its former hydrology is in place; this may be all that is necessary for its plant community to return
- Creating a wetland that resembles a locally indigenous wetland community in species composition and physiognomy on sites that have been altered

The success of wetland restoration depends, in part, on the degree of disturbance at the project site and the condition of the surrounding landscape at the beginning of the project. Success is more likely in areas with little or short-term disturbance and where the landscape is generally in its natural condition. The most difficult wetlands to restore are those in very degraded sites, such as the salt marshes of southern California and the Hackensack River Meadowlands of New Jersey (U.S. National Research Council 1992). Wetlands in urbanized areas or in many developing countries are also difficult to restore due to intense human pressures (Helfield and Diamond 1997; Walters 1997, 2000a, b; see Case Study 9.A, Integrating Wetland Restoration with Human Uses of Wetland Resources).

To determine the success of restoration, a monitoring plan is usually part of the project. Deciding whether or not a restoration project has been successful is often based on the structure of the plant community or on an ecosystem function such as primary productivity. In some cases, the presence or absence of indicator species can reveal whether a project is successful (see Case Study 9.B, Restoring the Habitat of an Endangered Bird in Southern California). Monitoring often includes comparing the restored wetland to nearby natural reference wetlands. Parameters that are compared include species diversity, plant productivity, stem density, sediment texture, sediment nutrient content, invertebrate populations, and wildlife use (Langis et al. 1991; Zedler 1993; Havens et al. 1995; Boyer and Zedler 1998; Walters 2000b). Throughout the monitoring period, it is important that the restoration plan remain flexible in order to respond to problems. A management strategy that adapts to problems and allows for changes is essential in many cases (Zedler 1993; Pastorok et al. 1997; Thom 1997).

The necessary length of the monitoring period varies with the type of wetland and the goals of the project. In many cases, success is assumed if the new wetland's community structure resembles that of reference wetlands. However, the establishment of food webs, the movement of carbon and energy, nutrient recycling, and other wetland functions may never be restored, or may take many years to develop (McKee and Faulkner 2000). For salt marshes, estimates of the time required for the success of plant community restoration vary from 3 to 10 years or even longer (Broome et al. 1988). Because of wide year-to-year variability, Zedler (1993) suggests that salt marsh restoration requires 20 years of monitoring along with a large data base from natural reference wetlands against which to compare. Forested wetlands may require much longer monitoring periods because of the long establishment time for trees. Given the correct hydrological conditions, restored mangrove forests may resemble natural communities within about 20 years of planting (Ellison 2000b). Mitsch and Wilson (1996) suggest that restored wetlands of all types should be given enough time for wetland functions to become established. They state that monitoring should continue for 15 to 20 years or even longer for specific types of wetlands (e.g., forested, coastal, and peatlands).

A. The Development of Plant Communities in Restored and Created Wetlands

Whether plants are carefully chosen and planted, arise from the seed bank, or arrive through natural dispersal mechanisms, the new wetland plant community is determined, to a large extent, by the environmental conditions found in the wetland. While some wetland restoration efforts include planting and managing for specific species, others have relied on volunteer plant species to colonize the site. Propagules arrive via wind, water, or animals. In some restored sites, wetland species already exist in the seed bank.

1. Environmental Conditions

One way to look at the assembly of wetland plant communities is as a series of filters, or environmental sieves, that strain species so that only the final assemblage remains (see Chapter 7, Section III.A.3, The Environmental Sieve Model; van der Valk 1981). Knowledge of each of the filters and how to manipulate them aids in restoring the desired community. Filters in wetlands include water levels, soil fertility, disturbance, salinity, competition, herbivory, and the accumulation of sediments that may bury seeds and propagules. Different wetland types may be more influenced by some filters than others. For example, species distribution in estuarine wetlands is heavily influenced by salinity, while plants in deltaic wetlands may be influenced most by the accumulation of sediments (Keddy 1999).

Organisms possess life-history traits that allow them to pass through different filters. A systematic method of predicting how a set of species might respond to a particular filter would be helpful in many cases (Shipley et al. 1989; Keddy 1999). Screening studies provide data that enable the researcher to predict how a set of species might respond to a particular filter. In order to screen wetland plants, a large number of species would need to be exposed to a certain filter or a set of filters. For example, in a salt marsh or mangrove, salinity levels provide a suitable filter to test, since the number of salt-tolerant plants is relatively low. In wetlands where there are multiple filters, screening might be more complex but still feasible, particularly if one or two filters, such as climate or water regime, can be used to filter out a large number of potential plant species (Keddy 1999).

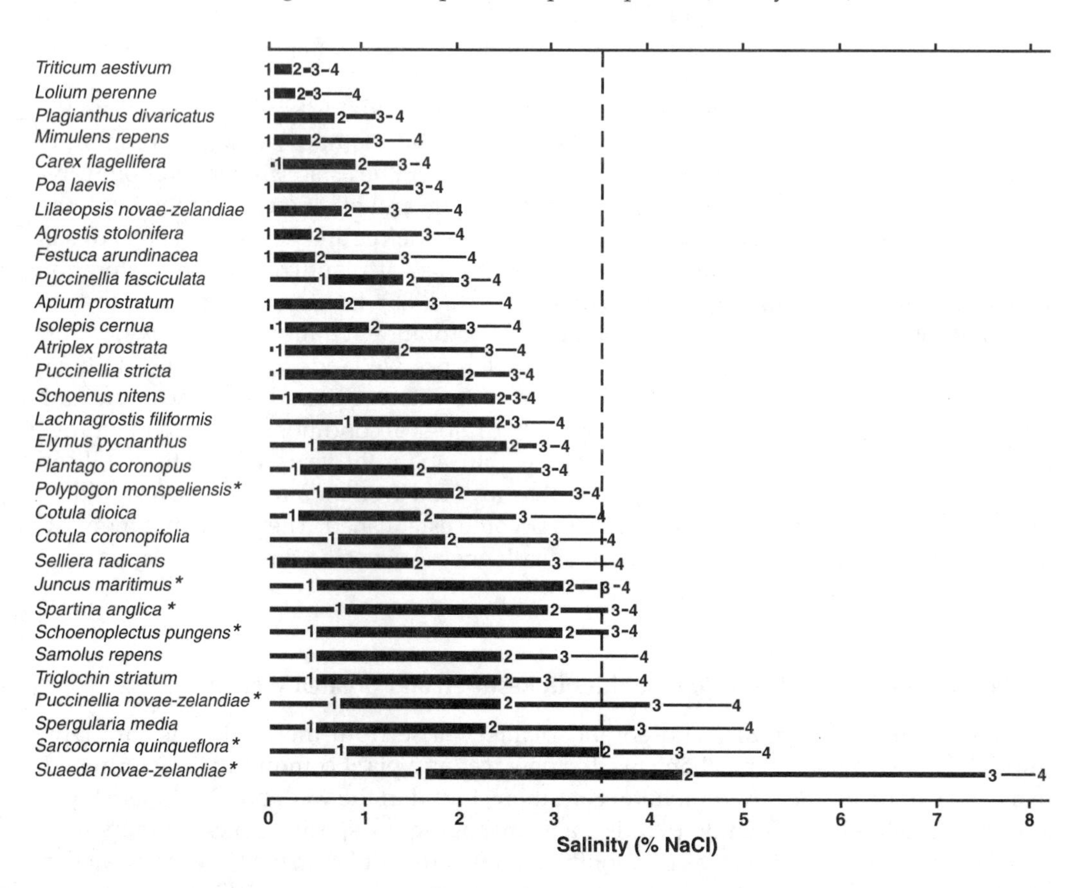

FIGURE 9.1
Growth parameters of salt-tolerant species from Otago, New Zealand salt marshes: 1 = salinity for maximum growth, 2 = half-growth salinity, 3 = salinity for death of plant parts, 4 = cessation of growth. The species are arranged in order based on cessation of growth. Asterisks indicate significant ($p = 0.05$) salt requirements for maximum growth. The thickness of the horizontal lines indicates the highest rates of growth and the vertical line, seawater salinity. (From Partridge, T.R. and Wilson, J.B. 1987. *New Zealand Journal of Botany* 25: 559–566. Reprinted with permission.)

Partridge and Wilson (1987) performed a screening experiment with 31 of the most frequently encountered species in the salt marshes of Otago, New Zealand. They measured the effects of salinity on survival and growth and found considerable differences among species (Figure 9.1). Most could not grow in seawater, which has a salinity of 35 ppt, although a small number could grow in hypersaline conditions of up to 75 ppt. Some species required some salt for maximum growth (e.g., *Suaeda novae-zelandiae*), although none required salt to survive. Most of the species grew best in fresh water. Similar knowledge of the salt tolerance, water level requirements, or other adaptations of a wide variety of plants would allow wetland restorationists to choose appropriate species for the environmental conditions of their site.

2. Self-Design and Designer Approaches

The designer approach and self-design are two general approaches to introducing vegetation to restored or constructed wetlands. The designer approach involves introducing and maintaining chosen plant species (and sometimes animals). In this approach, the wetland restorationist needs an understanding of the life history of the species involved, including their dispersal, germination, and establishment requirements (Middleton 1999). In the second approach, called 'self-design,' the self-organization capacity of natural systems is emphasized (Mitsch and Wilson 1996). In this approach, species may arrive as volunteers through wind, water, or animal dispersal. Species might also be introduced to the wetland, but their ultimate survival depends on the ecosystem's conditions, which filter out species not adapted to the conditions at hand. The assemblage of plants, microbes, and animals that is best adapted to the existing conditions will persist, while all other species will disappear from the system or not become established.

Although the introduction of plants is often required in order to comply with a mitigation or restoration plan, it may not always be ecologically necessary. When specific plants are chosen and carefully planted, their establishment and survival are ultimately a function of the abiotic filters in the wetland. When volunteer species arrive, as long as they are not invasives or otherwise undesirable, their presence is usually welcome in restored wetlands in which the self-design principle is at work. The self-design approach may, in some instances, be more sustainable than the close maintenance required in the designer approach (Mitsch et al. 1998). However, when a restoration site has a poor seed bank and limited possibilities for seed or propagule dispersal, planting may result in a more rapidly vegetated wetland (Middleton 1999). If the goal is to enhance the population of a specific species or set of species, wetland managers must ensure those species' survival and intervene with adaptive management approaches when necessary (Zedler 1993; 2000b).

The extent and rate of revegetation by natural dispersal can be unpredictable and depend on many interacting (and little understood) variables, including the availability of upstream or upwind seed sources, soil temperature and moisture regimes, streamflow regimes, slopes, soil fertility, and disturbance patterns (Goldner 1984; Day et al. 1988). In general, where there are nearby natural wetlands, more recovery of local flora might be expected, especially for species that are dispersed by wind or waterfowl. Species with poor dispersal capabilities may have to be reintroduced during restoration (Leck 1989; van der Valk and Pederson 1989; Reinartz and Warne 1993; Keddy 1999).

Some studies have shown that when initial conditions are suitable in constructed and restored wetlands, plant species arrive and new plant communities form, often without any human intervention (but see Case Study 9.C, Vegetation Patterns in Restored Prairie Potholes). In four constructed freshwater marshes in Illinois (from 1.9 to 3.4 ha in size; Figure 9.2) plant diversity increased with time (Fennessy et al. 1994a). During the first 4

FIGURE 9.2
An aerial photograph of four constructed marshes at the Des Plaines River Wetlands Demonstration Project in Illinois. The marshes were built for research purposes. The water source, the Des Plaines River, has relatively high levels of suspended solids and nutrients from agricultural sources. Researchers tested the capacity of the marshes to ameliorate water quality and they examined wetland plant community development. (Sanville and Mitsch 1994; photo courtesy of D. Hey, Wetlands Research, Inc.)

years of the wetlands' existence, the number of wetland taxa (obligate and facultative wetland species) increased from 2 to 19 in the first marsh, from 14 to 28 in the second, from 13 to 17 in the third, and from 12 to 22 in the last. Only one species was introduced, and it was only planted in the first marsh; all of the others arrived as volunteers.

In two 1-ha constructed marshes in Ohio, an experiment to test the effects of planting on species diversity began in 1994, when one of the marshes was planted with 13 species while the second was left unplanted. By the beginning of the fourth growing season, the plant cover in the unplanted wetland (58%) slightly exceeded the plant cover in the planted wetland (51%; Mitsch et al. 1998). By the end of the 1998 growing season, the number of wetland plants (obligate and facultative wetland species) in the planted wetland had increased from the 13 introduced species to 55 species. The number of species in the unplanted wetland increased from 0 to 45 species. The planted wetland has more species because many of the original planted species have become established there (Bouchard and Mitsch 1999).

In both the Ohio and Illinois studies, rivers adjacent to the study site were the main source of water for the constructed wetlands. Riverine wetlands may be more likely to revegetate naturally than isolated wetlands because the river water carries seeds and propagules from upstream wetlands (Middleton 1999).

Early introduction of a diversity of wetland plants may enhance the ultimate diversity of vegetation in constructed and restored wetlands. Reinartz and Warne (1993) examined the colonization of 5 constructed freshwater marshes that were seeded with 22 native species. They compared these to 11 unseeded constructed marshes. The diversity of native wetland species increased with wetland age, wetland size, and with proximity to the nearest established wetland. After 3 years, the unseeded wetlands had an average of 22 species. In contrast, the 5 seeded wetlands had an average of 42 species; 17 of the 22 planted species became established. *Typha latifolia* and *T. angustifolia* became the most dominant species in the unseeded wetlands; their cover increased from 15 to 55% during the 3-year study. The extent of the *Typha* cover was lower in the seeded sites with an average of 22% cover in the second year. Cover by the seeded species accounted for the difference in the *Typha* cover.

3. Seed Banks in Restored Wetlands

Seed banks may be present in restored wetlands from prior periods of wetland plant growth. The seeds of most herbaceous wetland species are capable of persisting more than a year in soil, and some persist for many years. Persistent species often have small seeds that respond positively to light, increased aeration, and/or alternating temperature. Herbaceous species dominate wetland seed banks, with graminoids usually constituting over half of the seed bank (Leck 1989).

In restoration projects, seed banks have been used to restore or establish native vegetation. Seed banks can be used only if suitable conditions can be established and maintained for the germination of the preferred species. Seed banks may not be the entire answer for the restoration of native vegetation because the desired species may not be represented or because the seeds of unwanted species are present (van der Valk and Pederson 1989). Seed banks in forested wetlands typically do not reflect the woody plant community. Rather, seeds are often from herbaceous species from nearby open areas. One cannot rely on the seed bank in forested wetland restoration projects, including mangrove forests (Leck 1989; Buckley et al. 1997; Walters 2000b).

The following are recommendations regarding the use of seed banks in restored wetlands:

- Before a management plan that relies on a seed bank is implemented, it is important to test the seed bank to determine the presence of viable seeds and the community composition (van der Valk and Pederson 1989). However, results of seed bank tests do not always reflect the species composition of the restored plant community. The hydrologic regime or soil organic matter of the restored site may allow for the germination of some species, but not others (van der Valk 1981; Wilson et al. 1993; ter Heerdt and Drost 1994).
- Relict seed banks can be used in the restoration of native vegetation, but their utility decreases with time because many seeds lose their viability. Sites where native vegetation has only recently been eliminated make the best candidates for restoration projects using the seed bank (van der Valk and Pederson 1989; Wienhold and van der Valk 1989; Galatowitsch and van der Valk 1994, 1995, 1996).
- Historical records of plant distribution at the site are useful because the seeds of desired species will be present where they had the densest growth in the past (Leck and Simpson 1987; Welling et al. 1988a).
- A period of drawndown conditions in which mudflats are exposed may enhance germination rates (van der Valk and Davis 1978; Siegley et al. 1988; Leck 1989; Willis and Mitsch 1995). However, if the purpose is to establish a maximum number of emergent seedlings, a 1-year drawdown may be sufficient. In a 2-year seed bank study in a Canadian marsh, recruitment of emergents occurred primarily during the first year. Many of the first-year seedlings died during the second year of drawndown conditions (Welling et al. 1988b).
- Knowledge of the desired plants' life history is necessary. If only certain species within the seed bank are desirable, then it is essential to know the conditions

required for germination (e.g., frost, aerobic conditions) as well as the plant's optimal hydroperiod (van der Valk 1981; van der Valk and Pederson 1989).
- The seed bank should not be covered with other sediments. For instance, 1 cm of sand can substantially reduce germination (Leck 1989).
- In general, germination rates in sand or sites with finely textured or highly organic soils are lower and these substrates should be avoided where possible (Leck 1989).
- The seeds of woody species are not common (as compared to herbaceous species), even in swamp seed banks, so for the restoration of forested or shrub wetlands, planting is necessary (Leck 1989).

Donor seed banks from other sites can be used in restoration projects, but they should be tested for species composition. Donor soils should be collected and carefully preserved in order to avoid a loss in seed viability. They should be used at the beginning of the growing season when germination would naturally occur (van der Valk and Pederson 1989). The uppermost portion of the soil contains the highest concentration of seeds and should be preserved. van der Valk and Pederson (1989) recommend that donor soils be collected to a depth no greater than 25 cm. If the soil layer is too thick, the seed bank is diluted and lower germination rates result (Putwain and Gillham 1990). Donor seed banks can enable the rapid development of diverse native vegetation and impede the establishment of unwanted species (van der Valk and Pederson 1989).

B. Planting Recommendations for Restoration and Creation Projects

The goal of many restoration projects is to produce a sustainable, diverse plant community with high percentages of desirable species that will attract wildlife. In some cases, particularly where the new wetland is close to natural ones, plants will arrive via natural dispersal mechanisms (Mitsch et al. 1998). When a specific community is desired, such as in the restoration of rare communities or a specific habitat type, or when natural dispersal may be unlikely, wetland restorationists must choose species for the site. The edaphic and hydrologic conditions of a site should be assessed in order to choose the right species and the best planting techniques (Imbert et al. 2000). Nichols (1991) suggests asking the following questions when considering species for restoration or construction projects:

- Does the species have the desired properties needed in the restoration? Does the plant provide good waterfowl food, desirable fish habitat, and aesthetic value? Is it able to withstand wind or waves?
- Does the species have weedy tendencies? Will it become a nuisance?
- Does the species have the potential to grow and reproduce well enough to maintain and increase its population?
- How large an initial population is needed to ensure a viable stand, taking into account losses from herbivores, pathogens, poor reproductive success, wind and wave action, and adverse climatic conditions?
- Is the physical and chemical habitat suitable for the desired species? Even if the species formerly grew in the area, the habitat might have been altered to the extent that it is no longer suitable.

Planting techniques have been developed for many species and the nursery or other plant source should always be consulted for planting instructions. The instructions may be quite specific and should be followed to ensure success. For example, the instructions for propagating *Spartina alterniflora* indicate that seeds should be harvested by hand or machine as near as possible to maturity or just prior to release from the plant. The seeds are threshed after being stored at 1° to 4°C for about 1 month. After threshing, the seeds are stored in covered containers filled with water with a salinity of 35 ppt at 2° to 4 °C. Seeds are broadcast from mid-April to mid-June, depending on the latitude. The seeds are incorporated into the substrate to a depth of 2 to 3 cm and the density of planting is 100 seeds m^{-2}. Seeding is only feasible in the upper half of the intertidal zone (Broome et al. 1988).

The timing of planting in both temperate and tropical latitudes is crucial. Mangrove seedlings, for example, may be best planted at the onset of the rainy season (July/August) to avoid drought. However, if the shoreline is poorly sheltered, planting may be done earlier (February/March) when the mean sea level is at a minimum (Imbert et al. 2000).

In general, when seeds are used, they may be broadcast or packed in mud balls before sowing. Whole plants or vegetative propagules can be placed directly in the sediments, or weighted with mesh bags and gravel and sown from the water surface. To plant emergents, it may be necessary to decrease the water level in order to expose the sediments and allow seeds to germinate (Nichols 1991).

Some wetland types pose unique challenges. For instance, in the restoration of sedge meadows, it is difficult to establish the dominant sedges, such as *Carex*, whose seeds are short-lived and do not usually remain viable within seed banks (Reinartz and Warne 1993; van der Valk et al. 1999). To maximize the probability that *Carex* will become established, the use of fresh seeds is necessary, preferably seeds produced earlier in the same growing season. The soil moisture must be kept as high as possible and the soil's organic matter content should be as high as that found in natural sedge meadows (van der Valk et al. 1999).

Wetland restoration often includes the careful choice of native plants; however, invasives may become established. Fast-growing species such as *Phragmites australis* (common reed), *Lythrum salicaria* (purple loosestrife), and *Typha* species may dominate sites that were intended for other vegetation. *Typha* is frequently found in freshwater marshes; it often outcompetes other species and creates dense monocultures with little variety in food or habitat. Extensive stands of *Typha* have become established in several freshwater marsh restoration projects (Reinartz and Warne 1993; Fennessy et al. 1994a; Bouchard and Mitsch 1999).

Weiher and others (1996) performed a 5-year mesocosm study using seeds from 20 wetland species under a range of environmental conditions. Although all of the species germinated, only six species were found in large numbers after 5 years. By the end of the study, most of the mesocosms were dominated by *Lythrum salicaria* while the other eudicot species were extirpated. *L. salicaria* establishment and dominance were minimal only under low fertility conditions and when the mesocosms were flooded in the spring and early summer to a depth of 5 cm. The growth of *Typha angustifolia* was poor on coarse substrates (particle size >4 mm). To inhibit the establishment of these fast-growing species, adverse conditions such as those noted in this study might be included in the restoration plan.

II. Treatment Wetlands

Because of their capacity to enhance water quality, hundreds of wetlands have been constructed around the world to treat liquid wastes in a number of forms, including domestic sewage (Figure 9.3; Hammer 1989; Kadlec and Knight 1996), livestock wastewater (Figures 9.4 and 9.5; Hammer 1994; Cronk 1996), nonpoint source pollution (Figure 9.2 Hammer

1992; Mitsch and Cronk 1992), landfill leachate (Mulamoottil et al. 1999), stormwater runoff (Figure 9.6; Livingston 1989; Strecker et al. 1992), mine drainage (Wieder 1989; Fennessy and Mitsch 1989; Hedin et al. 1994; Nairn et al. 2000), and other industrial discharges (Kadlec and Knight 1996; Odum et al. 2000). In addition, many riparian wetlands have been restored in an effort to intercept sediment- and nutrient-laden runoff from agricultural fields (Vought et al. 1994; Fennessy and Cronk 1997).

FIGURE 9.3
Winter at one of several wetland cells at the Mayo, Maryland wastewater treatment wetlands. The wetlands treat septic tank effluent in a town of about 2000 residents. The vegetation in this marsh is dominated by *Phalaris arundinacea* (reed canary grass). The wastewater was sprayed out of pipes spread throughout the wetland's area in an effort to aerate it. (Photo by J. Cronk.)

FIGURE 9.4
Two densely vegetated marshes were constructed in the depression shown in the middle of the photo. They were built to treat wastewater from a dairy farm in Montgomery County, Maryland. (Photo by J. Cronk.)

FIGURE 9.5
This rectangular, newly planted marsh treats irrigation water from a dairy farm near Sequim, Washington. (Photo by H. Crowell.)

FIGURE 9.6
This small marsh, vegetated with *Phragmites australis* (common reed), was constructed adjacent to a parking lot at the University of Maryland in an effort to filter stormwater runoff before it entered a tributary of the Patuxent River. (Photo by J. Cronk.)

While early studies of wastewater treatment wetlands were performed using natural wetlands such as *Taxodium distichum* (bald cypress) swamps in Florida (Odum et al. 1977) and peatlands in Michigan (Kadlec and Kadlec 1979), today wetlands are constructed specifically for the purpose of wastewater treatment. Wastewater treatment wetlands include surface flow marshes, vegetated subsurface flow beds (found mostly in Europe, and vegetated with *Phragmites australis*), submerged aquatic beds, and beds of floating plants such as *Eichhornia crassipes* (water hyacinth), as well as other types (Kadlec and

Knight 1996). Treatment wetlands have become widespread because, in general, they are effective for the reduction of suspended solids (SS), biochemical oxygen demand (BOD), nitrogen, phosphorus, and some metals. Constructed wetlands provide a low-energy, low-technology solution to many wastewater problems (Brix 1986; Kadlec and Knight 1996).

A. Removal of Wastewater Contaminants

The contaminants in domestic and animal wastewater and in agricultural runoff consist mostly of plant macronutrients (e.g., phosphorus and nitrogen), solids, and pathogens. Although nutrients are necessary for plant growth, an excess of nutrients in water bodies leads to adverse conditions for aquatic life. The removal of excess nutrient loadings is essential to the health of aquatic ecosystems. In treatment wetlands, nutrient and solids removal is facilitated by shallow water (which maximizes the sediment to water interface), high primary productivity, the presence of aerobic and anaerobic sediments, and the accumulation of litter (Mitsch and Gosselink 2000). Slow water flow causes SS to settle from the water column in wetlands. BOD is reduced by the settling of organic matter and through the decomposition of BOD-causing substances. We focus our discussion on removal processes for nitrogen, phosphorus, and pathogens in domestic and animal wastewater treatment wetlands and the role of plants in these processes. We also briefly describe the uptake of metals in treatment wetlands and in contaminated sites.

1. Nitrogen Removal

Nitrogen enters treatment wetlands in either an organic or inorganic form. As organic nitrogen is mineralized, it enters the inorganic nitrogen cycle. The inorganic forms are nitrate (NO_3^-), nitrite (NO_2^-), ammonia (NH_3), and ammonium (NH_4^+). Most of the inorganic nitrogen entering wastewater treatment wetlands is in the form of ammonia and ammonium. Ammonia may be volatilized or taken up by plants or microbes. Under aerobic conditions, it may be transformed into nitrate in the *nitrification* process. Similarly, ammonium may be taken up in biota or transformed into nitrate (see Chapter 3, Section III.A.1.a, Nitrogen). In addition, because of its positive charge, ammonium can be sorbed onto negatively charged soil particles that can be deposited as sediment.

In wetlands, nitrification (the oxidation of ammonia and ammonium to nitrate and nitrite) occurs in oxidized areas of the substrate or water column. Oxygen is present at the soil surface and in the root zone, where it enters the soil via diffusion from plant roots (see Chapter 4, Section II.A.5, Radial Oxygen Loss). As nitrate diffuses into anaerobic areas in the soil, it is reduced by bacteria to nitrous oxide (N_2O) or dinitrogen gas (N_2), in a process called *denitrification*. Both N_2O and N_2 are released to the atmosphere (Gambrell and Patrick 1978; see Chapter 3, Section III.A.1.a, Nitrogen). The occurrence of both aerobic and anaerobic soils in wetlands provides ideal conditions for nitrogen conversions. Since denitrification results in the removal of nitrogen from the aqueous system, it is the most important removal pathway for nitrogen in most wetlands (Faulkner and Richardson 1989). Because the transformations of nitrogen involve microbial processes, nitrogen removal is enhanced during the growing season when high temperatures stimulate microbial population growth (Gambrell and Patrick 1978). Low temperatures or acidic soil conditions inhibit denitrification (Engler and Patrick 1974; Schipper et al. 1993).

Uptake and incorporation into plant and algal biomass are another mechanism by which nitrogen is removed. This may or may not represent a permanent loss. Nitrogen and other nutrients that accumulate in tissues may be leached back into the water column or interstitial water upon plant senescence. Alternatively, nutrients may become permanently buried in undecomposed plant litter. Vegetative uptake of nutrients shows seasonal variation in temperate climates (see Section II.B.3, Nutrient Uptake).

2. Phosphorus Retention

Many treatment wetlands have been shown to be successful at retaining phosphorus. Reviews of phosphorus uptake at a wide variety of treatment wetlands in different climates receiving different loadings reveal that most function as net phosphorus sinks (Kadlec and Knight 1996; Reddy et al. 1999). The same is not necessarily true in natural wetlands, where there may be a seasonal release of phosphorus (Lee et al. 1975).

Phosphorus is retained within wetlands through biotic uptake, sorption onto soil particles, and accretion of wetland soils over time. Biotic uptake is considered to provide short-term removal (days to a few years), while the other two retention pathways provide longer-term removal (Kadlec 1995, 1997; Reddy et al. 1999).

a. Biotic Uptake of Phosphorus

Phosphorus enters treatment wetlands as organic or inorganic phosphorus. A portion of the inorganic phosphorus is bioavailable. Organic and other non-available forms can be broken down and transformed into bioavailable forms within the wetland. The proportion of the phosphorus that is bioavailable varies with the source of wastewater. Bioavailable phosphorus is taken up by macrophytes, algae, and microbes. Phytoplankton and periphyton are able to rapidly assimilate phosphorus and often respond to new inputs with rapid growth. Algal productivity has been observed to be higher near treatment wetland inflows than near outflows, probably because high levels of nutrients stimulate high assimilation rates (Cronk and Mitsch 1994a, b). Greater phosphorus retention during the growing season at wastewater treatment wetlands has been attributed to biotic uptake (Gearhart et al. 1989).

The amount of phosphorus stored in plant tissue depends on the type of vegetation and its rate of growth, the season and the climate (with more taken up during the growing season and in warmer climates), litter decomposition rates, leaching of phosphorus from detrital tissue, and translocation of phosphorus from aboveground to belowground parts (see Section II.B.3, Nutrient Uptake). At the end of the growing season in temperate areas, or as shoots die and are replaced throughout the year in subtropical and tropical areas, a portion of the aboveground plant tissues is decomposed and phosphorus is released. Some of the plant's nutrients are translocated to belowground parts where they aid the plant in overwintering and spring growth. Translocation can account for a high amount of phosphorus retention within the plant. In *Typha glauca*, for example, approximately 45% of the shoot phosphorus was translocated to roots and rhizomes at the end of the growing season (Davis and van der Valk 1983).

b. Sorption onto Soil Particles

Inorganic forms of phosphorus may become chemically bound with suspended solids and sediments in a process called *sorption*. As suspended solids settle, the sorbed phosphorus is removed from the water column. Phosphorus sorbs to oxides and hydroxyoxides of iron and aluminum and to calcium carbonate. There is a finite supply of these minerals in the sediments, and inorganic phosphorus must come in direct contact with the sediments before it can be retained there. Once the sorption sites are saturated (which occurs more readily in sites where phosphorus loadings have been high in the past or in sites with low levels of clay mineral surfaces), the capacity for the soil to release phosphorus increases (Kadlec 1985). Under oxidized conditions, phosphorus is held more tightly to soil particles than under reduced conditions. Under reduced conditions, phosphorus is released due to the reduction of ferric (Fe^{3+}) phosphate compounds to more soluble ferrous (Fe^{2+}) forms. If the soil is not vegetated, this released phosphorus diffuses back to surface waters. When plants are present, they assimilate the released phosphorus (or a portion of it) and prevent its movement out of the sediments (Reddy et al. 1999).

As phosphorus inputs to a constructed wetland continue over a period of several years, sorption sites in the sediments may become increasingly unavailable (Kadlec 1985). Incoming phosphorus is often rapidly removed from the water column very close to the inlet, through soil sorption or plant uptake (Figure 9.7; Mitsch et al. 1995; Kadlec 1999). For this reason, one way to enhance phosphorus sorption is to increase the surface area of initial contact by distributing the inflow along the length of a pipe (with severals outlet points rather than just one, as seen in Figure 9.3; Hammer 1992). Adding aluminum to the substrate can also enhance phosphorus removal (James et al. 1992) since phosphorus sorption is positively correlated to aluminum content in the substrate (Richardson 1985). Periodic draining can allow oxidation and recharge sorption sites for greater phosphorus removal than under permanently reduced conditions (Faulkner and Richardson 1989).

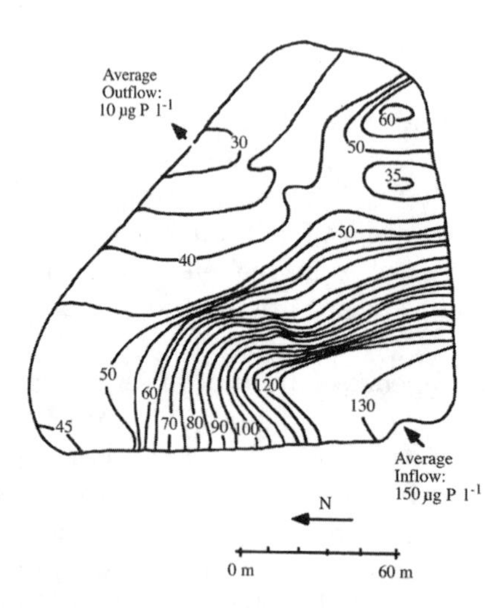

FIGURE 9.7
Because incoming phosphorus is often rapidly removed from the water column through soil sorption or plant uptake, the phosphorus concentration in treatment wetlands often exhibits a stable decreasing gradient from inlet to outlet. This figure shows a schematic of Experimental Wetland 5 at the Des Plaines River Wetland Demonstration Project in Illinois. Total phosphorus concentrations were measured monthly at 16 sites between the inflow and the outflow from May to September 1991. The concentration of phosphorus decreased steadily across the wetland from an average of 150 µg l^{-1} at the inflow to 10 µg l^{-1} at the outflow. (From Cronk 1992.)

c. Accretion of Wetland Soils

Sediment accretion by the accumulation of organic matter represents a long-term, sustainable phosphorus removal pathway (Kadlec 1997). The accumulation of litter is generally on the order of a few millimeters per year. A portion of the plant's biomass remains on or in the sediments and decomposes relatively slowly. Over time, the storage of phosphorus in plant litter becomes increasingly significant (Kadlec 1995, 1999).

In the Houghton wastewater treatment wetlands in Michigan, sorption sites became saturated during the first 3 years of operation. During the first 9 years, the formation of new biomass (vascular plants, algae, bacteria, and other organisms) had a significant effect on phosphorus removal. Thereafter, soil accretion (at the rate of 2 to 3 mm yr^{-1}) was the principal mechanism for phosphorus removal (Figure 9.8; Kadlec 1997).

3. Pathogen Removal

Wastewater, both human and animal, may be contaminated with pathogens. Most wastewater-related diseases in North America are caused by bacteria and viruses rather than

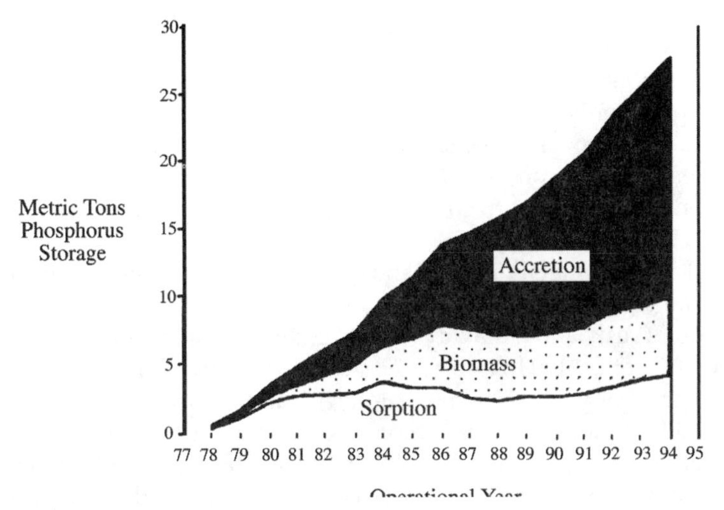

FIGURE 9.8
Storages of phosphorus within the treatment zone of the Houghton Lake, Michigan wastewater treatment wetland. Sediment accretion accounts for the greatest level of phosphorus storage. (From Kadlec, R.H. 1997. *Ecological Engineering* 8: 145–172. Reprinted with permission from Elsevier Science.)

worms and protozoa, so the treatment wetlands literature deals primarily with these two groups of organisms. Total or fecal coliforms are generally the only measured pathogen indicators in wastewater treatment wetlands. Coliforms are reduced within wetlands through exposure to sunlight, predation, and competition for resources. In addition, they may be buried beneath sediments or adsorbed. In two cases in which constructed wetlands were used as tertiary treatment systems for domestic wastewater, bacterial and viral indicators were 90 to 99% removed (Gersberg et al. 1989). If the wastewater has not been pretreated, additional disinfection through chlorination or exposure to ultraviolet radiation may be necessary.

4. Metal Removal

Some metals are essential micronutrients for both plants and animals, but in wastewaters they may be found in concentrations that are toxic to sensitive organisms. Biomagnification through the food chain occurs with a number of metals. For this reason it is essential that metals be removed from wastewater flows before they enter natural waters (Knight 1997). Kadlec and Knight (1996) and Odum and others (2000) report the removal of several metals in treatment wetlands, including aluminum, arsenic, cadmium, chromium, copper, iron, lead, manganese, mercury, nickel, selenium, silver, and zinc.

Metals are removed in treatment wetlands by three major mechanisms (Kadlec and Knight 1996):

- Binding to soils, sediments, particulates, and soluble organics by cation exchange and chelation
- Precipitation as insoluble salts, principally sulfides and oxyhydroxides
- Uptake by plants, including algae, and by bacteria

While the first two mechanisms along with microbial uptake are the predominant pathways of metal removal in treatment wetlands, we focus on the uptake of metals by vascular plants.

a. Plant Uptake of Metals

When plants accumulate metals, the roots and rhizomes generally show greater concentrations than the shoots (Sinicrope et al. 1992). Plants' effectiveness in removing metals is seasonal, with uptake only during the growing season (Simpson et al. 1983b). The accumulation of metals in plants may be short-lived since a portion of the metals are released upon senescence. The undecomposed portion of the litter may be a longer-term storage although data on metal release from wetland plant litter are not available (Kadlec and Knight 1996).

The accumulation of metals in wetland plants has been studied primarily in three species: *Eichhornia crassipes, Typha latifolia,* and *Phragmites australis. E. crassipes* accumulates copper, lead (Vesk and Allaway 1997), cadmium, chromium, mercury, zinc (as reviewed in Schmitz et al. 1993), and silver (Rai et al. 1995). *T. latifolia* accumulates high concentrations of nickel with no signs of toxicity, and up to 80 μg of copper g^{-1} before showing reduced leaf elongation and biomass production (Taylor and Crowder 1983). *T. latifolia* has also been shown to accumulate low levels of lead, zinc, and cadmium in the roots and it is reported to be tolerant of relatively high levels of these metals (Shutes et al. 1993; Ye et al. 1997a). *P. australis* accumulates iron, lead, zinc, cadmium, and copper in the roots and rhizomes, and in some cases it appears to impede their translocation to the shoots (Larsen and Schierup 1981; Peverly et al. 1995; Ye et al. 1997b; Wójcik and Wójcik 2000).

The oxygenation of the rhizosphere by wetland plants may play a role in the removal of some metals in wetlands (Otte et al. 1995). Arsenic and zinc have a high binding affinity for iron oxyhydroxides and were found to accumulate in the iron plaque on roots of *Aster tripolium*. In a salt marsh, arsenic and zinc levels were higher in the rhizosphere of *Halimione portulacoides* and *Spartina anglica* because they were associated with the oxidized iron found there.

b. Phytoremediation

The use of wetland plants in phytoremediation is a matter of current study. Phytoremediation is the "use of living green plants for *in situ* risk reduction of contaminated soil, sludge, sediments, and ground water through contaminant removal, degradation, or containment" (U.S. Environmental Protection Agency 1998). The basis of phytoremediation is that all plants extract nutrients, including metals, from soil and water. Some plants have the ability to store large amounts of metals, even some that are not required for plant function. In order for the metals to be removed from the system, the plants need to be harvested frequently and processed to reclaim the metals.

Phytoremediation is different from treatment wetland technology because it is used to clean up areas that have been contaminated by past use rather than a steady flow of wastewater. While most phytoremediation is of soils or groundwater, the use of wetland plants may be feasible when shallow water is contaminated (Miller 1996; U.S. Environmental Protection Agency 1998).

A number of wetland plants have been studied for potential use in phytoremediation. *Scapania undulata* (a liverwort from forested streams; Samecka-Cymerman and Kempers 1996), *Ceratophyllum demersum, Bacopa monnieri,* and *Hygrorrhiza aristata* appear to be *hyperaccumulators* of some metals (Cu, Cr, Fe, Mn, Cd, Pb; Rai et al 1995; Zayed et al. 1998). Hyperaccumulators take up metals into their roots, translocate them to the shoots, and sequester the metals within the shoots (Brown et al. 1994). The thresholds of metal content that define hyperaccumulation were derived from studies of terrestrial plants and may not be completely applicable to wetland plants (Zayed et al. 1998). Terrestrial plants with a metal content above 10,000 mg of the metal per kg dry weight (1%) for Zn and Mn, 1000 mg kg^{-1} (0.1%) for Ni, Co, Cu, Cr, and Pb, and 100 mg kg^{-1} (0.01%) for Cd and Se are considered to be hyperaccumulators. Some metal accumulators may take up several metals while others may only take up one or two specific metals. Examples of specific collectors among wetland plants are *Salvinia natans,* which accumulates mercury, and *Spirodela polyrrhiza,* which accumulates zinc (Rai et al. 1995; Sharma and Gaur 1995; Zayed et al. 1998).

In phytoremediation studies, metal content is reported using percentages (percent of dry weight), weight per dry weight, and bioconcentration factors. For this reason, it is somewhat difficult to compare the performances of different species. In addition, the maximum uptake capacity is seldom reported. Results for the floating plants, *Lemna minor* (duckweed) and *Azolla pinnata* (water velvet), have shown maximum concentrations of iron and copper up to 78 times their concentration in the wastewater (Rai et al. 1995).

B. The Role of Vascular Plants in High-Nutrient Load Treatment Wetlands

Plants in treatment wetlands serve several functions in wastewater treatment. They provide the conditions for physical filtration of wastewater: dense macrophyte stands can decrease water velocity causing solids to settle. Plants provide a large surface area for microbial growth, as well as a source of carbohydrates for microbial consumption (Brix 1997). Plants take up nutrients and incorporate them into their tissues. Although some of these nutrients are released when plants senesce and decompose, some remain in the undecomposed litter that accumulates in wetlands, building organic sediments (Kadlec 1995). Wetland plant roots leak oxygen into the sediments creating a zone in which aerobic microbes persist and in which chemical oxidation can occur. Macrophytes also provide wildlife habitat and make wastewater treatment wetlands aesthetically pleasing (Knight 1997).

For these reasons, vegetated treatment wetlands are more efficient at removing BOD, SS, nitrogen, and phosphorus than unvegetated wetlands (Table 9.2; Radoux 1982; Gersberg et al. 1986; Karnchanawong and Sanjitt 1995; Ansola et al. 1995; Tanner et al. 1995a, b; Sikora et al. 1995; Zhu and Sikora 1995; Heritage et al. 1995; Drizo et al. 1997). The removal of fecal coliforms is not affected by the presence of plants (Karnchanawong and Sanjitt 1995; Ansola et al. 1995; Tanner et al. 1995a, b), probably because, for the most part, fecal coliforms are removed by exposure to sunlight.

1. Vegetation as a Growth Surface and Carbon Source for Microbes

Perhaps the most important role of plants in wastewater treatment wetlands is that their submerged and buried parts provide surface area for the growth of bacteria, algae, and protozoa which take up nutrients or transform them in oxidation/reduction reactions. This 'biofilm' behaves somewhat like the trickling filters of traditional wastewater treatment facilities by breaking down dissolved organic matter. Microbes on submerged plant surfaces and in the rhizosphere are responsible for the majority of the microbial processing that occurs in wetlands (Nichols 1983; Brix 1997).

Denitrifying bacteria require carbon as an energy source. When sufficient carbon is available for microbial metabolism (as is the case in most saturated soils, where organic matter accumulates), denitrification is enhanced (Groffman and Tiedje 1989). The roots and root exudates of wetland plants release organic carbon in the soil profile. This link between vegetation and carbon availability is one of the ecologically critical features of effective treatment of high nitrogen loadings (Sedell et al. 1991).

TABLE 9.2
The Percent Removal of Wastewater Contaminants in Unplanted and Planted Treatment Wetlands

Wetland Type	Location	Species Used	BOD	SS	TN	NH_4^+	NO_3^-	TP	Source
Surface flow									
Constructed marshes	Belgium	*Carex acuta*			87–92			73–77	Radoux 1982
		Phragmites australis							
		Typha latifolia							
		Unplanted			69			65	
Plastic-lined beds, gravel substrate	California	*P. australis*	81			80			Gersberg et al. 1986
		Scirpus tabernaemontani	96			96			
		T. latifolia	74			28			
		Unplanted	69			12			
Constructed marshes (dairy waste)	New Zealand	*S. tabernaemontani*	50–80	75–80	48–75			37–74	Tanner et al. 1995a, b
		Unplanted	20	75–80	12–41			12–36	
Lined beds	India	*Ipomoea aquatica*	45–72		4–76				Karnchanawong and Sanjitt 1995
		Unplanted	22–44		4–64				
Subsurface flow									
Vertical flow	Australia	*T. orientalis*	96–98	81–100	58–78			34–96	Heritage et al. 1995
		S. tabernaemontani	92–97	84–100	25–66			3–57	
		Baumea articulata	98–100	97–99	79–97			54–95	
		Cyperus involucratus	97–99	95–98	81–91			55–71	
		Unplanted	87–95	67–91	22–34			11–56	
Horizontal flow shale substrate	Scotland	*P. australis*				99	85–95		Drizo et al. 1997
		Unplanted				45–75	45–75		

Note: In some of the studies, several species were planted together, and for these only one set of results is given. In others, species were planted separately and separate results for each species are shown. Ranges in percent removal reflect a range of loading rates.

2. Physical Effects of Vegetation

The presence of macrophyte stands reduces water velocity and allows for the filtering and settling of organic particulate matter, other suspended solids, and associated nutrients. With decreased water velocity, the contact time between the wastewater and the sediments and plant surface area is increased, thus adding to the potential treatment of the waste by adsorption or microbial processes (Carpenter and Lodge 1986). In two constructed freshwater marshes in Illinois with low water inflow (8 cm wk^{-1}), stands of *Typha latifolia* and *T. angustifolia* were shown to decrease water velocity. Sedimentation was highest within the stands of *Typha* during a 3-month study period (Brueske and Barrett 1994). If macrophyte stands are very dense, however, wastewater may be routed around them rather than through them (Johnston et al. 1984; Bowmer 1987; Fennessy et al. 1994b).

Stands of vegetation reduce the risk of erosion since the plants serve as a buffer against wind, waves, and flowing water. Macrophytes' dense roots impede the formation of erosion channels. Wind velocity is reduced near the soil when macrophytes are present, and this reduces the resuspension of settled material (Nichols 1983; Ward et al. 1984; Stevenson et al. 1988; Brix 1997).

In treatment wetlands that use floating plants such as *Eichhornia crassipes,* the plants act as a filter, straining wastewater and retaining solids in their dense roots. In a study comparing *E. crassipes* beds to unplanted areas, turbidity decreased up to 30% in the vegetated areas while it increased 22% in the unplanted ones. With a decrease in turbidity came a reduction in suspended organic matter, which resulted in a decrease in BOD (Reddy et al. 1983).

3. Nutrient Uptake

Plant nutrient uptake is usually not the major pathway of nitrogen and phosphorus removal in high-nutrient treatment wetlands and in many cases nutrient uptake accounts for only 1 to 4% of nutrient removal (as reviewed by Nichols 1983 and Brix 1997). For example, in a natural wetland receiving runoff from a peat mining operation in Finland for 6 years, the average decreases in both nitrogen and phosphorus were 55%. The plants accounted for 4% of the nitrogen removal and were actually a source of phosphorus rather than a sink. Thus, the retention of nutrients was mainly the result of processes other than plant uptake (Huttunen et al. 1996). In greenhouse wastewater treatment wetlands with a nitrogen loading of 15.6 g N m^{-2} d^{-1}, 4% of the nitrogen removal was by plant uptake. In a subsystem at the same treatment facility, nitrogen loadings were 5.2 g N m^{-2} d^{-1} and 38% was removed, 1% via plant uptake (Peterson and Teal 1996).

Plants make a greater contribution to the percent nutrient removal in treatment wetlands under low-load conditions than under high loads. In systems with a high load, the plants may take up higher amounts of nutrients, but as a percentage of the incoming loadings, the uptake is small. Peterson and Teal (1996) compared plant uptake in wastewater treatment wetlands with high loads of nitrogen (3.2 to 15.6 g N m^{-2} d^{-1}) to wetlands with lower loads (0.4 to 2.0 g N m^{-2} d^{-1}). In the heavily loaded system, the plants assimilated only 1 to 4% of the nitrogen. In the lightly loaded system, plant uptake accounted for 18 to 30% of nitrogen removal.

Plant uptake varies by season, latitude, and certain attributes of each species, such as growth rate and maximum biomass. In temperate climates, wetland plants are seasonally effective at incorporating nutrients into biomass. Most temperate herbaceous species show a maximum rate of uptake early in the growing season which slows considerably after flowering (Boyd 1970, 1978) or peak biomass (Peverly 1985).

Ultimately, nutrient storage in live plant tissues is temporary. A portion of the nutrients sequestered by wetland plants is released through tissue sloughing, plant senescence, and

TABLE 9.3
Ranges of Phosphorus and Nitrogen Content (mg g^{-1}) of Some Wetland Plant Species under High Nutrient Loads

Plant	P Content			N Content		
	Leaf	Root	Rhizome	Leaf	Root	Rhizome
Emergent species						
Cyperus involucratus[5,13]	2–5	1–7	2–7	15–43	11–45	5–21
Phragmites australis[7,10,11,13]	2–4	1–3	1–3	10–40	15–31	5–31
Typha spp.[7,8,10,13]	1–5	2–7	1–7	5–32	4–52	2–40
Scirpus tabernaemontani[12,13]	2–4	2–8	2–7	6–25	4–21	9–18
Bolboschoenus spp.[12,13]	1–5	2–7	4–6	2–15	2–15	13–19
Baumea articulata[11,13]	1–9	2–8	2–7	11–18	8–25	8–19
Floating Species						
Eichhornia crassipes[7]	1–12			10–40		
Salvinia molesta[7,9,13]	2–9			20–48		
Lemna spp.[6,7,9,13]	4–18			25–59		
Pistia stratiotes[7,9,13]	2–12			12–40		
Floating-Leaved Species						
Alternanthera philoxeroides[7]	2–9			15–35		
Ludwigia peploides[13]	4–6			25–45		
Marsilea mutica[13]	5–7			23–36		
Hydrocleys nymphoides[13]	5–10			14–50		
Hydrocotyle umbellata[7]	2–13			15–45		
Nymphoides indica[13]	5–12			15–35		

Submerged Species				
Ceratophyllum demersum[6,13]	10–14		35–42	
Elodea canadensis[6]	7–11		40–41	
Potamogeton crispus[6]	6–10		35–40	
P. pectinatus[6,13]	4–7	4–8	27–31	27–31
Trees	**Leaves**	**Stem wood**	**Leaves**	**Stem wood**
Acer rubrum[2,4]	2–3		10–22	
Magnolia virginiana[2,4]	1–2	<1	19–25	4
Nyssa sylvatica[4]	1	<1	19	1
N. sylvatica var. *biflora*[2]	1–2		20–24	
Taxodium distichum[1,3]	1–3	<1	<1	<1
T. ascendens[4]	1–2		15–18	

Note: Where only one set of data is given, the whole plant was analyzed and not broken down into leaf, root, and rhizome

Data for herbaceous species from authors 5 and 7–12 compiled in Greenway 1997; data for tree species from authors 1–4 compiled in Reddy and DeBusk 1987; additional data from Peverly 1985, Reddy and DeBusk 1987, and Greenway 1997.

[1]Schlesinger 1978; [2]Reynolds et al. 1979; [3]Brown 1981; [4]DeBusk 1984; [5]Hocking 1985; [6]Peverly 1985; [7]Reddy and DeBusk 1987; [8]Breen 1990; [9]Tripathi et al. 1991; [10]Gumbricht 1993; [11]Adcock and Ganf 1994; [12]Tanner 1996; [13]Greenway 1997.

decomposition. Temporary nutrient storage does provide benefits, however. The decrease in biologically available nutrients during the growing season protects downstream areas from eutrophication (Howarth and Fisher 1976; Vincent and Downes 1980; Nichols 1983).

a. Tissue Nutrient Content of Wetland Plants

Concentrations of nitrogen and phosphorus in wetland vegetation at peak biomass range from 1 to 3% of dry weight for nitrogen and 0.1 to 0.3% of dry weight for phosphorus in both emergent species and the leaves of woody vegetation. Woody structures themselves have much lower concentrations, averaging 0.4% nitrogen and 0.01% phosphorus dry weight in tree boles and roots (Johnston 1991).

In part, the growth habit of macrophytes determines their capacity to assimilate nutrients. Perennial emergents are the most widely used plant type in wastewater treatment wetlands. Many emergents are tolerant of a range of substrate types, grow quickly, and thrive in a wide variety of wastewaters. Emergents have a large network of roots and rhizomes and they store nutrients in perennial tissues. Emergents take up nutrients from the soil porewater and this uptake can establish a gradient between the water column and the soil, thus improving overall nutrient retention (Reddy et al. 1999).

Commonly used emergents are in the genera *Phragmites, Cyperus, Typha, Scirpus, Juncus, Pontederia* and *Sagittaria*. Early in the growing season they have rapid nutrient uptake and tend to have a high peak biomass (often >3000 g m^{-2}). While their nutrient content may be lower than submerged, floating, or floating-leaved plants in some instances (Table 9.3), their large size translates to high amounts of nutrient storage (Nichols 1983; Reddy and DeBusk 1987). The uptake capacity of emergent macrophytes (i.e., the amount that can be removed if the vegetation is harvested) ranges from 30 to 150 kg P ha^{-1} yr^{-1}, and from 200 to 2500 kg N ha^{-1} yr^{-1} (Brix 1997).

Rooted submerged plants can sequester nutrients from both the sediments and the water column. Under natural conditions, most of their nutrients come from the sediments (Barko and Smart 1981b). However, submerged plants are able to take advantage of high nutrient levels in the water column and take up more from the water column when more is available. Submerged plants tend to have relatively high tissue nutrient content (Table 9.3); however, large amounts of nutrients are released to the water when the plants die (Nichols 1983). The uptake capacity of submerged macrophytes is up to 100 kg P ha^{-1} yr^{-1} and 700 kg N ha^{-1} yr^{-1} (Brix 1997).

Floating plants absorb nutrients directly from the water column. Their turnover is rapid and when the plants or plant parts decompose, nutrients are released into the water. Some of the senescing tissues settle in the treatment wetland and this represents a transfer of nutrients from the water to the soil. *Eichhornia crassipes* has been used extensively for wastewater treatment because of its high uptake capacity (350 to 1125 kg P ha^{-1} yr^{-1} and 1950 to 5850 kg N ha^{-1} yr^{-1}; Reddy and DeBusk 1987). Other floating plants that have rapid growth and high nutrient assimilative capacity are *Hydrocotyle umbellata, Pistia stratiotes,* and various species of *Lemna* and *Salvinia* (Reddy and DeBusk 1987; Greenway 1997).

Trees tend to have lower nutrient content than herbaceous species, particularly of phosphorus. In addition, their annual net primary productivity is usually less than that of emergents. Their rate of nutrient uptake may be insignificant in a system used for wastewater treatment. The uptake capacity of *Taxodium distichum* is about 200 kg N ha^{-1} yr^{-1} and ranges from 3 to 23 kg P ha^{-1} yr^{-1}. Since the standing crop of forests is large, the total nutrient storage in forested wetlands is generally greater than in herbaceous wetlands. Forest uptake of nutrients, particularly in the woody parts of trees, represents a long-term storage of excess nutrients (Reddy and DeBusk 1987).

b. Factors Affecting Nutrient Uptake

Plants tend to accumulate more nutrients than are needed for growth when supplemental nutrients are available (luxury uptake). Therefore, plant nutrient content is greater under high nutrient loads than under natural, or background, levels of nutrients. Greenway (1997) analyzed eight common wetland plant emergents and floating-leaved species from both high-nutrient load treatment wetlands and from control wetlands. Plant phosphorus levels in the treatment wetlands averaged 2 mg P g^{-1} dry weight more than in the control wetlands. Nitrogen levels averaged 7 mg N g^{-1} more than in control wetlands (calculated from data in Greenway 1997).

Plants near the inflow of wastewater treatment wetlands are subjected to higher nutrient levels than those near the outflow and, as a result, they tend to have a higher tissue nutrient content. In an Australian surface flow wastewater treatment wetland, plants near the inflow had a phosphorus content that was twice as high as plants near the outflow. Stems and roots near the inlet had a phosphorus content of 4.3 and 5.8 mg g^{-1}, respectively. At the outflow, the phosphorus content was 1.9 mg g^{-1} in the stems and 2.9 mg g^{-1} in the roots. The plants' nitrogen content was also higher at the inflow than at the outflow, though the difference was not as great as for phosphorus (stems: 16.9 mg g^{-1} at the inflow and 9.9 mg g^{-1} at the outflow; roots: 18.1 mg g^{-1} at the inflow and 17.7 mg g^{-1} at the outflow; Adcock et al. 1995).

Trees also accumulate luxury levels of nutrients near wastewater outfalls. Fail and others (1987) measured higher tissue nutrient concentrations in trees (*Acer rubrum, Liriodendron tulipfera,* and *Nyssa sylvatica*) growing adjacent to pigpens compared with those from reference sites. Trees near the swine operation had 25% higher average nitrogen concentrations and 45% higher average phosphorus concentrations in woody tissue than trees in reference areas. Leaf tissue accumulated approximately 30% more nitrogen in the enriched area and up to twice as much phosphorus.

The nutrient content of rhizomes tends to be highest during the non-growing season in plants of temperate areas. The same plants grown in tropical areas do not display seasonal variation and have lower tissue nutrient concentrations. In a study of wetland plants in tropical Australia (Greenway 1997), tissue nutrient concentrations were lower than in studies of the same plants in temperate Australia (Hocking 1985; Breen 1990; Adcock and Ganf 1994) because rhizomes were less likely to function as storage organs for nutrients.

c. The Accretion of Organic Sediments

The longest-term nutrient storage associated with herbaceous wetland plant growth and tree leaf production is the process of organic soil development (Nichols 1983; Kadlec and Knight 1996). The undecomposed fraction of the litter that remains within the wetland accumulates and sediment accretion results in long-term nutrient storage. The amount of nutrients that remains in the litter depends on how much is released during plant senescence. In general, the death of wetland vegetation is typically followed by the rapid release to the water of 35 to 75% of plant tissue phosphorus and smaller but still substantial amounts of nitrogen (Nichols 1983). The nutrient content of mixed litter (composed primarily of *Phragmites australis, Typha orientalis,* and *Echinochloa crus-galli*) in a wastewater treatment wetland in Australia was found to be almost as high as the nutrient content for live tissue for these species, containing from 1.6 to 3.0 mg P g^{-1} and from 11.4 to 13.2 mg N g^{-1} (Table 9.3; Adcock et al. 1995). In a natural wetland in New York State, senescent plants (*Typha latifolia, Phalaris arundinacea, Acer rubrum, Populus deltoides, Salix nigra*) released from 10 to 100% of the nutrients originally taken up. However, this amounted to only 1 to 2% of the total annual stream load (Peverly 1985).

The rates at which nitrogen and phosphorus accumulate in the substrate range between 0.1 and 4.7 g N m^{-2} yr^{-1} and between 0.005 and 0.22 g P m^{-2} yr^{-1} in moderate to cold climates, and up to 10.0 g N m^{-2} yr^{-1} and 0.5 g P m^{-2} yr^{-1} in warm, highly productive areas (as reviewed by Nichols 1983).

4. Vegetation as a Source of Rhizospheric Oxygen

Wetland plants translocate oxygen from their shoots to their belowground parts via aerenchyma. Some of the oxygen in the roots and rhizomes diffuse into the soil, creating an oxidized rhizosphere (see Chapter 4, Section II.A.5, Radial Oxygen Loss). Radial oxygen loss often supplies enough oxygen so that reduced elements become oxidized near the roots. For example, nitrate is formed in the oxidized root zone of freshwater wetland sediments. As a result of higher nitrate levels in the rhizosphere, denitrification is enhanced (Laanbroek 1990).

An oxidized rhizosphere stimulates the decomposition of oxygen-demanding organic waste (BOD). In a study comparing water quality improvement in unvegetated beds and beds planted with three emergent species (*Scirpus tabernaemontani, Phragmites australis,* and *Typha latifolia*), Gersberg and others (1986) found that both nitrogen and BOD removal was higher in the vegetated beds than in the unplanted ones. BOD removal averaged 98% in the *S. tabernaemontani* bed, 81% in the *P. australis* bed, 74% with *T. latifolia,* and 69% in the unplanted bed. The authors surmised that the plants' ability to translocate oxygen from the shoots to the roots (and the subsequent diffusion of oxygen into the soil) resulted in enhanced decomposition of organic compounds. Among the plants, removal was greatest in the *S. tabernaemontani* and *P. australis* beds. Both of these species had deeper root systems than *T. latifolia,* and so were able to oxygenate the soil to a greater depth.

Tests of three floating species (*Hydrocotyle umbellata, Eichhornia crassipes, Pistia stratiotes*) and six emergent species (*Canna flaccida, Pontederia cordata, Sagittaria latifolia, Scirpus tabernaemontani, S. pungens, Typha latifolia*) showed that the plants' capacity to transport oxygen into the root zone created an oxidized microenvironment and stimulated carbon and nitrogen transformations. The presence of plants brought about significant reductions in BOD and NH_4^+ and increased the dissolved oxygen in the water column from <1 to 6 mg O_2 l^{-1} (Reddy et al. 1989).

In the same study, oxygen movement through the plants was compared to mechanical aeration. After 20 days, both vegetated and mechanically aerated systems removed all of the incoming BOD. In the mechanical aeration treatment, complete conversion of NH_4^+ to nitrate occurred after 12 days; however, the nitrate was not removed. In the vegetated treatment, 65 to 100% of the ammonium was converted to nitrate and the nitrate was entirely removed from the system via denitrification. The success of BOD and nitrogen removal in the vegetated tests was attributed to the capacity of the plants to transport oxygen into the root zone, and the subsequent use of the excess oxygen during microbial respiration.

In subsurface flow wastewater treatment wetlands, the diffusion of oxygen from the plants into the rhizosphere may be of minimal importance. Brix (1990) measured the air entering *Phragmites australis* plants and found that most of the air went through the cavities of dead standing culms. He found that the roots' and rhizomes' respiratory consumption of oxygen consumed nearly all of the oxygen that entered the plants, leaving only 0.02 g O_2 m^{-2} d^{-1}, an amount he surmised to be insufficient to have an effect in the rhizosphere. Other studies contradict these findings, however, and have shown that *P. australis* can oxygenate the rhizosphere at a rate of 15 g O_2 m^{-2} d^{-1} (Armstrong et al. 1992), but an extensive root system takes 3 years to develop (Biddlestone et al. 1991), so system effectiveness may be delayed.

5. Wildlife Habitat and Public Recreation

Wastewater treatment wetlands can be planned with the ancillary goal of attracting wildlife (Figures 9.9a, b). In general, greater plant diversity results in greater faunal diversity. Wastewater high in nutrients encourages primary productivity. Consumer populations of invertebrates, fish, reptiles, and amphibians can be found in abundance where plant productivity is high. These populations serve in turn to attract birds and mammals. Plants can be chosen for high productivity (or other features) to maximize the wildlife habitat in treatment wetlands (Knight 1997).

Variations in hydrology that include open water, deep and shallow water, and emergent marsh areas can result in a variety of animal species inhabiting the wetland. Attractiveness to waterfowl may be enhanced by a 1:1 ratio of marsh to open water. Wading birds require shallow littoral areas or perching surfaces near open water. Treatment wetlands can include a shallow (20 to 30 cm) emergent marsh area to encourage wading birds. Living and dead trees or nesting boxes in or near the wetlands provide habitat for wood ducks, owls, and bats. A study of bird populations in six treatment wetlands across the U.S. showed that the number of bird species observed at each site ranged from 33 to 63. The densities of wading birds at two central Florida constructed wetland treatment systems (0.29 to 0.38 birds ha^{-1}) were as high or higher than densities in nearby natural marshes (as reviewed in Kadlec and Knight 1996).

Water quality enhancement may not be compatible with habitat enhancement in urban environments, or wherever there are complex mixtures of metals and organic compounds. These contaminants may accumulate in wetland sediments and bioaccumulate in wetland biota. Wildlife habitat enhancement should be kept separate from the treatment of compounds that are potentially harmful to animals (Helfield and Diamond 1997).

Treatment wetlands are often used for recreational activities such as walking, running, biking, photography, and birdwatching. These activities can promote public support for the site and they can be encouraged by including walkways, bike paths, birdwatching blinds, and interpretive signs.

FIGURE 9.9a
This wastewater treatment wetland at the southern tip of the Florida Everglades treats a portion of the wastewater from the National Park's facilities. (Photo by H. Crowell.)

FIGURE 9.9b
This wastewater treatment wetland at the southern tip of the Florida Everglades is also among the most popular bird-watching sites in the park. Shown among the edge vegetation of the marsh are a Little blue heron (*Florida caerulea*) and an American coot (*Fulica americana*). (Photo by H. Crowell.)

C. Species Commonly Used in Treatment Wetlands

The appropriate species for wastewater treatment wetlands depend on local conditions, the water depth, the design (e.g., surface or subsurface flow), and characteristics of the wastewater. Studies of wetland plants' survival and effectiveness in treatment wetlands (Reddy and DeBusk 1987; Hammer 1993, 1994) have led to a list of general requirements for suitable plants (Tanner 1996):

- Ecological acceptability; i.e., no significant weed risks or danger to the ecological or genetic integrity of surrounding natural ecosystems
- Tolerance of local climatic conditions, pests, and diseases
- Tolerance of pollutants and hypertrophic waterlogged conditions
- Ready propagation, and rapid establishment, spread and growth (perennial habit)
- High pollutant removal capacity, either through direct assimilation and storage, or indirectly by enhancement of microbial transformations such as nitrification (via root-zone oxygen release) and denitrification (via production of carbon substrates)

In order to meet the first two criteria, managers may want to match plant species that are found in nearby natural wetlands. In areas where some of the commonly used species (such as those in Table 9.4) are not locally found, local species should be tested for survivability and effectiveness and used in preference to non-indigenous species (Maschinski et al. 1999). Many other plants besides those mentioned in Table 9.4 are used around the world, and specific guidelines for each region should be consulted.

The best strategy in choosing wetland plants is probably to use a variety of well-adapted species rather than to choose only one or two. Monocultures may be more susceptible to pests, or the chosen plant may not become established. In any event, managers

TABLE 9.4

Examples of Plant Species Used in High-Nutrient Load Wastewater Treatment Wetlands

Latin Name	Common Name	Source
Emergents		
Sagittaria spp.	Arrowhead	NRCS 1991, Hammer 1993
Scirpus spp.	Bulrush	NRCS 1991, Hammer 1993, Surrency 1993
Canna flaccida	Canna lily	NRCS 1991, Tanner 1996
Typha spp.	Cattail	NRCS 1991, Hammer 1993
Colocasia esculenta	Elephant ear	NRCS 1991
Zizaniopsis milacea	Giant cutgrass	NRCS 1991, Surrency 1993
Iris versicolor, I. pseudacorus	Iris	NRCS 1991, Hammer 1993
Panicum hemitomon	Maidencane	NRCS 1991
Pontederia cordata	Pickerelweed	NRCS 1991
Alisma spp.	Plantain	Hammer 1993
Phragmites australis	Common reed	NRCS 1991, Biddlestone et al. 1991
Juncus spp.	Rush	NRCS 1991, Hammer 1993
Cyperus spp.	Sedges	NRCS 1991, Hammer 1993
Fimbristylis spp.		
Eleocharis spp.		
E. dulcis	Water chestnut	NRCS 1991
Zizania latifolia	Wild rice	Tanner 1996
Baumea articulata		Tanner 1996
Hydrocotyle umbellata		Tanner 1996
Glyceria maxima	Manna grass	Tanner 1996
Submerged		
Ceratophyllum demersum	Hornwort	Hammer 1993
Najas spp.	Naiad	Hammer 1993
Potamogeton spp.	Pondweed	Hammer 1993
Elodea canadensis	Water weed	Hammer 1994
Vallisneria americana	Wild celery	Hammer 1994
Floating		
Spirodela spp.	Big duckweed	Koles et al. 1987
Lemna spp.	Duckweed	Koles et al. 1987
Eichhornia crassipes	Water hyacinth	Tanner 1996
Pistia stratiotes	Water lettuce	Tanner 1996
Salvinia spp.	Water fern	Tanner 1996
Rooted Floating-Leaved		
Nelumbo lutea	American water lily	Hammer 1993
Nymphoides spp.	Gentian	Hammer 1993
Nymphaea spp.	Water lily	Hammer 1993

Note: The species from Tanner 1996 have been used in New Zealand; some are of Asian origin. The species from Biddlestone et al. 1991 and Koles et al. 1987 have been used in Europe. The remaining species have been used in the U.S. (NRCS 1991 = U.S. Natural Resources Conservation Service 1991).

should be willing to introduce new species or even replant and start again if the growth and spread of the original plants are not successful. Quite frequently, volunteer species become established in treatment wetlands and unless they are invasives or detrimental in some way, they should probably be left in place since they are adapted to the growth conditions of the wetland.

While local plants are desirable, non-indigenous species have been used successfully in wastewater treatment and their use can be justified where their escape is unlikely. For example, *Eichhornia crassipes* has high nutrient assimilative capacity, grows rapidly, filters suspended solids, and is relatively easily harvested since it is free-floating. It may be desirable in contained systems such as greenhouse wastewater treatment systems (Peterson and Teal 1996).

Plants with high pollutant removal capacity tend to have high rates of primary productivity because they assimilate a greater quantity of nutrients. Tanner (1996) showed that peak biomass is positively correlated to total nitrogen removal in ammonium-rich dairy farm wastewater (average total nitrogen was nearly 100 mg N l^{-1}). The same plants that achieved a high peak biomass in his study, *P. australis* and *Zizania latifolia*, also showed high root-zone aeration, which may have increased the oxygenated area in which nitrification could occur (about 87% of the nitrogen was in the form of ammonia at the wetlands' inflow).

The selection of macrophytes with high levels of radial oxygen loss is critical in optimizing the treatment of domestic and animal wastewater. Plants that have been used in wastewater treatment wetlands and judged to have high radial oxygen loss include the floating plants, *Hydrocotyle umbellata, Eichhornia crassipes,* and *Pistia stratiotes,* and the emergents, *Pontederia cordata, Sagittaria latifolia, Canna flaccida, Scirpus tabernaemontani* (Reddy et al. 1989), *Zizania latifolia, Phragmites australis, Baumea articulata,* and *Glyceria maxima* (Tanner 1996). Floating plants tend to diffuse more oxygen through their roots than emergents. In laboratory tests, the average oxygen loss from floating roots was 13 to 26 mg O_2 g^{-1} root d^{-1} and for emergents, 2 to 14 mg O_2 g^{-1} root d^{-1} (Reddy et al. 1989). Whether the oxygen is diffused into the water column or the sediments, it stimulates the decomposition of organic matter and the growth of nitrifying bacteria.

In animal wastewater treatment wetlands, ammonia is often implicated in plant death; therefore a tolerance of high ammonia concentrations is necessary unless reliable pretreatment methods are used to control ammonia. In field trials, *Scirpus californicus* (giant bulrush) and *S. tabernaemontani* survived ammonia concentrations greater than 200 mg NH_3-N l^{-1}. Both *S. californicus* and *S. tabernaemontani* rapidly produce rhizomes and continue to grow past the end of the growing season for other plants. *Zizaniopsis milacea* (giant cutgrass) has also been observed to survive and grow rapidly in animal wastewater treatment wetlands (Surrency 1993). In a greenhouse experiment in which *Typha latifolia* was exposed to several concentrations of ammonia, it was shown to thrive at 100 mg NH_3-N l^{-1}, and to have reduced biomass above 200 NH_3-N l^{-1} (Clarke 1999). *Lemna minor* was a volunteer species in a dairy farm wetland in Maryland and it flourished despite ammonia concentrations as high as 160 mg NH_3-N l^{-1} (Cronk et al. 1994).

Plants of all the wetland growth forms, i.e., emergent, submerged, floating, floating-leaved, as well as trees, have been used in wastewater treatment wetlands (Table 9.4). Emergent genera that are widespread, able to tolerate a range of conditions, and have been shown to be effective in wastewater treatment wetlands include a variety of perennial monocots: *Phalaris, Spartina, Carex, Juncus, Typha, Phragmites,* and *Scirpus* (Guntenspergen et al. 1989). Under the right conditions (i.e., enough sunlight, an appropriate substrate, and sufficient pretreatment of the waste), species in these emergent genera become quickly established. *Phragmites australis* is commonly used in subsurface flow wetlands in Europe; however, its use is discouraged in the U.S. because it is an aggressive species that invades and dominates native plant communities (Hammer 1993).

Some submerged genera, such as *Potamogeton, Ceratophyllum,* and *Najas,* are suitable for wastewater treatment because they are rapid colonizers and can grow in deeper water than emergents (Table 9.4). Submerged species used along with emergents may help optimize nutrient uptake because they are able to assimilate nutrients directly from the water column (Guntenspergen et al. 1989).

Free-floating species are commonly used in wastewater treatment wetlands. *Eichhornia crassipes, Pistia stratiotes,* and species of *Salvinia* have high nutrient uptake capacity (Table 9.3) and they grow rapidly. Their use is limited to warm environments. *Lemna minor* as well as other Lemnaceae are attractive for treatment wetlands because they are widely distributed, adapted to a range of environmental conditions, and have high nutrient content. In addition, *L. minor* has a high protein content (140 to 410 g kg dry weight^{-1}) and may therefore be useful as livestock feed. The U.S. Natural Resources Conservation Service (1991) discourages planting *L. minor* and submerged plants in animal wastewater treatment wetlands, arguing that they may need to be harvested to effect permanent pollutant removals.

Lemna treatment ponds have been found to be effective in the treatment of domestic and animal wastewater with good BOD and SS removal (Koles et al. 1987; Whitehead et al. 1987). However, a dense cover of *Lemna* can shade the water column, rendering it anoxic and decreasing the effectiveness of the treatment (Cathcart et al. 1994). In temperate areas, *Lemna* is only effective for wastewater treatment during the growing season. *Lemna* treatment ponds may be particularly appropriate in small communities that need supplemental waste treatment during the summer months (Bonomo et al. 1997).

The use of rooted floating-leaved plants is relatively uncommon and there is little information on their use in treatment wetlands. Most treatment wetlands are relatively shallow, making them more suitable for emergents or floating species. In marsh–pond–marsh systems in which shallow marshes are at the beginning and end of the treatment system and a deeper pond is in the middle, floating-leaved plants may be suitable.

Trees have been used in some wastewater treatment wetlands. Florida *Taxodium distichum* swamps were among the first wastewater treatment wetlands to be tested for their effectiveness in reducing nutrient concentrations (Odum et al. 1977; Boyt et al. 1977; Ewel and Odum 1984). However, in both surface and subsurface flow wetlands, trees are not recommended. They take a long time to become established and they generally have lower productivity and nutrient uptake capacity than emergents.

On the other hand, various species of the tree genus *Melaleuca* are considered to be prime candidates for use in treatment wetlands in Australia (Bolton and Greenway 1997). The trees tolerate periodic inundation and extreme environmental conditions such as salinity, alkalinity, acidity, and high metal concentrations. They have a high growth rate and are therefore a high potential sink for nutrients. In addition, *Melaleuca* species concentrate excess phosphorus in their leaves. The trees have high rates of litter fall but low rates of decomposition. When the leaves fall, the portion that is not decomposed provides a long-term sink for phosphorus.

Trees may be best used in the restoration of riparian areas, particularly where nitrate is a water quality concern. Studies of riparian buffer strips that are managed to decrease nutrient loads to streams and rivers in agricultural landscapes indicate that forested systems are generally more efficient than herbaceous systems in intercepting the flux of agricultural chemicals. Haycock and Pinay (1993) compared the effectiveness of grassed and forested buffer strips to determine the effect of vegetation types on nitrate removal.

Forested buffer strips were more efficient in removing nitrate under all flow conditions, removing essentally 100%, while the grassed strips removed an average of 84%.

Correll (1991) compared the relative effectiveness of forested and herbaceous vegetation in controlling the loss of nitrogen and phosphorus from a pair of watersheds. He found that the discharge from a watershed containing forested buffer strips had an N:P ratio 11 times lower than the discharge from a watershed with herbaceous buffer strips: the forested buffer strips removed a greater proportion of nitrate. In forested and grassed buffer strips in Sweden, Vought and others (1991) conducted a series of experiments and found that subsurface-flow forested strips removed more nitrogen (76%), than surface-flow herbaceous strips (50%). It may be that the root system of trees provides more biomass (carbon) at greater depths in the soil profile, thus enhancing denitrification. In contrast, in one study comparing a grassed riparian strip to a newly reforested one, the grassed area had significantly higher denitrification rates (Lowrance et al. 1995). The outcome may have been influenced by the young age of the trees.

Phosphorus retention in buffer strips may be maximized when there is a combination of dense herbaceous and woody vegetation. In Sweden, a combination shrub/grass buffer strip retained significantly higher total phosphorus than did a grassed or a forested buffer strip. Differences in phosphorus retention efficiency were explained by differences in stem density. Increased hydraulic roughness translated to higher and more uniform patterns of sediment deposition from surface flows (Vought et al. 1994). In Illinois, forested buffer strips were less effective at reducing phosphorus than were grassed strips and both types of strips released phosphorus to the groundwater during the non-growing season (Osborne and Kovacic 1993).

D. The Establishment and Management of Plants in Wastewater Treatment Wetlands

The optimal conditions for the initial planting of most seeds or cut materials are created by shallow flooding, followed by draining to the surface of the soil. This leaves soft, moist soil conditions for planting (Hammer 1994). Plants may survive best if they are planted before waste enters the wetlands and then allowed to establish under wet conditions. Hammer (1994) recommends planting emergents when the wetlands are dry and adding 1 to 2.5 cm of clean water per week until the desired depth is reached (without covering the tops of plants). When using submerged or floating-leaved plants, water levels should be increased slowly but should be sufficient to support the plants' stems and avoid drying. The goal is to inhibit the growth of upland plants while still allowing young wetland plants to become firmly established. At the end of 6 or 7 weeks, wastewater can be added. Plant survival may be enhanced because the initial flush of wastewater is diluted by the accumulated fresh water. McCaskey and others (1994) suggest that plants should be allowed to establish for two growing seasons before wastewater is applied; however, a delay of this length before treatment may not be an option.

A number of wetland plant nurseries offer seeds, seedlings, tubers, roots, rhizomes, and mature plants. The nurseries' directions for successful planting (i.e., optimal distance between plants and other criteria) should be strictly followed. Plant stocks should be obtained immediately before planting is to begin. They should be stored for less than 2 days and kept cool and moist. Seeds are available for many wetland plants and they may be less expensive than rootstocks, but germination may not be successful. Germination rates of many wetland seeds are less than 5% per year, so large quantities of seed must be collected and distributed (Hammer 1994).

In some treatment wetlands, harvesting is a common practice. In intensely managed systems such as greenhouse facilities (Peterson and Teal 1996) or animal wastewater treatment/fish ponds in China (Yuan et al. 1993), harvesting is necessary to maintain rapid plant growth. In China, harvested plants from treatment wetlands are used for animal feed and composted and used as fertilizer (Yan et al; 1993, Yuan et al. 1993; Ma et al. 1993).

Harvesting permanently removes the nutrients in plant biomass from the treatment system (Reddy and DeBusk 1987; Brix 1997). However, it is a time-consuming process that may in fact be detrimental to the function of most wastewater treatment wetlands (Kadlec and Knight 1996). As plant litter accumulates, the substrate and carbon available for microbes increase. Since water quality improvement in wetlands is largely a function of well-developed microbial populations, removing plants does not improve the performance of most systems. The amount of nutrients removed by harvesting is generally insignificant compared to the loadings in wastewater treatment wetlands. Harvesting plants removes the organic material that would otherwise bring about an accretion of the sediments, providing a long-term sink for a portion of the nutrients.

Summary

In an effort to reduce the global loss of wetland area, wetlands are being restored and constructed around the world. The plants in restored and constructed wetlands may be planted, arrive through natural dispersal mechanisms, or arise from the seed bank. Abiotic filters such as the hydrologic regime and soil nutrient status determine the plants that will be able to persist. Several features of plant species are important considerations when choosing plants for a site. The species should have the desired properties for the restoration, such as good wildlife value and good growth and reproduction, so that the population will be sustained. In addition, the species should be indigenous to the area.

Wetlands are also constructed to treat wastewater of many types, including domestic and animal wastewater, industrial discharges, mine drainage, stormwater runoff, nonpoint source pollution, and landfill leachate. Treatment wetlands have been shown to effectively remove or reduce biochemical oxygen demand, suspended solids, nitrogen, phosphorus, and heavy metals from wastewater. BOD and SS are removed primarily through settling and decomposition of organic wastes. Nitrogen is removed by denitrification and plant uptake. Phosphorus is removed by sorption onto sediment particles, plant uptake, and the accretion of soils due to the accumulation of detritus.

Vegetation plays an important role in wastewater treatment wetlands. Plants provide a substrate for microbes, which are the most important processors of wastewater contaminants. Plants also provide microbes with a source of carbon. Stands of vegetation reduce water velocity, allowing solids to settle out of the water column. Plants assimilate nutrients, keeping them out of downstream water bodies during the growing season. As plants senesce, some nutrients are released back into the water. A portion of the nutrients is retained in the undecomposed fraction of the plant litter and accumulates in the sediments. Plants oxygenate the root zone by diffusion of oxygen out of their roots. The oxygenated rhizosphere provides aerobic microbes a habitat within the substrate and enhances nitrification and the removal of BOD. Vegetated treatment wetlands have been shown to be more effective than unvegetated wetlands at the removal of pollutants.

Wetland species of all growth forms have been used in treatment wetlands. Treatment wetland plants should not be invasive species; they should be tolerant of the local environmental conditons and of pollutants, and they should be easily established and maintained. They should also have a high pollutant removal capacity through the assimilation of contaminants or by root-zone oxygenation.

Case Studies

9.A. Integrating Wetland Restoration with Human Uses of Wetland Resources

In the Philippines, as elsewhere in the tropics, mangrove forests are under intense human pressure (Walters 1997, 2000a, b; Ellison 2000a). The rates of forest destruction and biodiversity loss from both upland and coastal mangrove areas are among the highest in the world. The trees are used for fuel and construction material, and the forests are cleared to make way for settlements and to create fish ponds. Under such circumstances, ecosystem restoration presents unique problems. In a few sites in the Philippines, mangroves have been replanted for decades by local residents in an effort to sustain the resources provided by mangroves. In the early 1990s, a team from Filipino and Canadian universities studied restoration strategies in Bais Bay Basin, a watershed of approximately 16,000 ha. They sought to understand how environmental restoration and protection could be integrated with local social and economic development. Their restoration efforts were modeled, in part, after existing local mangrove planting strategies. This study illuminates ways in which social factors can be crucial to the success of restoration projects.

In the upland areas of the watershed, native trees were planted along streams to reduce soil erosion and flooding and to restore some of the native tree biodiversity. Eight small villages were targeted for this reforestation. All of the natural forests in six of the eight villages had been harvested and only degraded secondary forest remained in the other two villages. The Bais Bay Program used a community leader approach in which three to five respected people from each village were given training and support with the thought that they would share the knowledge. These cooperators (who were paid a small wage) were to construct and manage tree nurseries, collect indigenous seedlings, plant and care for seedlings in riparian areas, and participate in soil conservation training. Unfortunately the villagers did not always regard the chosen individuals as leaders, and their cooperation was of mixed success. In some villages many of the seedlings were never planted and in one, the cooperator kept the seedlings for his own lands. In all, about 8000 seedlings were planted along 10 km of streams, with more trees planted by some villages than others. Flash floods, road building, crowding from neighboring agricultural crops, dessication, and grazing caused widespread seedling mortality. The restoration effort was most successful where the cooperators were committed to the project, technically competent, and able to gain the support of the other villagers.

The upland restoration project also included the introduction of soil and water conservation techniques such as hedgerows, rock walls, drainage canals, sediment traps, and check dams. This program was open to all and voluntary participation in workshops was encouraged through local networks. These group training sessions were more successful, on the whole, than selecting and training a few individuals as was done in the tree planting effort.

The mangrove forests of Bais Bay are located at lower elevations, near the coast. They have been reduced in area by about 60% and only about 300 ha of mostly degraded mangroves remain. The goal of the Bais Bay Program here was to restore biodiversity and increase protection of the Talabong Mangrove Sanctuary, which constitutes two thirds of the remaining mangrove forest (Figures 9.A.1 to 9.A.4). A nursery was established in cooperation with a local college to investigate the best ways to propagate and grow several mangrove species. With the help of government officials (who were allotting 25-year leases to households with mangrove areas adjacent to their homes), program staff encouraged

FIGURE 9.A.1
A 5-month-old *Rhizophora stylosa* plantation on Banacon Island, Philippines. The woman in the photograph is collecting shellfish. (Photo by B. Walters.)

FIGURE 9.A.2
Rhizophora mucronata in North Bais Bay, Philippines, originally planted 40 years ago to protect the adjacent home from storm winds and waves. (Photo by B. Walters.)

FIGURE 9.A.3
Newly planted *Rhizophora stylosa* in the foreground, with extensive older plantations in the background, Banacon Island, Philippines. Plantations here are used primarily for construction wood. (Photo by B. Walters.)

FIGURE 9.A.4
Fish pond owner stands on an earthen dike in North Bais Bay, Philippines. The dike is well protected from wave damage by an extensive stand of *Rhizophora mucronata* and *R. stylosa.* (Photo by B. Walters.)

people living near the shore to plant mangroves. Meetings concerning the importance of the mangroves were held for large groups of nearshore residents and officials. School children were also encouraged to help and they planted 4000 mangrove propagules. The mangrove reforestation met with mixed results, however. Walters (1997) attributes this, in part, to insufficient time devoted to extension by program staff. Specifically, a large share of the staff's time was spent studying technical and ecological issues such as the growth rates and survival of different species, leaving little time to understand the knowledge, motives, and interests of the local people.

Walters (1997) suggests that in developing countries the local people must be involved in and fully understand the restoration effort. He asserts that restoration is more likely to be successful when it is viewed by the local people as offering economic benefits, when the restoration is compatible with local land and resource use, and when local social and political groups are mobilized to support the restoration activities. It is important to know about local land use and societal values before attempting a restoration project. Since many impoverished people depend on the direct exploitation of natural resources for their livelihood, it is imperative that restoration efforts benefit these people.

9.B. Restoring the Habitat of an Endangered Bird in Southern California

In southern California, a number of salt marshes have historically been habitat for an endangered bird, the light-footed clapper rail (*Rallus longirostris levipes*). Due to tidal closures, road construction, and other pressures from the ever-expanding human population along the California coast, many of the birds' nesting sites have been lost (Zedler 1993, 2000b).

The clapper rail usually nests in tall, dense stands of the grass *Spartina foliosa* (intertidal cordgrass). In southern California salt marshes, *S. foliosa* growth is patchy and variable from year to year. It is usually <1 m in height and occurs only in wetlands with good tidal flushing. To provide the clapper rail with new habitat areas, three San Diego Bay marshes were planted with *S. foliosa* in the 1980s. The clapper rail failed to nest in the newly restored sites. To determine why the bird failed to adopt the new habitat, Zedler (1993) closely examined the birds' needs and compared the conditions in the new wetlands to three marshes that did support clapper rails.

The typical clapper rail nest is hidden in stands of tall, dense *S. foliosa* and constructed primarily of dead *S. foliosa* stems. The nest's platform is built up from the ground, supported in the cordgrass. The average height of the nest's rim is 45 cm off the ground. The nest is covered by live stems that are pulled over and entwined, making it undetectable from above. The *S. foliosa* platform allows the nest to float upward during high tide. A ramp of dead stems leads from the nest's rim to the ground (Figure 9.B.1).

Data regarding the biomass, stem density, mean stem height, and maximum height were gathered for both the reference wetlands and the restored ones. The total stem length per unit area (TSL; m m^{-2}) was calculated as stem density × mean stem height, a parameter that integrated stem density and height. In the reference wetlands, suitable habitat areas had a stem density of at least 100 stems m^{-2} with most of the stems (90 of them) taller than 60 cm and at least 30 of those taller than 90 cm. During the first year of sampling, stem density in some sampling areas of the restored marshes reached 100 stems m^{-2}; however, only a few plants grew to over 60 cm in height. The following year a larger proportion of the stems were tall (26% of stems were >90 cm), but the stem density decreased substantially, to only 53 stems m^{-2}. In one of the restored wetlands, *S. foliosa* biomass was significantly less than in reference wetlands (190 g m^{-2} vs. 450 g m^{-2}). The average TSL in the

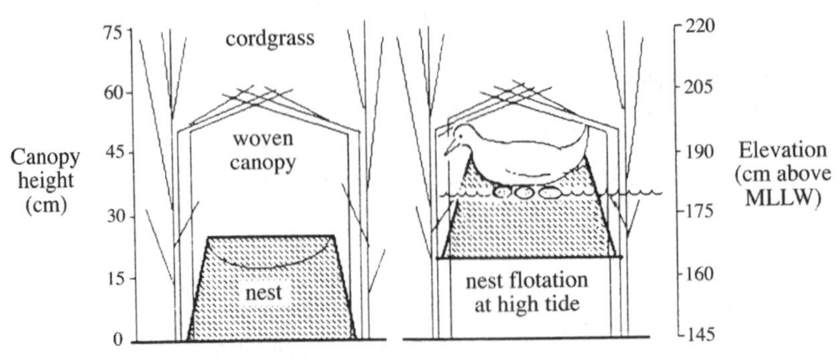

FIGURE 9.B.1
A diagram of a typical clapper rail nest in cordgrass 60 to 90 cm in height at low tide (left) and with 20 cm of flotation during high tide (right; MLLW = mean lower low water = zero tidal datum). (From Zedler, J.B. 1993. *Ecological Applications* 31: 123–138. Reprinted with permission.)

restored wetlands was 64, while in the natural wetlands it was 106, indicating the lower stem height and stem density in the restored marshes.

While planting *S. foliosa* appeared to offer a solution to declining clapper rail populations, the situation proved to be more complex than initially thought. The clapper rail requires plant stands with specific conditions, which probably offer optimal nesting, hiding, and foraging.

In a separate study, plants in the constructed marshes were found to be nitrogen limited (Covin and Zedler 1988). The coarse dredge-spoil substrate in the constructed marshes may have been too coarse to retain nutrients. It contained only one third to one fourth the total nitrogen found in the fine-textured soils of a nearby natural marsh (Boyer and Zedler 1998). When nitrogen was added in the constructed marsh, *S. foliosa* responded with increased stem length and foliar nitrogen, reaching the desired stem density of 100 stems m^{-2} with more than 30 stems exceeding 90 cm in height. This level of growth was sustained during the following growing season only in plots where nitrogen additions continued. Without long-term fertilization, the *S. foliosa* stands returned to sparser growth. Boyer and Zedler (1998) recommend that future marsh restoration projects begin with fine soils that promote nitrogen retention and support tall canopies of *S. foliosa*. In another study, *S. foliosa* was shown to grow best along tidal creek edges where it was not in competition with other species (Boyer and Zedler 1999). Therefore, in order to maximize *S. foliosa* habitat area, a further recommendation is to maximize tidal creek edges. Both of these changes in the restoration plan would likely result in self-sustaining *S. foliosa* stands.

9.C. Vegetation Patterns in Restored Prairie Potholes

Galatowitsch and van der Valk (1994, 1995, 1996) suggest that the definitive test of whether a wetland has been successfully restored is to compare the vegetation in the new wetland with that of nearby natural wetlands. They examined 62 reflooded depressional wetlands in the prairie pothole region of Iowa, Minnesota, and South Dakota. They compared the vegetation of the reflooded wetlands to that of natural wetlands to determine if there were patterns in the plant community structure and if these patterns could be related to historical factors such as the method of drainage or to environmental conditions.

Prairie potholes have distinct vegetation zones defined by water depth and length of flooding. The deepest area is in the center of the pothole. As one moves toward the outer edges, the elevation increases and the length of time of flooding decreases. The deepest area typically has submerged plants such as *Potamogeton* species and some emergents such as *Scirpus*. To the outside of the deep zone are shallow emergents such as *Sagittaria* and *Sparganium*. To the outside of the emergent zone are sedge meadows and wet prairies. If a pothole is shallow, it may consist of only the sedge meadow and wet prairie communities.

In the southern prairie pothole region, about 90% of the wetlands were drained for agriculture, most between 25 and 75 years ago. A number of townships restored the hydrology of prairie potholes between 1987 and 1991 by removing drainage tiles for agriculture (Figure 9.C.1). The basins were planted with upland cover crops such as *Bromus inermis* (brome) or *Panicum virgatum* (switchgrass). Revegetation of native wetland plants was dependent upon refugial adult plant populations, remnant seed banks, and the dispersal of seeds and other propagules to the site.

The wetlands were sampled for 3 years. The vegetation of the reflooded wetlands did not resemble that of the natural wetlands in several respects. The reflooded wetlands, with

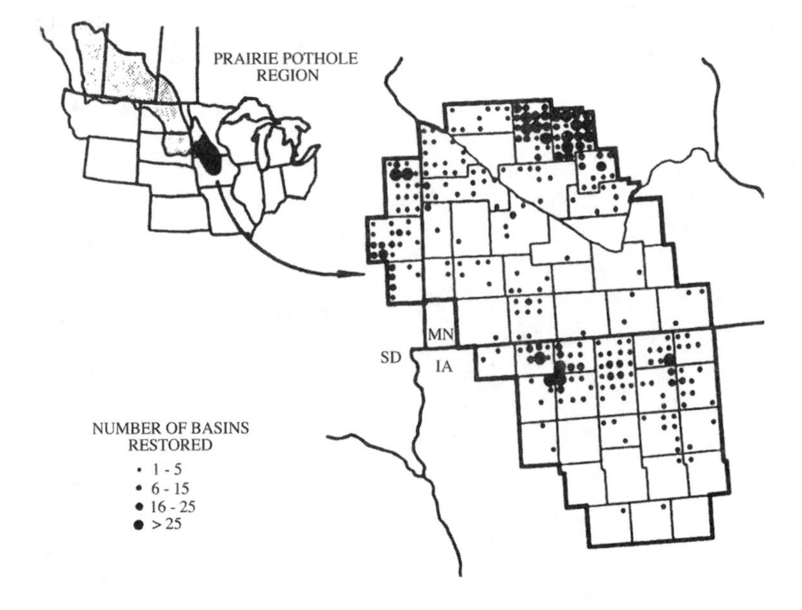

FIGURE 9.C.1
The prairie pothole region of the U.S. and Canada and the location of wetlands that were restored between 1987 and 1991 in the southern prairie pothole region. (From Galatowitsch, S.M. and van der Valk, A.G. 1995. *Restoration of Temperate Wetlands*. B.D. Wheeler, S.C. Shaw, W.J. Fojt, and R.A. Robertson, Eds. Chichester. John Wiley & Sons. Reprinted with permission.)

an average of 27 species per basin, had about half as many species as the natural wetlands, which had, on average, 46 species. Of the 106 species observed, 43% of them were found only in the natural wetlands. Only two of the four vegetation zones recovered in the reflooded wetlands. The submerged and emergent species recovered relatively well, but the sedge meadow and wet prairie species did not, perhaps due to a lack of viable seeds. Of the 46 species found only in the natural wetlands, 16 were wet prairie perennials and 21 were sedge meadow species.

The seed banks of most of the reflooded wetlands were probably too old to contribute many viable seeds. The mean length of time the wetlands were drained was 32 years. The seed banks of prairie potholes that have been drained for 30 years have been observed to have less than 10% of the seed density and about 50% of the number of species of existing wetlands. Many species are not represented in the seed banks of even recently drained wetlands (<20 years). *Carex* species, whose seeds are short-lived, are notably absent (Weinhold and van der Valk 1989). Because the sedge meadow and wet prairie communities did not recover, these two zones would need to be replanted for a complete restoration. Without replanting and careful management, fast-growing plants such as *Phalaris arundinacea* (reed canary grass) and *Leersia oryzoides* (rice-cut grass) may outcompete the desired species.

Recolonization of the reflooded wetlands occurred mainly through the dispersal of submerged, emergent, and floating plants from other wetlands. Proximity to natural wetlands probably made the dispersal of these plants possible. Over half (51%) of the reflooded wetlands were within 2 km of a natural wetland, 22% were within 5 km, and none were more than 13 km from a natural wetland. The authors did not find that the reflooded basins closest to a natural wetland had more species than the ones farther away; however, distance to the nearest seed source has been found to have a strong effect in other restored wetlands (Leck 1989; Reinartz and Warne 1993).

10

Wetland Plants as Biological Indicators

I. Introduction

There is a strong ecological basis for using vegetation to identify and characterize wetlands and delineate their boundaries. The relationship between soil saturation, either continuous or recurrent, and the development of wetland plant communities is well documented and there is a long history of using hydrophytic vegetation to identify wetlands (Hall and Penfound 1939; Sculthorpe 1967; U.S. Army Corps of Engineers 1987; Tiner 1991; U.S. National Research Council 1995). As has been well established, flooding and soil saturation foster conditions that the majority of plants cannot tolerate. Reed (1997) estimates that nearly 70% of all plant species found in the U.S. or its territories do not occur in wetlands. This fact has led to the use of wetland plants as indicators of the presence of wetlands, and where these communities give way to upland species, the wetland's boundary.

The composition of a wetland's plant community has also been shown to serve as a practical indicator of ecological stress. Changes in vegetation represent a community level response that integrates the effects of a wide range of ecological stressors. Predictable changes in community composition, species abundance, productivity, and other ecosystem properties have been observed as environmental conditions shift (Lopez and Fennessy in press; Carlisle et al. 1999). This idea has a long history in plant ecological studies. Clements (1935) is notable as one of the first to observe that taking specific measurements of environmental conditions, such as water or soil chemistry, or hydrology, may yield far less information than using the performance of the organisms themselves (in Keddy et al. 1993). Vegetation can integrate the temporal, spatial, chemical, physical, and biological dynamics of the system.

The focus of this chapter is on the use of wetland plants as indicators of ecological conditions including the existence of wetlands, and as a tool for the assessment of their biological integrity. Both of these approaches integrate community dynamics and the landscape context of wetlands, and they represent an application of wetland plant ecology.

II. Wetland Plants as Indicators of Wetland Boundaries

Using wetland plants to delineate the boundary of a wetland is based on consistently observed changes in vegetation composition that occur as a function of environmental gradients, such as elevation and moisture (Carter 1996). For instance, as elevations increase and soils dry, wetlands give way to uplands, and plant community composition changes in response. The shift in species composition forms the basis for using plants to identify wetland boundaries.

Wetland delineation, the set of techniques and procedures used to identify wetland boundaries, was designed to support wetland protection and management efforts in the U.S, and is used to establish the areal extent of government jurisdiction (Carter 1996; Tiner 1996). Wetland delineation is based on the three-parameter approach, namely, that hydrophytic vegetation, hydric soils, and wetland hydrology must be present for a wetland to be present (except in specified exceptions). The U.S. Army Corps of Engineers, which developed the three-parameter approach, has published the technical field procedures in what is known as the *delineation manual* (U.S. Army Corps of Engineers 1987).

In the delineation procedure, determining whether or not a plant community is hydrophytic is a pivotal decision (Wakely and Lichvar 1997). The identification of individual species as hydrophytic is made by using a compilation of plant species that ranks the probability of occurrence of each species in wetland habitats. The ratings for wetland plants are found in the "National List of Plant Species that Occur in Wetlands" (Reed 1988, 1997), which is a list of the indicator status of all plants know to occur in U.S. wetlands. There are currently about 7500 species on the list, each of which has been assigned an indicator status for the regions in which it occurs. All species on the list are assigned one of four wetland indicator status categories based on the probability that the species will be found in a wetland. These are obligate wetland (OBL), facultative wetland (FACW), facultative (FAC), and facultative upland (FACU; Table 10.1). Obligate species occur in wetlands more that 99% of the time, while facultative species are just as likely to be found in uplands as in wetlands. Species not found on the list are considered to be obligate upland (UPL) species. The indicator status assigned to species in the FACW, FAC, and FACU categories can be refined by assigning a "+" or "-" to the designation. Addition of a "+" indicates that, within its indicator status category, the species is more likely to be found in wetlands, while a "-" indicates it will more likely be found in uplands. Thus, a FAC+ species is more likely to be found in wetlands than a FAC- species.

TABLE 10.1

Wetland Indicator Status Categories for Plant Species

Wetland Indicator Status	Probability of Occurrence in Wetlands (%)	Probability of Occurrence in Non-Wetlands (%)	Weight[a]
Obligate wetland (OBL)	>99	<1	1
Facultative wetland (FACW)	67–99	1–33	2
Facultative (FAC)	24–66	34–66	3
Facultative upland (FACU)	1–33	67–99	4
Upland (UPL)	<1	>99	5

[a] Weights used for calculating weighted averages (prevalence index) from Wentworth et al. 1988.

According to Reed 1997.

The nearly 7500 species on the list represent approximately one third of the U.S. flora (estimated to be 22,500 vascular plant species). Of the listed species, Tiner (1991) estimated that 27% are obligate species. He considers obligate hydrophytes to be the best vegetative indicators of wetlands because they are almost never found in any other habitat (Tiner 1996). Table 10.2 presents some examples of OBL, FACW, and FAC hydrophytes. Because FACW and FAC species make up nearly two thirds of the species on the list, and they are able to grow in both wetland and upland environments, the delineation procedure cannot

TABLE 10.2
Examples of Common Wetland Plant Species in the U.S. with Indicator Status of OBL, FACW, and FAC

Obligate Species (OBL)
Alisma subcordatum (water plantain)
Caltha palustris (marsh marigold)
Cephalanthus occidentalis (buttonbush)
Chamaecyparis thyoides (Atlantic white cedar)
Elodea spp. (waterweeds)
Gleditsia aquatica (water locust)
Juncus militaris (bayonet rush)
Leersia oryzoides (rice cutgrass)
Lemna spp. (duckweeds)
Lonicera oblongifolia (swamp honeysuckle)
Nuphar spp. (pond lilies)
Nymphaea spp. (water lilies)
Nyssa aquatica (water tupelo)
Osmunda regalis (royal fern)
Rhizophora mangle (red mangrove)
Scirpus americanus (three square bulrush)
Typha latifolia (broad-leaved cattail)
Taxodium distichum (bald cypress)
Vallisneria americana (wild celery)
Zizania aquatica (wild rice)

Facultative Wetland Species (FACW)
Bidens frondosa (Spanish needles)
Cyperus odoratus (sedge)
Eleocharis tenuis (spike rush)
Helianthus giganteus (swamp sunflower)
*Ilex decidu*a (holly)
Impatiens capensis (impatiens)
Juncus torreyi (torrey's rush)
Leersia virginica (cut grass)
Mentha arvensis (field mint)
Onoclea sensibilis (sensitive fern)
Phalaris caroliniana (canary grass)
Quercus palustris (pin oak)
Salix lucida (shining willlow)
Spartina patens (salt marsh hay)

Facultative Species (FAC)
Acer rubrum (red maple)
Eupatorium purpureum (joe-pye weed)
Lonicera hirsuta (hairy honeysuckle)
Nyssa sylvatica (black gum)
Oenothera perennis (primrose)
Quercus macrocarpa (burr-oak)
Ranunculus hispidus (hispid buttercup)
*Rosa virginian*a (Virginia rose)
Scutellaria nervosa (skullcap)
Smilax rotundifolia (catbriar)
Solanum dulcamara (bittersweet nightshade)
Ulmus rubra (slippery elm)

Based on Tiner 1999; indicator status from Reed 1997.

be based on vegetation indicators alone. This fact led to development of the three-parameter approach (Tiner 1996, 1999).

The first draft of the hydrophyte list was completed by P. B. Reed in 1976 and he remains its "custodian" (in U.S. National Research Council 1995). The impetus for development of the list came in the mid-1970s from the U.S. Fish and Wildlife Service who needed it to define wetlands in the field. The list was compiled through a search of nearly 300 regional and state floras, regional wetland manuals, and information from the Fairchild Tropical Gardens in Miami. The final list contained 5244 species. Extensive interagency peer review was conducted in 1983 to 1984 by representatives of the U.S. Fish and Wildlife Service, the U.S. Army Corps of Engineers, the U.S. Environmental Protection Agency, and the U.S. Department of Agriculture's Natural Resources Conservation Service. Thirteen regional sub-lists were established using the geographic regions previously established by the U.S. Department of Agriculture (1982) for the "National List of Plant Names" (Figure 10.1). The indicator status of a given species sometimes varies between regions due to the ecotypic variation in the different populations. The interagency peer groups designated a regional panel for each region and gave them responsibility for assigning the wetland indicator status to as many species as possible (U.S. National Research Council 1995). The final list was published in 1988, and a revised edition was issued in 1997.

The first delineation manual to assist those charged with delineating wetlands was adopted in 1987 by the U.S. Army Corps of Engineers who make final jurisdictional determinations on wetland delineations and authorize certain activities in wetlands under

FIGURE 10.1
Map indicating regions used to identify the wetland indicator status of U.S. plant species. Many species' indicator statuses change across their range. (From Reed, P.B. 1988. National List of Plant Species that Occur in Wetlands: 1988 National Summary. Biological Report 88 (24). Washington, D.C. U.S. Department of the Interior, U.S. Fish and Wildlife Service. Reprinted with permission.)

Section 404 of the Clean Water Act (U.S. Army Corps of Engineers 1987). The manual is a technical document designed to provide methods to apply the definition of wetlands on the ground (U.S. National Research Council 1995). In this way wetlands, which are protected by the Clean Water Act, can be identified and protected through the proper regulatory process. Other federal agencies, including the U.S. EPA, U.S. NRCS, and U.S. FWS, also developed delineation manuals. Subsequently, the four agencies joined forces and developed a uniform manual that all would use (Federal Interagency Committee for Wetland Delineation 1989). Critics charged that the 1989 manual was too inclusive, causing many non-wetland areas to be regulated as wetlands. The result was a revised delineation manual (Proposed Revisions 1991). The controversy surrounding the 1991 manual prevented it from being adopted and it has not been used in the field. All agencies, save the U.S. NRCS, were directed to use the 1987 manual. Currently, the NRCS uses a modified delineation method detailed in the Food Security Act of 1985 (FSA; amended in 1990). Since then, the 1987 manual has been updated by a series of memoranda issued by the U.S. Army Corps of Engineers. A comparison of how the three manuals use wetland vegetation in delineation is shown in Table 10.3.

TABLE 10.3

A Comparison of Vegetation Criteria Used in Different U.S. Delineation Methods

	Year of 4-Agency Manual		
Vegetation Characteristic	**1987**	**1989**	**1991**
Use of the National Hydrophyte list to evaluate indicator status (OBL, FACW, FAC, FACU, UPL)	yes	yes	yes
Use of + and – to modify the indicator status	yes	no	no
Use of 50% rule to determine hydrophytic vegetation, where >50% of dominant species are OBL, FACW, or FAC	yes	yes	no
Use of prevalence index to determine hydrophytic vegetation, where the prevalence index <3.0 (a)	no[a]	yes	yes
Some secondary indicators of hydrophytic vegetation used (morphologic or physiologic adaptations, literature documentation)	yes[b]	no	no
Use of the FAC-neutral test	yes[c]	no	no

[a] Allowed under the updated 1987 manual.

[b] Not used under the updated 1987 manual.

[c] Now used as a secondary indicator of hydrology.

From U.S. National Research Council. 1995. Wetlands: Characteristics and Boundaries, p. 69. Committee on Characterization of Wetlands, Water Science, and Technology Board, Board on Environmental Studies and Toxicology, Commission on Geosciences, Environment, and Resources. Washington, D.C. National Academy Press. Reprinted with permission.

A. Hydrophytic Vegetation as a Basis for Delineation

In part, the identification and delineation of a wetland center on whether the plant community is hydrophytic. Several field indicators of hydrophytic vegetation have been used, some of which are detailed in Table 10.3. Currently, the most basic criterion is in use, which states that when more than 50% of the dominant species from all strata are hydrophytic (i.e., OBL, FACW, or FAC), then the plant community is considered to be hydrophytic. The U.S. NRCS uses the prevalence index under the Food Security Act, which is allowable under the 1987 manual.

Two widely used methods that have been used to determine the presence of hydrophytic vegetation include the *dominance ratio* and the *prevalence index*. The dominance ratio is calculated using the "50/20 rule" in which the dominant species in each stratum are defined as the species whose cumulative cover makes up >50% of the total cover of the stratum, plus any individual species that was at least 20% of the total cover in the stratum (Federal Interagency Committee for Wetland Delineation 1989). Vegetation is designated hydrophytic by this method if >50% of dominant species across all strata have an indicator status of OBL, FACW, or FAC (excluding FAC-). The prevalence index is a weighted average of the wetland indicator status of all plants present (Wentworth et al. 1988; Table 10.1). In this method, each plant along a transect must be identified. Each plant is given a score (OBL = 1.0, FACW = 2.0, FAC = 3.0, FACU = 4.0, and UPL = 5.0). The scores are summed and the average is the score for that plot. Plots that score <3.0 are considered to be wetland and those >3.0 are designated upland. The Federal Interagency Committee for Wetland Delineation (1989) presented these two approaches as alternative, but equivalent, methods, although there had been little study to confirm this.

In a study designed to test the parity between the two methods, Wakeley and Lichvar (1997) sampled 338 vegetation plots at sites throughout the U.S. and calculated the dominance ratio and the prevalence index for each. They found a 16% rate of disagreement on the decision to classify a given plot as hydrophytic, and disagreement tended to increase as vegetation complexity increased (Figure 10.2). The prevalence index averaged 2.65 (± 0.80) for all plots, while the mean score for the plots where results disagreed was 3.01

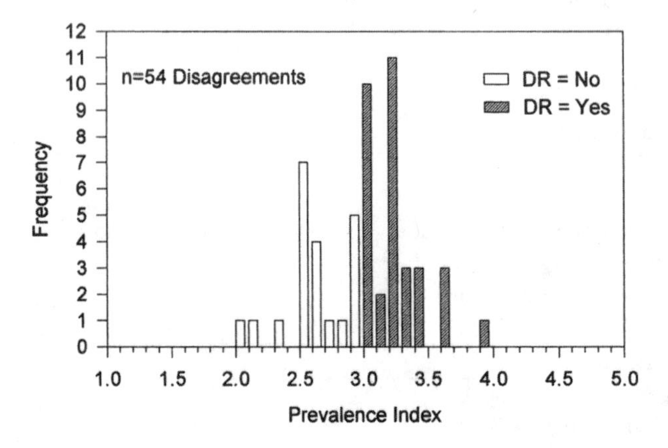

FIGURE 10.2
Data from 338 plots located in sites across the U.S. showing the frequency of disagreement between the assessment of whether or not hydrophytic vegetation is present (yes or no) based on two different methods, the dominance ratio and prevalence index. (From Wakeley, J.S. and Lichvar, R.W. 1997. *Wetlands* 17: 301–309. Reprinted with permission.)

(± 0.39). The authors concluded that the methods cannot be considered equivalent and should not be interpreted as such. Current research does not indicate which method might be more reliable, although the data support the recommendation of Wentworth and others (1988) that for scores between 2.5 and 3.5, vegetation data should be confirmed with soil and hydrology indicators.

Where a significant discontinuity exists in a landscape, such as where riparian wetlands give way to river terraces, boundary identification is relatively straightforward. However, where environmental conditions change gradually, locating a boundary is difficult and somewhat arbitrary (Johnson et al. 1992). This is particularly true where environmental gradients are gradual and species composition changes gradually as a result, or where vegetation is tolerant of both wet and dry conditions (Carter 1996). An extensive study conducted in the Great Dismal Swamp, an 84,000-ha forested wetland on the Virginia–North Carolina border, provides an example of the difficulty delineation sometimes poses. Carter and others (1994) collected data on vegetation, soils, and hydrology along transects (400 to 625 m) on the western side of the swamp to study the boundary identification. The elevation gradient along this west–east axis is only 19 cm km^{-1} and the transition zone between the Great Dismal Swamp and adjacent uplands is dominated by FAC species including *Acer rubrum* (red maple), *Nyssa sylvatica* (black tupelo), and *Liquidambar styraciflua* (sweet gum). These factors conspired to make the wetland edge obscure, and different teams of researchers delineated the boundary at different locations. Carter and others (1994) also tested the use of the prevalence index. Depending on how the weighted average was calculated (based on all species, by individual stratum, eliminating ubiquitous species, or using weights based on the best professional judgment of the investigators), different results were obtained on the boundary location. The researchers concluded that three zones actually exist at the western edge of the Great Dismal Swamp: the wetland itself, an ecotone (transition zone), and the upland. The vegetation of the transition zone contains species common to both the wetland and adjacent upland.

B. Wetland Boundaries and Wetland Functions

Concern has been raised about the degree to which delineated boundaries coincide with the functional properties of wetlands. Ideally the wetland boundary will be located where the functional properties of wetlands diminish rapidly. Because site-specific information on the functional capacity of wetlands is difficult to collect and often subjective, structural attributes, such as species composition, are commonly used instead (Holland 1996). This issue has received relatively little study, but several authors report discontinuities between measured functions and boundary identification, and the preoccupation with the wetland "edge" that has developed in the U.S. has led to unrealistic assumptions that habitat values and ecosystem processes coincide with that boundary. For instance, the inclusion of fauna in delineation often results in a much larger wetland area since many species (e.g., amphibians) require adjacent buffer areas or uplands to complete their life cycle (Hapley and Milne 1996). In another example, Groffman and Hanson (1997) report finding a poor relationship between the spatial and temporal patterns of denitrification (a key biogeochemical process important to the maintenance of water quality) and the location of hydrophytic vegetation or hydric soils. In an earlier study designed to quantify the ability of riparian wetlands to remove nitrate from shallow groundwater, Haycock and Pinay (1993) found that the most intense zone of nitrate removal due to denitrification was just upslope of the floodplain (wetland) boundary.

If the goal of delineation is to identify wetlands so that both their diversity and functions can be preserved, then protecting the wetland only as far as its border (leaving all land beyond that point to be converted) may not achieve this goal. The values that are placed on wetland functions often accrue at different scales, such as stream reach, watershed, or landscape. Modest proposals include designing wetland protection measures to offer protection for a buffer zone around the wetland (i.e., an area surrounding the boundary that is a transition zone between the wetland and surrounding uplands) that will lead to the preservation of more functionally intact systems. These studies offer support for including a buffer around the wetland as part of the measures designed to protect wetlands.

C. The Use of Remotely Sensed Data in Wetland Identification and Classification

Considerable emphasis has been placed on the use of vegetation to identify wetlands in the field; however, wetlands may also be identified remotely using aerial photographs or satellite imagery (Lehmann and Lachavanne 1997; Narumalani et al. 1997; Malthus and George 1997; Williams and Lyon 1997). The interpretation of aerial photographs has been used extensively to map and inventory wetlands, including state coastal wetland maps (e.g., New Jersey, New York, Ohio, South Carolina) and the mapping of inland wetlands (Maine, New York, Wisconsin). These photographs have also been the basis for the creation of the U.S. National Wetlands Inventory (NWI), a national mapping effort in the U.S. which uses the Cowardin Classification System (Cowardin et al. 1979). Satellite imagery has also been used to monitor wetlands and evaluate the suitability of their plant communities as habitat for wildlife, particularly for waterfowl (Tatu et al. 1999).

The success of wetland photointerpretation is a function of several factors including the quality of the photography, the season in which the photo was taken, and the photographic scale, which sets a limit on what can be interpreted (e.g., the minimum mapping unit; Tiner 1999). Large-scale photographs (such as 1:24,000) are better when the goal is to locate changes in vegetation communities precisely, when small wetlands must be identified, or when discrete plant communities must be located. On the other hand, small-scale photography (such as 1:58,000) works best for inventories over large regions (Tiner 1999).

The seasonal changes in vegetation are a primary consideration in photointerpretation. For example, deciduous forested wetlands are easiest to interpret when they have dropped their leaves since standing water (if present) is visible. Early spring photographs are ideal for this purpose since standing water is more likely to be present than it is following leaf-fall in the autumn. Evergreen forested wetlands are among the most difficult to discern from aerial photos, in part because their foliage is always present and because evergreen stands occur in both wetlands and adjacent uplands. In this case, Tiner (1999) recommends using the height of the canopy to assess the difference in wetness. For example, evergreens growing in Alaskan forested wetlands tend to be shorter than upland evergreens (18 vs. 30 m tall), a difference that can be detected in aerial photos. In another example in which aerial photographs were used to identify submerged aquatic vegetation in coastal wetlands, the timing was found to be critical. The most useful photos were taken at peak biomass (when the plants are most conspicuous in the water column) and within 2 h of low tide early in the morning when the sun angle was low (Dobson et al. 1995).

III. Wetland Plants as Indicators of Ecological Integrity

In the U.S., the national goal of achieving no overall net loss of wetlands created a preoccupation with the amount of wetland area that remains on the landscape. Several reports issued on the status and trends in wetland area document the extent of wetland loss (e.g., Dahl 1990; Dahl and Johnson 1991). Wetland impacts, and requirements to mitigate those impacts, are typically based on the acreage lost. However, recognition that many existing wetlands have become degraded has led to an increasing focus on the quality or condition of those wetlands, as well as on the quality of those that are being created or restored. The U.S. Clean Water Act provides a broad definition of water quality that includes all aspects of the ecological health or integrity of the nation's waters. This includes their "chemical, physical and biological integrity." Karr (1991) refers to this as "the quality of the water resource," and expresses this concept as "ecological integrity." Ecological integrity (sometimes referred to as "health") is defined as (Karr and Dudley 1981):

> ... the capability of supporting and maintaining a balanced, integrated, adaptive community of organisms having a species composition, diversity and functional organization comparable to that of natural habitats of the region.

The Clean Water Act mandates that biological integrity be restored in all degraded aquatic ecosystems, including wetlands. While substantial progress has been made to develop and implement methods to assess the condition of rivers and lakes, research in this area for wetlands has lagged (Danielson 1998). The goal of maintaining ecological integrity cannot be accomplished without monitoring the condition of wetlands. Conventional water quality monitoring has relied on the chemical analysis of water, an approach that misses many physical or biological stressors to the system. The most robust monitoring and assessment programs not only rely on measures of chemical water quality, but also include biological monitoring (Karr 1991; Yoder and Rankin 1995). The biological assessment of wetlands requires assessment methods that can quickly and reliably detect ecosystem changes in response to human activities (Karr and Chu 1997; Galatowitsch et al. 1999b). Wetland plants have the capacity to indicate the cumulative response of the ecosystem to a wide array of chemical, physical, and biological alterations.

Because one of the ultimate goals of preserving the environment is to preserve biological diversity, the most direct approach to measuring the quality of a wetland is to assess its biota. The goal of biological assessment is to identify biological attributes that provide reliable information on wetland condition. Standard terms have been defined to describe the certainty with which a given measure reflects ecosystem integrity. These include (Karr and Chu 1997):

- *Attribute*, a quantifiable component of a biological system
- *Metric*, an attribute that has been shown to change in value along a gradient of human influence
- *Multimetric index*, an index that integrates several biological metrics into a single number to indicate the condition of a site (e.g., an *Index of Biotic Integrity*, IBI; see Section III.D, Vegetation-Based Indicators)
- *Biological assessment*, using information (metrics, etc.) collected from species assemblages to evaluate site condition

Several factors have provided impetus in the development of biological indicators. One is the need for tools to monitor the quality of our wetland resources. Without such tools, it is difficult to determine whether current or potential ecosystem problems are increasing, or whether current environmental policies are effective in maintaining quality. Metrics can be used to describe the overall condition of an ecosystem, diagnose probable causes where conditions are poor, and identify human activities that are contributing to these causes (Messer et al. 1991). Another factor in the development of biological indicators stems from the need to monitor wetland restoration, both to understand the factors that might limit recovery and to set quantifiable goals to determine if a restoration project is successful. Previously developed techniques used to assess wetlands (e.g., WET; Adamus et al. 1987) are not as suitable for monitoring restoration projects because they provide estimates of a wetland's ability to perform certain functions including those, such as sediment trapping, that can lead to wetland degradation (Galatowitsch et al. 1999b).

The choice of biological indicators must reflect both policy goals and scientific issues. Factors that dictate the choice of indicator include the sensitivity of the metric, the replicability of its response, and cost (Adamus 1992). While indicators of ecosystem quality have historically been based on chemical or physical characteristics, indicators that include biological measures tend to be more informative because of their ability to reflect the totality of environmental conditions. The ability to be diagnostic (i.e., to begin to explain how plant communities respond to stress) initially requires both biological and physical/chemical data. Ultimately the goals of wetland biological assessment include (Adamus 1992; Galatowitsch et al. 1999b):

- To determine if wetland condition is changing and, if so, in what direction
- To assess the degree of disturbance a wetland has sustained and use that information to set priorities for restoration or mitigation
- To evaluate the success of wetland restoration and mitigation projects
- To define management approaches to protection and/or manage wetlands
- To diagnose the cause of wetland degradation
- To increase our understanding of wetland ecosystem science

In all cases, monitoring wetland recovery or restoration requires assessment techniques that can reliably discern changes in ecosystems.

A. An Operational Definition of Ecological Integrity

Ecosystem integrity or "health" was originally defined in the same terms that describe human health (Schaeffer 1991). The term *ecosystem integrity* has come to be used more generally to indicate the ecological condition of an ecosystem and its response to human-induced stressors. As human disturbance increases over time, the ecological integrity of the wetland is diminished due to changes in processes such as nutrient cycling, photosynthesis, hydrology, competition, or predation (Karr 1993). At the community level, anthropogenic disturbance tends to decrease species richness and alter community composition. Because ecosystems are complex, made up of interacting elements that are controlled by, and may control, elements from other trophic or organizational levels, metrics developed for biological assessment must be selected in light of these ecosystem relationships (Schaeffer et al. 1988).

Choosing indicators of ecosystem integrity would be relatively straightforward if the science of ecology were able to supply simple, robust models to predict the response of ecosystems to stress, i.e., identify which state variables are important to monitor when assessing wetland condition (Keddy et al. 1993). Since these models are not available, operational definitions, based on the collection and interpretation of field data, are being developed. We can approach a practical definition of ecological integrity empirically by comparing the community structure and functions of the wetland of interest to unimpacted or 'minimally disturbed' wetland reference areas.

Plant species composition reflects both current and historical environmental conditions, and for this reason changes in species composition over time indicate environmental change. Plant communities have been called one of the best indicators of the "unique combination of climatic and hydrogeologic factors that shape wetlands within their landscape" (Bedford 1996). For example, the rate of attrition of conservative plant species (species adapted to a specific, narrow set of biotic and abiotic factors) tends to escalate with increasingly rapid and severe disturbance. Disturbance is a natural element of all ecological systems and ecosystems have adapted to the natural disturbance regime. However, the magnitude and rapidity of disturbance associated with human activities today has resulted in the reduction and/or extirpation of numerous conservative plant species (Wilhelm and Ladd 1988; Vitousek 1994; Andreas and Lichvar 1995).

Wetland plants are particularly useful as biological indicators because they are a universal component of wetland ecosystems. Plants, both vascular and nonvascular, are common and present in sufficient diversity to provide clear and robust signals of human disturbance. They have been used effectively to distinguish among environmental stressors including hydrologic alterations, excessive siltation, nutrient enrichment, and other types of disturbance (Wilcox 1995). Some of the advantages and disadvantages of using plants as biological indicators are shown in Table 10.4.

TABLE 10.4

Some Advantages and Disadvantages of Using Vegetation in Wetland Biological Assessment

Characteristics of Plants That Confer an Advantage in Assessing Wetland Integrity
They are a universal component of wetland ecosystems.
They are immobile (except for a few free-floating species), thus they integrate the temporal, spatial, chemical, physical, and biological dynamics of the system. They are also indicative of long-term, chronic stress to a system.
Their taxonomy is well known, and excellent field guides are available. For many plants, identification to genus or species is relatively easy by experienced field biologists.
There is a great diversity of species.
Because the ecological tolerances for many species are known, changes in community composition might be used to diagnose the stressor responsible. For example, plant responses to changing hydrology are reasonably predictable.
Sampling techniques are well developed and extensively documented.
Similar methods can be used in both freshwater and saltwater systems.
Functionally or structurally based vegetation guilds have been proposed for some regions.
Characteristics of Plants That Confer a Disadvantage in Assessing Wetland Integrity
There may be a lag in the response time to stressors, particularly in long-lived species.
For some plants, identification to species level is laborious, or restricted to narrow periods during the field season; results may vary with field personnel. Some assemblages, such as the grasses and sedges, may be particularly difficult to identify to species.
Some assemblages, such as the submerged species, are difficult to sample.
Vegetation sampling may be limited to the growing season.
Many species appear to be insensitive to insecticides and heavy metals (Adamus 1995).
Research or literature on plant species responses to different stressors is not well developed. Adamus (in press) estimated that only 17% of all wetland plant species have been the subject of detailed studies.

U.S. EPA Biological Assessment of Wetlands Workgroup (BAWWG, plant subgroup)

B. Wetland Plant Community Composition as a Basis for Indicator Development

As wetlands continue to be exploited and degraded, attention has turned to understanding the response of the plant community. Nutrient enrichment, sediment loading, hydrological changes, other changes in the physical and chemical environment, and invasion by exotic species are common types of disturbance (Niering 1990; Leach 1995; Toner and Keddy 1997; Hulot et al. 2000). One of the most common disturbances in wetlands is land use changes in the watershed, including conversion to agricultural and urban land use (Gusewell et al. 1998). A high proportion of global wetland loss is due to drainage for agriculture (Mitsch and Gosselink 2000). Many remaining wetlands are substantially smaller and more isolated than in the past. For instance, in five agricultural counties in Ohio, 75% of the wetlands are less than 0.4 ha in size (Fennessy et al. 1998b; data from the Ohio Wetlands Inventory). Habitat fragmentation on this scale has had a profound impact on ecosystem diversity and function.

The response of a given plant species to disturbance is a function of its autecological tolerance to different environmental conditions. Species with specialized requirements or those that are not tolerant of disturbance tend to be displaced (Table 10.5). Recognition that species intolerant of disturbance tend to be eliminated from disturbed areas led to an early approach to biological assessment based on identifying the "most sensitive species," and using its presence (or absence) as an indicator of the quality of a given environment. This is a problematic approach and has been dubbed by John Cairns, Jr. as "the myth of the most sensitive species" (Cairns 1986). Although his critique was set in the context of toxicity testing, the flawed assumptions inherent in the most sensitive species approach, listed below, also apply to vegetation-based assessments of wetland disturbance:

- Choosing the most sensitive species from a whole community means that the response of that species must correspond to that of all members of the community. In fact, species respond individually to different stressors.
- The response of the indicator species is assumed to be more sensitive than any other level of biological organization. As will be discussed below, community-level responses are often more sensitive.
- A species shown to be most sensitive to a limited array of disturbance types is assumed to respond in the same way to any type of disturbance. An example of the individualistic nature of species responses is seen in *Cirsium arvense* (swamp thistle), which has been shown to be relatively tolerant of sedimentation (Wardrop and Brooks 1998) but is less tolerant of hydrologic alterations.

TABLE 10.5
Predicted Effects of Anthropogenic Disturbance on the Plant Community Composition in Wetlands

The proportion of r-strategists increases
The mean size of plant species decreases
Mean life span of plants or plant parts (leaves) decreases
Food chains shorten (reduced energy flow to higher trophic levels)
Species diversity decreases
Increasing dominance by fewer species

Adapted from Odum 1985.

Rather than focus on the response of one or a few sensitive species, recent developments in the use of wetland plants as biological indicators has focused on changes in the structure and composition of the community as a whole. Changes in vegetation composition are community level responses that integrate the effects of a wide range of ecological stress (Carlisle et al. 1999).

The U.S. EPA's national Environmental Monitoring and Assessment Program (EMAP) has proposed a list of potential indicators for use in the EMAP program based on different taxonomic groups (Table 10.6). Few studies have compared the usefulness of different taxonomic groups as a basis for the development of biological indicators, although many groups have been evaluated including amphibians, birds, mammals, and plants (Kooser and Garono 1993; Garono and Kooser 1994; Danielson 1998).

TABLE 10.6

Physical and Biological Indicators of Wetland Integrity Proposed for Use in the EMAP Wetlands Program

Physical Indicators
The diversity of wetland types
Wetland pattern in the landscape
Hydroperiod
Sediment/organic matter accretion
Chemical contamination of sediments, plant and animal tissues
Biological Indicators
Vegetation
Species composition
Percent cover
Spectral greenness
Birds
Species composition
Bioaccumulation
Amphibians
Species composition
Bioaccumulation

Note: Some taxonomic groups other than plants demonstrate biological characteristics that commonly indicate stress.

U.S. Environmental Protection Agency 1996.

C. General Framework for Wetland Biological Indicator Development

A good framework for developing biological indicators based on wetland vegetation should include several key components. Effective sampling using an appropriate sampling protocol is required. Sufficient expertise is required to identify plant species to an appropriate taxonomic level. A general framework includes the following:

- *Wetland classification.* Classification allows wetlands to be grouped so that only "like-kind" wetlands are compared. In this way, wetlands that are similar in the absence of human disturbance and that respond similarly to disturbance are grouped together (Karr and Chu 1997). One of the goals of classification is to reduce variability in the data collected within a group of wetland sites. Natural variability is reduced within groups while being maximized between groups. Minimizing variability within a group of wetlands facilitates detection of human-induced variability. There are several well-established wetland classification schemes including the hydrogeomorphic approach (HGM; see Chapter 2, Section III, Broad Types of Wetland Plant Communities; Brinson 1993a), which has proven to be useful in classifying wetlands for biological assessment (Fennessy et al. 1998b; Wardrop and Brooks 1998), and the classification system employed in the National Wetland Inventory (Cowardin et al. 1979). Ecoregions have also been used to classify wetlands (Omernik 1995).

- *Establishment of reference sites and reference conditions.* A crucial component of a biological assessment program is the careful selection of least-impacted reference sites. *Reference sites* are wetlands of the same class that are used to define the best possible condition for that class. Reference sites serve as the standard against which other sites are judged, and so provide a baseline for the evaluation of all wetlands in a given class (Karr and Chu 1997). As such, reference sites set the standard for ecological integrity for that class (also referred to as the *best attainable condition*). Because reference wetlands are used to define the best attainable condition, they should be as undisturbed as possible and be representative of the wetland class for which they will serve as models (Yoder and Rankin 1995).
- *The inclusion of other wetlands that represent the full gradient of disturbance.* Selected sites should represent the full range of human influence for each wetland group. There is no standard method to quantify disturbance, so many projects have relied on surrogate measures including the percent impervious surface (Richter and Azous 1995) or the percent agricultural land use in the watershed. Another approach is to use a qualitative index of disturbance based on dominant land use surrounding the wetland, buffer characteristics, and the degree of hydrologic alteration to the site (Fennessy et al. 1998a; Lopez and Fennessy, in press). Sites should include the full range of human influence, from severely degraded to least impacted wetlands. Sampling at sites with different intensities of disturbance can indicate which attribute is sensitive to human activity (Figure 10.3). This makes it possible to evaluate the response of the wetland plant community to increasing "doses" of human activity (Karr and Chu 1997).
- *Establishment of standard field sampling methods, laboratory and analytical methods.* Standardized field methods must be adopted, tested, and refined to ensure that an equal and consistent sampling effort is made at each site. Consideration must be given to the type of data that will be collected (species inventory, cover, stem counts, etc.), the time of year or *sampling window* that will be used to characterize the vegetation, the sampling technique that will be employed, the number of samples that will be collected, and retention of voucher specimens. The same must be done for any analyses that will be performed subsequently in the laboratory.
- *The choice of metrics.* In writing about how to choose appropriate metrics, Karr and Chu (1997) have said that "a bewildering variety of biological attributes can be measured, but only a few provide useful signals about the impact of human activities." Only a few attributes will show a consistent response to disturbance; those that do have been termed 'metrics.' Certain attributes of wetland plant communities have been shown to vary consistently and systematically with human influence (discussed below). The goal of using metrics across a gradient

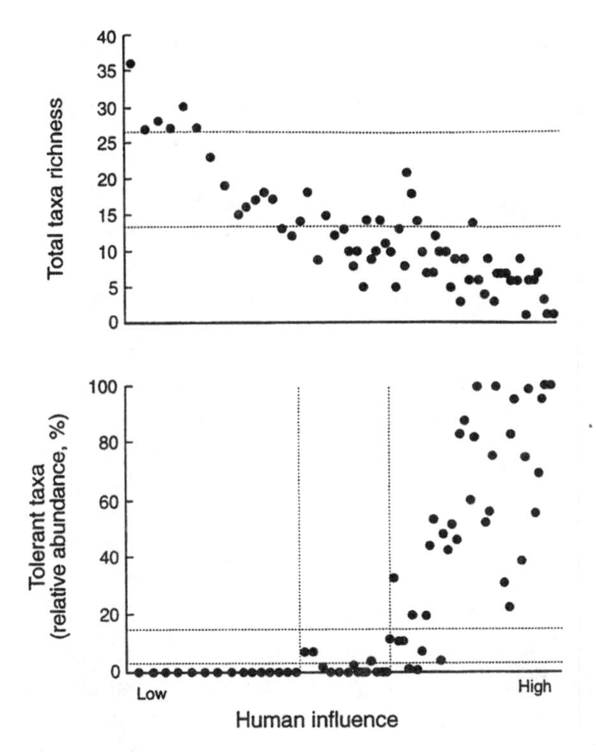

FIGURE 10.3
Graphs showing two sample metrics illustrating the ways that they differ in their response to human disturbance. (From Karr, J.R. and Chu, E.W. 1997. *Biological Monitoring and Assessment: Using Multimetric Indexes Effectively*. Seattle. University of Washington. Reprinted with permission.)

of disturbance regimes is to establish a dose–response curve that plots the response of the plant community to increasing levels of human activity.

D. Vegetation-Based Indicators

Keddy and others (1993) provide guidelines to consider when developing indicators of biological integrity, stating that appropriate indicators should:

- Be ecologically meaningful, i.e., closely related to key ecosystem processes (primary productivity) or environmental variables (hydrological fluctuations)
- Operate at the macro-scale by indicating changes in the entire community rather than for selected species
- Be general enough to be used on different community types
- Be sensitive enough that they respond quickly to stress and disturbance
- Be simple to measure

One method to assess the biological condition of wetlands is to develop an index of biotic integrity (IBI) based on proven metrics (i.e., those atrributes that respond predictably to chemical, physical and biological disturbance gradients in wetlands). An IBI combines several metrics into a composite index value that can be compared to values obtained at reference sites (see Case Study 10.A, The Development of a Vegetation IBI). A well-constructed IBI can be sensitive to disturbance and offer diagnostic capabilities (i.e., indicate the type of stress). One of the most effective ways to interpret the "biological signal" that attributes provide is to plot them as a function of disturbance level, as described below. In this way useful metrics can be identified among the many possible community attributes.

The use of plant functional guilds has also been proposed as a technique to identify useful indicators. Plant functional guilds are groups of species that show similar responses to disturbance through similar mechanisms (Hobbs 1997). Because of this they can collectively serve as useful indicators (Adamus 1992). Guilds are simplifications of the real world that allow variation to be interpreted more easily. It is possible to use functional guilds to evaluate environmental change (Hobbs 1997; Wardrop 1997). Species can be grouped on the basis of ecosystem function or response to environmental variables in a number of ways, including (from Hobbs 1997):

- Resource use
- Ecosystem functions such as primary production, decomposition, N-fixation
- Response to different types of disturbance
- Reproductive strategies
- Tolerance to different types of stressors
- Physiological types (e.g., C_3 vs. C_4 species)
- Physiognomic types

Wardrop and Brooks (1998) used this approach in Pennsylvania wetlands. Because many existing schemes to identify plant guilds do not group species by their characteristic response to stress, the authors constructed guilds (dubbed 'tolerance groups') based on species responses to sedimentation and hydrologic stress (e.g., in terms of germination success). Field data were collected for presence/absence and percent cover data for over 500 plant species in 800 plots across many wetland sites. Sediment tolerance groups were created by calculating the average percent cover of each species (when present) with measured sedimentation levels for each plot. Species were then grouped as being very tolerant, moderately tolerant, slightly tolerant, and intolerant to sedimentation. As disturbance due to sediment loading increased, species that were grouped as very or moderately tolerant tended to increase in dominance (percent cover). Hydrologic tolerance groups were established using water level data collected from groundwater monitoring wells installed at 27 sites. The hydrologic measures included median depth to the water table; percent time the water level was in the top 30 cm of the soil profile; the percent time the upper 30 cm of soil was saturated, inundated, or dry; and the percent time the upper 10 cm was saturated, inundated, or dry. The hydrologic groups were created by calculating the average percent cover of individual species, when present, within each of the five hydrologic groups. The authors found that the wetland plant indicator status (OBL, FACW, FAC, etc.) was an "extremely poor predictor of an individual species' hydrologic regime." An example of how data were tabulated for the sedimentation group is shown in Figure 10.4. These tolerance groups, which show strong correlation with the degree of human impact at a site, will form the basis for the development of a plant-based Index of Biological Integrity.

E. The Floristic Quality Assessment Index for Wetland Assessment

The Floristic Quality Assessment Index (FQAI) is a vegetation-based index that can be used to assess ecological integrity (Wilhelm and Ladd 1988; Andreas and Lichvar 1995; Fennessy et al. 1998a, b). The method was originally developed by Wilhelm and Ladd (1988) for the Chicago region, although the floristic quality lists necessary for its use have

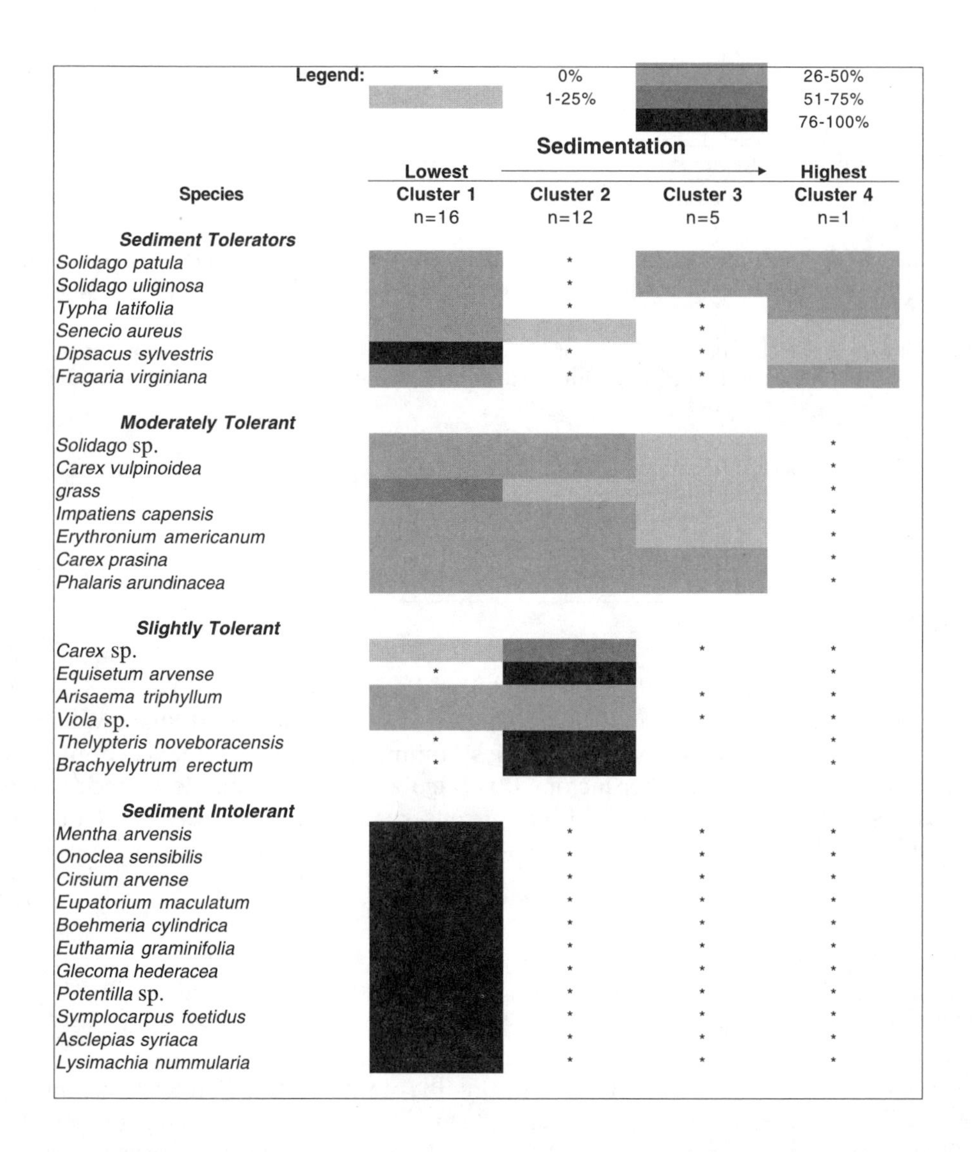

FIGURE 10.4
Sediment tolerance groups of wetland plants in slope wetlands of central Pennsylvania. These data were the basis for creation of plant sensitivity guilds. (Figure by Wardrop, D.H. and Brooks, R.P.)

been compiled in other areas including Ohio (Andreas and Lichvar 1995) and Michigan (Herman et al. 1996). The FQAI provides information about the condition of an ecosystem because it accounts for both the presence of exotic species and the degree of fidelity of each native species for specific environmental conditions (i.e., the ability of a native species to persist in the ecosystem as conditions change over time). Because it is a sensitive measure of changes in plant species composition, the FQAI can be used to assess an area based on the balance between ecologically conservative and highly tolerant species. The FQAI is based on the *Coefficient of Conservatism,* a rating of the tolerance of each species in the community to varying environmental conditions (Table 10.7). The Coefficients of Conservatism

TABLE 10.7

Definitions Used to Assign Coefficients of Conservatism to Plant Species for Use in Calculating the FQAI

Coefficient of Conservatism	Species Characteristics
Value of 0	Opportunistic native invasive species and all non-native species
Values of 1–3	Species that are widespread and are not an indicator of a particular community, that tolerate moderate disturbance, and are found in a variety of communities
Values of 4–6	Species that are typical of a successional phase of some native community; species that display fidelity to a particular community, but tolerate moderate disturbance
Values of 7–8	Species that are typical of relatively stable conditions; typical of well-established communities that have sustained only minor disturbance
Values of 9–10	Species that exhibit high degrees of fidelity to a narrow set of ecological conditions

From Andreas and Lichvar 1995.

are taken from a floristic checklist with pre-determined values for each species. The FQAI is calculated based on the community composition at a site irrespective of the proportional representation (evenness) of any given species, community type, abundance, dominance, growth form, showiness, or other factors. It is calculated using a complete species inventory. Each species is assigned its coefficient of conservatism and the index is calculated as follows:

$$FQAI = R/\,N^{1/2} \tag{10.1}$$

where
R = the sum of all the coefficients of conservatism for the community
N = the number of native species

Several projects have been completed testing the ability of the FQAI to reflect the relative level of disturbance in a given wetland. In a pilot study conducted by the Ohio Environmental Protection Agency, the FQAI was tested in ten forested riparian wetlands in eastern Ohio. Sites were selected along a gradient of disturbance from the least impacted to those highly disturbed by human activities (Fennessy et al. 1998a). Sites were assigned a qualitative disturbance rank based upon the land use surrounding the site, its land use history (e.g., had it been plowed at any point), and the degree of observed hydrological modification to the wetland and adjacent stream. All plant species were identified. A strong correlation was found between the relative disturbance rank of the site and its FQAI value ($R^2 = 0.92$; $p < 0.01$; Figure 10.5).

The FQAI was also tested in 22 depressional wetlands (both forested and emergent), again using sites chosen along a gradient of disturbance (Fennessy et al. 1998b). Project objectives included determining the most suitable sampling window, and linking the biological data with quantitatively measured stressors to the wetlands as well as with ecosystem processes. Stressors included nutrient and metal concentrations in the water column and sediments, and the proportion of human-dominated land use in the area surrounding

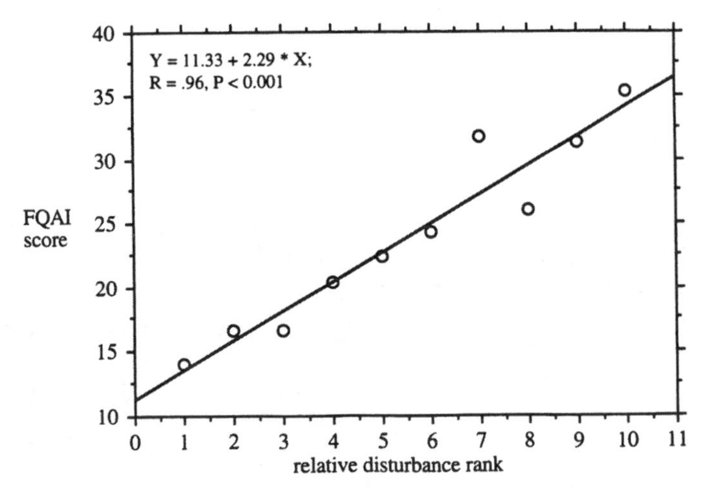

FIGURE 10.5
A regression analysis of FQAI scores as a function of relative disturbance ranks in ten forested wetlands in Ohio (Fennessy et al. 1998a).

the wetland. The FQAI was not correlated with differences in surface water chemistry ($p = 0.05$), but was positively correlated with soil total organic carbon ($p = 0.01$), phosphorus, and calcium ($p = 0.05$). Biomass is a measure that integrates the effects of many other processes as well as the trophic status of the wetland (Keddy et al. 1993). FQAI scores showed a negative correlation with biomass production (Figure 10.6), indicating that the FQAI index reflects ecosystem condition in an ecologically realistic way. As diversity declines and FQAI scores drop, the herbaceous community tends to become dominated by invasive exotic species capable of producing large amounts of biomass.

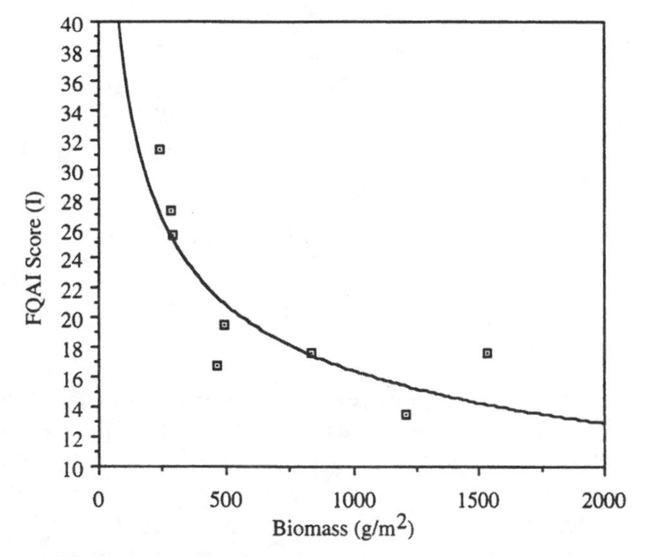

FIGURE 10.6
The relationship between standing crop (peak biomass) and FQAI scores in 15 depressional marshes in Ohio (Fennessy et al. 1998b).

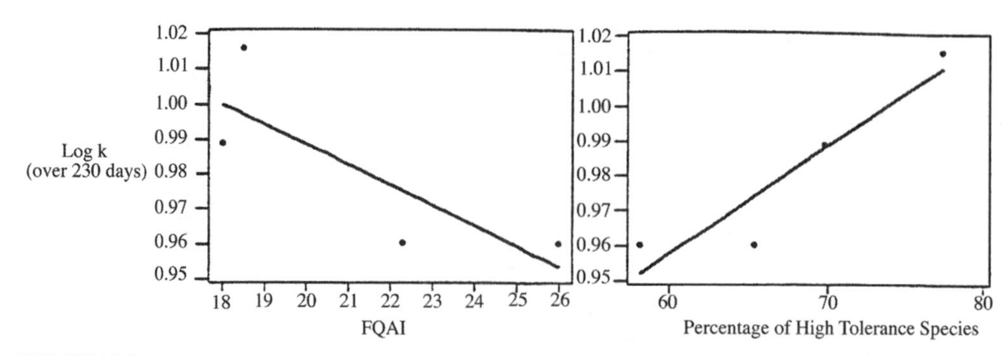

FIGURE 10.7
Relationship between litter mass lost and the plant-based biological indicators, the FQAI, and the percent high tolerant species (PHTS), defined as those with a coefficient of conservatism of 0, 1, 2, or 3. (From Bush and Fennessy, unpublished data.)

In a subsequent study, ecosystem processes such as primary productivity and decomposition rates were measured in a number of emergent sites in order to examine whether the FQAI was correlated with function (Bush and Fennessy, unpublished data). Decomposition rates and nutrient loss rates (P, N) from decomposing plant litter were also calculated. Decomposition rates (Figure 10.7) in the month following senescence, as well as the loss of P from that litter (Table 10.8), were significantly correlated with FQAI score. The results indicated that degraded wetlands (those with lower FQAI scores) were more productive, had higher rates of decomposition, and lost nutrients at a faster rate from decomposing tissue than less degraded systems.

TABLE 10.8

FQAI Scores and Nutrient Losses from Macrophyte Litter after 1 and 8 Months of Decomposition

Site	FQAI Score	% Decrease in Tissue P after 1 Month (mg/g)	% Decrease in Tissue P after 8 Months (mg/g)	% Decrease in Tissue N after 1 Month (mg/g)	% Decrease in Tissue N after 8 Months (mg/g)
Wetland A	18	56.6	72.9	19.8	11.8
Wetland B	18.5	55.7	72.5	21.1	16
Wetland C	22.6	18.3	66.6	4.6	2.5
Wetland D	27	14.8	65	26.8	16.9

Bush and Fennessy, unpublished data.

F. Using Biological Indicators to Assess Risk

Biological monitoring is a central tool in assessing ecological risk. Ecological risk assessment represents a shift from traditional risk assessment in which the focus is human health (and generally the effects of a single toxic substance) to a focus on ecological effects, or the impact on ecosystem integrity. The classical framework of risk assessment is being used to systematically study the effects of stress on ecosystem structure and function (Schaeffer 1991). Typically, risk assessment is done in a series of steps including measuring the characteristics of the ecosystem, assessing its ecological integrity, and identifying the stressors responsible for causing ecosystem degradation (Stevenson 1998). The U.S. Environmental

Protection Agency (1996) has established a framework of questions that are used in ecological risk assessment:

- Is there a problem?
- What is the nature of the problem (characterize exposure and any ecological effects)?
- What can we do about it (risk management)?

Risk assessment for wetlands can be assessed using the vegetation-based indices described above. Abiotic measures can also be used including water and sediment chemistry, temperature, basin morphometry, or hydrology. Measures of ecosystem processes may also be useful including primary productivity, decomposition rates, or nutrient cycling (Stevenson 1998).

Summary

Plant community composition can provide information on wetlands, including the location of their boundaries and the extent of anthropogenic disturbance that has occurred. Efforts to protect wetlands in the U.S. have focused on delineation of the wetland edge based on changes in vegetation composition that occur as a function of gradients in elevation, soil type, and wetness (Carter 1996). The wetland delineation procedure, which refers to the set of procedures used to identify wetland boundaries, is used to establish the extent of government jurisdiction. Delineation is based on the three-parameter approach, namely, that hydrophytic vegetation, hydric soils, and wetland hydrology must be present for an area to be considered a wetland. In the delineation procedure, ascertaining whether or not a plant community is hydrophytic is a pivotal decision. The identification of individual species as hydrophytes is made by using a compilation of plant species that ranks the probability of occurrence of each species in wetland habitats. For the U.S., these ratings for wetland plants are found in the "National List of Plant Species that Occur in Wetlands" (Reed 1997).

The composition of the plant community and the predictable changes that are the result of anthropogenic disturbance can also serve as sensitive biological indicators of the 'health' or ecological integrity of the wetland. Because communities consist of a diverse assemblage of species with different adaptations, ecological tolerances, and life history strategies, their composition can reflect (often with great sensitivity) the ecological integrity of the wetland. Plants are particularly useful biological indicators because they are a universal component of wetland ecosystems and occur with sufficient richness to provide clear and robust signals of human disturbance. They have been used effectively to distinguish environmental stressors including hydrologic alterations, excessive siltation, nutrient enrichment, and other types of disturbance. The use of biological data to evaluate ecosystem health (known as biological monitoring and assessment) is a powerful tool to measure and interpret the consequences of human activities on wetland ecosystems.

Case Study

10.A. The Development of a Vegetation IBI

The state of Minnesota has developed a vegetation-based multimetric assessment index known as the Index of Vegetative Integrity (IVI; Gernes 2000). This index is modeled after the Index of Biotic Integrity (IBI), a bioassessment technique used extensively in stream studies (Karr 1991). The IVI is composed of ten metrics based on different aspects of the plant community: two are based on taxa richness, four on life-form guilds, two on sensitive and tolerant species, and two on community structure. It is planned that the IVI will be used to monitor the status and trends of wetland conditions in the state.

The IVI is calculated based on the collection of field data. The final index score is the sum of individual scores for the ten metrics. Each metric is scored by assigning a value of 5, 3, or 1. The scoring system for each metric was calibrated based on data collected at least-impacted reference wetlands as well as wetlands selected to represent the full gradient of human disturbance. Because different metrics are measured in different units, and because some metrics increase in response to disturbance while others decrease, the scoring system for this type of index is done by comparing the value to that obtained at reference sites (Karr and Chu 1997). The range of values obtained for each metric was divided into three categories. Values that are closest to values obtained at reference sites receive a score of 5, while values that deviate the most from the reference condition receive a score of 1. Metrics that deviate somewhat from the reference sites receive a score of 3. A list of the metrics, the range of values obtained in field sampling for each, and the criteria for scoring each are shown in Table 10.A.1.

In both the development and use of this IVI, vegetation sampling is done using a 100-m^2 relevè plot in the emergent portion of the wetland being evaluated. According to the relevè method all species within the plot are inventoried, and estimates of their cover are made. These data are then used to score the site based on the ten metrics that make up the IVI. The metrics and a rationale for each are given below.

1. Vascular Genera Metric

Predicted response to human influence: a decrease in value of metric.

The vascular genera metric is based on the number of native vascular plant genera in the relevè plot. If a genus contains both native and non-native species it is included in the count. *Typha* is one such example. It contains two common wetland species, *T. latifolia,* which is native, and *T. angustifolia,* which is not. Non-wetland species (i.e., those with a wetland indicator status of FAC-, FACU, and UPL; Reed 1997) are excluded from the count. Care should be used when applying this metric to plant communities that are naturally low in diversity, because the results may indicate a low value when in fact the number of genera is naturally low. Examples of wetland plant communities with naturally low species numbers include lake sedge (*Carex lacustris*) fringe communities, wild rice (*Zizania palustris*) beds, and hardstem bulrush (*Scirpus acutus*) communities.

2. Nonvascular Taxa Metric

Predicted response to human influence: a decrease in value of metric.

The nonvascular taxa metric is based on the number of nonvascular taxa including liverworts, mosses, lichen taxa, and the macroscopic algae *Chara* and *Nitella*. In the Minnesota study, the maximum number of nonvascular taxa observed at any site was two.

TABLE 10.A.1
The Scoring Criteria Used for Ten Wetland Vegetation Metrics in the Minnesota IVI

			No. Sites That Scored in Each Category	
Metric	**Range of Values**	**Score Assigned**	**Reference Sites**	**Impaired Sites**
Number of vascular genera	>14	5	3	3
	9–14	3	3	7
	<9	1	0	10
Number of nonvascular taxa	>1	5	4	2
	1	3	2	7
	0	1	0	11
Sum of all *Carex* species (sum of cover)	>3	5	2	4
	0.1–2.9	3	4	7
	0	1	0	9
Taxa whose presence decreases with disturbance	>4	5	3	0
	1–3	3	3	9
	0	1	0	11
Proportion of tolerant taxa to total taxa	<25	5	4	0
	25.1–60	3	2	10
	>60	1	0	10
Number of grass, sedge, and rush species	>8	5	2	2
	2–7	3	4	10
	<2	1	0	8
Proportion of monocarpic species cover and richness divided by cove	>2.6	5	4	3
	2–2.59	3	2	9
	<2	1	0	8
Number of aquatic guild species	>6	5	4	1
	3–5	3	2	7
	<3	1	0	12
Equitability of cover within the sample	<0.07	5	3	4
	0.08–0.2	3	3	8
	>0.2	1	0	8
Sum of taxa with persistent litter (sum of cover)	<1	5	4	2
	1–5.9	3	2	11
	>6	1	0	7

Note: The range of values obtained in field sampling, the score assigned for each range, and the number of sites tested which fell into each category are shown.

From Gernes 2000.

Most of the sites affected by human activities such as agricultural or stormwater runoff contained no non-vascular taxa.

3. *Carex* Cover Metric

Predicted response to human influence: a decrease in value of metric.

The *Carex* cover metric is based on the sum of the percent cover for all *Carex* taxa sampled in the 100-m^2 plot. *Carex* are an important structural component of shallow emergent wetlands, and are known to be adversely affected by such environmental stressors as excessive siltation, hydrologic alteration, and nutrient enrichment (Wilcox 1995).

4. Grass-Like Species Metric

Predicted response to human influence: a decrease in value of metric.

The grass-like species metric is calculated by summing the number of grass (Poaceae), sedge (Cyperaceae), and rush (Juncaceae) species. This metric is based on the fact that this group of species is structurally very similar and on reports in the literature (Wilcox 1995) that native taxa in these plant families are frequently among the first to begin decreasing following human disturbance.

5. Monocarpic Species Metric

Predicted response to human influence: an increase in value of metric.

Monocarpic species are defined as those that flower only once in their life cycle, i.e., annual and biennial species. This metric is calculated by summing the number of native monocarpic species and their cover class values and dividing by the total cover of native monocarpic species. Gernes found that this metric responded strongly to hydrologic fluctuations, making it a sensitive indicator of sites with severe hydrologic disturbance.

6. Aquatic Guild Metric

Predicted response to human influence: a decrease in value of metric.

Working in Minnesota, Galatowitsch and McAdams (1994) recognized six distinct guilds of aquatic plants. Four of these, rooted submerged aquatics, unrooted submerged aquatics, floating perennials, and floating annuals, were used to develop the aquatic guild metric which is calculated by counting the number of native aquatic guild species. It was expected that the aquatic guild metric would be most responsive to water quality, although a regression analysis between the metric and the disturbance index was not significant.

7. Sensitive Taxa Metric

Predicted response to human influence: a decrease in value of metric.

This metric is based on the fact that some species are more sensitive to human disturbance than others. It is calculated based on the prevalence of species that are considered to be most susceptible to human disturbance. No non-native species are considered to be sensitive. The final list of sensitive species includes: *Asclepias incarnata, Dulichium arundinaceum, Eriophorum gracile, Scirpus tabernaemontani, Iris versicolor, Iris* sp., *Scutellaria galericulata, Utricularia macrorhiza, Calamagrostis canadensis, C. stricta, Glyceria striata, G. borealis, Polygonum sagittatum, P. scandens, P. amphibium, Riccia fluitans, Spiraea alba,* and *Rubus occidentallis*. Gernes found this to be among the metrics most highly correlated with the disturbance index.

8. Tolerant Species Metric

Predicted response to human influence: an increase in value of metric.

The tolerant species metric is based on the presence of species that are tolerant of disturbance (e.g., hydrologic fluctuation, the influx of toxic substances, nutrient inputs, or heavy sedimentation rates). The list of tolerant taxa used to calculate this metric was based on a literature review of the responses of plant species to human disturbance (including Wilcox 1995; Weisner 1993; Squires and van der Valk 1992) as well as field observations. All non-native species are considered to be tolerant. The percent of the community comprised of tolerant species values is calculated as follows:

$$\frac{\text{number of tolerant species} \times 100}{\text{total number of all taxa}}$$

Results show that between 11.1 and 100% of species at a site are in the tolerant group. Reference wetlands had proportionately fewer tolerant species than the impaired sites. This metric gave the strongest response (as measured by the fit of a least-squares regression analysis) of the ten vegetation metrics.

9. Dominance Metric

Predicted response to human influence: an increase in value of metric.

The dominance metric is based on the percent cover relative to the taxa richness for native species within the wetland, thus it is a measure of evenness. The formula for calculating dominance was taken from Odum (1971):

$$D = \Sigma(ni/N)^2$$

where ni = the percent cover for each taxa and N = the sum of all cover values for all taxa within the sampling plot.

The mathematical range of this function is between 0 and 1 with a more biologically diverse wetland scoring near 0 and more monotypic sites scoring near 1. This metric is considered to be only moderately reliable.

10. Persistent Litter Metric

Predicted response to human influence: increase in value of metric.

Persistent litter is defined as being resistant to decomposition. As such, it does not have the same quantity of available nutrients or detrital energy as readily decomposable litter. Taxa recognized as having persistent litter include: *Phragmites* (common reed), *Scirpus* (bulrushes), *Polygonum* (smartweeds), *Typha* (cattails), *Sparganium* (burreeds), and *Lythrum salicaria* (purple loosestrife). Although *Carex* is reported to have persistent litter, it was not counted as contributing persistent standing litter in this index. For this metric, the higher the cover of persistent litter, the lower the score. Results show that there was an increase in the amount of persistent litter at sites in response to human disturbance, and as such this measure was judged to be a reliable metric.

References

A.P.H.A. 1995. Standard Methods for the Examination of Water and Wastewater. American Association of Public Health, American Water Works Association, and the Water Pollution Control Federation. A. E. Greenberg, R. R. Trussell, and L. S. Clesceri, Eds. 1268 pp. Washington, D.C. American Public Health Association.

Abernethy, V.J., Sabbatini, M.R., and Murphy, K.J. 1996. Response of *Elodea canadensis* Michx. and *Myriophyllum spicatum* L. to shade, cutting and competition in experimental cultures. *Hydrobiologia* 340: 219–224.

Ackerman, J.D. 1995. Convergence of filiform pollen morphologies in seagrasses: functional mechanisms. *Evolutionary Ecology* 9: 139–153.

Ackerman, J.D. 1997a. Submarine pollination in the marine angiosperm *Zostera marina* (Zosteraceae). I. The influence of floral morphology on fluid flow. *American Journal of Botany* 84: 1099–1109.

Ackerman, J.D. 1997b. Submarine pollination in the marine angiosperm *Zostera marina* (Zosteraceae). II. Pollen transport in flow fields and capture by stigmas. *American Journal of Botany* 84: 1110–1119.

Adams, M.S. and McCracken, M.D. 1974. Seasonal production of the *Myriophyllum* component of the littoral of Lake Wingra, *Wisconsin Journal of Ecology* 62: 457–465.

Adamus, P.R. 1992. Choices in monitoring wetlands. In *Ecological Indicators.* D.H. McKenzie, D.L. Hyatt, and V.J. McDonald, Eds. pp. 571–592. New York. Elsevier Applied Science.

Adamus, P.R. 1996. Bioindicators for Assessing Ecological Integrity of Prairie Wetlands. EPA/600/R-96/082. Corvallis, OR. U.S. Environmental Protection Agency, National Health and Environmental Effects Research Laboratory, Western Ecology Division.

Adamus, P.R. in review. National Database of Wetland Plant Tolerances. EPA/600/R-xx/xxx. Corvallis, OR. U.S. Environmental Protection Agency, National Health and Environmental Effects Research Laboratory, Western Ecology Division.

Adamus, P.R. and Stockwell, L.T. 1983. A Method for Wetland Functional Assessment: Vol. I, Critical Review and Evaluation Concepts, 176 pp. Washington, D.C. Federal Highway Administration, U.S. Department of Transportation.

Adamus, P.R., Clairain, E.J., Smith, R.D., and Young, R.E. 1987. Wetland Evaluation Technique (WET): Vol. II, Methodology. Vicksburg, MS. U.S. Army Corps of Engineers, Waterways Experiment Station.

Adcock, P.W. and Ganf, G.G. 1994. Growth characteristics of three macrophyte species growing in a natural and constructed wetland system. *Water Science and Technology* 29: 95–102.

Adcock, P.W., Ryan, G.L., and Osborne, P.L. 1995. Nutrient partitioning in a clay-based surface flow wetland. *Water Science and Technology* 32: 203–209.

Agami, M. and Waisel, Y. 1984. Germination of *Najas marina* L. *Aquatic Botany* 19: 37–44.

Agami, M. and Waisel, Y. 1986. The role of mallard ducks (*Anas platyrhynchos*) in distribution and germination of seeds of the submerged hydrophyte *Najas marina* L. *Oecologia* 68: 473–475.

Agami, M. and Waisel, Y. 1988. The role of fish in distribution and germination of seeds of the submerged macrophytes *Najas marina* L. and *Ruppia maritima* L. *Oecologia* 76: 83–88.

Agami, M., Beer, S., and Waisel, Y. 1986. The morphology and physiology of turions in *Najas marina* L. in Israel. *Aquatic Botany* 26: 371–376.

Aiken, S.G., Newroth, P.R., and Wile, I. 1979. The biology of Canadian weeds. 34. *Myriophyllum spicatum* L. *Canadian Journal of Plant Science* 59: 201–215.

Albert, V.A., Williams, S.E., and Chase, M.W. 1992. Carnivorous plants: phylogeny and structural evolution. *Science* 257: 1491–1495.

Allen, E.D. and Spence, D.H.N. 1981. The differential ability of aquatic plants to utilize the inorganic carbon supply in fresh waters. *New Phytologist* 87: 269–283.

Allen, J.A., Chambers, J.L., and McKinney, D. 1994. Intraspecific variation in the response of *Taxodium distichum* seedlings to salinity. *Forest Ecology and Management* 70: 203–214.

Allen, K.R. 1951. The Horokiwi Stream: a study of a trout population. *Fisheries Bulletin of Wellington, New Zealand* 10: 1–238.

Aloi, J. 1990. A critical review of recent freshwater periphyton methods. *Canadian Journal of Fisheries and Aquatic Science* 47: 656–670.

Alongi, D.M. 1998. *Coastal Ecosystem Processes*, 419 pp. Boca Raton, FL. CRC Press.

Anderson, C.E. 1974. A review of structure in several North Carolina salt marsh plants. In *Ecology of Halophytes*. R. J. Reimold and W. H. Queen, Eds. pp. 307–344. New York. Academic Press.

Anderson, C.M. and Treshow, M. 1980. A review of environmental and genetic factors that affect height in *Spartina alterniflora* Loisel. (salt marsh cord grass). *Estuaries* 3: 168–176.

Anderson, D. 1982. Plant Communities of Ohio: A Preliminary Classification and Description, 183 pp. Columbus. Ohio Department of Natural Resources.

Andreas, B.K. and Lichvar, R.W. 1995. Floristic Index for Establishing Assessment Standards: A Case Study for Northern Ohio. Wetlands Research Program Technical Report WRP-DE-8. Vicksburg, MS. U.S. Army Corps of Engineers, Waterways Experiment Station.

Angiosperm Phylogeny Group. 1998. An ordinal classification for the families of flowering plants. *Annals of the Missouri Botanical Garden* 85: 531–553.

Ansola, G., Fernandez, C., and de Luis, E. 1995. Removal of organic matter and nutrients from urban wastewater by using an experimental emergent aquatic macrophyte system. *Ecological Engineering* 5: 13–19.

Anten, N.P.R., Schieving, F., Medina, E., Werger, M.J.A., and Schuffelen, P. 1995. Optimal leaf area indices in C_3 and C_4 mono- and dicotyledonous species at low and high nitrogen availability. *Physiologia Plantarum* 95: 541–550.

Ap Rees, T. and Wilson, P.M. 1984. Effects of reduced supply of oxygen on the metabolism of roots of *Glyceria maxima* and *Pisum sativum*. *Zeitschrift für Pflanzenphysiologie* 114: 493–503.

Arenovski, A.L. and Howes, B.L. 1992. Lacunal allocation and gas transport capacity in the salt marsh grass *Spartina alterniflora*. *Oecologia* 90: 316–322.

Armstrong, J. and Armstrong, W. 1991. A convective through-flow of gases in *Phragmites australis* (Cav.) Trin. ex Steud. *Aquatic Botany* 39: 75–88.

Armstrong, J., Armstrong, W., and Beckett, P. 1992. *Phragmites australis*: Venturi- and humidity induced pressure flows enhance rhizome aeration and rhizosphere oxidation. *New Phytologist* 120: 197–207.

Armstrong, J., Armstrong, W., Wu, Z., and Afreen-Zobayed, F. 1996. A role for phytotoxins in the *Phragmites* die-back syndrome? *Folia Geobotanica Phytotaxonomica* 31: 127–142.

Armstrong, W. 1978. Root aeration in the wetland condition. In *Plant Life in Anaerobic Environments*. D.D. Hook and R.M.M. Crawford, Eds. pp. 269–297. Ann Arbor, MI. Ann Arbor Science Publishers.

Armstrong, W. 1979. Aeration in higher plants. In *Advances in Botanic Research*. H.W. Woolhouse, Ed. pp. 225–332. London. Academic Press.

Armstrong, W., Armstrong, J., and Beckett, P. 1996. Pressurized aeration in wetland macrophytes: some theoretical aspects of humidity-induced convection and thermal transpiration. *Folia Geobotanica Phytotaxonomica* 31: 25–36.

Arteca, R.N. 1997. Flooding. In *Plant Ecophysiology*. M. N. V. Prasad, Ed. pp. 151–171. New York. John Wiley & Sons.

Ashton, P.J. and Mitchell, D.S. 1989. Aquatic plants: patterns and modes of invasion, attributes of invading species and assessment of control programmes. In *Biological Invasions—A Global Perspective*. J.A. Drake, H.A. Mooney, F. di Castri, R.H. Groves, F.J. Kruger, M. Rejmanek, and M. Williamson, Eds. pp. 111–154. Chichester. John Wiley & Sons.

Avery, T. 1967. *Forest Measurements*, 290 pp. New York. McGraw-Hill.

Badger, K.S. and Ungar, I.A. 1994. Seed bank dynamics in an inland salt marsh, with special emphasis on the halophyte *Hordeum jubatum* L. *International Journal of Plant Science* 155: 66–72.

Baldwin, A.H. and I.A. Mendelssohn. 1998. Effects of salinity and water level on coastal marshes: an experimental test of disturbance as a catalyst for vegetation change. *Aquatic Botany* 61: 255–268.

Barclay, A.M. and Crawford, R.M.M. 1982. Plant growth and survival under strict anaerobiosis. *Journal of Experimental Botany* 33: 541–549.

Barko, J.W. and Smart, R.M. 1978. The growth and biomass distribution of two emergent freshwater plants, *Cyperus esculentus* and *Scirpus validus*, on different sediments. *Aquatic Botany* 5: 109–117.

Barko, J.W. and Smart, R.M. 1979. The nutritional ecology of *Cyperus esculentus*, an emergent aquatic plant, grown on different sediments. *Aquatic Botany* 6: 13–28.

Barko, J.W. and Smart, R.M. 1980. Mobilization of sediment phosphorus by submersed freshwater macrophytes. *Freshwater Biology* 10: 229–238.
Barko, J.W. and Smart, R.M. 1981a. Comparative influences of light and temperature on the growth and metabolism of selected submersed freshwater macrophytes. *Ecological Monographs* 51: 219–235.
Barko, J.W. and Smart, R.M. 1981b. Sediment-based nutrition of submersed macrophytes. *Aquatic Botany* 10: 339–352.
Barko, J.W. and Smart, R.M. 1983. Effects of organic matter additions to sediment on the growth of aquatic plants. *Journal of Ecology* 71: 161–175.
Barko, J.W., Murphy, P.G., and Wetzel, R.G. 1977. An investigation of primary productivity and ecosystem metabolism in a Lake Michigan dune pond. *Archivs für Hydrobiologie* 81: 155–187.
Barko, J.W. and Smart, R.M. 1986. Sediment-related mechanisms of growth limitation in submersed macrophytes. *Ecology* 67: 1328–1340.
Barko, J.W., Adams, M.S. and Clesceri, N.L. 1986. Environmental factors and their consideration in the management of submersed aquatic vegetation: a review. *Journal of Aquatic Plant Management* 24: 1–10.
Barko, J.W., Smart, R.M., McFarland, D.G., and Chen, R.L. 1988. Interrelationships between the growth of *Hydrilla verticillata* L.F. Royal and sediment nutrient availability. *Aquatic Botany* 32: 205–216.
Barko, J.W., Gunnison, D., and Carpenter, S.R. 1991. Sediment interacters with submersed macrophyte growth and community dynamics. *Aquatic Botany* 41: 41–65.
Barko, J.W., Smith, C.S., and Chambers, P.A. 1994. Perspectives on submersed macrophyte invasions and declines. *Lake and Reservoir Management* 10: 1–3.
Barnabas, A.D. 1996. Casparian band-like structures in the root hypodermis of some aquatic angiosperms. *Aquatic Botany* 55: 217–225.
Barrat-Segretain, M.H. 1996. Strategies of reproduction, dispersion, and competition in river plants: a review. *Vegetatio* 123: 13–37.
Barrett, S.C.H. 1980a. Sexual reproduction in *Eichhornia crassipes* (water hyacinth). I. Fertility of clones from diverse regions. *Journal of Applied Ecology* 17: 101–112.
Barrett, S.C.H. 1980b. Sexual reproduction in *Eichhornia crassipes* (water hyacinth). II. Seed production in natural populations. *Journal of Applied Ecology* 17: 113–124.
Barrett, S.C.H., Eckert, C.G., and Husband, B.C. 1993. Evolutionary processes in aquatic plant populations. *Aquatic Botany* 44: 105–145.
Bartley, M.R. and Spence, D.H.N. 1987. Dormancy and propagation in helophytes and hydrophytes. *Archivs für Hydrobiologie* 27: 127–155.
Bartsch, I. and Moore, T.R. 1985. A preliminary investigation of primary production and decomposition in four peatlands near Schefferville, Quebec. *Canadian Journal of Botany* 63: 1241–1248.
Baskin, Y. 1994. California's ephemeral vernal pools may be a good model for speciation. *BioScience* 44: 384–388.
Bayley, S.E., Zoltek, J., Hermann, T., Dolan, T., and Tortora, L. 1985. Experimental manipulation of nutrients and water in a freshwater marsh: effects on biomass, decomposition, and nutrient accumulation. *Limnology and Oceanography* 30: 500–512.
Bazzaz, F.A. 1986. Life history of colonizing plants: some demographic, genetic, and physiological features. In *Ecology of Biological Invasions of North America and Hawaii*. H.A. Mooney and J.A. Drake, Eds. pp. 96–110. New York. Springer-Verlag.
Beadle, N.C.W. 1981. *The Vegetation of Australia*, 690 pp. London. Cambridge University Press.
Bedford, B. 1996. The need to define hydrologic equivalence at the landscape scale for freshwater wetland mitigation. *Ecological Applications* 6: 57–68.
Bedford, B.L. 1999. Cumulative effects on wetland landscapes: links to wetland restoration in the United States and southern Canada. *Wetlands* 19: 775–788.
Beeftink, W.G. 1977. The coastal salt marshes of Western and Northern Europe: an ecological and phytosociological approach. In *Ecosystems of the World 1: Wet Coastal Ecosystems*. V.J. Chapman, Ed. pp. 109–149. New York. Elsevier Science Publishing.

Beer, S. and Wetzel, R.G. 1981. Photosynthetic carbon metabolism in the submerged aquatic angiosperm *Scirpus subterminalis*. *Plant Science Letter* 21: 199–207.
Bellamy, D.J. 1968. An ecological approach to the classification of the lowland mires of Europe. In *Proceedings of the Third International Peat Congress*, pp. 74–79. Quebec, Canada. National Research Council, Ottawa.
Bertness, M.D. 1985. Fiddler crab regulation of *Spartina alterniflora* production on a New England salt marsh. *Ecology* 66: 1042–1055.
Bertness, M.D. 1991a. Zonation of *Spartina patens* and *Spartina alterniflora* in a New England salt marsh. *Ecology* 71: 138–148.
Bertness, M.D. 1991b. Interspecific interactions among high marsh perennials in a New England salt marsh. *Ecology* 72: 125–137.
Bertness, M.D. and Ellison, A.M. 1987. Determinants of pattern in a New England salt marsh plant community. *Ecological Monographs* 57: 129–147.
Best, E.P. 1988. The phytosociological approach to the description and classification of aquatic macrophyte vegetation. In *Vegetation of Inland Waters*. J.J. Symoens, Ed. pp. 155–182. Dordrecht. Kluwer Academic Publishers.
Biddlestone, A.J., Gray, K.R., and Thayanithy, K. 1991. A botanical approach to the treatment of wastewaters. *Journal of Biotechnology* 17: 209–220.
Bird, K.T. 1993. Comparisons of herbicide toxicity using *in vitro* cultures of *Myriophyllum spicatum*. *Journal of Aquatic Plant Management* 31: 43–45.
Bischoff, W.D., Paterson, V.L., and Mackenzie, F.T. 1984. Geochemical mass balance for sulfur- and nitrogen-bearing acid components: eastern United States. In *Geological Aspects of Acid Deposition*, Vol. 7. O.P. Bricker, Ed. pp. 1–21. Boston. Butterworth.
Blanken, P.D. and Rouse, W.R. 1996. Evidence of water conservation mechanisms in several subarctic wetland species. *Journal of Applied Ecology* 33: 842–850.
Bliss, S.A. and Zedler, P.H. 1998. The germination process in vernal pools: sensitivity to environmental conditions and effects on community structure. *Oecologia* 113: 67–73.
Blossey, B. and Schat, M. 1997. Performance of *Galerucella calmariensis* (Coleoptera: Chrysomelidae) on different North American populations of purple loosestrife. *Entomological Society of America* 26: 439–445.
Blum, U., Seneca, E.D., and Stroud, L.M. 1978. Photosynthesis and respiration of *Spartina* and *Juncus* salt marshes in North Carolina: some models. *Estuaries* 1: 228–238.
Bolen, E.G., Smith, L.M., and Schramm, H.L., Jr. 1989. Playa lakes: prairie wetlands of the southern high plains. *BioScience* 39: 615–623.
Bolton, J.K. and Brown, R.H. 1980. Photosynthesis of grass species differing in carbon dioxide fixation pathway. V. Response of *Panicum maximum, Panicum miloides* and tall fescue (*Festuca arundinacea*) to nitrogen nutrition. *Plant Physiology* 66: 97–100.
Bolton, K.G.E. and Greenway, M. 1997. A feasibility study of *Melaleuca* trees for use in constructed wetlands in subtropical Australia. *Water Science and Technology* 35: 247–254.
Bonomo, L., Pastorelli, G., and Zambon, N. 1997. Advantages and limitations of duckweed-based wastewater treatment systems. *Water Science and Technology* 35: 239–246.
Bornette, G., Amoros, C., and Lamourous, N. 1998. Aquatic plant diversity in riverine wetlands: the role of connectivity. *Freshwater Biology* 39: 267–283.
Bos, R. 1997. The human health impact of aquatic weeds. In International Water Hyacinth Consortium. Washington, D.C. U.S. Department of Agriculture website: http://www.sidney.ars.usda.gov/scientists/nspencer/water_h/consortium.htm.
Bouchard, V. and Mitsch, W.J. 1999. Plant richness and community establishment after five growing seasons in the two experimental wetland basins. In *Olentangy River Wetland Research Park at The Ohio State University*. W.J. Mitsch and V. Bouchard. Eds. pp. 43–51. Columbus. Ohio State University.
Bouny, J.M. and Saglio, P.H. 1996. Glycolytic flux and hexokinase activities in anoxic maize root tips acclimated by hypoxic pretreatment. *Plant Physiology* 111: 187–194.
Bowes, G. and Salvucci, M.E. 1984. *Hydrilla*: inducible C4–type photosynthesis without Kranz anatomy. In *Advances in Photosynthesis Research*, Vol. III. C. Sybesma, Ed. pp. 829–832. The Hague. Dr. W. Junk Publishers.

Bowes, G. and Salvucci, M.E. 1989. Plasticity in the photosynthetic carbon metabolism of submersed aquatic macrophytes. *Aquatic Botany* 34: 233–266.

Bowmer, K.H. 1987. Nutrient removal from effluents by an artificial wetland: influence of the rhizosphere aeration and preferential flow studied using bromide and dye tracers. *Water Resources* 21: 591–599.

Boyd, C.E. 1970. Production, mineral accumulation and pigment concentrations in *Typha latifolia* and *Scirpus americanus*. *Ecology* 51: 285–290.

Boyd, C.E. 1971. Further studies on productivity, nutrient and pigment relationships in *Typha latifolia* populations. *Bulletin of the Torrey Botanical Club* 98: 144–150.

Boyd, C.E. 1978. Chemical composition of wetland plants. In *Freshwater Wetlands: Ecological Processes and Management Potential*. R.E. Good, D.F. Whigham, and R.L. Simpson, Eds. pp. 155–167. New York. Academic Press.

Boyd, C.E. and Vickers, D.H. 1971. Relationships between production, nutrient accumulation, and chlorophyll synthesis in an *Eleocharis quadrangulata* population. *Canadian Journal of Botany* 49: 883–888.

Boyer, K.E. and Zedler, J.B. 1998. Effects of nitrogen additions on the vertical structure of a constructed cordgrass marsh. *Ecological Applications* 8: 692–705.

Boyer, K.E. and Zedler, J.B. 1999. Nitrogen addition could shift plant community composition in a restored California salt marsh. *Restoration Ecology* 7: 74–85.

Boyt, F.L., Bayley, S.E., and Zoltek, J. 1977. Removal of nutrients from treated wastewater by wetland vegetation. *Journal of the Water Pollution Control Federation* 49: 789–799.

Bradbury, I.K. and Grace, J.B. 1993. Primary production in wetlands. In *Mires: Swamp, Bog, Fen and Moor. Ecosystems of the World*, Vol. 4A. A. Gore, Ed. pp. 285–310. Amsterdam. Elsevier Science.

Bradbury, I.K. and Hofstra, G. 1976. Vegetation death and its importance in primary production measurements. *Ecology* 57: 209–211.

Bradley, P.M. and Dunn, E.L. 1989. Effects of sulfide on the growth of three salt marsh halophytes of the southeastern United States. *American Journal of Botany* 76: 1701–1713.

Bradley, P.M and Morris, J.T. 1990. Influence of oxygen and sulfide concentration on nitrogen uptake kinetics in *Spartina alterniflora*. *Ecology* 71: 282–287.

Bradley, P.M. and Morris, J.T. 1991a. Relative importance of ion exclusion, secretion and accumulation in *Spartina alterniflora* Loisel. *Journal of Experimental Botany* 42: 1525–1532.

Bradley, P.M. and Morris, J.T. 1991b. The influence of salinity on the kinetics of NH_4^+ uptake in *Spartina alterniflora*. *Oecologia* 85: 375–380.

Bradshaw, W.E. and Creelman, R.A. 1984. Mutualism between the carnivorous purple pitcher plant and its inhabitants. *The American Midland Naturalist* 112: 294–304.

Braendle, R. and Crawford, R.M.M. 1987. Rhizome anoxia tolerance and habitat specialization in wetland plants. In *Plant Life in Aquatic and Amphibious Habitats*. R.M.M. Crawford, Ed. pp. 397–410. Oxford. Blackwell Scientific.

Bray, J.R. 1963. Root production and the estimation of net productivity. *Canadian Journal of Botany* 41: 65–71.

Breen, C.M., Rogers, K.H., and Ashton, P.J. 1988. Vegetation processes in swamps and flooded plains. In *Vegetation of Inland Waters*. J.J. Symoens, Ed. pp. 223–247. Dordrecht. Kluwer Academic Publishers.

Breen, P. 1990. A mass balance method for assessing the potential of artificial wetlands for wastewater treatment. *Water Research* 24: 689–697.

Brewer, J.S. 1996. Site differences in the clone structure of an emergent sedge, *Cladium jamaicense*. *Aquatic Botany* 55: 79–91.

Bridgham, S.D. and Richardson, C.J. 1993. Hydrology and nutrient gradients in North Carolina peatlands. *Wetlands* 13: 207–218.

Bridgham, S.D., Pastor, J.A., Janssens, J., Chapin, C., and Malterer, T.J. 1996. Multiple limiting gradients in peatlands: a call for a new paradigm. *Wetlands* 16: 45–65.

Brinson, M.M. 1990. Riverine forests. In *Forested Wetlands Ecosystems of the World*. A.E. Lugo, M.M. Brinson, and S. Brown, Eds. pp. 87–141. Amsterdam. Elsevier Science.

Brinson, M.M. 1993a. A Hydrogeomorphic Classification for Wetlands. Technical Report WRP-DE-4. Washington, D.C. U.S. Army Corps of Engineers.

Brinson, M.M. 1993b. Changes in the functioning of wetlands along environmental gradients. *Wetlands* 13: 65–74.

Brinson, M.M., A.E. Lugo, and S. Brown. 1981. Primary productivity, decomposition, and consumer activity in freshwater wetlands. *Annual Review of Ecology and Systematics* 12: 123–161.

Bristow, J.M. and Whitcombe, M. 1971. The role of roots in the nutrition of aquatic vascular plants. *American Journal of Botany* 58: 8–13.

Britton, R.H. and Crivelli, A.J. 1993. Wetlands of southern Europe and North Africa: Mediterranean wetlands. In *Wetlands of the World I: Inventory, Ecology and Management*. D.F. Whigham, D. Dykyjova, and S. Hejny, Eds. pp. 129–194. Dordrecht. Kluwer Academic Publishers.

Brix, H. 1986. Treatment of wastewater in the rhizosphere of wetland plants—the root-zone method. *Water Science and Technology* 19: 107–118.

Brix, H. 1990. Gas exchange through the soil-atmosphere interphase and through dead culms of *Phragmites australis* in a constructed reed bed receiving domestic sewage. *Water Resources* 24: 259–266.

Brix, H. 1993. Macrophyte-mediated oxygen transfer in wetlands: transport mechanisms and rates. In *Constructed Wetlands for Water Quality Improvement*. G.A. Moshiri, Ed. pp. 393–398. Boca Raton, FL. Lewis Publishers.

Brix, H. 1997. Do macrophytes play a role in constructed treatment wetlands? *Water Science and Technology* 35: 11–17.

Brix, H., Sorrell, B.K., and Orr, P.T. 1992. Internal pressurization and convective gas flow in some emergent freshwater macrophytes. *Limnology and Oceanography* 37: 1420–1433.

Brix, H., Sorrell, B.K., and Schierup, H.H. 1996. Gas fluxes achieved by *in situ* convective flow in *Phragmites australis*. *Aquatic Botany* 54: 151–163.

Brodrick, S.J., Cullen, P., and Maher, W. 1988. Denitrification in a natural wetland receiving secondary treated effluent. *Water Resources* 22: 431–439.

Broome, S.W., Seneca, E.D., and Woodhouse, W.W., Jr. 1983. The effects of source, rate and placement of nitrogen and phosphorus fertilizers on growth of *Spartina alterniflora*. *Estuaries* 6: 212–226.

Broome, S.W., Seneca, E.D., and Woodhouse, W.W., Jr. 1988. Tidal salt marsh restoration. *Aquatic Botany* 32: 1–22.

Brown, M.T. 1985. Managing landscapes for humanity and nature: the role of wetlands in regional nutrient dynamics. In *Proceedings of Wetlands of the Chesapeake: Protecting the Future of the Bay*. CFW-85-12. pp. 63–75. Gainesville. Center for Wetlands, University of Florida.

Brown, S. 1981. A comparison of the structure, primary productivity, and transpiration of cypress ecosystems in Florida. *Ecological Monographs* 51: 403–427.

Brown, S. 1990. Structure and dynamics of basin forested wetlands in North America. In *Forested Wetlands Ecosystems of the World*. A.E. Lugo, M.M. Brinson, and S. Brown, Eds. pp. 171–199. Amsterdam. Elsevier Science.

Brown, S. and Peterson, D.L. 1983. Structural characteristics and biomass production of two Illinois bottomland forests. *The American Midland Naturalist* 110: 107–117.

Brown, S.L., Chaney, R.L., Angle, J.S., and Baker, A.J.M. 1994. Phytoremediation potential of *Thlaspi caerulescens* and bladder campion for zinc- and cadmium-contaminated soil. *Journal of Environmental Quality* 23: 1151–1157.

Brueske, C.C. and Barrett, G.W. 1994. Effects of vegetation and hydrologic load on sedimentation patterns in experimental wetland ecosystems. *Ecological Engineering* 3: 429–447.

Buckley, G.P., Howell, R., and Anderson, M.A. 1997. Vegetation succession following ride edge management in lowland plantations and woods. 2. The seed bank resource. *Biological Conservation* 82: 305–316.

Burchett, M.D., Clarke, C.J., Field, C.D., and Pulkownik, A. 1989. Growth and respiration in two mangrove species at a range of salinities. *Physiologia Plantarum* 75: 299–303.

Burdick, D.M. and Mendelssohn, I.A. 1990. Relationship between anatomical and metabolic responses to waterlogging in the coastal grass *Spartina patens*. *Journal of Experimental Botany* 41: 223–228.

Burg, M.E., Tripp, D.R., and E.S. Rosenberg. 1980. Plant associations and primary productivity of the Nisqually salt marsh of southern Puget Sound, Washington. *Northwest Science* 54: 222–236.

Busch, D.E. 1992. Analyses of the Structure and Function of Lower Colorado River Riparian Plant Communities, Ph.D. dissertation. 241 pp. Las Vegas. University of Nevada.

Busch, D.E. and Smith, S.D. 1995. Mechanisms associated with the decline of woody species in riparian ecosystems of the southwestern U.S. *Ecological Monographs* 65: 347–370.

Bush, C.B. and Fennessy, M.S. Unpublished data.

Cahoon, D.R. and Stevenson, J.C. 1986. Production, predation, and decomposition in a low-salinity *Hibiscus* marsh. *Ecology* 67: 1341–1350.

Cairns, J. Jr. 1986. The myth of the most sensitive species: multispecies testing can provide valuable evidence for protecting the environment. *BioScience* 36: 670–672.

Callaway, J.C. and Zedler, J.B. 1997. Interactions between a salt marsh native perennial (*Salicornia virginica*) and an exotic annual (*Polypogon monspeliensis*) under varied salinity and hydroperiod. *Wetlands Ecology and Management* 5: 179–194.

Callaway, R.M. and King, L. 1996. Temperature-driven variation in substrate oxygenation and the balance of competition and facilitation. *Ecology* 77: 1189–1195.

Callaway, R.M., Jones, S., Ferren W.F., Jr., and Parikh, A. 1990. Ecology of a mediterranean-climate estuarine wetland at Carpinteria, California: plant distributions and soil salinity in the upper marsh. *Canadian Journal of Botany* 68: 1139–1146.

Campbell, B.D. and Grime, J.P. 1992. An experimental test of plant strategy theory. *Ecology* 73:15–29.

Cantelmo, A.J. and Ehrenfeld, J.G. 1999. Effects of microtopography on mycorrhizal infection in Atlantic white cedar (*Chamaecyparis thyoides* (L.) Mills). *Mycorrhiza* 8: 175–180.

Cargill, S.M. and Jefferies, R.L. 1984. Nutrient limitation of primary production in a sub-arctic salt marsh. *Journal of Applied Ecology* 21: 657–668.

Carignan, R. and Kalff, J. 1980. Phosphorus sources for aquatic weeds: water or sediments. *Science* 207: 987–989.

Carlisle, B.K., Hicks, A.L., Smith, J.P., Garcia, S.R., and Largay, B.G. 1999. Plants and aquatic invertebrates as indicators of wetland biological integrity in Waquoit Bay watershed, Cape Cod. *Environment Cape Cod* 2: 30–60.

Carpenter, S.R. 1980a. The decline of *Myriophyllum spicatum* in a eutrophic Wisconsin lake. *Canadian Journal of Botany* 58: 527–535.

Carpenter, S.R. 1980b. Estimating net shoot production by a hierarchical cohort method of herbaceous plants subject to high mortality. *The American Midland Naturalist* 104: 163–175.

Carpenter, S.R. and Lodge, D.M. 1986. Effects of submersed macrophytes on ecosystem processes. *Aquatic Botany* 26: 341–370.

Carpenter, S.R., Elser, J.J., and Olson, K.M. 1983. Effects of roots of *Myriophyllum verticillatum* L. on sediment redox conditions. *Aquatic Botany* 17: 243–249.

Carpenter, S.R., Fisher, S.G., Grimm, N.B., and Kitchell, J.B. 1992. Global change and freshwater ecosystems. *Annual Review of Ecology and Systematics* 23: 119–139.

Carr, G.M., Duthie, H.C., and Taylor, W.D. 1997. Models of aquatic plant productivity: a review of the factors that influence growth. *Aquatic Botany* 59: 195–215.

Carrick, H.J. and Lowe, R.L. 1988. Response of Lake Michigan benthic algae to *in situ* enrichment with Si, N, and P. *Canadian Journal of Fisheries and Aquatic Science* 45: 271–279.

Carter, C., Gammon, P.T., and Garrett, M.K. 1994. Ecotone dynamics and boundary determination in the Great Dismal Swamp. *Ecological Applications* 4: 189–203.

Carter, M.F. and Grace, J.B. 1990. Relationships between flooding tolerance, life history, and short-term competitive performance in three species of *Polygonum*. *American Journal of Botany* 77: 381–387.

Carter, M.L., Burns, T., Cavinder, K., Dugger, P., Fore, D., Revells, H., and Schmidt, T. 1973. Ecosystem Analysis of the Big Cypress Swamp and Estuaries, EPA 904/9–74–002, Region IV, Atlanta, GA. U.S. Environmental Protection Agency.

Carter, V. 1996. Environmental gradients, boundaries and buffers: an overview. In *Wetlands: Environmental Gradients, Boundaries and Buffers*. G. Mulamoottil, B.G. Warner, and E.A. McBean, Eds. pp. 9–17. New York. Lewis Publishers.

Carter, V. and Rybicki, N. 1986. Resurgence of submersed aquatic macrophytes in the tidal Potomac River, Maryland, Virginia, and the District of Columbia. *Estuaries* 9: 368–375.

Carter, V., Bedinger, M.S., Novitzki, R.P., and Wilen, W.O. 1979. Water resources and wetlands. In *Wetland Functions and Values: The State of Our Understanding*. P.E. Greeson, J.R. Clark, and J.E. Clark, Eds. pp. 344–376. Minneapolis, MN. American Water Resources Association.

Carter, V., Gammon, P.T., and Garrett, M.K. 1994. Ecotone dynamics and boundary determination in the Great Dismal Swamp, Virginia and North Carolina. *Ecological Applications* 4:189–203.

Castellanos, E.M., Figueroa, M.E., and Davy, A.J. 1994. Nucleation and facilitation in saltmarsh succession: interactions between *Spartina maritima* and *Arthrocnemum perenne*. *Journal of Ecology* 82: 239–248.

Cathcart, T.P., Hammer, D.A., and Triyono, S. 1994. Performance of a constructed wetland—vegetated strip system used for swine waste treatment. In *Proceedings of a Workshop on Constructed Wetlands for Animal Waste Management*. P.J. DuBowy and R.P. Reaves, Eds. pp. 9–22. Lafayette, IN. Purdue University.

Cavalieri, A.J. and Huang, H.C. 1981. Accumulation of proline and glycinebetaine in *Spartina alterniflora* Loisel. in response to NaCl and nitrogen in the marsh. *Oecologia* 49: 224–228.

Center, T.D., Dray, F.A., and Vandiver V.V., Jr. 1998a. Biological Control with Insects: The Water Hyacinth Moth, Vol. 1999. Department of Agronomy, Florida Cooperative Extension Service, Institute of Food and Agricultural Sciences, University of Florida. Website: http://hammock.ifas.ufl.edu/txt/fairs/2123.

Center, T.D., Dray, F.A., and Vandiver V.V., Jr. 1998b. Biological Control with Insects: the Water Weevils, Vol. 1999. Department of Agronomy, Florida Cooperative Extension Service, Institute of Food and Agricultural Sciences, University of Florida. Website: http://hammock.ifas.ufl.edu/txt/fairs/2123.

Chambers, P.A. 1987. Light and nutrients in the control of aquatic plant community structure. II. *In situ* observations. *Journal of Ecology* 75: 621–628.

Chambers, P.A. and Kalff, J. 1985. Depth distribution and biomass of submersed aquatic macrophyte communities in relation to Secchi depth. *Canadian Journal of Fisheries and Aquatic Science* 42: 701–709.

Chambers, P.A. and Prespas, E.E. 1988. Underwater spectral attenuation and its effect on the maximum depth of angiosperm colonization. *Canadian Journal of Fisheries and Aquatic Science* 45: 1010–1017.

Chambers, P.A., Prespas, E.E., Hamilton, H.R., and Bothwell, M.L. 1991. Current velocity and its effect on aquatic macrophytes in flowing waters. *Ecological Applications* 1: 249–257.

Chambers, P.A., Barko, J.W., and Smith, C.S. 1993. Evaluation of submersed aquatic macrophytes. *Journal of Aquatic Plant Management* 31: 218–220.

Chambers, R.M., Smith, S.V., and Hollibaugh, J. 1994. An ecosystem-level context for tidal exchange studies in salt marshes of Tomales Bay, California, USA. In *Global Wetlands: Old World and New*. W. J. Mitsch, Ed. pp. 265–276. Amsterdam. Elsevier Science.

Chambers, R.M., Meyerson, L.A., and Saltonstall, K. 1999. Expansion of *Phragmites australis* into tidal wetlands of North America. *Aquatic Botany* 64: 261–273.

Chapin, C.T. and Pastor, J. 1995. Nutrient limitations in the northern pitcher plant *Sarracenia purpurea*. *Canadian Journal of Botany* 73: 728–734.

Chapman, V. 1974. Salt marshes and salt deserts of the world. In *Ecology of Halophytes*. R. Reimold and W. Queen, Eds. pp. 3–19. New York. Academic Press.

Chapman, V.J. 1976. *Mangrove Vegetation*, 447 pp. Vaduz. J. Crammer Publisher.

Charudattan, R. 1997. Bioherbicides for the control of water hyacinth: feasibility and needs. In International Water Hyacinth Consortium. Washington, D.C. U.S. Department of Agriculture. Website: http://www.sidney.ars.usda.gov/scientists/nspencer/water_h/consortium.htm.

Chen, R.L. and Barko, J.W. 1988. Effects of freshwater macrophytes on sediment chemistry. *Journal of Freshwater Ecology* 4: 279–289.

Chiang, C., Craft, C.B., Rogers, D.W., and Richardson, C.J. 2000. Effects of 4 years of nitrogen and phosphorus additions on Everglades plant communities. *Aquatic Botany* 65: 61–78.

Chow, V.T. 1964. *Handbook of Applied Hydrology*, 1453 pp. New York. McGraw–Hill.

Christensen, K.K. and Wigand, C. 1998. Formation of root plaques and their influence on tissue phosphorus content in *Lobelia dortmanna*. *Aquatic Botany* 61: 111–122.

Christopher, S.V. and Bird, K.T. 1992. The effects of herbicides on development of *Myriophyllum spicatum* L. cultured *in vitro*. *Journal of Environmental Quality* 21: 203–207.

Chung, C.H. 1993. Thirty years of ecological engineering with *Spartina* plantations. *Ecological Engineering* 2: 261–290.

Cicerone, R.J. and Shetter, J.D. 1981. Sources of atmospheric methane: measurements in rice paddies and a discussion. *Journal of Geophysical Research* 86: 7203–7209.

Cintrón, G., Lugo, A.E., Pool, D.J., and Morris, G. 1978. Mangroves of arid environments in Puerto Rico and adjacent islands. *Biotropica* 10: 110–121.

Cipollini, D.F., Newell, S.J. and Nastase, A.J. 1994. Total carbohydrates in nectar of *Sarracenia purpurea* L. (northern pitcher plant). *The American Midland Naturalist* 131: 374–377.

Cizkova, H., Strand, J.A., and Lukavska, J. 1996. Factors associated with reed decline in a eutrophic fishpond, Rozmberk (South Bohemia, Czech Republic). *Folia Geobotanica Phytotaxonomica* 31: 73–84.

Clark, W.R. 1994. Habitat selection by muskrats in experimental marshes undergoing succession. *Canadian Journal of Zoology* 72: 675–680.

Clarke, E. 1999. Constructed Wetlands for Treating Dairy Wastewater: Suitability of Plant Species and Treatment Efectiveness, Master's thesis. Department of Marine, Estuarine, and Environmental Science, 100 pp. College Park. University of Maryland.

Clayton, J.S. and Bagyaraj, D.J. 1984. Vesicular-arbuscular mycorrhizas in submerged aquatic plants of New Zealand. *Aquatic Botany* 19: 251–262.

Clements, F.E. 1904. The development and structure of vegetation. *Botanical Survey of Nebraska* 7: 5–175.

Clements, F.E. 1916. *Plant Succession*, 512 pp. Carnegie Institute of Washington Publ. 242.

Clevering, O.A. 1998. Effects of litter accumulation and water table on morphology and productivity of *Phragmites australis. Wetlands Ecology and Management* 5: 275–287.

Cleverly, J.R., Smith, S.D., Sala, A., and Devitt, D.A. 1997. Invasive capacity of *Tamarix ramosissima* in a Mojave Desert floodplain: the role of drought. *Oecologia* 111: 12–18.

Clewell, A.F. and Lea, R. 1990. Creation and restoration of forested wetland vegetation in the southeastern United States. *In Wetland Creation and Restoration*. J.A. Kusler and M.E. Kentula, Eds. pp. 195–231. Washington, D.C. Island Press.

Clinton, W.J. 1999. Executive Order: Invasive Species, February 3, 1999. White House Press Release. Washington, D.C. U.S. Government Printing Office.

Clymo, R.S. 1970. The growth of *Sphagnum*: methods of measurement. *Journal of Ecology* 58: 13–49.

Clymo, R.S. and Hayward, P.M. 1982. The ecology of *Sphagnum*. In *Bryophyte Ecology*. A.J.E. Smith, Ed. pp. 229–289. London. Chapman & Hall.

Cockburn, W. 1985. Variation in photosynthetic acid metabolism in vascular plants: CAM and related phenomena. *New Phytologist* 101: 3–24.

Colinvaux, P. 1993. *Ecology 2*, 688 pp. New York. John Wiley & Sons.

Confer, S.R. and Niering, W.A. 1992. Comparison of created and natural freshwater emergent wetlands in Connecticut. *Wetlands Ecology and Management* 2: 143–156.

Connell, J.H. 1961. The influence of interspecific competition and other factors on the distribution of the barnacle *Chthamalus stellatus. Ecology* 42: 710–723.

Connell, J.H. 1990. *Apparent versus "Real" Competition in Plants. Perspectives on Plant Competition.* J.B. Grace and D. Tilman, Eds. pp. 9–26. San Diego, CA. Academic Press.

Connell, J.H. and Slatyer, R.O. 1977. Mechanisms of succession in natural communities and their role in community stability and organization. *The American Naturalist* 111: 1119–1144.

Connell, J.H., Noble, I.R., and Slatyer, R.O. 1987. On the mechanisms producing successional change. *Oikos* 50: 136–137.

Conner, W.H. and Day, J.W., Jr. 1976. Productivity and composition of a bald cypress-water tupelo site and a bottomland hardwood site in a Louisiana swamp. *American Journal of Botany* 63: 1354–1364.

Conner, W.H. and Day, J.W., Jr. 1982. The ecology of forested wetlands in the southeastern United States. In *Wetlands: Ecology and Management*. B. Gopal, R.E. Turner, R.G. Wetzel and D.F. Whigham, Eds. pp. 69–87. Jaipus, India. National Institute of Ecology and International Scientific Publications.

Conner, W.H. and Day, J.W., Jr., 1992. Diameter growth of *Taxodium distichum* (L.) Rich. and *Nyssa aquatica* L. from 1979–1985 in four Louisiana swamp stands. *The American Midland Naturalist* 127: 290–299.

Conner, W.H., Gosselink, J.G., and Parrondo, R.T. 1981. Comparison of the vegetation of three Louisiana swamp sites with different flooding regimes. *American Journal of Botany* 68: 320–331.

Constable, J.V.H. and Longstreth, D.J. 1994. Aerenchyma carbon dioxide can be assimilated in *Typha latifolia* L. leaves. *Plant Physiology* 106: 1065–1072.

Constable, J.V.H., Grace, J.B., and Longstreth, D.J. 1992. High carbon dioxide concentrations in aerenchyma of *Typha latifolia. American Journal of Botany* 79: 415–418.

Conway, V.M. 1949. The bogs of central Minnesota. *Ecological Monographs* 19: 173–206.

Cook, C.D.K. 1982. Pollination mechanisms in the Hydrocharitaceae. In *Studies on Aquatic Vascular Plants*. J.J. Symoens, S.S. Hooper, and P. Compere, Eds. pp. 1–15. Brussels. Royal Botanical Society of Belgium.

Cook, C.D.K. 1985. Range extensions of aquatic vascular plant species. *Journal of Aquatic Plant Management* 23: 1–6.

Cook, C.D.K. 1988. Wind pollination in aquatic angiosperms. *Annals of the Missouri Botanical Garden* 75: 768–777.

Cook, C.D.K. 1993. Origin, autecology, and spread of some of the world's most troublesome aquatic weeds. In *Aquatic Weeds: The Ecology and Management of Nuisance Aquatic Vegetation*. A. Pieterse and K. Murphy, Eds. pp. 31–38. Oxford. Oxford University Press.

Cook, C.D.K. 1996. *Aquatic Plant Book.* 228 pp. The Hague. SPB Academic Publishing/Backhuys Publishers.

Cook, S.A. and Johnson, M.P. 1968. Adaptation to heterogeneous environments. I. Variation in heterophylly in *Ranunculus flammula* L. *Evolution* 22: 496–516.

Cooke, G.D. 1980. Lake level drawdown as a macrophyte control technique. *Water Resources Bulletin* 16: 317–322.

Cooke, J.C. and Lefor, M.W. 1998. The mycorrhizal status of selected plant species from Connecticut wetlands and transition zones. *Restoration Ecology* 6: 214–222.

Cooper, A. 1982. The effects of salinity and waterlogging on the growth and cation uptake of salt marsh plants. *New Phytologist* 90: 263–275.

Copeland, B.J. and Duffer, W.R. 1964. Use of a clear plastic dome to measure gaseous diffusion rates in natural waters. *Limnology and Oceanography* 9: 494–499.

Corlett, R.T. 1986. The mangrove understory: some additional observations. *Journal of Tropical Ecology* 2: 93–94.

Correll, D.L. 1991. Human impact on the functioning of landscape boundaries. In *Ecotones: The Role of Landscape Boundaries in the Management and Restoration of Changing Environments*. M.M. Holland, P.G. Risser, and R.J. Naimer, Eds. pp. 90–109. New York. Chapman & Hall.

Couee, I., Defontaine, S., Carde, J.P., and Pradet, A. 1992. Effects of anoxia on mitochondrial biogenesis in rice shoots. *Plant Physiology* 98: 411–421.

Covin, J.D. and Zedler, J.B. 1988. Nitrogen effects on *Spartina foliosa* and *Salicornia virginica* in the salt marsh at Tijuana Estuary, California. *Wetlands* 8: 51–65.

Cowardin, L.M., Carter, V., Golet, F.C., and LaRoe, E.T. 1979. Classification of Wetlands and Deepwater Habitats of the United States, 103 pp. Washington, D.C. U.S. Department of the Interior, U.S. Fish and Wildlife Service.

Cox, P.A. 1983. Search theory, random motion, and the convergent evolution of pollen and spore morphology in aquatic plants. *The American Naturalist* 121: 9–31.

Cox, P.A. 1988. Hydrophilous pollination. *Annual Review of Ecology and Systematics* 19: 261–280.

Cox, P.A. 1993. Water-pollinated plants. *Scientific American* 269: 68–74.

Cox, P.A. and Knox, R.B. 1988. Pollination postulates and two-dimensional pollination in hydrophilous monocotyledons. *Annals of the Missouri Botanical Garden* 75: 811–818.

Craft, C.B. and Richardson, C.J. 1993. Peat accretion and phosphorus accumulation along a eutrophication gradient in the northern Everglades. *Biogeochemistry* 22:133–156.

Craft, C.B., Vymazal, J., and Richardson, C.J. 1995. Response of Everglades plant communities to nitrogen and phosphorus additions. *Wetlands* 15: 258–271.

Cranford, P.J., Gordon, D.C., and Jarvis, C.M. 1989. Measurement of cordgrass, *Spartina alterniflora,* production in a macrotidal estuary, Bay of Fundy. *Estuaries* 12: 27–34.

Crawford, R.M.M. 1978. Metabolic adaptations to anoxia. In *Plant Life in Anaerobic Environments.* D.D. Hook and R.M.M. Crawford, Eds. pp. 119–136. Ann Arbor, MI. Ann Arbor Science Publishers.

Crawford, R.M.M. 1982. Physiological responses to flooding. In *Physiological Plant Ecology II.* O.L. Lange, P.S. Nobel, C.B. Osmond, and H. Ziegler, Eds. pp. 453–477. Berlin. Springer-Verlag.

Crawford, R.M.M. 1992. Oxygen availability as an ecological limit to plant distribution. *Advances in Ecological Research* 23: 93–185.

Crawford, R.M.M. 1993. Root survival in flooded soils. In *Mires: Swamp, Bog, Fen and Moor. Ecosystems of the World,* Vol. 4A. A. Gore, Ed. pp. 257–283. Amsterdam. Elsevier Science.

Crawford, R.M.M. and Tyler, P.D. 1969. Organic acid metabolism in relation to flooding tolerance in roots. *Journal of Ecology* 57: 235–244.

Creed, R.P. 1998. A biogeographic perspective on Eurasian watermilfoil declines: additional evidence for the role of herbivorous weevils in promoting declines? *Journal of Aquatic Plant Management* 36: 16–22.

Creed, R.P. and Sheldon, S.P. 1993. The effect of feeding by a North American weevil, *Euhrychiopsis lecontei,* on Eurasian watermilfoil (*Myriophyllum spicatum*). *Aquatic Botany* 45: 245–256.

Creed, R.P. and Sheldon, S.P. 1994. The effect of two herbivorous insect larvae on Eurasian watermilfoil. *Journal of Aquatic Plant Management* 32: 21–26.

Creswell, J.E. 1991. Capture rates and composition of insect prey of the pitcher plant *Sarracenia purpurea. The American Midland Naturalist* 125: 1–9.

Creswell, J.E. 1993. The morphological correlates of prey capture and resource parasitism in pitchers of the carnivorous plant *Sarracenia purpurea. The American Midland Naturalist* 129: 35–41.

Cronk, J.K. 1992. Spatial Water Quality and Aquatic Metabolism in Four Newly Constructed Freshwater Riparian Wetlands. Ph.D. dissertation, 293 pp. Columbus, OH. Environmental Biology Department, Ohio State University.

Cronk, J.K. 1996. Constructed wetlands to treat wastewater from dairy and swine operations: a review. *Agriculture, Ecosystems & Environment* 58: 97–114.

Cronk, J.K. and Mitsch, W.J. 1994a. Aquatic metabolism in four newly constructed freshwater wetlands with different hydrologic inputs. *Ecological Engineering* 3: 449–468.

Cronk, J.K. and Mitsch, W.J. 1994b. Periphyton productivity on artificial and natural surfaces in four constructed freshwater wetlands under different hydrologic regimes. *Aquatic Botany* 48: 325–342.

Cronk, J.K., Shirmohammadi, A., and Kodmur, V. 1994. An evaluation of wetlands for the treatment of dairy effluent: results from the first year of operations, 6 pp. International Winter Meeting. American Society of Agricultural Engineers. Paper No. 94–2600. Chicago, IL.

Cronk, Q.C.B. and Fuller, J.L. 1995. *Plant Invaders: The Threat to Natural Ecosystems,* 241 pp. London. Chapman & Hall.

Crooks, E. 1997. Insight into the Pitcher Plant (*Sarracenia purpurea*): The Anatomical and Mutualistic Adaptations for Survival in Nutrient-Poor Environments and the Importance of Carnivory, 10 pp. Unpublished paper, Allendale, MI. Grand Valley State University.

Crowder, A.A. and Macfie, S.M. 1986. Seasonal deposition of ferric hydroxide plaque on roots of wetland plants. *Canadian Journal of Botany* 64: 2120–2124.

Crozier, C.R., Devai, I., and DeLaune, R.D. 1995. Methane and reduced sulfur gas production by fresh and dried wetland soils. *Soil Science Society of America Journal* 59: 277–284.

Crum, H. 1992. *A Focus on Peatlands and Peat Mosses,* 306 pp. Ann Arbor, MI. The University of Michigan Press.

D'Angelo, E.M. and Reddy, K.R. 1993. Ammonium oxidation and nitrate reduction in sediments of a hypereutrophic lake. *Soil Science Society of America Journal* 57: 1156–1163.

D'Antonio, C.M. and Vitousek, P.M. 1992. Biological invasions by exotic grasses, the grass-fire cycle and global change. *Annual Review of Ecology and Systematics* 23: 63–87.

Dacey, J.W.H. 1980. Internal winds in water lilies: an adaptation for life in anaerobic sediments. *Science* 210: 1017–1019.

Dacey, J.W.H. 1981. Pressurized ventilation in the yellow waterlily. *Ecology* 62: 1137–1147.

Dacey, J.W.H. and Klug, M.J. 1982. Tracer studies of gas circulation in *Nuphar*: $^{18}O_2$ and $^{14}CO_2$ transport. *Physiologia Plantarum* 56: 361–366.

Daehler, C.C. 1998. The taxonomic distribution of invasive angiosperm plants: ecological insights and comparison to agricultural weeds. *Biological Conservation* 84: 167–180.

Dahl, T.E. 1990. Wetland Losses in the United States 1780s to 1980s, 13 pp. Washington, D.C. U.S. Department of the Interior, U.S. Fish and Wildlife Service.

Dahl, T.E. and Johnson, C.E. 1991. Status and Trends of Wetlands in the Conterminous United States, Mid-1970's to Mid-1980's, 28 pp. Washington, D.C. U.S. Department of the Interior, U.S. Fish and Wildlife Service.

Dai, T. and Wiegert, R.G. 1996. Ramet population dynamics and net aerial primary productivity of *Spartina alterniflora*. *Ecology* 77: 276–288.

Dame, R.F. and Kenny, P.D. 1986. Variability of *Spartina alterniflora* primary production in the euhaline North Inlet estuary. *Marine Ecology Progress Series* 32: 71–80.

Damman, A.W.H. and French, T.W. 1987. The Ecology of Peat Bogs of the Glaciated Northeastern United States: a Community Profile, 100 pp. U.S. Department of the Interior, U.S. Fish and Wildlife Service.

Danielson, T.J. 1998. Indicators for Monitoring and Assessing Biological Integrity of Inland, Freshwater Wetlands: A Survey of Technical Literature (1989–1996). Washington, D.C. EPA843-R-98-002. U.S. Environmental Protection Agency, Office of Wetlands, Oceans, and Watersheds.

Dansereau, P. and Segadas-Vianna, F. 1952. Ecological study of the peat bogs of eastern North America. I. Structure and evolution of vegetation. *Canadian Journal of Botany* 30: 490–520.

Daoust, R.J. and Childers, D.L. 1998. Quantifying aboveground biomass and estimating net aboveground primary production for wetland macrophytes using a non-destructive phenometric technique. *Aquatic Botany* 62: 115–133.

Daoust, R.J. and Childers, D.L. 1999. Controls on emergent macrophyte composition, abundance, and productivity in freshwater Everglades wetland communities. *Wetlands* 19: 262–275.

Daubenmire, R.F. 1968. *Plant Communities: A Textbook of Plant Synecology*, 300 pp. New York. Harper & Row.

Davies, D.D. 1980. Anaerobic metabolism and production of organic acids. In *The Biochemistry of Plants: A Comprehensive Treatise*, Vol. 2. P.K. Stumpf and E.E. Conn, Eds. pp. 581–611. New York. Academic Press.

Davis, A.F. 1993. Rare wetland plants and their habitats in Pennsylvania. *Proceedings of the Academy of Natural Sciences of Philadelphia* 114: 254–262.

Davis, C.B. and van der Valk, A.G. 1983. Uptake and release of nutrients by living and decomposing *Typha glauca* Godr. tissues at Eagle Lake, Iowa. 16: 75–89.

Davis, F.W. 1985. Historical changes in submerged macrophyte communities of upper Chesapeake Bay. *Ecology* 66: 981.

Davis, J.H. 1940. *The Ecology and Geologic Role of Mangroves in Florida*, pp. 303–412. Washington, D.C. Carnegie Institution.

Davis, S.M. 1994. Phosphorus inputs and vegetation sensitivity in the Everglades. In *Everglades: The Ecosystem and Its Restoration*. S.M. Davis and J.C. Ogden, Eds. pp. 357–378. Delray Beach, FL. St. Lucie Press.

Davis, T.J. 1993. Towards the Wise Use of Wetlands. Report of the Ramsar Convention Wise Use Project. Gland, Switzerland.

Day, J.W., Jr., Coronado-Molina, C., Vera-Herrera, F., Twilley, R., Rivera-Monroy, V.H., Alvarez-Guillen, H., Day, R., and Conner, W. 1996. A 7 year record of above-ground net primary production in a southeastern Mexican mangrove forest. *Aquatic Botany* 55: 39–60.

Day, R.T., Keddy, P.A., McNeill, J., and Carleton, T. 1988. Fertility and disturbance gradients: a summary model for riverine marsh vegetation. *Ecology* 69: 1044–1054.

DeBusk, W.F. 1984. Nutrient Dynamics in a Cypress Strand Receiving Municipal Wastewater Effluent, Thesis. 103 pp. Gainesville. University of Florida.

De Leeuw, J., Olff, H., and Bakker, J. 1990. Year-to-year variation in peak above-ground biomass of six salt-marsh angiosperm communities as related to rainfall deficit and inundation frequency. *Aquatic Botany* 36: 139–151.

Deloach, C.J. 1997. Biological control of weeds in the United States and Canada. In *Assessment and Management of Plant Invasions*. J.O. Luken and J.W. Thieret, Eds. pp. 172–194. New York. Springer.

Dennison, W.C., Orth, R.J., Moore, K.A., Stevenson, J.C., Carter, V., Kollar, S., Bergstrom, P.W., and Batiuk, R.A. 1993. Assessing water quality with submersed vegetation. *BioScience* 43: 86–94.

Denny, P. 1980. Solute movement in submerged angiosperms. *Biological Review* 55: 65–92.

Denny, P. 1993. Wetlands of Africa: introduction. In *Wetlands of the World I: Inventory, Ecology and Management*. D.F. Whigham, D. Dykyjova, and S. Hejny, Eds. pp. 1–31. Dordrecht. Kluwer Academic Publishers.

de Winton, M.D. and Clayton, J.S. 1996. The impact of invasive submerged weed species on seed banks in lake sediments. *Aquatic Botany* 53: 31–45.

Dibble, E., Killgore, K.J., and Dick, G.O. 1996. Measurement of plant architecture in seven aquatic plants. *Journal of Freshwater Ecology* 11: 311–318.

Dickerman, J.A., Stewart, A.J., and Wetzel, R.G. 1986. Estimates of net annual aboveground production: sensitivity to sampling frequency. *Ecology* 67: 650–659.

Dierberg, F.E. and Brezonik, P.L. 1983. Nitrogen and phosphorus mass balances in natural and sewage-enriched cypress domes. *Journal of Applied Ecology* 20: 323–337.

Dierberg, F.E., Straub, P.A., and Hendry, C.D. 1986. Leaf-to-twig transfer conserves nitrogen and phosphorus in nutrient poor and enriched cypress swamps. *Forest Science* 32: 900–913.

Dobson, J.E., Bright, E.A., Ferguson, R.L., Field, D.W., Wood, K.D., Haddad, K., Iredale, H., III, Jensen, J., Klemas, V., Orth, R.J., and Thomas, J.P. 1995. NOAA Coastal Change Analysis Program (C-CAP): Guidance for Regional Implementation. NOAA Technical Report NMFS 123. Seattle, WA. U.S. Department of Commerce.

Dolan, T.J., Bayley, S.E., Zoltek, J., and Hermann, A. 1981. Phosphorus dynamics of a Florida freshwater marsh receiving treated wastewater. *Journal of Applied Ecology* 18: 205–219.

Drake, B.G. and Read, M. 1981. Carbon dioxide assimilation, photosynthetic efficiency, and respiration of a Chesapeake Bay salt marsh. *Journal of Ecology* 69: 405–423.

Drizo, A., Frost, C.A., Smith, K.A., and Grace, J. 1997. Phosphate and ammonium removal by constructed wetlands with horizontal subsurface flow, using shale as a substrate. *Water Science and Technology* 35: 95–102.

Dunn, C.P. and Stearns, F. 1987. A comparison of vegetation and soils in floodplain and basin forested wetlands of southeastern Wisconsin. *The American Midland Naturalist* 118: 375–384.

Eastman, J. 1995. *The Book of Swamp and Bog: Trees, Shrubs, and Wildflowers of Eastern Freshwater Wetlands*, 237 pp. Mechanicsburg, PA. Stackpole Books.

Edwards, D. 1969. Some effects of siltation upon macrophytic vegetation in rivers. *Hydrobiologia* 34: 29–37.

Egler, F. 1952. Vegetation science concepts. I. Initial floristic composition a factor in old-field vegetation development. *Vegetation* 4: 412–417.

Ehleringer, J.R. and Monson, R.K. 1993. Evolutionary and ecological aspects of photosynthetic pathway variation. *Annual Review of Ecology and Systematics* 24: 411–439.

Ehrenfeld, J.G. 1983. The effects of changes in land-use on swamps of the New Jersey pine barrens. *Biological Conservation* 25: 353–375.

Ehrenfeld, J.G. 2000. Evaluating wetlands within an urban context. *Ecological Engineering* 15: 253–265.

Ehrenfeld, J.G. and Schneider, J.P. 1991. *Chamaecyparis thyoides* wetlands and suburbanization: effects on hydrology, water quality and plant community composition. *Journal of Applied Ecology* 28: 467–490.

Ehrenfeld, J.G. and Schneider, J.P. 1993. Response of forested wetland vegetation to perturbations of water chemistry and hydrology. *Wetlands* 13: 122–129.

Ehrlich, P.R. and Raven, P.H. 1969. Differentiation of populations. *Science* 165: 1228–1231.

Eichler, L.W., Bombard, R.T., Sutherland, J.W., and Boylen, C.W. 1993. Suction harvesting of Eurasian watermilfoil and its effect on native plant communities. *Journal of Aquatic Plant Management* 31: 144–148.

Eilers, H.P. 1979. Production ecology in an Oregon coastal salt marsh. *Estuarine and Coastal Marine Science* 8: 399–410.

Elakovich, S.D. and Wooten, J.W. 1991. Allelopathic potential of *Nuphar lutea* (L.) Sibth. & Sm. (Nymphaeaceae). *Journal of Chemical Ecology* 17: 701–714.

Elakovich, S.D. and Yang, J. 1996. Structures and allelopathic effects of *Nuphar* alkaloids: nupharolutine and 6,6′-dihydroxythiobinupharidine. *Journal of Chemical Ecology* 22: 2209–2219.

Ellison, A.M. 2000a. Restoration of mangrove ecosystems. *Restoration Ecology* 8: 217–218.

Ellison, A.M. 2000b. Mangrove restoration: do we know enough? *Restoration Ecology* 8: 219–229.

Engel, S. and Nichols, S. 1984. Lake sediment alteration for macrophyte control. *Journal of Aquatic Plant Management* 22: 38–41.

Engler, R.M. and Patrick, W.H., Jr. 1974. Nitrate removal from floodwater overlying flooded soils and sediments. *Journal of Environmental Quality* 3: 409–413.

Ernst, W.H.O. 1990. Ecophysiology of plants in waterlogged and flooded environments. *Aquatic Botany* 38: 73–90.

Ervin, G.N. and Wetzel, R.G. 2000. Allelochemical autotoxicity in the emergent wetland macrophyte *Juncus effusus* (Juncaceae). *American Journal of Botany* 87: 853–860.

Ewel, J.J. 1986. Invasability: lessons from South Florida. In *Ecology of Biological Invasions of North America and Hawaii*. H.A. Mooney and J.A. Drake, Eds. pp. 214–230. New York. Springer-Verlag.

Ewel, K.C. 1984. Effects of fire and wastewater on understory vegetation in cypress domes. In *Cypress Swamps*. K.C. Ewel and H.T. Odum, Eds. pp. 119–126. Gainesville. University Presses of Florida.

Ewel, K.C. 1990a. Multiple demands on wetlands. *BioScience* 40: 660–666.

Ewel, K.C. 1990b. Swamps. In *Ecosystems of Florida*. R.L. Myers and J.J. Ewel, Eds. pp. 281–323. Orlando, FL. University of Central Florida Press.

Ewel, K.C. and Mitsch, W.J. 1978. The effects of fire on species composition in cypress dome ecosystems. *Florida Scientist* 41: 25–31.

Ewel, K.C. and Odum, H.T. 1984. *Cypress Swamps*, 472 pp. Gainseville. University Presses of Florida.

Ewel, K.C. and Wickenheiser, L.P. 1988. Effect of swamp size on growth rates of cypress (*Taxodium distichum*) trees. *The American Midland Naturalist* 120: 362–370.

Faber, P.A., Keller, E., Sands, A., and Massey, B.M. 1989. The Ecology of Riparian Habitats of the Southern California Coastal Region: A Community Profile, 152 pp. Washington, D.C. U.S. Department of the Interior, U.S. Fish and Wildlife Service.

Faegri, K. and van der Pijl, L. 1979. *The Principles of Pollination Ecology*, 244 pp. Oxford. Pergamon Press.

Fail, J.L., Haines, B.L., and Todd, R.L. 1987. Riparian forest communities and their role in nutrient conservation in an agricultural watershed. *American Journal of Alternative Agriculture* 2: 114–121.

Fan, W.M., Higashi, R.M., and Lane, A.N. 1988. An *in vivo* ^{1}H and ^{31}P NMR investigation of the effect of nitrate on hypoxic metabolism in maize roots. *Archives of Biochemistry and Biophysics* 266: 592–606.

Farney, R.A. and Bookhout, T.A. 1982. Vegetation changes in a Lake Erie marsh (Winous Point, Ottawa County, Ohio) during high water years. *Ohio Journal of Science* 82: 103–107.

Farnsworth, E. and Ellison, A. 1997. The global conservation status of mangroves. *Ambio* 26: 328–334.

Fassett, N.C. 1957. *.A Manual of Aquatic Plants,* 405 pp. Madison. University of Wisconsin Press.

Faulkner, S.P. and Richardson, C.J. 1989. Physical and chemical characteristics of freshwater wetland soils. In *Constructed Wetlands for Wastewater Treatment*. D.A. Hammer, Ed. pp. 41–72. Chelsea, MI. Lewis Publishers.

Federal Interagency Committee for Wetland Delineation. 1989. Federal Manual for Identifying and Delineating Jurisdictional Wetlands, 76 pp. Cooperative Technical Publication. Washington, D.C. U.S. Army Corps of Engineers, U.S. Environmental Protection Agency, U.S. Fish and Wildlife Service, and U.S.D.A. Soil Conservation Service.

Fennessy, M.S. 1991. Ecosystem Development in Restored Riparian Wetlands, Ph.D. dissertation. 201 pp. Columbus, OH. Environmental Biology Department, Ohio State University.

Fennessy, M.S. and Cronk, J.K. 1997. The effectiveness and restoration potential of riparian ecotones for the management of nonpoint source pollution, particulary nitrate. *Critical Reviews in Environmental Science and Technology* 27: 285–317.

Fennessy, M.S. and Mitsch, W.J. 1989. Treating coal mine drainage with an artificial wetland. Journal Water Pollution Control Federation 61: 1691–1701.

Fennessy, M.S., Cronk, J.K., and Mitsch, W.J. 1994a. Macrophyte productivity and community development in newly restored wetlands. *Ecological Engineering* 3: 469–484.

Fennessy, M.S., Brueske, C.C., and Mitsch, W.J. 1994b. Sediment deposition patterns in restored freshwater wetlands using sediment traps. *Ecological Engineering* 3: 409–468.

Fennessy, M.S., Geho, R., Elifritz, B., and Lopez, R. 1998a. Testing the Floristic Quality Assessment Index as an Indicator of Riparian Wetland Quality, Final Report to U.S. Environmental Protection Agency. Columbus. Ohio Environmental Protection Agency, Division of Surface Water.

Fennessy, M.S., Gray, M.A., and Lopez, R.D. 1998b. An ecological assessment of wetlands using reference sites Volume 1: Final Report, Volume 2: Appendices. Final Report to U.S. Environmental Protection Agency. Columbus. Ohio Environmental Protection Agency, Division of Surface Water.

Field, C.D. 1995. Impact of expected climate change of mangroves. *Hydrobiologia* 295: 75–81.

Finlayson, C. and Von Oertzen, I. 1993. Wetlands of Australia: northern (tropical) Australia. In *Wetlands of the World I: Inventory, Ecology and Management*. D.F. Whigham, D. Dykyjova, and S. Hejny, Eds. pp. 195–304. Dordrecht. Kluwer Academic Publishers.

Fish, D. and Hall, D.W. 1978. Succession and stratification of aquatic insects inhabiting the leaves of the insectivorous pitcher plant, *Sarracenia purpurea*. *The American Midland Naturalist* 99: 172–183.

Fitter, A.H. and Hay, R.K.M. 1987. *Environmental Physiology of Plants*, 423 pp. London. Academic Press.

Flack, S.R. and Benton, N.B. 1998. Invasive species and wetland biodiversity. *National Wetlands Newsletter* 20: 7–11.

Flowers, T.J., Troke, P.F., and Yeo, A.R. 1977. The mechanism of salt tolerance in halophytes. *Annual Review of Plant Physiology* 28: 89–121.

Flowers, T.J., Hajibagheri, M.A., and Clipson, N.J.W. 1986. Halophytes. *Quarterly Review of Biology* 61: 313–337.

Folkerts, G.W. 1982. The Gulf coast pitcher plant bogs. *American Scientist* 70: 260–267.

Fontaine, T.D. and Ewel, K.C. 1981. Metabolism of a Florida lake ecosystem. *Limnology and Oceanography* 26: 754–763.

Forno, I.W. and Cofrancesco, A.F. 1993. New frontiers in biocontrol. *Journal of Aquatic Plant Management* 31: 222–224.

Fromard, F., Puig, H., Mougin, E., Marty, G., Betoulle, J.L., and Cadamuro, L. 1998. Structure, above-ground biomass and dynamics of mangrove ecosystems: new data from French Guiana. *Oecologia* 115: 39–53.

Gabor, T.S., Haagsma, T., Murkin, H.R., and Armson, E. 1995. Effects of triclopyr amine on purple loosestrife and non-target wetland plants in southeastern Ontario, Canada. *Journal of Aquatic Plant Management* 33: 48–51.

Galatowitsch, S.M. and McAdams, T.V. 1994. Distribution and Requirements of Plants on the Upper Mississippi River: Literature Review. Ames. Iowa Cooperative Fish and Wildlife Research Unit.

Galatowitsch, S.M. and van der Valk, A. G. 1994. *Restoring Prairie Wetlands — An Ecological Approach*, 246 pp. Ames. Iowa State University Press.

Galatowitsch, S.M. and van der Valk, A.G. 1995. Natural revegetation during restoration of wetlands in the southern prairie pothole region of North America. In *Restoration of Temperate Wetlands*. B.D. Wheeler, S.C. Shaw, W.J. Fojt, and R.A. Robertson, Eds. pp. 129–142. Chichester. John Wiley & Sons.

Galatowitsch, S.M. and van der Valk, A.G. 1996. The vegetation of restored and natural prairie wetlands. *Ecological Applications* 6: 102–112.

Galatowitsch, S.M., Anderson, N.O., and Ascher, P.D. 1999a. Invasiveness in wetland plants in temperate North America. *Wetlands* 19: 733–755.

Galatowitsch, S.M., Whited, K.C., and Tester, J.R. 1999b. Development of community metrics to evaluate recovery of Minnesota wetlands. *Journal of Aquatic Ecosystem Stress and Recovery* 6: 217–234.

Gallagher, J.L. 1974. Sampling macro-organic matter profiles in salt marsh plant root zones. *Soil Science Society of America Proceedings* 38: 154–155.

Gallagher, J.L. 1975. Effect of an ammonium nitrate pulse on the growth and elemental composition of natural stands of *Spartina alterniflora* and *Juncus roemerianus*. *American Journal of Botany* 62: 644–648.

Gallagher, J.L. and Daiber, F.C. 1974. Primary production of edaphic algal communities in a Delaware salt marsh. *Limnology and Oceanography* 19: 390–395.

Gallagher, J.L. and Plumley, F.G. 1979. Underground biomass profiles and productivity in Atlantic coastal marshes. *American Journal of Botany* 66: 156–161.

Gallagher, J.L., Reimold, R.J., Linthurst, R.A., and Pfeiffer, W.J. 1980. Aerial production, mortality, and mineral accumulation-export dynamics in *Spartina alterniflora* and *Juncus roemerianus* plant stands in a Georgia salt marsh. *Ecology* 61: 303–312.

Gallagher, J.L., Somers, G.F., Grant, D.M., and Seliskar, D.M. 1988. Persistent differences in two forms of *Spartina alterniflora*: a common garden experiment. *Ecology* 69: 1005–1008.

Gambrell, R.P. and Patrick, W.H. 1978. Chemical and microbiological properties of anaerobic soils and sediments. In *Plant Life in Anaerobic Environments*. D.D. Hook and R.M.M. Crawford, Eds. pp. 375–423. Ann Arbor, MI. Ann Arbor Science Publishers.

Garono, R.J. and Kooser, J.G. 1994. Ordination of wetland insect populations: evaluation of a potential mitigation monitoring tool. In *Global Wetlands: Old World and New*. W.J. Mitsch, Ed. pp. 509–516. Amsterdam. Elsevier Science.

Gates, F.C. 1942. The bogs of lower Michigan. *Ecological Monographs* 12: 213–254.

Gaudet, C.L. and Keddy, P.A. 1988. A comparative approach to predicting competitive ability from plant traits. *Nature* 334: 242–243.

Gaudet, C.L. and Keddy, P.A. 1995. Competitive performance and species distribution in shoreline plant communities: a comparative approach. *Ecology* 76: 280–291.

Gearhart, R.A., Klopp, F., and Allen, G. 1989. Constructed free surface wetlands to treat and receive wastewater: pilot project to full scale. In *Constructed Wetlands for Wastewater Treatment*. D.A. Hammer, Ed. pp. 121–138. Chelsea, MI. Lewis Publishers.

Gerloff, G.C. and Krombholz, P.H. 1966. Tissue analysis as a measure of nutrient availability for the growth of angiosperm aquatic plants. *Limnology and Oceanography* 11: 529–537.

Germain, V., Ricard, B., Raymond, P., and Saglio, P.H. 1997. The role of sugars, hexokinase, and sucrose synthase in the determination of hypoxically induced tolerance to anoxia in tomato roots. *Plant Physiology* 114: 167–175.

Gernes, M.C. 2000. *Using Plants in Wetland Biological Indexes*, 10 pp. Minneapolis, MN. Pollution Control Agency.

Gernes, M.C. and Helgen, J.C. 1999. Indexes of Biotic Integrity (IBI) for Wetlands: Vegetation and Invertebrate IBI's, Final Report to U.S. Environmental Protection Agency, Assistance Number CD995525-01, April 1999. St. Paul, MN. Minnesota Pollution Control Agency, Environmental Outcomes Division.

Gersberg, R.M., Elkins, B.V., Lyon, S.R., and Goldman, C.R. 1986. Role of aquatic plants in wastewater treatment by artificial wetlands. *Water Resources* 20: 363–368.

Gersberg, R.M., Gearhart, R.A., and Ives, I. 1989. Pathogen removal in constructed wetland. In *Constructed Wetlands for Wastewater Treatment*. D.A. Hammer, Ed. pp. 431–445. Chelsea, MI. Lewis Publishers.

Gholz, H.L. 1982. Environmental limits on aboveground net primary production, leaf area, and biomass in vegetation zones of the Pacific Northwest. *Ecology* 63: 469–481.

Gibbs, J., Morrell, S., Valdez, A., Setter, T.L., and Greenway, H. 2000. Regulation of alcoholic fermentation in coleoptiles of two rice cultivars differing in tolerance to anoxia. *Journal of Experimental Botany* 51: 785–796.

Gibson, T.C. 1991. Differential escape of insects from carnivorous plant traps. *The American Midland Naturalist* 125: 55–62.

Gilman, K. 1994. *Hydrology and Wetland Conservation*, 101 pp. Chichester. John Wiley & Sons.

Glaser, P.H. 1987. The Ecology of Patterned Boreal Peatlands of Northern Minnesota: A Community Profile, 98 pp. Washington, D.C. U.S. Department of the Interior, U.S. Fish and Wildlife Service.

Glaser, P.H. 1992. Raised bogs in eastern North America — regional controls for species richness and floristic assemblages. *Journal of Ecology* 80: 535–554.

Gleason, H.A. 1917. The structure and development of the plant association. *Bulletin of the Torrey Botanical Club* 44: 463–481.

Glenn, E., Thompson, T.L., Frye, R., Riley, J., and Baumgarter, D. 1995. Effects of salinity on growth and evapotranspiration of *Typha domingensis* Pers. *Aquatic Botany* 52: 75–91.
Glooschenko, W.A. 1978. Above-ground biomass of vascular plants in a subarctic James Bay salt marsh. *Canadian Field-Naturalist* 92: 30–37.
Glooschenko, W.A., Tarnocai, C., Zoltai, S., and Glooschenko, V. 1993. Wetlands of Canada and Greenland. In *Wetlands of the World I: Inventory, Ecology and Management*. D.F. Whigham, D. Dykyjova, and S. Hejny, Eds. pp. 415–514. Dordrecht. Kluwer Academic Publishers.
Goldner, B.H. 1984. Riparian restoration efforts associated with structurally modified flood control channels. In *California Riparian Systems: Ecology, Conservation and Productive Management*. California Resource Report No. 55. R.E. Warner and K.M. Hendrix, Eds. 445 pp. Berkeley. University of California Press.
Goldsborough, L.G. and Hickman, M. 1991. A comparison of periphytic algal biomass and community structure on *Scirpus validus* and on morphologically similar artificial substratum. *Journal of Phycology* 27: 196–206.
Golet, F.C. and Larson, J.S. 1974. Classification of Freshwater Wetlands in the Glaciated Northeast, 56 pp. U.S. Department of the Interior, U.S. Fish and Wildlife Service.
Golley, F.C., Odum, H.T., and Wilson, R.F. 1962. The structure and metabolism of a Puerto Rican red mangrove forest in May. *Ecology* 43: 9–19.
Gopal, B. and Goel, U. 1993. Competition and allelopathy in aquatic plant communities. *The Botanical Review* 59: 155–210.
Gopal, B. and Krishnamurthy, K. 1993. Wetlands of South Asia. In *Wetlands of the World I: Inventory, Ecology and Management*. D.F. Whigham, D. Dykyjova, and S. Hejny, Eds. pp. 345–414. Dordrecht. Kluwer Academic Publishers.
Gordon, D.C. and Cranford, P.J. 1994. Export of organic matter from macrotidal salt marshes in the upper Bay of Fundy, Canada. In *Global Wetlands: Old World and New*. W.J. Mitsch, Ed. pp. 257–264. Amsterdam. Elsevier Science.
Gordon, D.R. 1998. Effects of invasive, non-indigenous plant species on ecosystem processes: lessons from Florida. *Ecological Applications* 8: 975–989.
Gorham, E., Eisenreich, S.J., Ford, J., and Santelmann, M.V. 1985. The chemistry of bog waters. In *Chemical Processes in Lakes*. W. Stumm, Ed. pp. 339–363. New York. John Wiley & Sons.
Gosselink, J.G. and Turner, R.E. 1978. The role of hydrology in freshwater wetland ecosystems. In *Freshwater Wetlands — Ecological Processes and Management Potential*. R.E. Good, D.F. Whigham, and R.L. Simpson, Eds. pp. 63–78. New York. Academic Press.
Grace, J.B. 1991. A clarification of the debate between Grime and Tilman. *Functional Ecology* 5: 583–587.
Grace, J.B. 1993. The adaptive significance of clonal reproduction in angiosperms: an aquatic perspective. *Aquatic Botany* 44: 159–180.
Grace, J.B. and Tilman, D., Eds. 1990. *Perspectives on Plant Competition*. 484 pp. San Diego, CA. Academic Press.
Grace, J.B. and Wetzel, R.G. 1978. The production biology of Eurasian watermilfoil (*Myriophyllum spicatum* L.): a review. *Journal of Aquatic Plant Management* 16: 1–11.
Grace, J.B. and Wetzel, R.G. 1981. Habitat partitioning and competitive displacement in cattails (*Typha*): experimental field studies. *The American Naturalist* 118: 463–474.
Grace, J.B. and Wetzel, R.G. 1982. Niche differentiation between two rhizomatous plant species: *Typha latifolia* and *Typha angustifolia*. *Canadian Journal of Botany* 60: 46–57.
Grace, J.R. and Wetzel, R.G. 1998. Long-term dynamics of *Typha* populations. *Aquatic Botany* 61: 137–146.
Greenway, M. 1997. Nutrient content of wetland plants in constructed wetland receiving municipal effluent in tropical Australia. *Water Science and Technology* 35: 135–142.
Grelsson, G. and Nilsson, C. 1991. Vegetation and seed-bank relationships on a lakeshore. *Freshwater Biology* 26: 199–207.
Grigal, D.F. 1985. *Sphagnum* production in forested bogs of northern Minnesota. *Canadian Journal of Botany* 63: 1204–1207.
Grigal, D.F. and Kernik, L.K. 1984. Generality of black spruce biomass estimation equations. *Canadian Journal of Forest Research* 11: 599–605.

Grigal, D.F., Buttleman, C.G., and Kernik, L.K. 1985. Biomass and productivity of the woody strata of forested bogs in northern Minnesota. *Canadian Journal of Botany* 63: 2416–2424.

Griggs, F.T. and Jain, S.K. 1983. Conservation of vernal pool plants in California, II. Population biology of a rare and unique grass genus *Orcuttia*. *Biological Conservation* 27: 171–193.

Grillas, P. and Duncan, P. 1986. On the distribution and abundance of submersed macrophytes in temporary marshes in the Camargue (S. France). In Proceedings of the European Weed Research Society 7th International Symposium on Aquatic Weeds, pp. 133–141. Loughborough, England.

Grime, J.P. 1977. Evidence for the existence of three primary strategies in plants and its relevance to ecological and evolutionary theory. *The American Naturalist* 111: 1169–94.

Grime, J.P. 1979. *Plant Strategies and Vegetation Processes*, 222 pp. Chichester, England. John Wiley & Sons.

Grodowitz, M.J. 1998. An active approach to the use of insect biological control for the management of non-native aquatic plants. *Journal of Aquatic Plant Management* 36: 57–61.

Groenendijk, A.M. 1984. Primary production of four dominant salt-marsh angiosperms in the SW Netherlands. *Vegetatio* 57: 143–152.

Groffman, P.M. and Hanson, G.C. 1997. Wetland denitrification: influence of site quality and relationships with wetland delineation protocols. *Soil Science Society of America Journal* 61: 323–329.

Groffman, P.M. and Tiedje, J.M. 1989. Denitrification in riparian wetlands receiving high and low groundwater nitrate inputs. *Journal of Environmental Quality* 23: 273.

Grootjans, A.P., Ernst, W.H.O., and Stuyfzand, P.J. 1998. European dune slacks: strong interactions of biology, pedogenesis and hydrology. *Tree* 13: 96–100.

Grosse, W. 1996. Pressurized ventilation in floating-leaved aquatic macrophytes. *Aquatic Botany* 54: 137–150.

Grosse, W. and Bauch, C. 1991. Gas transfer in floating-leaved plants. *Vegetatio* 97: 185–192.

Grosse, W., Armstrong, J., and Armstrong, W. 1996. A history of pressurized gas-flow studies in plants. *Aquatic Botany* 54: 87–100.

Grubb, P.J. 1977. The maintenance of species-richness in plant communities: the importance of the regeneration niche. *Biological Reviews of the Cambridge Philosophical Review* 52: 107–145.

Gumbricht, T. 1993. Nutrient removal processes in freshwater submersed macrophyte systems. *Ecological Engineering* 2: 1–30.

Guntenspergen, G.R., Stearns, F., and Kadlec, J.A. 1989. Wetland vegetation. In *Constructed Wetlands for Wastewater Treatment*. D.A. Hammer, Ed. pp. 73–88. Chelsea, MI. Lewis Publishers.

Gusewell, S., and Klotzli, F. 1998. Abundance of common reed (*Phragmites australis*), site conditions and conservation value of fen meadows in Switzerland. *Acta Botanica Neerlandica* 47: 113–129.

Gusewell, S.W., Koerselman, W., and Verhoeven, J.T.A. 1998. The N:P ratio and the nutrient limitation of wetland plants. *Bulletin of the Geobotanical Institute* ETH 64:77–90.

Hacker, S.D. and Bertness, M.D. 1995. Morphological and physiological consequences of a positive plant interaction. *Ecology* 76: 2165–2175.

Hagemeyer, J. 1997. Salt. In *Plant Ecophysiology*. M.N.V. Prasad, Ed. pp. 173–206. New York. John Wiley & Sons.

Hall, C.A.S. and Moll, R.A. 1975. Methods of assessing aquatic primary productivity. In *Primary Productivity of the Biosphere*. H. Lieth and R.H. Whittaker, Eds. pp. 19–53. New York. Springer-Verlag.

Hall, J.V., Frayer, W.E., and Wilen, B.O. 1994. Status of Alaska Wetlands, 33 pp. Alaska Region. U.S. Department of the Interior, U.S. Fish and Wildlife Service.

Hall, T.F. and Penfound, W.T. 1939. A phytosociological study of a *Nyssa biflora* consocies in southeastern Louisiana. *American Midland Naturalist* 23: 369–375.

Haller, W.T., Sutton, D.L., and Barlowe, W.C. 1974. Effects of salinity on growth of several aquatic macrophytes. *Ecology* 55: 891–894.

Halsey, L., Vitt, D., and Zoltai, S. 1997. Climatic and physiographic controls on wetland type and distribution in Manitoba, Canada. *Wetlands* 17: 243–262.

Hammer, D.A. 1989. *Constructed Wetlands for Wastewater Treatment*, 381 pp. Chelsea, MI. Lewis Publishers.

Hammer, D.A. 1992. Designing constructed wetland systems to treat agricultural nonpoint source pollution. *Ecological Engineering* 1: 49–82.

Hammer, D.A. 1993. Constructed wetland for wastewater teatment: an overview of a low cost technology. In Constructed Wetlands for Wastewater Treatment Conference. Middleton, County Cork, Ireland.

Hammer, D.A. 1994. Guidelines for design, construction and operation of constructed wetlands for livestock wastewater treatment. In proceedings of a workshop on Constructed Wetlands for Animal Waste Management. P.J. DuBowy and R.P. Reaves, Eds. pp. 155–181. Lafayette, IN.

Hann, B.J. 1991. Invertebrate grazer-periphyton interactions in a eutrophic marsh pond. *Freshwater Biology* 26: 87–96.

Hanslin, H.M. and Karlsson, P.S. 1996. Nitrogen uptake from prey and substrate as affected by prey capture level and plant reproductive status in four carnivorous plant species. *Oecologia* 106: 370–375.

Hapley, P.J. and Milne, R.J. 1996. The use of avian fauna in delineating wetlands in the Baldwin wetland complex, southern Ontario. In *Wetlands: Environmental Gradients, Boundaries, and Buffers.* G. Mulamoottil, B.G. Warner, and E.A. McBean, Eds. pp. 163–176. Boca Raton, FL. Lewis Publishers.

Hardisky, M.A. 1980. A comparison of *Spartina alterniflora* primary production estimated by destructive and nondestructive techniques. In *Estuarine Perspectives.* V. Kennedy, Ed. pp. 223–234. New York. Academic Press.

Harley, K.L.S., Julien, M.H., and Wright, A.D. 1996. Water hyacinth: a tropical worldwide problem and methods for its control. In Second International Weed Control Congress. Copenhagen, Denmark. U.S. Department of Agriculture. Website: http://www/sidny.ars.usda.gov/scientists/nspencer/water_h/appendix7.htm.

Harper, J.L. 1977. *Population Biology of Plants.* 892 pp. London. Academic Press.

Harris, P. 1988. Environmental impact of weed-control insects. *BioScience* 38: 542–548.

Haslam, S.M. 1978. *River Plants: The Macrophytic Vegetation of Watercourses.* 396 pp. Cambridge. Cambridge University Press.

Havens, K.J., Varnell, L.M., and Bradshaw, J.G. 1995. An assessment of ecological conditions in a constructed tidal marsh and two natrual reference tidal marshes in coastal Virginia. *Ecological Engineering* 4: 117–141.

Havill, D.C., Ingold, A., and Pearson, J. 1985. Sulphide tolerance in coastal halophytes. *Vegetatio* 62: 279–285.

Haworth-Brockman, M.J., Murkin, H.R., and Clay, R.T. 1993. Effects of shallow flooding on newly established purple loosestrife seedlings. *Wetlands* 13: 224–227.

Haycock, N.E. and Pinay, G. 1993. Nitrate retention in grass and poplar vegetated riparian buffer strips during winter. *Journal of Environmental Quality* 22: 273–278.

Haynes, R.R. 1988. Reproductive biology of selected aquatic plants. *Annals of the Missouri Botanical Garden* 75: 805–810.

Heard, S.B. 1994. Pitcher-plant midges and mosquitoes: a processing chanin commensalism. *Ecology* 75: 1647–1660.

Heddle, E.M. and Specht, R.L. 1975. Dark island heath (Ninety-Mile Plain, South Australia). VIII. The effect of fertilizers on composition and growth 1950–1972. *Australian Journal of Botany* 23: 151–164.

Hedin, R.S., Nairn, R.W., and Kleinmann, R.L.P. 1994. Passive Treatment of Coal Mine Drainage, 37 pp. U.S. Bureau of Mines Information Circular 9389.

Heinselman, M.L. 1963. Forest sites, bog processes, and peatland types in the glacial Lake Agassiz region, Minnesota. *Ecological Monographs* 33: 327–374.

Heinselman, M.L. 1975. Boreal peatlands in relation to environment. In *Coupling of Land and Water Systems.* A.D. Hasler, Ed. pp. 83–103. New York. Springer-Verlag.

Helfield, J.M. and Diamond, M.L. 1997. Use of constructed wetlands for urban stream restoration: a critical analysis. *Environmental Management* 21: 329–341.

Henry, C.P., Amoros, C., and Bornette, G. 1996. Species traits and recolonization processes after flood disturbances in riverine macrophytes. *Vegetatio* 122: 13–27.

Henzi, T. and Braendle, R. 1993. Long-term survival of rhizomatous species under oxygen deprivation. In *Interacting Stresses on Plants in a Changing Climate.* M.B. Jackson and C.R. Black, Eds. pp. 305–314. Berlin. Springer-Verlag.

Herbst, M. and Kappen, L. 1999. The ratio of transpiration versus evaporation in a reed belt as influenced by weather conditions. *Aquatic Botany* 63: 113–125.

Herdendorf, C.E. 1987. The Ecology of the Coastal Marshes of Western Lake Erie: A Community Profile, 171 pp. Washington, D.C. U.S. Department of the Interior, U.S. Fish and Wildlife Service.

Heritage, A., Pistillo, P., Sharma, K.P., and Lantzke, I.R. 1995. Treatment of primary-settled urban sewage in pilot-scale vertical flow wetlands filters: comparison of four emergent macrophyte species over a 12 month period. *Water Science and Technology* 32: 295–304.

Herman, K.D., Masters, L.A., Penskar, M.R., Reznicek, A.A., Wilhelm, G.S., and Brodowicz, W.W. 1996. Floristic Quality Assessment with Wetland Categories and Computer Application Programs for the State of Michigan. Michigan Department of Natural Resources, Wildlife Division, Natural Heritage Program.

Hight, S.D. 1993. Control of the ornamental purple loosestrife (*Lythrum salicaria*) by exotic organisms. In *Biological Pollution: The Control and Impact of Invasive Exotic Species*. B. McKnight, Ed. pp. 147–148. Indianapolis. Indiana Academy of Science.

Hik, D.S. and Jefferies, R.L. 1990. Increases in the net above-ground primary production of a salt-marsh forage grass: a test of the predictions of the herbivore-optimization model. *Journal of Ecology* 78: 180–195.

Hik, D.S., Jefferies, R.L., and Sinclair, A.R.E. 1992. Foraging by geese, isostatic uplift and asymmetry in the development of salt-marsh communities. *Journal of Ecology* 80: 395–406.

Hill, G., Waage, J., and Phiri, G. 1997. The water hyacinth problem in tropical Africa. In International Water Hyacinth Consortium. Washington, D.C. U.S. Department of Agriculture. Website: http://www.sidney.ars.usda.gov/scientists/nspencer/water_h/appendix5.htm.

Hobbs, R.J. 1997. Can we use plant functional types to describe and predict responses to environmental change? In *Plant Functional Types: Their Relevance to Ecosystem Properties and Global Change*. T.M. Smith, H.H. Shugart, and F.I. Woodward, Eds. pp. 66–90. Cambridge. Cambridge University Press.

Hocking, P.J. 1985. Responses of *Cyperus involucratus* Rottb. to nitrogen and phosphorus with reference to wastewater reclamation. *Water Research* 19: 1379–1386.

Hodges, J.D. 1997. Development and ecology of bottomland hardwood sites. *Forest Ecology and Management* 90: 117–125.

Hofstra, D.E., Clayton, J., Green, J.D., and Auger, M. 1999. Competitive performance of *Hydrilla verticillata* in New Zealand. *Aquatic Botany* 63: 305–324.

Hogenbirk, J.C. and Wein, R.W. 1991. Fire and drought experiments in northern wetlands: a climate change analogue. *Canadian Journal of Botany* 69: 1991–1997.

Holland, M.M. 1996. Wetlands and environmental gradients. In *Wetlands: Environmental Gradients, Boundaries and Buffers*. G. Mulamoottil, B.G. Warner, and E.A. McBean, Eds. pp. 19–44. New York. Lewis Publishers.

Hollis, G.E., Fennessy, M.S., and Thompson, J. 1993. The Hydrology of the Isle of Sheppey Grazing Marshes (A249 Iwade to Queensborough, UK). Final Report to Ove Arup for the United Kingdom Department of Transport. London.

Hopkinson, C.S. and Schubauer, J.P. 1984. Static and dynamic aspects of nitrogen cycling in the salt marsh graminoid *Spartina alterniflora*. *Ecology* 65: 961–969.

Hopkinson, C.S., Gosselink, J.G., and Parrondo, R.T. 1978. Aboveground production of seven marsh plant species in coastal Louisiana. *Ecology* 59: 760–769.

Hopkinson, C.S., Gosselink, J.G., and Parrondo, R.T. 1980. Production of coastal Louisiana marsh plants calculated from phenometric techniques. *Ecology* 61: 1091–1098.

Hostrup, O. and Wiegleb, G. 1991. Anatomy of leaves of submerged and emergent forms of *Littorella uniflora* (L.) Ascherson. *Aquatic Botany* 39: 195–209.

Hotchkiss, N. 1972. *Common Marsh, Underwater and Floating-Leaved Plants of the United States and Canada*, 124 pp. New York. Dover Publications.

Houghton, R.A. 1985. The effect of mortality on estimates of net above-ground production by *Spartina alterniflora*. *Aquatic Botany* 22: 121–132.

Houghton, R.A. and Woodwell, G.M. 1980. The Flax Pond ecosystem study: exchanges of CO_2 between a salt marsh and the atmosphere. *Ecology* 61: 1434–1445.

Howard, G.W. and Harley, K.L.S. 1998. How do floating aquatic weeds affect wetland conservation and development? How can these effects be minimised? *Wetlands Ecology and Management* 5: 215–225.
Howard-Williams, C. 1985. Cycling and retention of nitrogen and phosphorous in wetlands: A theoretical and applied perspective. *Freshwater Biology* 15: 391–431.
Howard-Williams, C. 1993. Processes of aquatic weed invasions: the New Zealand example. *Journal of Aquatic Plant Management* 31: 17–22.
Howarth, R. and Fisher, S. 1976. Carbon, nitrogen, and phosphorus dynamics during leaf decay in nutrient-enriched stream microcosms. *Freshwater Biology* 6: 221–228.
Howarth, R.W. and Teal, J.M. 1979. Sulfate reduction in a New England salt marsh. *Limnology and Oceanography* 24: 999–1013.
Howes, B.L. and Teal, J.M. 1994. Oxygen loss from *Spartina alterniflora* and its relationship to salt marsh oxygen balance. *Oecologia* 97: 431–438.
Howes, B.L., Howarth, R.W., Teal, J.M. and Valiela, I. 1981. Oxidation-reduction potentials in a salt marsh: spatial patterns and interactions with primary production. *Limnology and Oceanography* 26: 350–360.
Howes, B.L., Dacey, J.W.H., and Teal, J.M., 1985. Annual carbon mineralization and belowground production of *Spartina alterniflora* in a New England salt marsh. *Ecology* 66: 595–605.
Howes, B.L., Dacey, J.W.H., and Goehringer, D.D. 1986. Factors controlling the growth form of *Spartina alterniflora*: feedbacks between above-ground production, sediment oxidation, nitrogen and salinity. *Journal of Ecology* 74: 881–898.
Hoyer, M.V. and Canfield, D.E., Jr. 1997. Aquatic Plant Management in Lakes and Reservoirs. Washington, D.C. North American Lake Management Society and Aquatic Plant Management Society for the U.S. Environmental Protection Agency.
Huenneke, L.F. and Sharitz, R.R. 1986. Microsite abundance and disribution of woody seedlings in a South Carolina cypress-tupelo swamp. *The American Midland Naturalist* 115: 328–335.
Huff, D.D. and Young, H.L. 1980. The effect of a marsh on runoff. I. A water budget model. *Journal of Environmental Quality* 9: 633–640.
Hughes, R.M. 1995. Defining acceptable biological status by comparing with reference conditions. In *Biological Assessment and Criteria: Tools for Water Resource Planning and Decision Making.* W.S. Davis and T.P. Simon, Eds. pp. 31–48. Boca Raton, FL. Lewis Publishers.
Hulot, F.D., Lacroix, G., Lescher-Moutoue, F., and Loreau, M. 2000. Functional diversity governs ecosystem response to nutrient enrichment. *Nature* 405: 340–344.
Husch, B., Miller, C.I., and Beers, T.W. 1993. *Forest Mensuration*, 402 pp. Malabar, FL. Krieger Publishing.
Hussain, M.I. and Khoja, T.M. 1993. Intertidal and subtidal blue-green algal mats of open and mangrove areas in the Farasan Archipelago (Saudi Arabia), Red Sea. *Botanica Marina* 36: 377–388.
Hutchinson, G.E. 1975. *A Treatise on Limnology III. Aquatic Macrophytes and Attached Algae*. New York. Wiley.
Huttunen, A., Heikkinen, K., and Ihme, R. 1996. Nutrient retention in the vegetation of an overland flow treatment system in northern Finland. *Aquatic Botany* 55: 61–73.
Imbert, D., Rousteau, A., and Scherrer, P. 2000. Ecology of mangrove growth and recovery in the Lesser Antilles: state of knowledge and basis for restoration projects. *Restoration Ecology* 8: 230–236.
Ingold, A. and Havill, D.C. 1984. The influence of sulphide on the distribution of higher plants in salt marshes. *Journal of Ecology* 72: 1043–1054.
Ingrouille, M. 1992. *Diversity and Evolution of Land Plants*. 340 pp. London. Chapman & Hall.
Jackson, D., Long, S.P., and Mason, C.F. 1986. Net primary production, decomposition and export of *Spartina anglica* on a Suffolk salt marsh. *Journal of Ecology* 74: 647–662.
Jackson, M.B. 1982. Ethylene as a growth promoting hormone under flooded conditions. In *Plant Growth Substances*. P.E. Wareing, Ed. pp. 291–301. London. Academic Press.
Jackson, M.B. 1989. Regulation of aerenchyma formation in roots and shoots by oxygen and ethylene. In *Cell Separation in Plants: Physiology, Biochemistry and Molecular Biology.* D.J. Osborne and M.B. Jackson, Eds. pp. 263–274. Berlin. Springer-Verlag.
Jackson, M.B. 1990. Hormones and developmental change in plants subjected to submergence or soil waterlogging. *Aquatic Botany* 38: 49–72.

Jackson, M.B. 1994. Hormone action and plant adaptations to poor aeration. *Proceedings of the Royal Society of Edinburgh* 102B: 391–405.

Jackson, M.B., Drew, M.C., and Giffard, S.C. 1981. Effects of applying ethlene to the root system of *Zea mays* on growth and nutrient concentration in relation to flooding tolerance. *Physiologia Plantarum* 52: 23–28.

Jackson, M.B., Hermann, B., and Goodenough, A. 1982. An examination of the importance of ethanol in causing injury to flooded plants. *Plant Cell Environment* 5: 163–172.

Jackson, M.B., Fenning, T.M., and Jenkins, W. 1985. Aerenchyma (gas-space) formation in adventitious roots of rice (*Oryza sativa* L.) is not controlled by ethylene or small partial pressures of oxygen. *Journal of Experimental Botany* 36: 1566–1572.

James, B.R., Rabenhorst, M.C., and Frigon, G.A. 1992. Phosphorus sorption by peat and sand amended with iron oxides or steel wool. *Water Environmental Research* 64: 699–705.

James, C.S., Easton, J.W., and Hardwick, K. 1999. Competition between three submerged macrophytes, *Elodea canadensis* Michx, *Elodea nuttallii* (Planch) St. John and *Lagarosiphon major* (Ridl.) moss. *Hydrobiologia* 415: 35–40.

Janzen, D.H. 1985. Mangroves: where's the understory? *Journal of Tropical Ecology* 1: 89–92.

Jefferies, R.L. 1977. The vegetation of salt marshes at some coastal sites in arctic North America. *Journal of Ecology* 65: 661–672.

Jefferies, R.L. and Rudmik, T. 1991. Growth, reproduction and resource allocation in halophytes. *Aquatic Botany* 39: 3–16.

Jervis, R.A. 1969. Primary productivity in the freshwater marsh ecosystem of Troy Meadows, New Jersey. *Bulletin of the Torrey Botanical Club* 96: 209–231.

Jespersen, D.N., Sorrell, B.K., and Brix, H. 1998. Growth and root oxygen release by *Typha latifolia* and its effects on sediment methanogenesis. *Aquatic Botany* 61: 165–180.

John, C.D., Limpinuntana, V., and Greenway, H. 1974. Adaptation of rice to anaerobiois. *Australian Journal of Plant Physiology* 1: 513–520.

Johnson, A.M. and Leopold, D.J. 1994. Vascular plant species richness and rarity across a minerotrophic gradient in wetlands of St. Lawrence County, New York, USA. *Biodiversity and Conservation* 3: 606–627.

Johnson, F.L. and Bell, D.T. 1976. Plant biomass and net primary production along a flood-frequency gradient in the streamside forest. *Castanea* 41: 156–165.

Johnson, N.C., Tilman, D., and Wedin, D. 1992. Plant and soil controls on mycorrhizal fungal communities. *Ecology* 73: 2034–2042.

Johnson, R.L., Gross, E.M., and Hairston, N.G. 1997. Decline of the invasive submersed macrophytes *Myriophyllum spicatum* (Haloragaceae) associated with herbivory by larvae of *Acentria ephemerella* (Lepidoptera). *Aquatic Ecology* 31: 273–282.

Johnston, C.A. 1991. Sediment and nutrient retention by freshwater wetlands: effects on surface water quality. *Critical Reviews in Environmental Control* 21: 491–565.

Johnston, C.A. 1994. Ecological engineering of wetlands by beavers. In *Global Wetlands: Old World and New*. W.J. Mitsch, Ed. pp. 379–384. Amsterdam. Elsevier Science.

Johnston, C.A. and Naiman, R.J. 1990. Aquatic patch creation in relation to beaver population trends. *Ecology* 71: 1617–1621.

Johnston, C.A., Bubenzer, G.D., Lee, G.B., Madison, F.W., and McHenry, J.R. 1984. Nutrient trapping by sediment deposition in a seasonally flooded lakeside wetland. *Journal of Environmental Quality* 13: 283–290.

Johnstone, I.M. 1983. Succession in zoned mangrove communities: where is the climax? In B*iology and Ecology of Mangroves*. H.J. Teas, Ed. pp. 131–139. The Hague. Dr. W. Junk Publishers.

Jones, M.B. 1987. The photosynthetic characteristics of papyrus in a tropical swamp. *Oecologia* 71: 355–359.

Jones, M.B. 1988. Photosynthetic responses of C_3 and C_4 wetland species in a tropical swamp. *Journal of Ecology* 76: 253–262.

Jones, R.H., Sharitz, R.R., and McLeod, K.W. 1989. Effects of flooding and root competition on growth of shaded bottomland hardwood seedlings. *The American Midland Naturalist* 121: 165–175.

Jordan, T.E., Whigam, D.F., and Correll, D.L. 1990. Effects of nutrient and litter manipulations on the narrow-leaved cattail, *Typha angustifolia* L. *Aquatic Botany* 36: 179–191.

Jordan, F., Jelks, H.L., and Kitchens, W.M. 1997. Habitat structure and plant community composition in a northern Everglades wetland landscape. *Wetlands* 17: 275–283.

Junk, W.J. and Welcomme, R.L. 1990. *Floodplains in Wetlands and Shallow Continental Water Bodies. Vol.1: Natural and Human Relationships.* B.C. Patten, Ed. pp. 491–524. The Hague. SPB Academic Publishing.

Justin, S. and Armstrong, W. 1987. The anatomical characteristics of roots and plant response to soil flooding. *New Phytologist* 106: 465–495.

Kadlec, R.H. 1985. Aging phenomenon in wastewater wetlands. In *Ecological Considerations in Wetland Treatment of Municipal Wastewaters*. P.J. Godfrey, Ed. pp. 338–350. New York. Van Nostrand Reinhold.

Kadlec, R.H. 1987. Northern natural wetland water treatment systems. In *Aquatic Plants for Water Treatment and Resource Recovery*. K.R. Reddy and W.H. Smith. Eds. Orlando, FL. Magnolia Publishing.

Kadlec, R.H. 1995. Overview: surface flow constructed wetlands. *Water Science and Technology* 32: 1–12.

Kadlec, R.H. 1997. An autobiotic wetland phosphorus model. *Ecological Engineering* 8: 145–172.

Kadlec, R.H. 1999. The limits of phosphorus removal in wetlands. *Wetlands Ecology and Management* 7: 165–175.

Kadlec, R.H. and Kadlec, J.A. 1979. Wetlands and water quality. In *Wetland Functions and Values: The State of Our Understanding*. P.E. Greeson, J.R. Clark, and J.E. Clark, Eds. pp. 436–456. Minneapolis, MN. American Water Resources Association.

Kadlec, R.H. and Knight, R.L. 1996. *Treatment Wetlands*, 893 pp. New York. Lewis Publishers.

Kangas, P.C. and Lugo, A.E. 1990. The distribution of mangroves and saltmarsh in Florida. *Tropical Ecology* 31: 32–39.

Kantrud, H.A., Krapu, G.L., and Swanson, G.A. 1989. *Prairie Basin Wetlands of the Dakotas: A Community Profile*, 116 pp. Washington, D.C. U.S. Department of the Interior, U.S. Fish and Wildlife Service.

Karlsson, P.S., Nordell, K.O., Carlsson, B.A., and Svensson, B.M. 1991. The effect of soil nutrient status on prey utilization in four carnivorous plants. *Oecologia* 86: 1–7.

Karnchanawong, S. and Sanjitt, J. 1995. Comparative study of domestic wastewater treatment efficiencies between facultative pond and water spinach pond. *Water Science and Technology* 32: 262–270.

Karr, J.R. 1991. Biological integrity: a long neglected aspect of water resource management. *Ecological Applications* 1: 66–84.

Karr, J.R. 1993. Measuring biological integrity: lessons from streams. In *Ecological Integrity and the Management of Ecosystems*. S. Woodley, J. Kay, and G. Francis, Eds. pp. 83–104. Delray Beach, FL. St. Lucie Press.

Karr, J.R. 1994. Biological monitoring: challenges for the future. In *Biological Monitoring of Aquatic Systems*. S.L. Loeb and A. Spacie, Eds. pp. 357–373. Boca Raton, FL. CRC Press.

Karr, J.R. and Dudley, D.R. 1981. Ecological perspective on water quality goals. *Environmental Management* 5: 55–68.

Karr, J.R. and Chu, E.W. 1997. *Biological Monitoring and Assessment: Using Multimetric Indexes Effectively*, 119 pp. Seattle. University of Washington.

Kaswadji, R.F., Gosselink, J.G., and Turner, R.E. 1990. Estimation of primary production using five different methods in a *Spartina alterniflora* salt marsh. *Wetlands Ecology and Management* 1: 57–64.

Kaul, R.D. 1976. Anatomical observations on floating leaves. *Aquatic Botany* 2: 215–234.

Kautsky, L. 1988. Life strategies of aquatic soft bottom macrophytes. *Oikos* 53:126–135.

Kawase, M. 1971. Causes of centrifugal root promotion. II. Ethylene. *HortScience* 6: 282.

Kay, S.H. and Rice, J.A. 1999. Using Grass Carp for Aquatic Weed Management in North Carolina, North Carolina State University and A&T State University. Website: http://www.ces.ncsu.edu/nreos/wild/aquatics/grass_carp.html.

Keammerer, W.R., Johnson, W.C., and Burgess, R.L. 1975. Floristic analysis of the Missouri River bottomland forests in North Dakota. *Canadian Field Naturalist* 89: 5–19.

Keast, A. 1984. The introduced macrophyte, *Myriophyllum spicatum*, as habitat for fish and their invertebrate prey. *Canadian Journal of Zoology* 62: 1289–1303.

Keddy, P.A. 1999. Wetland restoration: the potential for assembly rules in the service of conservation. *Wetlands* 19: 716–732.

Keddy, P.A. 2000. Wetland Ecology: Principles and Conservation. In *Cambridge Studies in Ecology* H.J.B. Birks and J.A. Wiens, Eds. 614 pp. Cambridge. Cambridge University Press.

Keddy, P.A. and Reznicek, A.A. 1986. Great Lakes vegetation dynamics: the role of fluctuating water levels and buried seeds. *Journal of Great Lakes Research* 12: 25–36.

Keddy, P.A., Lee, H.T., and Wisheu, I.C. 1993. Choosing indicators of ecosystem integrity: wetlands as a model system. In *Ecological Integrity and the Management of Ecosystems.* S. Woodley, J. Kay, and G. Francis, Eds. Ottawa. St. Lucie Press.

Keeley, J.E. 1979. Population differentiation along a flood frequency gradient: physiological adaptation to flooding in *Nyssa sylvatica. Ecological Monographs* 49: 89–108.

Keeley, J.E. 1980. Endomycorrhizae influence growth of blackgum seedlings in flooded soils. *American Journal of Botany* 67: 6–9.

Keeley, J.E. 1988. Anaerobiosis as a stimulus to germination in two vernal pool grasses. *American Journal of Botany* 75: 1086–1089.

Kellison, R.C. and Young, M.J. 1997. The bottomland hardwood forest of the southern United States. *Forest Ecology and Management* 90: 101–115.

Kemp, W.M. and Boynton, W. 1980. Influence of biological and physical processes on dissolved oxygen dynamics in an estuarine system: implications for measurement of community metabolism. *Estuarine and Coastal Marine Science* 2: 407–431.

Kemp, W.M., Lewis, M.R., and Jones, T.W. 1986. Comparison of methods for measuring production by the submersed macrophyte, *Potamogeton perfoliatus* L. *Limnology and Oceanography* 31: 1322–1334.

Kennedy, R.A., Rumpho, M.E., and Fox, T.C. 1987. Germination physiology of rice and rice weeds: metabolic adaptations to anoxia. In *Plant Life in Aquatic and Amphibious Habitats.* R.M.M. Crawford, Ed. pp. 193–203. Oxford. Blackwell Scientific.

Kennedy, R.A., Rumpho, M.E., and Fox, T.C. 1992. Anaerobic metabolism in plants. *Plant Physiology* 100: 1–6.

Kentula, M.E. 1997. A comparison of approaches to prioritizing sites for riparian restoration. *Restoration Ecology* 5(Suppl.): 69–74.

Kevern, N.R., Wilhm, J.L., and Dyne, G.M.V. 1966. Use of artificial substrata to estimate the productivity of periphyton communities. *Limnology and Oceanography* 11: 499.

Kilham, P. 1982. The biogeochemistry of bog ecosystems and the chemical ecology of *Sphagnum. The Michigan Botanist* 21: 159–168.

King, G.M., Klug, M.J., Wiegert, R.G., and Chalmers, A.G. 1982. Relation of soil water movement and sulfide concentration to *Spartina alterniflora* production in a Georgia salt marsh. *Science* 218: 61–63.

Kiorboe, T. 1980. Distribution and production of submerged macrophytes in Tipper Grund (Rinkobing Fjord, Denmark), and the impact of waterfowl grazing. *Journal of Applied Ecology* 17: 675–687.

Kirby, C.J. and Gosselink, J.G. 1976. Primary production in a Louisiana Gulf Coast *Spartina alterniflora* marsh. *Ecology* 57: 1052–1059.

Kirk, J.T.O. 1994. *Light and Photosynthesis in Aquatic Ecosystems*, 509 pp. Cambridge. Cambridge University Press.

Kirkman, L.K., Goebel, P.C., West, L., Drew, M.B., and Palik, B.J. 2000. Depressional wetland vegetation types: a question of plant community development. *Wetlands* 20: 373–385.

Kistritz, R.U., Hall, K.J., and Yesaki, I. 1983. Productivity, detritus flux, and nutrient cycling in a *Carex lyngbyei* tidal marsh. *Estuaries* 6: 227–236.

Klopatek, J.M. and Stearns, F.W. 1978. Primary productivity of emergent macrophytes in a Wisconsin freshwater marsh ecosystem. *The American Midland Naturalist* 100: 320–332.

Kludze, H.K. and DeLaune, R.D. 1994. Methane emissons and growth of *Spartina patens* in response to soil redox intensity. *Soil Science Society of America Journal* 58: 1838–1845.

Kludze, H.K. and DeLaune, R.D. 1996. Soil redox intensity effects on oxygen exchange and growth of cattail and sawgrass. *Soil Science Society of America Journal* 60: 616–621.

Kludze, H.K., DeLaune, R.D., and Patrick, W.H., Jr. 1993. Aerenchyma formation and methane and oxygen exchange in rice. *Soil Science Society of America Journal* 57: 386–391.

Kludze, H.K., Pezeshki, S.R., and DeLaune, R.D. 1994. Evaluation of root oxygenation and growth in baldcypress in response to short-term soil hypoxia. *Canadian Journal of Forest Research* 24: 804–809.

Knight, R.L. 1997. Wildlife habitat and public use benefits of treatment wetlands. *Water Science and Technology* 35: 35–43.

Knight, R.L., McKim, T.W., and Kohl, H.R. 1987. Performance of a natural wetland treatment system for wastewater management. *Journal Water Pollution Control Federation* 59: 746–754.

Knight, S.E. 1992. Costs of carnivory in the common bladderwort, *Utricularia macrorhiza. Oecologia* 89: 348–355.

Knight, S.E. and Frost, T.M. 1991. Bladder control in *Utricularia macrorhiza*: lake-specific variation in plant investment in carnivory. *Ecology* 72: 728–734.

Knopf, F.L. and Samson, F.B. 1994. Scale perspectives on avian diversity in western riparian ecosystems. *Conservation Biology* 8: 669–676.

Koch, M.S. and Mendelssohn, I.A. 1989. Sulfide as a soil phytotoxin: differential responses of two marsh species. *Journal of Ecology* 77: 565–578.

Koch, M.S. and Reddy, K.R. 1992. Distribution of soil and plant nutrients along a trophic gradient in the Florida Everglades. *Soil Science Society of America* 56: 1492–1499.

Koch, M.S., Mendelssohn, I.A., and McKee, K.L. 1990. Mechanism for the hydrogen sulfide-induced growth limitation in wetland macrophytes. *Limnology and Oceanography* 35: 399–408.

Koles, S.M., Petrell, R.J., and Bagnall, L.O. 1987. Duckweed culture for reduction of ammonia, phosphorus and suspended solids from algal-rich water. In *Aquatic Plants for Water Treatment and Resource Recovery*. K.R. Reddy and W.H. Smith, Eds. pp. 769–774. Orlando, FL. Magnolia Publishing.

Koncalova, H. 1990. Anatomical adaptations to waterlogging in roots of wetland graminoids: limitations and drawbacks. *Aquatic Botany* 38: 127–134.

Koncalova, H., Pokorny, J., and Kvet, J. 1988. Root ventilation in *Carex gracilis* Curt.: diffusion or mass flow? *Aquatic Botany* 30: 149–155.

Kooser, J.G. and Garono, R.J. 1993. Ordination of wetland insect populations: continued evaluation of a potential mitigation monitoring tool. *The Ohio Journal of Science* 93: 44.

Krabel, D., Eschrich, W., Gamelei, Y.V., Fromm, J., and Ziegler, H. 1995. Acquisition of carbon in *Elodea canadensis* Michx. *Journal of Plant Physiology* 145: 50–56.

Kubin, P. and Melzer, A. 1997. Chronological relationship between eutrophication and reed decline in three lakes of southern Germany. *Folia Geobotanica Phytotaxonomica* 32: 15–23.

Kuhn, N.L. and Zedler, J.B. 1997. Differential effects of salinity and soil saturation on native and exotic plants of a coastal salt marsh. *Estuaries* 20: 391–403.

Kushlan, J.A. 1990. Freshwater marshes. In *Ecosystems of Florida*. R.L. Myers and J.J. Ewel, Eds. pp. 324–363. Orlando. University of Central Florida Press.

Laan, P., Berrevoets, M.J., Lythe, S., Armstrong, W., and Blom, C.W.P.M. 1989. Root morphology and aerenchyma formation as indicators of the flood-tolerance of *Rumex* species. *Journal of Ecology* 77: 693–703.

Laanbroek, H.J. 1990. Bacterial cycling of minerals that affect plant growth in waterlogged soils: a review. *Aquatic Botany* 38: 109–125.

Lafleur, P.M. 1990a. Focus: aspects of the physical geography of wetlands. *The Canadian Geographer* 34: 79–88.

Lafleur, P.M. 1990b. Evapotranspiration from sedge-dominated wetland surfaces. *Aquatic Botany* 37: 341–353.

Laing, H.E. 1940. The composition of the internal atmosphere of *Nuphar advenum* and other water plants. *American Journal of Botany* 27: 861–868.

Lambers, H., Chapin, F.S., and Pons, T.L. 1998. *Plant Physiological Ecology*, 540 pp. New York. Springer.

Lamberti, G.A. and Resh, V.H. 1985. Comparability of introduced tiles and natural substrates for sampling lotic bacteria, algae, and macroinvertebrates. *Freshwater Biology* 15: 21.

Langeland, K.A. 1993. *Hydrilla* response to Mariner applied to lakes. *Journal of Aquatic Plant Management* 31: 175–178.

Langeland, K.A. and Laroche, F.B. 1994. Persistence of bensulfuron methyl and control of hydrilla in shallow ponds. *Journal of Aquatic Plant Management* 32: 12–14.

Langis, R., Zalejko, M., and Zedler, J.B. 1991. Nitrogen assessments in a constructed and a natural salt marsh of San Diego Bay. *Ecological Applications* 1: 40–51.

Larsen, J.A. 1982. *Ecology of the Northern Lowland Bogs and Conifer Forests*, 307 pp. New York. Academic Press.

Larsen, V.J. and Schierup, H.H. 1981. Macrophyte cycling of zinc, copper, lead and cadmium in the littoral zone of a polluted and a non-polluted lake. II. Seasonal changes in heavy metal content of aboveground biomass and decomposing leaves of *Phragmites australis* (Cav.) Trin. *Aquatic Botany* 11: 211–230.

Laurentius, A.C., Voesenek, J., Banga, M., Rijnders, J.G.H.M., Visser, E.J.W., and Blom, C.W.P.M. 1996. Hormone sensitivity and plant adaptations to flooding. *Folia Geobotanica Phytotaxonomica* 31: 47–56.

Lauridsen, T.L., Jeppesen, E., and Sondergaard, M. 1994. Colonization and succession of submerged macrophytes in shallow Lake Vaeng during the first five years following fish manipulation. *Hydrobiologia* 275–276: 233–242.

Laushman, R.H. 1993. Population genetics of hydrophilous angiosperms. *Aquatic Botany* 44: 147–158.

Laws, E.A. 1993. *Aquatic Pollution: An Introductory Text*, 611 pp. New York. John Wiley & Sons.

Leach, J.H. 1995. Non-indigenous species in the Great Lakes: were colonization and damage to ecosystem health predictable? *Journal of Aquatic Ecosystem Health* 4: 117–128.

Leakey, R.R.B. 1981. Adaptive biology of vegetatively regenerating weeds. *Advances in Applied Biology* 6: 57–90.

Leck, M.A. 1989. Wetland seed banks. In *Ecology of Soil Seed Banks*, pp. 283–305. New York. Academic Press.

Leck, M.A. and Simpson, R.L. 1987. Seed bank of a freshwater tidal wetland: turnover and relationship to vegetation change. *American Journal of Botany* 74: 360–370.

Leck, M.A. and Simpson, R.L. 1994. Tidal freshwater wetland zonation: seed and seedling dynamics. *Aquatic Botany* 47: 61–75.

Leck, M.A. and Simpson, R.L. 1995. Ten-year seed bank and vegetation dynamics of a tidal freshwater marsh. *American Journal of Botany* 82: 1547–1557.

Lee, G.F., Bentley, E., and Amundsen, R. 1975. Effects of marshes on water quality systems. In *Coupling of Land and Water Systems*. A.D. Hasler, Ed. pp. 105–127. New York. Springer-Verlag.

Lee, R.W., Kraus, D.W., and Doeller, J.E. 1999. Oxidation of sulfide by *Spartina alterniflora* roots. *Limnology and Oceanography* 44: 1155–1159.

Lefeuvre, J.C. and Dame, R. 1994. Comparative studies of salt marsh processes in the new and old worlds: an introduction. In *Global Wetlands: Old World and New*. W.J. Mitsch, Ed. pp. 169–179. Amsterdam. Elsevier Science.

Lehmann, A. and Lachavanne, J.B. 1997. Geographic information systems and remote sensing in aquatic botany. *Aquatic Botany* 58: 195–207.

Lentz, K.A. and Dunson, W.A. 1999. Distinguishing characteristics of temporary pond habitat of endangered northeastern bulrush, *Scirpus ancistrochaetus*. *Wetlands* 19: 162–167.

Les, D.H. 1988. Breeding systems, population structure, and evolution in hydrophilous angiosperms. *Annals of the Missouri Botanical Garden* 75: 819–835.

Les, D.H. and Schneider, E.L. 1995. The Nymphaeales, Alismatidae, and the theory of an aquatic monocotyledon origin. In *Monocotyledons: Systematics and Evolution*. P.J. Rudall, P.J. Cribb, D.F. Cutler, and C.J. Humphries, Eds. pp. 23–42. Kew. Royal Botanic Gardens.

Les, D.H., Garvin, D.K., and Wimpee, C.F. 1991. Molecular evolutionary history of ancient aquatic angiosperms. *Proceedings of the National Academy of Science, U.S.A.* 88: 10119–10123.

Les, D.H., Cleland, M.A., and Waycott, M. 1997. Phylogenetic studies in Alismatidae. II. Evolution of marine angiosperms (seagrasses) and hydrophily. *Systematic Botany* 22: 443–463.

Les, D.H., Mehrhoff, L.J., Cleland, M.A., and Gabel, J.D. 1997. *Hydrilla verticillata* (Hydrocharitaceae) in Connecticut. *Journal of Aquatic Plant Management* 35: 10–14.

Lesica, P. 1992. Autecology of the endangered plant *Howellia aquatilis;* implications for management and reserve design. *Ecological Applications* 2: 411–421.

Levine, J., Brewer, J.S., and Bertness, M.D. 1998. Nutrients, competition and plant zonation in a New England salt marsh. *Journal of Ecology* 86: 285–292.

Li, M.R., Wedin, D.A., and Tieszen, L.L. 1999. C_3 and C_4 photosynthesis in *Cyperus* (Cyperaceae) in temperate eastern North America. *Canadian Journal of Botany* 77: 208–218.

Li, M.S. and Lee, S.Y. 1997. Mangroves of China: a brief review. *Forest Ecology and Management* 96: 241–259.

Lind, O.T. 1985. *Handbook of Common Methods in Limnology*, 199 pp. Dubuque, IA. Kendall/Hunt.

Lindeman, R.L. 1941. The developmental history of Cedar Creek Bog, Minnesota. *The American Midland Naturalist* 25: 101–112.

Lindsey, A.A., Petty, R.O., Sterlin, D.K., and Asdall, W.V. 1961. Vegetation and environment along the Wabash and Tippecanoe rivers. *Ecological Monographs* 31: 105–156.

Linhart, Y.B. and Baker, G. 1973. Intra-population differentiation of physiological response to flooding in a population of *Veronica peregrina* L. *Nature* 242: 275.

Linthurst, R.A. and Blum, U. 1981. Growth modifications of *Spartina alterniflora* Loisel. by the interaction of pH and salinity under controlled conditions. *Journal of Experimental Marine Biology and Ecology* 55: 207–218.

Linthurst, R.A. and Reimold, R.J. 1978a. Estimated net aerial primary productivity for selected estuarine angiosperms in Maine, Delaware, and Georgia. *Ecology* 59: 945–955.

Linthurst, R.A. and Reimold, R.J. 1978b. An evaluation of methods for estimating the net aerial primary production of estuarine angiosperms. *Journal of Applied Ecology* 15: 919–931.

Linthurst, R.A. and Seneca, E.D. 1981. Aeration, nitrogen and salinity as determinants of *Spartina alterniflora* Loisel. growth response. *Estuaries* 4: 53–63.

Lissner, J. and Schierup, H.H. 1997. Effects of salinity on the growth of *Phragmites australis. Aquatic Botany* 55: 247–260.

Little, E. 1980. *The Audobon Society Field Guide to North American Trees*, 714 pp. New York. Alfred A. Knopf, Inc.

Livingston, E.H. 1989. Use of wetlands for urban stormwater management. In *Constructed Wetlands for Wastewater Treatment*. D.A. Hammer, Ed. pp. 253–264. Chelsea, MI. Lewis Publishers.

Lloyd, F.E. 1942. *The Carnivorous Plants*, 352 pp. New York. Ronald Press.

Lodge, D.M. 1991. Herbivory on freshwater macrophytes. *Aquatic Botany* 41: 195–224.

Lodge, D.M. and Lorman, J.G. 1987. Reductions in submersed macrophyte biomass and species richness by the crayfish *Orconectes rusticus. Canadian Journal of Fisheries and Aquatic Science* 44: 591–597.

Lomnicki, A., Bandola, E., and Jankowska, K. 1968. Modification of the Wiegert-Evans method for estimation of net primary production. *Ecology* 49: 147–149.

Longstreth, D.J. 1989. Photosynthesis and photorespiration in freshwater emergent and floating plants. *Aquatic Botany* 34: 287–299.

Longstreth, D.J. and Strain, B.R. 1977. Effects of salinity and illumination on photosynthesis and water balance of *Spartina alterniflora* Loisel. *Oecologia* 31: 191–199.

Longstreth, D.J., Bolanos, J.A., and Smith, J.E. 1984. Salinity effects on photosynthesis and growth in *Alternanthera philoxeroides* (Mart.) Griseb. *Plant Physiology* 75: 1044–1047.

Lopez, R.D. and Fennessy, M.S. in press. Testing the floristic quality assessment index as an indicator of wetland condition along gradients of human influence. *Ecological Applications*.

Loveless, C.M. 1959. A study of the vegetation in the Florida Everglades. *Ecology* 40: 1–9.

Lovett-Doust, L. and Lovett-Doust, J. 1995. Wetland management and conservation of rare species. *Canadian Journal of Botany* 73: 1019–1028.

Lowrance, R.R., Vellidis, G., and Hubbard, R.K. 1995. Denitrification in a restored riparian forest wetland. *Journal of Environmental Quality* 24: 808.

Lowrie, A. 1998. *Carnivorous Plants of Australia*, Vol. 3, 288 pp. Perth. University of Western Australia Press.

Lucas, W.J. 1983. Photosynthetic assimilation of exogenous HCO_3^- by aquatic plants. *Annual Review of Plant Physiology* 34: 71–104.

Lugo, A.E. 1980. Mangrove ecosystems: successional or steady state? *Biotropica* 12: 65–72.

Lugo, A.E. 1986. Mangrove understory: an expensive luxury? *Journal of Tropical Ecology* 2: 287–288.
Lugo, A.E. 1990. Introduction. In *Forested Wetlands Ecosystems of the World*. A.E. Lugo, M.M. Brinson, and S. Brown, Eds. pp. 1–14. Amsterdam. Elsevier Science.
Lugo, A.E. 1994. Maintaining an open mind on exotic species. In *Principles of Conservation Biology*. G. Meffe and C. Carroll, Eds. pp. 218–220. Sunderland, MA. Sinauer Associates, Inc.
Lugo, A.E. 1997. Old-growth mangrove forests in the United States. *Conservation Biology* 11: 11–20.
Lugo, A.E. and Snedaker, S.C. 1974. The ecology of mangroves. *Annual Review of Ecology and Systematics* 5: 39–64.
Lugo, A.E., Gonzalez-Liboy, J.A., Cintron, B., and Dugger, K. 1978. Structure, productivity and transpiration of a subtropical dry forest in Puerto Rico. *Biotropica* 10: 278–291.
Lugo, A.E., Brown, S., and Brinson, M.M. 1988. Forested wetlands in freshwater and saltwater environments. *Limnology and Oceanography* 33: 894–909.
Luken, J.O. 1994. Valuing plants in natural areas. *Natural Areas Journal* 14: 295–299.
Luken, J.O. 1997. Management of plant invasions: implicating ecological succession. In *Assessment and Management of Plant Invasions*. J.O. Luken and J.W. Thieret, Eds. pp. 133–144. New York. Springer.
Ma, X., Liua, X., and Wang, R. 1993. China's wetlands and agro-ecological engineering. *Ecological Engineering* 2: 291–302.
Maberly, S.C. and Spence, D.H.N. 1989. Photosynthesis and photorespiration in freshwater organisms: amphibious plants. *Aquatic Botany* 34: 267–286.
MacDonald, G.E., Shilling, D.G., Doong, R.L., and Haller, W.T. 1993. Effects of fluridone on hydrilla growth and reproduction. *Journal of Aquatic Plant Management* 31: 195–198.
MacDonald, K.B. and Barbour, M.G. 1974. Beach and salt marsh vegetation of the North American Pacific coast. In *Ecology of Halophytes*. R.J. Reimold and W.H. Queen, Eds. pp. 175–233. New York. Academic Press.
Mack, J.J., Micacchion, M., Augusta, L.D., and Sablak, G.R. 2000. Vegetation Indices of Biotic Integrity (VIBI) for Wetlands and Calibration of the Ohio Rapid Assessment Method for Wetlands. Grant CD95276. Final Report to U.S. Environmental Protection Agency. Columbus, OH. Ohio Environmental Protection Agency, Wetlands Unit, Division of Surface Water.
MacKay, W.P., Zak, J., and Whitford, W. 1992. Litter decomposition in a Chihuahuan desert playa. *The American Midland Naturalist* 128: 89–94.
Madsen, J.D. 1991. Resource allocation at the individual plant level. *Aquatic Botany* 41: 67–86.
Madsen, J.D. 1997. Methods for management of nonindigenous aquatic plants. In *Assessment and Management of Plant Invasions*. J.O. Luken and J.W. Thieret, Eds. pp. 145–171. New York. Springer.
Madsen, J.D. 1998. Predicting invasion success of Eurasian watermilfoil. *Journal of Aquatic Plant Management* 36: 28–32.
Madsen, J.D. and Adams, M.S. 1988. The seasonal biomass and productivity of the submerged macrophytes in a polluted Wisconsin stream. *Freshwater Biology* 20: 41–50.
Madsen, J.D., Hartleb, C.F., and Boylen, C.W. 1991. Photosynthetic characteristics of *Myriophyllum spicatum* and six submersed aquatic macrophyte species native to Lake George, New York. *Freshwater Biology* 26: 233–240.
Madsen, T. and Sand-Jensen, K. 1991. Photosynthetic carbon assimilation in aquatic macrophytes. *Aquatic Botany* 41: 5–40.
Madsen, T. and Sondergaard, M. 1983. The effect of current velocity on the photosynthesis of *Callitriche stagnalis* Scop. *Aquatic Botany* 15: 187–193.
Madsen, T., Enevoldsen, H.O., and Jorgensen, T.B. 1993. Effects of water velocity on photosynthesis and dark respiration in submerged stream macrophytes. *Plant, Cell and Environment* 16: 317–322.
Maehr, D.S. 1997. *The Florida Panther: Life and Death of a Vanishing Carnivore*, 320 pp. Washington, D.C. Island Press.
Mahall, B.E. and Park, R.B. 1976. The ecotone between *Spartina foliosa* Trin. and *Salicornia virginica* L. in salt marshes of northern San Francisco Bay. I. Biomass and production. *Journal of Ecology* 64: 421–433.
Mal, T., Lovett-Doust, J., Lovett-Doust, L., and Mulligan, G. 1992. The biology of Canadian weeds. 100. *Lythrum salicaria*. *Canadian Journal of Plant Science* 72: 1305–1330.

Malecki, R.A. and Rawinski, T.J. 1985. New methods for controlling purple loosestrife. *New York Fish and Game Journal* 32: 9–19.

Malecki, R.A., Lassoie, J.R., Rieger, E., and Seamans, T. 1983. Effects of long-term artificial flooding on a northern bottomland hardwood forest community. *Forest Science* 29: 535–544.

Malecki, R.A., Blossey, B., Hight, S.D., Schroeder, D., Kok, L.T., and Coulson, J.R. 1993. Biological control of purple loosestrife. *BioScience* 43: 680–686.

Maltby, E. 1986. *Waterlogged Wealth,* 200 pp. Washington, D.C. Earthscan.

Malthus, T.J. and George, D.G. 1997. Airborne remote sensing of macrophytes in Cefni Reservoir, Anglesey, UK. *Aquatic Botany* 58: 317–332.

Mann, C.J. and Wetzel, R.G. 1999. Photosynthesis and stomatal conductance of *Juncus effusus* in a temperate wetland ecosystem. *Aquatic Botany* 63: 127–144.

Marcum, K.B. 1999. Salinity tolerance mechanisms of grasses in the subfamily Chloridoideae. *Crop Science* 39: 1153–1160.

Martyn, R.D. 1985. Waterhyacinth decline in Texas caused by *Cercospora piaropi. Journal of Aquatic Plant Management* 23: 29–32.

Maschinski, J., Southam, G., Hines, J., and Strohmeyer, S. 1999. Efficiency of a subsurface constructed wetland system using native southwestern U.S. plants. *Journal of Environmental Quality* 28: 225–231.

Mathews, C.P. and Westlake, D.F. 1969. Estimation of production by populations of higher plants subject to high mortality. *Oikos* 20: 156–160.

Mathias, M.E. and Moyle, P. 1992. Wetland and aquatic habitats. *Agriculture, Ecosystems and Environment* 42: 165–176.

Matthews, G.V.T. 1993. *The Ramsar Convention; Its History and Development.* Gland, Switzerland. Ramsar Convention Bureau.

McCann, J.A., Arkin, L.N., and Williams, J.D. 1996. Nonindigenous aquatic and selected terrestrial species of Florida. Orlando. University of Florida, Center for Aquatic Plants. Website: http://aquat1/ifas.ufl.edu/mctitle.html.

McCaskey, T.A., Britt, S.N., Hannah, T.C., Eason, J.T., Payne, V.W.E., and Hammer, D.A. 1994. Treatment of swine lagoon effluent by constructed wetlands operated at three loading rates. In *Proceedings of a Workshop on Constructed Wetlands for Animal Waste Management.* P.J. DuBowy and R.P. Reaves, Eds. pp. 23–33. Lafayette, IN. Purdue University.

McCreary, N.J. 1991. Competition as a mechanism of submersed macrophyte community structure. *Aquatic Botany* 41: 177–193.

McJannet, C.L., Keddy, P.K., and Pick, F.R. 1995. Nitrogen and phosphorus tissue concentrations in 41 wetland plants: a comparison across habitats and functional groups. *Functional Ecology* 9: 231–238.

McKee, K.L. and Faulkner, P.L. 2000. Restoration of biogeochemical function in mangrove forests. *Restoration Ecology* 8: 247–259.

McKee, K.L., Mendelssohn, I.A., and Hester, M.W. 1988. Reexamination of pore water sulfide concentrations and redox potentials near the aerial roots of *Rhizophora mangle* and *Avicennia germinans. American Journal of Botany* 75: 1352–1359.

McLay, C.L. 1974. The distribution of duckweed *Lemna perpusilla* in a small southern California lake: an experimental approach. *Ecology* 55: 262–276.

McManmon, M. and Crawford, R.M.M. 1971. A metabolic theory of flooding tolerance: the significance of enzyme distribution and behaviour. *New Phytologist* 70: 299–306.

Meffe, G.K. and Carroll, C.R. 1994. *Principles of Conservation Biology,* 600 pp. Sunderland, MA. Sinauer Associates, Inc.

Megonigal, J.P. and Day, F.P. 1992. Effects of flooding on root and shoot production of bald cypress in large experimental enclosures. *Ecology* 73: 1182–1193.

Mendelssohn, I.A. 1979. Nitrogen metabolism in the height forms of *Spartina alterniflora* in North Carolina. *Ecology* 60: 574–584.

Mendelssohn, I.A. and McKee, K.L. 1987. Root metabolism response of *Spartina alterniflora* to hypoxia. In *Plant Life in Aquatic and Amphibious Habitats.* R.M.M. Crawford, Ed. pp. 239–253. Oxford. Blackwell Scientific.

Mendelssohn, I.A. and McKee, K.L. 1988. *Spartina alterniflora* die-back in Louisiana: time-course investigation of soil waterlogging effects. *Journal of Ecology* 76: 509–521.

Mendelssohn, I.A. and Morris, J.T. 2000. Eco-physiological constraints on the primary productivity of *Spartina alterniflora*. In *Concepts and Controversies of Tidal Marsh Ecology*. M.P. Weinstein and D.A. Kreeger, Eds. pp 59–80. Dordrecht. Kluwer Academic Publishers.

Mendelssohn, I.A., McKee, K.L., and Patrick, W.H., Jr. 1981. Oxygen deficiency in *Spartina alterniflora* roots: metabolic adaptation to anoxia. *Science* 214: 439–441.

Menegus, F., Cattaruzza, L., Chersi, A., and Fronza, G. 1989. Differences in the anaerobic lactate-succinate production and in the changes of cell sap pH for plants with high and low resistance to anoxia. *Plant Physiology* 90: 29–32.

Menegus, F., Cattaruzza, L., Mattana, M., Beffagna, N., and Ragg, E. 1991. Response to anoxia in rice and wheat seedlings. *Plant Physiology* 95: 760–767.

Menges, E.S. and Waller, D.M. 1983. Plant strategies in relation to elevation and light in floodplain herbs. *American Naturalist* 122: 454–473.

Messer, J.J., Linthurst, R.A., and Overton, W.S. 1991. An EPA program for monitoring ecological status and trends. *Environmental Monitoring and Assessment* 17: 67–78.

Messmore, N.A. and Knox, J.S. 1997. The breeding system of the narrow endemic, *Helenium virginicum* (Asteraceae). *Journal of the Torrey Botanical Society* 124: 318–321.

Metcalfe, W.S., Ellison, A.M., and Bertness, M.D. 1986. Survivorship and spatial development of *Spartina alterniflora* Loisel. (Gramineae) seedlings in a New England salt marsh. *Annals of Botany* 58: 249–258.

Meyer, J.L. and Edwards, R.T. 1990. Ecosystem metabolism and turnover of organic carbon along a blackwater river continuum. *Ecology* 71: 668–677.

Middleton, B. 1999. *Wetland Restoration, Flood Pulsing, and Disturbance Dynamics*, 384 pp. New York. John Wiley & Sons.

Miles, D.H., Tunsuwan, K., Chittawong, V., Kokpol, U., Choudhary, M.I., and Clardy, J. 1993. Boll weevil antifeedants from *Arundo donax*. *Phytochemistry* 34: 1277–1279.

Miller, R.R. 1996. *Phytoremediation*, 11 pp. Pittsburgh, PA. Ground-Water Remediation Technologies Analysis Center.

Mills, E.L., Leach, J.H., Carlton, J.T., and Secor, C.L. 1993. Exotic species in the Great Lakes: a history of biotic crises and anthropogenic introductions. *Journal of Great Lakes Research* 19: 1–54.

Milner, C. and Hughes, R.E. 1968. Methods for the measurement of the primary production of grasslands. In *IBP Handbook No. 6*. Oxford. Blackwell Scientific.

Misra, S., Choudhury, A., Ghosh, A., and Dutta, J. 1984. The role of hydrophobic substances in leaves in adaptation of plants to periodic submersion by tidal water in a mangrove ecosystem. *Journal of Ecology* 72: 621–625.

Mitsch, W.J. 1988. Productivity-hydrology-nutrient models of forested wetlands. In *Wetland Modelling*. W.J. Mitsch, M. Straskraba, and S.E. Jorgensen, Eds. pp. 115–132. Amsterdam. Elsevier Science.

Mitsch, W.J. 1992. Landscape design and the role of created, restored, and natural riparian wetlands in controlling nonpoint source pollution. *Ecological Engineering* 1: 22–47.

Mitsch, W.J. and Cronk, J.K. 1992. Creation and restoration of wetlands: some design considerations for ecological engineering. In *Advances in Soil Science*, Vol. 17. R. Lal and B.A. Stewart, Eds. pp. 217–259. New York. Springer-Verlag.

Mitsch, W.J. and Ewel, K.C. 1979. Comparative biomass and growth of cypress in Florida wetlands. *The American Midland Naturalist* 101: 417–426.

Mitsch, W.J. and Gosselink, J.G. 2000. *Wetlands*, 3rd edition, 920 pp. New York. John Wiley & Sons.

Mitsch, W.J. and Kaltenborn, K.S. 1980. Effects of copper sulfate application on diel dissolved oxygen and metabolism in the Fox Chain of Lakes. *Transcripts of the Illinois State Academy of Science* 73: 55–64.

Mitsch, W.J. and Reeder, B.C. 1991. Modelling nutrient retention of a freshwater coastal wetland: estimating the roles of primary productivity, sedimentation, resuspension and hydrology. *Ecological Modelling* 54: 151–187.

Mitsch, W.J. and Wilson, R.F. 1996. Improving the success of wetland creation and restoration with know-how, time, and self-design. *Ecological Applications* 6: 77–83.

Mitsch, W.J. and Wu, X. 1995. Wetlands and global change. In *Soil Management and Greenhouse Effect.* R. Lal, J. Kimble, E. Levine, and B.A. Stewart, Eds. pp. 205–230. Boca Raton, FL. Lewis Publishers.

Mitsch, W.J., Taylor, J.R., and Benson, K.B. Eds. 1991. Estimating primary productivity of forested wetland communities in different hydrologic landscapes. *Landscape Ecology* 5: 75–92.

Mitsch, W.J., Mitsch, R.H., and Turner, R.E. 1994. The Wetlands of the Old and New Worlds: ecology and management. In *Global Wetlands: Old World and New.* W.J. Mitsch, Ed. pp. 3–53. Amsterdam. Elsevier Science.

Mitsch, W.J., Cronk, J.K., Wu, X., Nairn, R.W., and Hey, D.L. 1995. Phosphorus retention in constructed freshwater riparian marshes. *Ecological Applications* 5: 830–845.

Mitsch, W.J., Wu, X., Nairn, R.W., Weihe, P.E., Wang, N., Deal, R., and Boucher, C. 1998. Creating and restoring wetlands: a whole-ecosystem experiment in self-design. *BioScience* 48: 1019–1030.

Moen, R.A. and Cohen, Y. 1989. Growth and competition between *Potomogenton pectinatus* L. and *Myriophyllum exalbescens* Fern. in experimental ecosystems. *Aquatic Botany* 33: 257–270.

Molles, M.C., Jr. 1999. *Ecology.* 509 pp. New York. McGraw-Hill.

Monk, L.S., Braendle, R., and Crawford, R.M.M. 1987. Catalase activity and post-anoxic injury in monocotyledonous species. *Journal of Experimental Botany* 38: 233–246.

Montague, T.G. and Givnish, T.J. 1996. Distribution of black spruce versus eastern larch along peatland gradients: relationship to relative stature, growth rate, and shade tolerance. *Canadian Journal of Botany* 74: 1514–1532.

Monteith, J.L. 1965. *Evaporation and Environment, Symposium of the Society for Experimental Biology XIX*, pp. 205–234. London. Cambridge University Press.

Mook, J.H. and van der Toorn, J. 1982. The influence of environmental factors and management on stands of *Phragmites australis.* H.: effects on yield and its relationship with shoot density. *Journal of Applied Ecology* 19: 501–517.

Moore, B.C., Lafer, J.E., and Funk, W.H. 1994. Influence of aquatic macrophytes on phosphorus and sediment porewater chemistry in a freshwater wetland. *Aquatic Botany* 49: 137–148.

Moore, P.D. and Bellamy, D.J. 1974. *Peatlands,* 221 pp. New York. Springer-Verlag.

Morin, P.J. 1999. *Community Ecology,* 320 pp. Oxford. Blackwell Scientific.

Morris, J.T. and Haskin, B. 1990. A 5-yr record of aerial primary production and stand characteristics of *Spartina alterniflora. Ecology* 71: 2209–2217.

Moss, B. 1988. *Ecology of Fresh Waters: Man and Medium,* 417 pp. Oxford. Blackwell Scientific.

Motzkin, G., Patterson, W.A., and Drake, N.E.R. 1993. Fire history and vegetation dynamics of a *Chamaecyparis thyoides* wetland on Cape Cod, Massachusetts. *Journal of Ecology* 81: 391–402.

Mozeto, A.A., De B. Nogueira, F. M., De O.E. Souza, M.H., and Victoria, R.L. 1996. C_3 and C_4 grasses distribution along soil moisture gradient in surrounding areas of the Lobo Dam (Sao Paulo, Brazil). *Anais da Academia Brasileira de Ciencias* 68: 113–121.

Mulamoottil, G., McBean, E.A., and Rovers, F. 1999. *Constructed Wetlands for the Treatment of Landfill Leachates,* 273 pp. Boca Raton, FL. Lewis Publishers.

Mulholland, M.M. and Otte, M.L. 2000. Effects of varying sulphate and nitrogen supply on DMSP and glycine betaine levels in *Spartina anglica. Journal of Sea Research* 43: 199–207.

Müller, N. 1995. River dynamics and floodplain vegetation and their alterations due to human impace. *Archivs für Hydrobiologie* 101: 477–511.

Murdock, N.A. 1994. Rare and endangered plants and animals of southern Appalachian wetlands. *Water Air and Soil Pollution* 77: 385–405.

Myers, R. 1984. Ecological compression of *Taxodium distichum* var. *nutans* by *Melaleuca quinquenervia* in Southern Florida. In *Cypress Swamps.* K.C. Ewel and H.T. Odum, Eds. pp. 358–364. Gainesville. University Presses of Florida.

Naidoo, G., McKee, K.L., and Mendelssohn, I.A. 1992. Anatomical and metabolic responses to waterlogging and salinity in *Spartina alterniflora* and *S. patens* (Poaceae). *American Journal of Botany* 79: 765–770.

Nairn, R.W., Mercer, M.N., and Lipe, S.A. 2000. Alkalinity generation and metals retention in vertical flow treatment wetlands. In *A New Era of Land Reclamation, Proceedings, 2000 Annual Meeting of the American Society for Surface Mining and Reclamation.* W.L. Daniels and S.G. Richardson, Eds. pp. 412–420. Tampa, FL. American Society for Surface Mining and Reclamation.

Narumalani, S., Zhou, Y., and Jensen, J.R. 1997. Application of remote sensing and geographic information systems to the delineation and analysis of riparian buffer zones. *Aquatic Botany* 58: 393–409.

Nastase, A.J., de la Rosa, C., and Newell, S.J. 1995. Abundance of pitcher-plant mosquitoes, *Wyeomyia smithii* (Coq.) (Diptera: Culcidae) and midges, *Metriocnemus knabi* Coq. (Diptera: Chironomidae), in relation to pitcher characteristics of *Sarracenia purpurea* L. *The American Midland Naturalist* 133: 44–51.

Neill, C. 1990. Effects of nutrients and water levels on emergent macrophyte biomass in a prairie marsh. *Canadian Journal of Botany* 68: 1007–1014.

Neill, C. and Deegan, L.A. 1986. The effect of Mississippi River delta lobe development on the habitat composition and diversity of Louisiana coastal wetlands. *The American Midland Naturalist* 116: 296–305.

Nelson, L.S., Netherland, M.D., and Getsinger, K.D. 1993. Bensulfuron methyl activity on Eurasian watermilfoil. *Journal of Aquatic Plant Managment* 31: 179–189.

Nestler, J. 1977. Interstitial salinity as a cause of ecophenic variation in *Spartina alterniflora*. *Estuarine and Coastal Marine Science* 5: 707–714.

Netherland, M.D. 1997. Turion ecology of hydrilla. *Journal of Aquatic Plant Management* 35: 1–10.

Netherland, M.D., Getsinger, K.D., and Turner, E.G. 1993. Fluridone concentration and exposure time requirements for control of Eurasian watermilfoil and hydrilla. *Journal of Aquatic Plant Management* 31: 189–194.

Newbould, P.J. 1970. Methods for estimating the primary production of forests. In *International Biological Programme*, Vol. No. 2, 60 pp. London. Blackwell Scientific.

Newman, S., Schuette, J., Grace, J.B., Rutchey, K., Fontaine, T., Reddy, K.R., and Pietrucha, M. 1998. Factors influencing cattail abundance in the northern Everglades. *Aquatic Botany* 60: 265–280.

Nichols, D.S. 1983. Capacity of natural wetlands to remove nutrients from wastewater. *Journal of the Water Pollution Control Federation* 55: 495–505.

Nichols, S.A. 1991. The interaction between biology and the management of aquatic macrophytes. *Aquatic Botany* 41: 225–252.

Niering, W.A. 1988. Endangered, threatened and rare wetland plants and animals of the continental United States. In *The Ecology and Management of Wetlands, Volume 1: Ecology of Wetlands*. D.D. Hook, Ed. pp. 227–238. Portland, OR. Timber Press.

Niering, W.A. 1990. Vegetation dynamics in relation to wetland creation. In *Wetland Creation and Restoration: The Status of The Science*. J.A. Kusler and M.E. Kentula, Eds. pp. 479–486. Washington, D.C. Island Press.

Niering, W.A. and Warren, R.S. 1980. Vegetation patterns and processes in New England salt marshes. *BioScience* 30: 301–307.

Nilsson, C. 1987. Distribution of stream-edge vegetation along a gradient of current velocity. *Journal of Ecology* 75: 513–522.

Nilsson, C., Ekblad, A., Dynesius, M., Backe, S., Gardfjell, M., Carlberg, B., Hellqvist, S., and Jansson, R. 1994. A comparison of species richness and traits of riparian plants between a main river channel and its tributaries. *Journal of Ecology* 82: 281–295.

Niswander, S.F., and Mitsch, W.J. 1995. Functional analysis of a two-year-old created in-stream wetland: hydrology, phosphorus retention, and vegetation survival and growth. *Wetlands* 15: 212–225.

Nixon, S.W. and Oviatt, C.A. 1973. Ecology of a New England salt marsh. *Ecological Monographs* 43: 463–498.

Notzold, R., Blossey, B., and Newton, E. 1998. The influence of below ground herbivory and plant competition on growth and biomass allocation of purple loosestrife. *Oecologia* 113: 82–93.

Nyvall, R.F. 1995. Fungi associated with purple loosestrife (*Lythrum salicaria*) in Minnesota. *Mycologia* 87: 501–506.

Nyvall, R.F. and Hu, A. 1997. Laboratory evaluation of indigenous North American fungi for biological control of purple loosestrife. *Biological Control* 8: 37–42.

Odum, E.P. 1969. The strategy of ecosystem development. *Science* 164: 262–270.

Odum, E.P. 1971. *Fundamentals of Ecology*, 574 pp. Philadelphia. W.B. Saunders.

Odum, E.P. 1985. Trends expected in stressed ecosystems. *BioScience* 35: 419–423.

Odum, H.T. 1956. Primary production in flowing waters. *Limnology and Oceanography* 1: 102–117.
Odum, H.T., Ewel, K.C., Mitsch, W.J., and Ordway, J.W. 1977. Recycling treated sewage through cypress wetlands in Florida. In *Wastewater Renovation and Reuse*. F.M. D'Itri, Ed. pp. 35–67. New York. Marcel Dekker.
Odum, H.T., Wojcik, W., Pritchard, L., Jr., Ton, S., Delfino, J.J., Woojcik, M., Leszczynski, S., Patel, J.D., Doherty, S.J., and Stasik, J. 2000. *Heavy Metals in the Environment: Using Wetlands for Their Removal*, 326 pp. Boca Raton, FL. Lewis Publishers.
Odum, W.E. and McIvor, C.C. 1990. Mangroves. In *Ecosystems of Florida*. R.L. Myers and J.J. Ewel, Eds. pp. 517–548. Orlando. University of Florida Press.
Odum, W.E., Fisher, J.S., and Pickral, J.C. 1979. Factors controlling the flux of particulate organic carbon from estuarine wetlands. In *Ecological Processes in Coastal and Marine Systems*. R.J. Livingston, Ed. pp. 69–80. New York. Plenum.
Odum, W.E., Smith, T.J., Hoover, J.K., and McIvor, C.C. 1984. The Ecology of Tidal Freshwater Marshes of the United States East Coast: A Community Profile. Technical Report FWS/OBS 81–24. Washington, D.C. U.S. Department of the Interior, U.S. Fish and Wildlife Service, Office of Biological Services.
Office of Technology Assessment. 1993. Harmful Non-Indigenous Species in the United States. Report Number OTA-F-565. Washington, D.C. U.S. Congress, U.S. Government Printing Office.
O'Hare, N.K. and Dalrymple, G.H. 1997. Wildlife in southern Everglades wetlands invaded by Melaleuca (*Melaleuca quinquenervia*). *Bulletin of the Florida Museum of Natural History* 41: 1–68.
Olff, H., Leeuw, J.D., Bakker, J.P., Platerink, R.J., Wijnen, H.J.V., and Munck, W.D. 1997. Vegetation succession and herbivory in a salt marsh: changes induced by sea level rise and silt deposition along an elevational gradient. *Journal of Ecology* 85: 799–814.
Olmsted, I. 1993. Wetlands of Mexico. In *Wetlands of the World I: Inventory, Ecology and Management*. D.F. Whigham, D. Dykyjova, and S. Hejny, Eds. pp. 637–677. Dordrecht. Kluwer Academic Publishers.
Omernik, J.M. 1995. Ecoregions: a spatial framework for environmental management. In *Biological Assessment and Criteria: Tools for Water Resource Planning and Decision Making*. W.S. Davis and T.P. Simon, Eds. pp. 49–62. Boca Raton, FL. Lewis Publishers.
Ornes, W.H. and Kaplan, D.I. 1989. Macronutrient status of tall and short forms of *Spartina alterniflora* in a South Carolina salt marsh. *Marine Ecology Progress Series* 55: 63–72.
Orth, R.J. and Moore, K.A. 1983. Chesapeake Bay: an unprecedented decline in submerged aquatic vegetation. *Science* 222: 51–53.
Orth, R.J. and Moore, K.A. 1984. Distribution and abundance of vegetation in Chesapeake Bay: an historical perspective. *Estuaries* 7: 531–540.
Osborne, L.L. and Kovacic, D.A. 1993. Riparian vegetated buffer strips in water quality restoration and stream management. *Freshwater Biology* 29: 243–258.
Osborne, P.L. 1993. Wetlands of Papua New Guinea. In *Wetlands of the World I: Inventory, Ecology and Management*. D.F. Whigham, D. Dykyjova, and S. Hejny, Eds. pp. 305–344. Dordrecht. Kluwer Academic Publishers.
Ostendorp, W. 1989. 'Die-back' of reed in Europe—a critical review of the literature. *Aquatic Botany* 35: 5–26.
Ostendorp, W., Iseli, C., Krauss, M., Krumscheid-Plankert, P., Moret, J., Rollier, M., and Schanz, F. 1995. Lake shore deterioration, reed management and bank restoration in some Central European lakes. *Ecological Engineering* 5: 51–75.
Ostrofsky, M.L. and Zettler, E.R. 1986. Chemical defences in aquatic plants. *Journal of Ecology* 74: 279–287.
Otte, M.L., Kearns, C.C., and Doyle, M.O. 1995. Accumulation of arsenic and zinc in the rhizosphere of wetland plants. *Bulletin of Environmental Contamination and Toxicology* 55: 154–161.
Owen, C.R. 1998. Hydrology and history: land use changes and ecological responses in the urban wetland. *Wetlands Ecology and Management* 6: 209–219.
Painter, D.S. and McCabe, K.J. 1988. Investigation into the disappearance of Eurasian watermilfoil from the Kawartha Lakes, Canada. *Journal of Aquatic Plant Management* 26: 3–12.
Parker, V.T. and Leck, M.A. 1985. Relationships of seed banks to plant distribution patterns in a freshwater tidal wetland. *American Journal of Botany* 72: 161–174.

Partridge, T.R. and Wilson, J.B. 1987. Salt tolerance of salt marsh plants of Otago, New Zealand. *New Zealand Journal of Botany* 25: 559–566.
Pastorok, R.A., MacDonald, A., Sampson, J.R., Wilber, P., Yozzo, D.J., and Titre, J.P. 1997. An ecological decision framework for environmental restoration projects. *Ecological Engineering* 9: 89–107.
Paul, B.J. and Duthie, H.C. 1988. Nutrient cycling in the epilithon of running waters. *Canadian Journal of Botany* 67: 2303–2309.
Pearce, D.M.E. and Jackson, M.B. 1991. Comparison of growth responses of barnyard grass (*Echinochloa oryzoides*) and rice (*Oryza sativa*) to submergence, ethylene, carbon dioxide, and oxygen shortage. *Annals of Botany* 68: 201–209.
Pearson, J. and Havill, D.C. 1988. The effect of hypoxia and sulphide on culture-grown wetland and non-wetland plants. I. Growth and nutrient uptake. *Journal of Experimental Botany* 39: 363–374.
Penfound, W.T. 1952. Southern swamps and marshes. *Botanical Review* 18: 413–436.
Penfound, W.T. 1956. Primary production of vascular aquatic plants. *Limnology and Oceanography* 1: 92–101.
Penfound, W.T. and Hathaway, E.S. 1938. Plant communities in the marshlands of southeastern Louisiana. *Ecological Monographs* 8: 1–56.
Penhale, P.A. and Wetzel, R.G. 1983. Structural and functional adaptations of eelgrass (*Zostera marina* L.) to the anaerobic sediment environment. *Canadian Journal of Botany* 61: 1421–1428.
Penman, H.L. 1948. Natural evaporation from open water, bare soil and grass. *Proceedings Royal Society of London*. A193: 120–146.
Pennings, S.C. and Callaway, R.M. 1992. Salt marsh plant zonation: the relative importance of competition and physical factors. *Ecology* 73: 681–690.
Perata, P. and Alpi, A. 1991. Ethanol induced injuries to carrot cells, the role of acetaldehyde. *Plant Physiology* 95: 748–752.
Perry, C.A. 2000. Significant Floods in the United States during the 20th Century—U.S.G.S. Measures a Century of Floods. Fact Sheet 024-00. Washington, D.C. U.S. Geological Survey. Website: http://ks.water.usgs.gov/kansas/pubs/fact-sheets/fs.024-00.html.
Peterson, S.B. and Teal, J.M. 1996. The role of plants in ecologically engineered wastewater treatment systems. *Ecological Engineering* 6: 137–148.
Peverly, J.H. 1985. Element accumulation and release by macrophytes in a wetland stream. *Journal of Environmental Quality* 14: 137–143.
Peverly, J.H., Surface, J.M., and Wang, T. 1995. Growth and trace metal absorption by *Phragmites australis* in wetlands constructed for landfill leachate treatment. *Ecological Engineering* 5: 21–35.
Pezeshki, S.R. 1994. Plant response to flooding. In *Plant-Environment Interactions*. R.E. Wilkinson, Ed. pp. 289–321. New York. Marcel Dekker.
Pezeshki, S.R., DeLaune, R.D., and Lindau, D.W. 1988. Interaction among sediment anaerobiosis, nitrogen uptake and photosynthesis of *Spartina alterniflora. Physiologia Plantarum* 74: 561–565.
Pezzolesi, T.P., Zartman, R.E., Fish, E.B., and Hickey, M.G. 1998. Nutrients in a playa wetland receiving wastewater. *Journal of Environmental Quality* 27: 67–74.
Philbrick, C.T. 1988. Evolution of underwater outcrossing from aerial pollination systems: a hypothesis. *Annals of the Missouri Botanical Garden* 75: 836–841.
Philbrick, C.T. 1991. Hydrophily: phylogenetic and evolutionary considerations. *Rhodora* 93: 36–50.
Philbrick, C.T. and Les, D.H. 1996. Evolution of aquatic angiosperm reproductive systems. *BioScience* 46: 813–826.
Philips, G.L., Eminson, D., and Moss, B. 1978. A mechanism to account for macrophyte decline in progressively eutrophicated freshwaters. *Aquatic Botany* 4: 103–126.
Pickett, S.T.A. and White, P.S. 1985. *The Ecology of Natural Disturbance and Patch Dynamics*. 472 pp. Orlando, FL. Academic Press.
Pickett, S.T.A., Kolasa, J., Armesto, J.J., and Collins, S.L. 1989. The ecological concept of disturbance and its expression at various hierarchical levels. *Oikos* 54: 129–136.
Pietropaolo, J. and Pietropaolo, P. 1986. *Carnivorous Plants of the World*, 206 pp. Portland, OR. Timber Press.
Pip, E. 1984. Ecogeographical tolerance range variation in aquatic macrophytes. *Hydrobiologia* 108: 37–48.

Pomeroy, L.R. 1959. Algae productivity in salt marshes of Georgia. *Limnology and Oceanography* 4: 386–397.

Pomeroy, L.R. and Wiegert, R.G. 1981. *The Ecology of a Salt Marsh*, 271 pp. New York. Springer-Verlag.

Ponnamperuma, F.N. 1984. Effects of flooding on soils. In *Flooding and Plant Growth*. T.T. Kozlowski, Ed. pp 9–45. New York. Academic Press.

Pool, D.J., Snedaker, S., and Lugo, A.E. 1977. Structure of mangrove forests in Florida, Puerto Rico, Mexico, and Costa Rica. *Biotropica* 9: 195–212.

Potts, M. 1979. Nitrogen fixation (acetylene reduction) associated with communities of heterocystous and non-heterocystous blue-green algae in mangrove forests of Sinai. *Oecologia* 39: 359–373.

Powell, S.W. and Day, F.P., Jr. 1991. Root production in four communities in the Great Dismal Swamp. *American Journal of Botany* 78: 288–297.

Prankevicius, A.B. and Cameron, D.M. 1991. Bacterial dinitrogen fixation in the leaf of the northern pitcher plant (*Sarracenia purpurea*). *Canadian Journal of Botany* 69: 2296–2298.

Pringle, C.M. 1987. Effects of water to substratum nutrient supplies on lotic periphyton growth: an integrated bioassay. *Canadian Journal of Fisheries and Aquatic Science* 44: 619–629.

Pringle, C.M. 1990. Nutrient spatial heterogeneity: effects on community structure, physiognomy, and diversity of stream algae. *Ecology* 71: 905–920.

Prins, H.B.A. and Elzenga, J.T.M. 1989. Bicarbonate utilization: function and mechanism. *Aquatic Botany* 34: 59–83.

Prins, H.B.A., O'Brien, J., and Zanstra, P.E. 1982a. Bicarbonate utilization in aquatic angiosperms, pH and CO_2 concentrations at the leaf surface. In *Studies on Aquatic Vascular Plants*. J.J. Symoens, S.S. Hooper, and P. Compere, Eds. pp. 112–119. Brussels. Royal Botanical Society of Belgium.

Prins, H.B.A., Snel, J.F.H., and Zanstra, P.E. 1982b. The mechanism of photosynthetic bicarbonate utilization. In *Studies on Aquatic Vascular Plants*. J.J. Symoens, S.S. Hooper, and P. Compere, Eds. pp. 120–126. Brussels. Royal Botanical Society of Belgium.

Proctor, M.C.F. 1995. The ombrogenous bog environment. In *Restoration of Temperate Wetlands*. B.D. Wheeler, S.C. Shaw, W.J. Fojt, and R.A. Robertson, Eds. pp. 287–303. Chichester. John Wiley & Sons.

Proposed Revisions. 1991 (to the Federal Interagency Committee for Wetland Delineation Manual, 1989). Washington, D.C. *Federal Register* 56 (40): 446.

Putwain, P.D. and Gillham, D.A. 1990. The significance of the dormant viable seed bank in the restoration of heathlands. *Biological Conservation* 52: 1–16.

Pysek, P. 1998. Is there a taxonomic pattern to plant invasions? *Oikos* 82: 282–294.

Quayyum, H.A., Mallik, A.U., and Lee, P.F. 1999. Allelopathic potential of aquatic plants associated with wild rice (*Zizania palustris*). I. Bioassay with plant and lake sediment samples. *Journal of Chemical Ecology* 25: 209–220.

Queen, W.H. 1974. Physiology of coastal halophytes. In *Ecology of Halophytes*. R.J. Reimold and W.H. Queen, Eds. pp. 345–353. New York. Academic Press.

Rachich, J. and Reader, R. 1999. Interactive effects of herbivory and competition on blue vervain (*Verbena hastata* L.: Verbenaceae). *Wetlands* 19: 156–161.

Radoux, M. 1982. Etude comparée des capacités épuratrices d'un lagunage et d'un marais artificiel miniatures recevant la même eau usée en zone rurale. In *Studies on Aquatic Vascular Plants*. J.J. Symoens, S.S. Hooper, and P. Compere, Eds. pp. 346–352. Brussels. Royal Botanical Society of Belgium.

Ragupathy, S., Mohankumar, V., and Mahadevan, A. 1990. Occurrence of vesicular-arbuscular mycorrhizae in tropical hydrophytes. *Aquatic Botany* 36: 287–291.

Rai, U.N., Sinha, S., Tripathi, R.D., and Chandra, P. 1995. Wastewater treatability potential of some aquatic macrophytes: removal of heavy metals. *Ecological Engineering* 5: 5–12.

Ramaprabhu, T., Kumaraiah, P., Parameswaran, S., Sukumaran, P.K., and Raghavan, S.L. 1987. Water hyacinth control by natural water level fluctuations in Byramangala Reservoir, India. *Journal of Aquatic Plant Management* 25: 63–64.

Ramsar Convention Bureau. 2000. Gland, Switzerland. Website: http://iucn.org/themes/ramsar.

Raskin, I. and Kende, H. 1985. Mechanism of aeration in rice. *Science* 228: 327–329.

Rattray, M.R., MacDonald, G., Shilling, D. and Bowes, G. 1993. The mechanism of action of bensulfuron–methyl on hydrilla. *Journal of Aquatic Plant Management* 31: 39–42.

Raven, J.A. 1984. *Energetics and Transport in Aquatic Plants. MBL Lectures in Biology*, Vol. 4, 587 pp. New York. Alan R. Liss.

Raven, P.H., Evert, R.F., and Eichhorn, S.E. 1999. *Biology of Plants*, 944 pp. New York. W.H. Freeman and Company.

Raymond, P., Saglio, P.H., and Ricard, B. 1995. Plant tissue responses to low oxygen concentration. *Cahiers Agricultures* 4: 343–350.

Rea, N. 1996. Water levels and *Phragmites*: decline from lack of regneration or dieback from shoot death. *Folia Geobotanica Phytotaxonomica* 31: 85–90.

Rea, N. 1998. Biological control: premises, ecological input and *Mimosa pigra* in the wetlands of Australia's Top End. *Wetlands Ecology and Management* 5: 227–242.

Rea, N. and Storrs, M.J. 1999. Weed invasions in wetlands of Australia's Top End: reasons and solutions. *Wetlands Ecology and Management* 7: 47–62.

Reader, R. and Stewart, J. 1971. Net primary productivity of bog vegetation in southeastern Manitoba. *Canadian Journal of Botany* 49: 1471–1477.

Reader, R.J. and Stewart, J.M. 1972. The relationship between net primary productivity and accumulation for a peatland in southeastern Manitoba. *Ecology* 53: 1024–1036.

Reddy, K.R. and DeBusk, W.F. 1987. Nutrient storage capabilities of aquatic and wetland plants. In *Aquatic Plants for Water Treatment*. K.R. Reddy and W.H. Smith, Eds. pp. 337–357. Orlando, FL. Magnolia Publishing.

Reddy, K.R., Sacco, P.D., Graetz, D.A., Campbell, K.L., and Porter, K.M. 1983. Effect of aquatic macrophytes on physico-chemical parameters of agricultural drainage water. *Journal of Aquatic Plant Management* 21: 1–7.

Reddy, K.R., D'Angelo, E.M., and DeBusk, T.A. 1989. Oxygen transport through aquatic macrophytes: the role in wastewater treatment. *Journal of Environmental Quality* 19: 261–267.

Reddy, K.R., Kadlec, R.H., Flaig, E., and Gale, P.M. 1999. Phosphorus retention in streams and wetlands: a review. *Critical Reviews in Environmental Science and Technology* 29: 83–146.

Redfield, A.C. 1972. Development of a New England salt marsh. *Ecological Monographs* 42: 201–237.

Reed, P.B. 1988. National List of Plant Species that Occur in Wetlands: 1988 National Summary. Biological Report 88 (24). Washington, D.C. U.S. Department of the Interior, U.S. Fish and Wildlife Service.

Reed, P.B. 1997. Revision of the National List of Plant Species that Occur in Wetlands, 209 pp. Washington, D.C. U.S. Department of the Interior, U.S. Fish and Wildlife Service.

Reeder, B.C. and Eisner, W.R. 1994. Holocene biogeochemical and pollen history of a Lake Erie, Ohio coastal wetland. *Ohio Journal of Science* 94: 87–93.

Reinartz, J.A. and Warne, E.L. 1993. Development of vegetation in small created wetlands in southeastern Wisconsin. *Wetlands* 13: 153–164.

Reiners, W.A. 1972. Structure and energetics of three Minnesota forests. *Ecological Monographs* 42: 71–94.

Rejmanek, M. 1989. Invasibility of plant communities. In *Biological Invasions: A Global Perspective.* J.A. Drake, H.A. Mooney, F. di Castri, R.H. Groves, F.J. Kruger, M. Rejmanek, and M. Williamson, Eds. pp. 369–388. Chichester. John Wiley & Sons.

Rey Benayas, J.M. and Scheiner, S.M. 1993. Diversity patterns of wet meadows along geochemical gradients in central Spain. *Journal of Vegetation Science* 4: 103–108.

Rey Benayas, J.M., Bernaldez, F.G., Levassor, C., and Peco, B. 1990. Vegetation of ground water discharge sites in the Douro basin, central Spain. *Journal of Vegetation Science* 1: 461–466.

Reynolds, P.E., Carlson, K.G., Fromm, T.W., Gigliello, K.A., and Kaminski, R.J. 1979. Phytosociology, biomass, productivity and nutrient budget for the tree stratum of a southern New Jersey hardwood swamp. In *Forest Resources Inventories: Workshop Proceedings*, Vol. II. W.E. Frayer, Ed. pp. 123–139. Fort Collins. Colorado State University.

Ricard, B., Couee, I., Raymond, P., Saglio, P.H., Saint-Ges, V., and Pradet, A. 1994. Plant metabolism under hypoxia and anoxia. *Plant Physiology and Biochemistry* 32: 1–10.

Rice, D., Rooth, J., and Stevenson, J.C. 2000. Colonization and expansion of *Phragmites australis* in Upper Chesapeake Bay tidal marshes. *Wetlands* 20: 280–299.

Richardson, C.J. 1985. Mechanisms controlling phosphorus retention capacity in freshwater wetlands. *Science* 228: 1424–1427.

Richardson, C.J., Ferrell, G.M., and Vaithiyanathan, P. 1999. Nutrient effects on stand structure, resorption efficiency, and secondary compounds in Everglades sawgrass. *Ecology* 80: 2182–2192.

Richter, K.O. and Azous, A.L. 1995. Amphibian occurrence and wetland characteristics in the Puget Sound Basin. *Wetlands* 15: 305–312.

Rickerl, D.H., Sancho, F.O., and Ananth, S. 1994. Vesicular-arbuscular endomycorrhizal colonization of wetland plants. *Journal of Environmental Quality* 23: 913–916.

Ridge, I. 1987. Ethylene and growth control in amphibious plants. In *Plant Life in Aquatic and Amphibious Habitats*. R.M.M. Crawford, Ed. pp. 53–76. Oxford. Blackwell Scientific.

Roberts, B.A. and Robertson, A. 1986. Salt marshes of Atlantic Canada: their ecology and distribution. *Canadian Journal of Botany* 64: 455–467.

Roberts, J.K.M. 1989. Cytoplasmic acidosis and flooding tolerance in crop plants. In *The Ecology and Management of Wetland Plants*, Vol. 1. D.D. Hook et al., Eds. pp. 392–397. Portland, OR. Timber Press.

Roberts, J.K.M., Callis, J., Jardetzky, O., Walbot, V., and Freeling, M. 1984. Cytoplasmic acidosis as a determinant of flooding intolerance in plants. *Proceedings of the National Academy of Science* 81: 6029–6033.

Roberts, J.K.M., Chang, K., Webster, C., Callis, J., and Walbot, V. 1989. Dependence of alcoholic fermentation, cytoplasmic pH regulation, and viability on the activity of alcoholic dehydrogenase in hypoxic maize root tips. *Plant Physiology* 89: 1275–1278.

Roberts, J.K.M., Hooks, M.A., Miaullis, A.P., Edwards, S., and Webster, C. 1992. Contribution of malate and amino acid metabolism to cytoplasmic pH regulation in hypoxic maize root tips studied using nuclear magnetic resonance spectroscopy. *Plant Physiology* 98: 480–487.

Robinson, G.G.C. 1983. Methodology: the key to understanding periphyton. In *Periphyton of Freshwater Ecosystems*. R.G. Wetzel, Ed. pp. 245–251. The Hague. Dr. W. Junk Publishers.

Rochefort, L., Vitt, D.H., and Bayley, S.E. 1990. Growth, production, and decomposition dynamics of *Sphagnum* under natural and experimentally acidified conditions. *Ecology* 71: 1986–2000.

Rogers, K.H. and Breen, C.M. 1980. Growth and reproduction of *Potamogeton crispus* in a South African lake. *Journal of Ecology* 68: 561–571.

Roman, C.T., Niering, W.A., and Warren, R.S. 1984. Salt marsh vegetation change in response to tidal restriction. *Environmental Management* 8: 141–150.

Rost, T.L., Barbour, M.G., Thornton, R.M., Weier, T.E., and Stocking, C.R. 1984. *Botany: A Brief Introduction to Plant Biology*, 342 pp. New York. John Wiley & Sons.

Rubec, C.D.A. 1994. Canada's federal policy on wetland conservation: a global model. In *Global Wetlands: Old World and New*. W.J. Mitsch, Ed. pp. 909–917. Amsterdam. Elsevier Science.

Ruesink, J.L., Parker, I.M., Groom, M.J., and Kareiva, P.M. 1995. Reducing the risks of nonindigenous species introductions: guilty until proven innocent. *Bioscience* 45: 465–477.

Rumpho, M.E. and Kennedy, R.A. 1981. Anaerobic metabolism in germinating seeds of *Echinochloa crus-galli* (barnyard grass) metabolite and enzyme studies. *Plant Physiology* 68: 165–168.

Rutzler, K. and Feller, I.C. 1996. Caribbean mangrove swamps. *Scientific American* March: 94–99.

Ryther, J.H. and Dunstan, W.M. 1971. Nitrogen, phosphorus and eutrophication in the coastal marine environment. *Science* 171: 1008–1013.

Sage, R.F. and Pearcy, R.W. 1987. The nitrogen use efficiency of C_3 and C_4 plants I. Leaf nitrogen, growth, and biomass partitioning in *Chenopodium album* (L.) and *Amaranthus retroflexus* (L.). *Plant Physiology* 84: 954–958.

Sage, R.F., Wedin, D.A., and Li, M.R. 1999. The biogeography of C_4 photosynthesis: patterns and controlling factors. In *The Biology of C_4 Photosynthesis*. R.F. Sage and R.K. Monson, Eds. pp. 313–375. San Diego, CA. Academic Press.

Saglio, P.H., Raymond, P., and Pradet, A. 1980. Metabolic activity and energy charge of excised maize root tips under anoxia. *Plant Physiology* 66: 1053–1039.

Saglio, P.H., Raymond, P., and Pradet, A. 1983. Oxygen transport and root respiration of maize seedlings. *Plant Physiology* 72: 1035–1039.

Saint-Ges, V., Roby, C., Bligny, R., Pradet, A., and Douce, R. 1991. Kinetic studies of the variation of cytoplasmic pH, nucleotide triphosphates (^{31}P-NMR) and lactate during normoxic and anoxic transitions in maize root tips. *European Journal of Biochemistry* 200: 477–482.

Salisbury, F.B. and Ross, C.W. 1985. *Plant Physiology*, 540 pp. Belmont, CA. Wadsworth Publishing.

Samecka-Cymerman, A. and Kempers, A.J. 1996. Bioaccumulation of heavy metals by aquatic macrophytes around Wroclaw, Poland. *Ecotoxicology and Environmental Safety* 35: 242–247.

Sand-Jensen, K. and Borum, J. 1991. Interactions among phytoplankton, periphyton, and macrophytes in temperate freshwaters and estuaries. *Aquatic Botany* 41: 137–175.

Sanville, W. and Mitsch, W.J. 1994. Creating freshwater marshes in a riparian landscape: research at the Des Plaines River Wetlands Demonstration Project. *Ecological Engineering* 3: 315–317.

Sather, J.H. and Smith, R.D. 1984. An Overview of Major Wetland Functions and Values, Report FWS/OBS–84/18. 68 pp. Washington, D.C. U.S. Department of the Interior, U.S. Fish and Wildlife Service, Western Energy Land Use Team.

Savage, J.M. 1995. Systematics and the biodiversity crisis. *Bioscience* 45: 673–679.

Schaeffer, D.J. 1991. A toxicological perspective on ecosystem characteristics to track sustainable development. *Ecotoxicology and Environmental Safety* 22: 225–239.

Schaeffer, D.J., Herricks, E.E., and Kerster, H.W. 1988. Ecosystem health: measuring ecosystem health. *Environmental Management* 5:187–190.

Schiemer, F. and Prosser, M. 1976. Distribution and biomass of submerged macrophytes in Neusiedlersee. *Aquatic Botany* 2: 289–307.

Schindler, D.W. and Fee, E.J. 1975. The roles of nutrient cycling and radiant energy in aquatic communities. In *Photosynthesis and Productivity in Different Environments.* J.P. Cooper, Ed. pp. 323–343. Cambridge. Cambridge University Press.

Schipper, L.A., Cooper, A.B., Harfoot, C.G., and Dyck, W.J. 1993. Regulators of denitrification in an organic riparian soil. *Soil Biology and Biochemistry* 25: 925.

Schlesinger, W.H. 1978. Community structure, dynamics and nutrient cycling in the Okefenokee cypress swamp-forest. *Ecological Monographs* 48: 43–65.

Schmitz, D.C., Schardt, J.D., Leslie, A.J., Dray, F.A., Osborne, J.A., and Nelson, B.V. 1993. The ecological impact and management history of three invasive alien aquatic plant species in Florida. In *Biological Pollution: The Control and Impact of Invasive Exotic Species.* B.N. McKnight, Ed. pp. 173–194. Indianapolis. Indiana Academy of Science.

Schneider, R.L. 1994. The role of hydrologic regime in maintaining rare plant communities of New York's coastal plain pondshores. *Biological Conservation* 68: 253–260.

Schneider, R.L. and Sharitz, R.R. 1988. Hydrochory and regeneration in a bald cypress-water tupelo swamp forest. *Ecology* 69: 1055–1063.

Schnell, D.E. 1976. *Carnivorous Plants of the United States and Canada*, 125 pp. Winston-Salem, NC. John F. Blair.

Scholander, P.F., van Dam, L., and Scholander, S.I. 1955. Gas exchange in the roots of mangroves. *American Journal of Botany* 42: 92–98.

Scholten, M.C.T. and Rozema, J. 1990. The competitive ability of *Spartina anglica* on Dutch salt marshes. In *Spartina anglica* – a Research Review. A.J. Gray and P.E.M. Nenham, Eds. London. Her Majesty's Stationery Office.

Schubauer, J.P. and Hopkinson, C.S. 1984. Above- and belowground emergent macrophyte production and turnover in a coastal marsh ecosystem, Georgia. *Limnology and Oceanography* 29: 1052–1065.

Schwartz, M.W. 1997. Defining indigenous species: an introduction. In *Assessment and Management of Plant Invasions.* J.O. Luken and J.W. Thieret, Eds. pp. 7–17. New York. Springer.

Schwintzer, C.R. 1983. Primary productivity and nitrogen, carbon, and biomass distribution in a dense *Myrica gale* stand. *Canadian Journal of Botany* 61: 2943–2948.

Scott, D.A. and Jones, T.A. 1995. Classification and inventory of wetlands: a global overview. *Vegetatio* 118: 3–16.

Sculthorpe, C.D. 1967. The *Biology of Aquatic Vascular Plants,* 610 pp. London. Edward Arnold Publishers.

Sebacher, D.I., Harriss, R.C., and Bartlett, K.B. 1985. Methane emissions to the atmosphere through aquatic plants. *Journal of Environmental Quality* 14: 40–46.

Sedell, J.R., Steedman, R., Regler, H., and Gregory, S. 1991. Restoration of human impacted land-water ecotones. In *Ecotones: The Role of Landscape Boundaries in the Management and Restoration of Changing Environments*. M.M. Holland, P.G. Risser, and R.J. Naimer, Eds. 110 pp. New York. Chapman & Hall.

Seliskar, D.M. 1987. The effect of soil moisture on structural biomass characteristics of four salt marsh plants. *Journal of Experimental Botany* 38: 1193–1202.

Setter, T.L., Ellis, M., Laureles, E.V., Ella, E.S., Senadhira, D., Mishra, S.B., Sarkarung, S., and Datta, S. 1997. Physiology and genetics of submergence tolerance in rice. *Annals of Botany* 79: 67–77.

Shaffer, G.P., Sasser, C.E., Gosselink, J.G., and Rejmanek, M. 1992. Vegetation dynamics in the emerging Atchafalaya Delta, Louisiana USA. *Journal of Ecology* 80: 677–687.

Sharma, S.S. and Gaur, J.P. 1995. Potential of *Lemna polyrrhiza* for removal of heavy metals. *Ecological Engineering* 4: 37–43.

Shaw, S.P. and Fredine, C.G. 1956. Wetlands of the United States, Their Extent, and Their Value for Waterfowl and Other Wildlife, 67 pp. Washington, D.C. U.S. Department of the Interior, U.S. Fish and Wildlife Service.

Shay, J.M. and Shay, C.T. 1986. Prairie marshes in western Canada, with specific reference to the ecology of five emergent macrophytes. *Canadian Journal of Botany* 64: 443–454.

Shay, J.M., de Geus, P.M.J., and Kapinga, M.R.M. 1999. Changes in shoreline vegetation over a 50-year period in the Delta Marsh, Manitoba in response to water levels. *Wetlands* 19: 413–425.

Shea, M.L., Warren, R.S., and Niering, W.A. 1975. Biochemical and transplantation studies of growth forms of *Spartina alterniflora* on Connecticut salt marshes. *Ecology* 56: 461–466.

Shearer, J.F. 1998. Biological control of hydrilla using an endemic fungal pathogen. *Journal of Aquatic Plant Management* 36: 54–56.

Shedlock, R.J., Wilcox, D.A., Thompson, T.A., and Cohen, D.A. 1993. Interactions between ground water and wetlands, southern shore of Lake Michigan, USA. *Journal of Hydrology* 141: 127–155.

Sheldon, S.P. 1997. Ecological approaches for biological control of the aquatic weed Eurasian water-milfoil: resource and interference competition, exotic and endemic herbivores and pathogens. In *Ecological Interactions and Biological Control*. D.A. Andow, D.W. Ragsdale, and R.F. Nyvall, Eds. pp. 53–70. Boulder, CO. Westview Press.

Sheldon, S.P. and Creed, R.P. 1995. Use of a native insect as a biological control for an introduced weed. *Ecological Applications* 5: 1122–1132.

Sheridan, R.P. 1991. Epicaulous, nitrogen–fixing microepiphytes in a tropical mangal community, Guadeloupe, French West Indies. *Biotropica* 23: 530–541.

Shew, D.M., Linthurst, R.A., and Seneca, E.D. 1981. Comparison of production computation methods in a southeastern North Carolina *Spartina alterniflora* salt marsh. *Estuaries* 4: 97–109.

Shipley, B., Keddy, P.A., Moore, D.R.J., and Lemky, K. 1989. Regeneration and establishment strategies of emergent macrophytes. *Journal of Ecology* 77: 1093–1110.

Shutes, R.B., Ellis, J.B., Revitt, D.M., and Zhang, T.T. 1993. The use of *Typha latifolia* for heavy metal pollution control in urban wetlands. In *Constructed Wetlands for Water Quality Improvement*. G.A. Moshiri, Ed. pp. 407–414. Boca Raton, FL. Lewis Publishers.

Siegel, D.I. and Glaser, P.H. 1987. Ground water flow in a bog-fen complex, Lost River Peatland, northern Minnesota. *Journal of Ecology* 75: 743–754.

Siegley, C.E., Boerner, R.E.J., and Reutter, J.M. 1988. Role of the seed bank in the development of vegetation on a freshwater marsh created from dredge spoil. *Journal of Great Lakes Research* 14: 267–276.

Sikora, F.J., Tong, Z., Behrends, L.L., Steinberg, S.L., and Coonrod, H.S. 1995. Ammonium removal in constructed wetlands with recirculating subsurface flow: removal rates and mechanisms. *Water Science and Technology* 32: 193–202.

Silvertown, J. 1987. *Introduction to Plant Population Ecology*, 229 pp. Essex. Longman Scientific and Technical.

Simberloff, D. 1996. Impacts of introduced species in the United States. *Consequences* 2: 13–22.

Simpson, R.L., Good, R.E., Leck, M.A., and Whigham, D.F. 1983a. The ecology of freshwater tidal wetlands. *BioScience* 33: 255–259.

Simpson, R.L., Good, R.E., Walker, R., and Frasco, B.R. 1983b. The role of Delaware River freshwater tidal wetlands in the retention of nutrients and heavy metals. *Journal of Environmental Quality* 12: 41–48.

Singer, A., Eshel, A., Agami, M., and Beer, S. 1994. The contribution of aerenchymal CO_2 to the photosynthesis of emergent and submerged culms of *Scirpus lacustris* and *Cyperus papyrus*. *Aquatic Botany* 49: 107–116.

Singh, J.S., Lauenroth, W.K., Hunt, H.W., and Swift, D.M. 1984. Bias and random errors in estimators of net root production: a simulation approach. *Ecology* 65: 1760–1764.

Sinicrope, T.L., Hine, P.G., Warren, R.S., and Niering, W.A. 1990. Restoration of an impounded salt marsh in New England. *Estuaries* 13: 25–30.

Sinicrope, T.L., Langis, R., Gersberg, R.M., Busnardo, M.J., and Zedler, J.B. 1992. Metal removal by wetland mesocosms subjected to different hydroperiods. *Ecological Engineering* 1: 309–322.

Sipple, W.S. 1988. Wetland Identification and Delineation Manual, Vol. I. Rationale, Wetland Parameters, and Overview of Jurisdictional Approach. Revised Interim Final. Washington, D.C. U.S. Environmental Protection Agency, Office of Wetlands Protection.

Sjors, H. 1950. On the relation between vegetation and electrolytes in north Swedish mires. *Oikos* 2: 243–258.

Slack, A. 1979. *Carnivorous Plants,* 240 pp. Cambridge, MA. MIT Press.

Smalley, A.E. 1959. The Role of Two Invertebrate Populations, *Littorina irrorata* and *Orchelimum fidicinium* in the Energy Flow of a Georgia Salt Marsh Ecosystem, Ph.D. dissertation. Athens. University of Georgia.

Smirnoff, N. and Crawford, R.M.M. 1983. Variation in the structure and response to flooding of root aerenchyma in some wetland plants. *Annals of Botany* 51: 237–249.

Smith, K.K., Good, R.E., and Good, N.F. 1979. Production dynamics for above and belowground components of a New Jersey *Spartina alterniflora* tidal marsh. *Estuarine and Coastal Marine Science* 9: 189–201.

Smith, S.G. and Yatskievych, G. 1996. Notes on the genus *Scirpus sensu lato* in Missouri. *Rhodora* 98: 168–179.

Smits, A.J.M., Laan, P., Thier, R.H., and van der Velde, G. 1990. Root aerenchyma, oxygen leakage patterns and alcoholic fermentation ability of the roots of some nymphaeid and isoetid macrophytes in relation to the sediment type of their habitat. *Aquatic Botany* 38: 3–17.

Snedaker, S.C. and Lahmann, E.J. 1988. Mangrove understory absence: a consequence of evolution? *Journal of Tropical Ecology* 4: 311–314.

Snow, A.A. and Vince, S.W. 1984. Plant zonation in an Alaskan salt marsh. *Journal of Ecology* 72: 669–684.

Soeda, K., Tada, M., and Murata, Y. 2000. Exogenous proline mitigates the inhibition of growth of *Nicotania tabacum* cultured cells under saline conditions. *Soil Science and Plant Nutrition* 46: 257–263.

Sondergaard, M. 1979. Light and dark respiration and the effect of the lacunal system on refixation of CO_2 in submerged aquatic plants. *Aquatic Botany* 6: 269–283.

Sondergaard, M. 1988. Photosynthesis of aquatic plants under natural conditions. In *Vegetation of Inland Waters.* J.J. Symoens, Ed. pp. 63–109. Dordrecht. Kluwer Academic Publishers.

Sondergaard, M. and Laegaard, S. 1977. Vesicular-arbuscular mycorrhiza in some aquatic vascular plants. *Nature* 268: 232–233.

Sorrell, B.K. and Boon, P.I. 1994. Convective gas-flow in *Eleocharis sphacelata* R. Br.: methane transport and release from wetlands. *Aquatic Botany* 47: 197–212.

Souch, C., Grimmond, C.S., and Wolfe, C.P. 1998. Evapotranspiration rates from wetlands with different disturbance histories: Indiana Dunes National Lakeshore. *Wetlands* 18: 216–229.

Spence, D.H.N. 1982. The zonation of plants in freshwater lakes. *Advances in Ecological Research* 12: 37–125.

Squires, J. and van der Valk, A.G. 1992. Water-depth tolerances of the dominant emergent macrophytes of the Delta marsh, Manitoba. *Canadian Journal of Botany* 70: 1860–1867.

Srivastava, D.S. and Jefferies, R.L. 1995. Mosaics of vegetation and soil salinity: a consequence of goose foraging in an arctic salt marsh. *Canadian Journal of Botany* 73: 75–83.

Srivastava, D.S. and Jefferies, R.L. 1996. A positive feedback: herbivory, plant growth, salinity, and the desertification of an Arctic salt–marsh. *Journal of Ecology* 84: 31–42.

Stefels, J. 2000. Physiological aspects of the production and conversion of DMSP in marine algae and higher plants. *Journal of Sea Research* 43: 183–197.

Steinke, W., von Willert, D. J., and Austenfeld, F.A. 1996. Root dynamics in a salt marsh over three consecutive years. *Plant and Soil* 185: 265–269.

Stengel, E. and Soeder, C.J. 1975. Control of photosynthetic production in aquatic ecosystems. In *Photosynthesis and Productivity in Different Environments.* J.P. Cooper, Ed. pp. 645–660. Cambridge. Cambridge University Press.

Stevens, K.J., Peterson, R.L., and Stephenson, G.R. 1997a. Vegetative propagation and the tissues involved in lateral spread of *Lythrum salicaria. Aquatic Botany* 56: 11–24.

Stevens, K.J., Peterson, R.L., and Stephenson, G.R. 1997b. Morphological and anatomical responses of *Lythrum salicaria* L. (purple loosestrife) to an imposed water gradient. *International Journal of Plant Science* 158: 172–183.

Stevenson, J.C. 1988. Comparative ecology of submersed grass beds in freshwater, estuarine, and marine environments. *Limnology and Oceanography* 33: 867–893.

Stevenson, J.C., Ward, L.G., and Kearney, M.S. 1988. Sediment transport and trapping in marsh systems: implications of tidal flux studies. *Marine Geology* 80: 37–59.

Stevenson, J.C., Staver, L.W., and Staver, K.W. 1993. Water quality associated with survival of submersed aquatic vegetation along an estuarine gradient. *Estuaries* 16: 346–361.

Stevenson, R.J. 1998. Diatom indicators of stream and wetland stressors in a risk management framework. *Environmental Monitoring and Assessment* 51: 107–118.

Steward, K.K. 1993. Seed production in monoecious and dioecious populations of Hydrilla. *Aquatic Botany* 46: 169–183.

Steward, K.K., Van, T.K., Carter, V., and Pieterse, A.H. 1984. *Hydrilla* invades Washington, D.C. and the Potomac. *American Journal of Botany* 71: 162–163.

Stewart, C.N. and Nilsen, E.T. 1992. *Drosera rotundifolia* growth and nutrition in a natural population with special reference to the significance of insectivory. *Canadian Journal of Botany* 70: 1409–1416.

Stewart, C.N. and Nilsen, E.T. 1993. Responses of *Drosera capensis* and *D. binata* var. *multifida* (Droseraceae) to manipulations of insect availability and soil nutrient levels. *New Zealand Journal of Botany* 31: 385–390.

Stewart, G.R. and Lee, J.A. 1974. The role of proline accumulation in halophytes. *Planta* 120: 279–289.

Stewart, R.E. and Kantrud, H.A. 1971. Classification of Natural Ponds and Lakes in the Glaciated Prairie Region, 57 pp. Washington, D.C. U.S. Department of the Interior, U.S. Fish and Wildlife Service.

Stocker, R.K. 1999. Mechanical harvesting of *Melaleuca quinquenervia* in Lake Okeechobee, Florida. *Ecological Engineering* 12: 373–386.

Strand, J.A. and Weisner, E.B. 1996. Wave exposure related growth of epiphyton: implications for the distribution of submerged macrophytes in eutrophic lakes. *Hydrobiologia* 325: 113–119.

Strecker, E.W., Kersnar, J.M., Driscoll, E.D., and Horner, R.R. 1992. *The Use of Wetlands for Controlling Stormwater Pollution,* 66 pp. Washington, D.C. Terrene Institute.

Streever, W.J. and Genders, A.J. 1998. Effect of improved tidal flushing and competitive interactions at the boundary between salt marsh and pasture. *Oceanographic Literature Review* 45: 999–1000.

Stromberg, J.C. and Patten, D.T. 1996. Instream flow and cottonwood growth in the eastern Sierra Nevada of California, USA. *Regulated Rivers: Research and Management* 12: 1–12.

Stuckey, R.L. 1971. Changes of vascular flowering plants during 70 years in Put-in-Bay Harbor, Lake Erie, Ohio. *Ohio Journal of Science* 71: 321–342.

Stuckey, R.L. 1980. Distributional history of *Lythrum salicaria* (purple loosestrife) in North America. *Bartonia* 47: 3–20.

Studer, C. and Braendle, R. 1987. Ethanol, acetaldehyde, ethylene release and ACC concentration of rhizomes from marsh plants under normoxia, hypoxia and anoxia. In *Plant Life in Aquatic and Amphibious Habitats.* R.M.M. Crawford, Ed. pp. 293–301. Oxford. Blackwell Scientific.

Sullivan, M.J. and Daiber, F.C. 1974. Response in production of cordgrass, *Spartina alterniflora,* to inorganic nitrogen and phosphorus fertilizer. *Chesapeake Science* 15: 121–123.

Summers, J.E., Ratcliffe, R.G., and Jackson, M.B. 2000. Anoxia tolerance in the aquatic monocot *Potamogeton pectinatus*: absence of oxygen stimulates elongation in association with an unusually large Pasteur effect. *Journal of Experimental Botany* 51: 1413–1422.

Surrency, D. 1993. Evaluation of aquatic plants for constructed wetlands. In *Constructed Wetlands for Water Quality Improvement.* G.A. Moshiri, Ed. pp. 349–357. Boca Raton, FL. Lewis Publishers.

Sutton, D.L. 1986. Growth of *Hydrilla* in established stands of spikerush and slender arrowhead. *Journal of Aquatic Plant Management* 24: 16–20.

Sutton, D.L. 1996. Depletion of turions and tubers of *Hydrilla verticillata* in the North New River Canal, Florida. *Aquatic Botany* 53: 121–130.

Svedang, M.U. 1990. The growth dynamics of *Juncus bulbosus* L. — a strategy to avoid competition? *Aquatic Botany* 37: 123–138.

Talbot, R.J. and Etherington, J.R. 1987. Comparative studies of plant growth and distribution in relation to waterlogging. XIII. The effect of Fe^{2+} on photosynthesis and respiration of *Salix caprea* and *S. cinerea* spp. *oleifolia*. *New Phytologist* 105: 575–583.

Talling, J.F. 1975. Primary production of aquatic plants—conclusions. In *Photosynthesis and Productivity in Different Environments*. J.P. Cooper, Ed. pp. 281–294. Cambridge. Cambridge University Press.

Tanner, C.C. 1996. Plants for constructed wetland treatment systems — a comparison of the growth and nutrient uptake of eight emergent species. *Ecological Engineering* 7: 59–83.

Tanner, C.C., Clayton, J.S., and Upsdell, M.P. 1995a. Effect of loading rate and planting on treatment of dairy farm wastewaters in constructed wetlands. I. Removal of oxygen demand, suspended solids and faecal coliforms. *Water Resources* 29: 17–26.

Tanner, C.C., Clayton, J.S., and Upsdell, M.P. 1995b. Effect of loading rate and planting on treatment of dairy farm wastewaters in constructed wetlands. II. Removal of nitrogen and phosphorus. *Water Resources* 29: 27–34.

Tatu, K., Kimothi, M., and Parihar, J.S. 1999. Remote sensing-based Habitat Availability Model (HAM): a tool for quick-look assessment of wetlands as waterbird habitats. *Indian Forester* 125: 1004–1017.

Taylor, B.W. 1959. The classification of lowland swamp communities in north-eastern Papua New Guinea. *Ecology* 40: 703–711.

Taylor, G.J. and Crowder, A.A. 1983. Uptake and accumulation of copper, nickel, and iron by *Typha latifolia* grown in solution culture. *Canadian Journal of Botany* 61: 1825–1830.

Teal, J.M. and Howes, B.L. 1996. Interannual variability of a salt–marsh ecosystem. *Limnology and Oceanography* 41: 802–809.

Teal, J.M. and Kanwisher, J.W. 1966. Gas transport in the marsh grass, *Spartina alterniflora*. *Journal of Experimental Botany* 17: 355–361.

ter Heerdt, G.N.J. and Drost, H.J. 1994. Potential for the development of marsh vegetation from the seed bank after a drawdown. *Biological Conservation* 67: 1–11.

Thibodeau, F.R. and Nickerson, N.H. 1985. Changes in a wetland plant association induced by impoundment and draining. *Biological Conservation* 33: 269–279.

Thibodeau, F.R. and Nickerson, N.H. 1986. Differential oxidation of mangrove substrate by *Avicennia germinans* and *Rhizophora mangle*. *American Journal of Botany* 73: 512–516.

Thom, R. M. 1997. System-development matrix for adaptive management of coastal ecosystem restoration projects. *Ecological Engineering* 8: 219–232.

Thompson, C.A., Bettis, E. A., and Baker, R.G. 1992. Geology of Iowa fens. *Journal of Iowa Academy of Science* 99: 53–59.

Thompson, D., Stuckey, R.L., and Thompson, E.B. 1987. Spread, Impact, and Control of Purple Loosestrife (*Lythrum salicaria*). Fish and Wildlife Research 2, 52 pp. Washington, D.C. U.S. Department of the Interior, U.S. Fish and Wildlife Service.

Thompson, D.J. and Shay, J.M. 1989. First-year response of a *Phragmites* marsh community to seasonal burning. *Canadian Journal of Botany* 67: 1448–1455.

Thompson, K. and Grime, J.P. 1979. Seasonal variation in the seed banks of herbaceous species in ten contrasting habitats. *Journal of Ecology* 67: 893–921.

Thompson, K., Hodgson, J.G., and Rich, T.C.G. 1995. Native and alien invasive plants: more of the same? *Ecography* 18: 390–402.

Thum, M. 1989a. The significance of opportunistic predators for the sympatric carnivorous plant species *Drosera intermedia* and *Drosera rotundifolia*. *Oecologia* 81: 397-400.

Thum, M. 1989b. The significance of carnivory for the fitness of *Drosera* in its natural habitat. II. The amount of captured prey and its effect on *Drosera intermedia* and *Drosera rotundifolia*. *Oecologia* 81: 401–411.

Tilman, D. 1982. *Resource Competition and Community Structure*, 296 pp. Princeton, NJ. Princeton University Press.

Tilman, D. 1985. The resource ratio hypothesis of succession. *American Naturalist* 125: 827–852.

Tilton, D.L. and Kadlec, R.H. 1979. The utilization of a freshwater wetland for nutrient removal from secondarily treated wastewater effluent. *Journal of Environmental Quality* 8: 328–334.

Tiner, R.W., Jr. 1988. Field Guide to Nontidal Wetland Identification. Annapolis, MD. Maryland Department of Natural Resources, U.S. Fish and Wildlife Service.

Tiner, R.W. 1991. The concept of a hydrophyte for wetland identification. *BioScience* 41: 236–247.

Tiner, R.W. 1996. Practical considerations for wetland identification and boundary delineation. In *Wetlands: Environmental Gradients, Boundaries, and Buffers*. G. Mulamoottil, B. G. Warner, and E. A. McBean, Eds. pp. 113–138. New York. Lewis Publishers.

Tiner, R.W. 1999. *Wetland Indicators: A Guide to Wetland Indentification, Delineation, Classification and Mapping*, 392 pp. Boca Raton, FL. CRC Press.

Titus, J. and Hoover, D.T. 1991. Toward predicting reproductive success in submersed freshwater angiosperms. *Aquatic Botany* 41: 111–136.

Titus, J. and Narayanan, V.K. 1995. The Probability of Sea Level Rise, U.S. Report, p. 138. Washington, D.C. U.S. Environmental Protection Agency.

Titus, J.G., Park, R., Leatherman, S., Weggle, J., Greene, M., et al. 1991. Greenhouse effect and sea level rise: the cost of holding back the sea. *Coastal Management* 19: 171–204.

Tjepkema, J. 1977. The role of oxygen diffusion from the shoots and nodule roots in nitrogen fixation by root nodules of *Myrica gale*. *Canadian Journal of Botany* 56: 1365–1371.

Toetz, D.W. 1974. Uptake and translocation of ammonia by freshwater hydrophytes. *Ecology* 55: 199–201.

Toledo, G., Bashan, Y., and Soeldner, A. 1995. Cyanobacteria and black mangroves in northwestern Mexico: colonization, and diurnal and seasonal nitrogen fixation on aerial roots. *Microbiology* 41: 999–1011.

Tolley, M.D., Delaune, R.D., and Patrick, W.H., Jr. 1986. The effect of sediment redox potential and soil acidity on nitrogen uptake, anaerobic root respiration, and growth of rice (*Oryza sativa*). *Plant and Soil* 93: 323–331.

Tomlinson, P.B. 1986. *The Botany of Mangroves*, 413 pp. London. Cambridge University Press.

Toner, M. and Keddy, P.A. 1997. River hydrology and riparian wetlands: a predictive model for ecological assembly. *Ecological Applications* 7: 236–246.

Topa, M.A. and McLeod, K.W. 1986. Aerenchyma and lenticel formation in pine seedlings: a possible avoidance mechanism to anaerobic growth conditions. *Physiologia Plantarum* 68: 540–550.

Trebitz, A.S., Nichols, S.A., Carpenter, S.R., and Lathrop, R.C. 1993. Patterns of vegetation change in Lake Wingra following a *Myriophyllum spicatum* decline. *Aquatic Botany* 46: 325–340.

Tripathi, B.D., Srivastava, J., and Misra, K. 1991. Nitrogen and phosphorus removal capacity of four chosen macrophytes in tropical freshwater ponds. *Environment Conservation* 12: 143-148.

Turner, R. 1976. Geographic variations in salt marsh macrophyte production: a review. *Contributions in Marine Science* 20: 47–68.

Turner, C.E., Center, T.D., Burrows, D.W., and Buckingham, G.R. 1998. Ecology and management of *Melaleuca quinquenervia*, an invader of wetlands in Florida, U.S.A. *Wetlands Ecology and Management* 5: 165–178.

Turner, S.D., Amon, J.P., Schneble, R.M., and Friese, C.F. 2000. Mycrorrhizal fungi associated with plants in ground-water fed wetlands. *Wetlands* 20: 200–204.

Twilley, R. 1988. Coupling mangroves to the productivity of estuarine and coastal waters. In *Coastal Offshore Interactions: Lecture Notes on Coastal and Estuarine Studies*, Vol. 22. pp. 155–180. B.O. Jansson, Ed. Berlin. Springer-Verlag.

Twilley, R., Kemp, W.M., Staver, K.W., Stevenson, J.C., and Boynton, W.R. 1985. Nutrient enrichment of estuarine submersed vascular plant communities. I. Algal growth and effects on production of plants and associated communities. *Marine Ecology Progress Series* 23: 179–191.

Twilley, R., Lugo, A.E., and Patterson-Zucca, C. 1986. Litter production and turnover in basin mangrove forests in southwest Florida. *Ecology* 67: 670–683.

Twolan-Strutt, L. and Keddy, P.A. 1996. Above- and belowground competition intensity in two contrasting wetland plant communities. *Ecology* 77: 259–270.
U.S. Army Corps of Engineers. 1987. U.S. Army Corps of Engineers Wetlands Delineation Manual, Technical Report Y-87-1. Vicksburg, MS. U.S. Army Corps of Engineers, Environmental Laboratory, Waterways Experiment Station.
U.S. Army Corps of Engineers, Jacksonville District.1999. Fact sheet on the Corps' three aquatic plant control programs. U.S. ACE. Website: http:/www.saj.usace.army.mil/conops/apc/.
U.S. Department of Agriculture. 1982. National List of Scientific Plant Names. Vol. I. List of Plant Names. Vol II. Synonymy. SCS-TP-159. Washington, D.C. Soil Conservation Service.
U.S. Environmental Protection Agency. 1996. Environmental Indicators of Water Quality in the United States. Report No. EPA 800-R-96-002. Washington, D.C.
U.S. Environmental Protection Agency. 1998. A Citizen's Guide to Phytoremediation. 6 pp. Report No. EPA 542-F-98-011. Washington, D.C. Office of Solid Waste and Emergency Response.
U.S. Fish and Wildlife Service. 1980. Whooping Crane Recovery Plan, 206 pp. Washington, D.C. U.S. Department of the Interior, U.S. Fish and Wildlife Service, Nebraska Game and Parks Commission, Texas Parks and Wildlife Department, National Audobon Society.
U.S. Fish and Wildlife Service. 1986. Part 17 — Endangered and Threatened Wildlife and Plants, 30 pp. Report No. 50 CFR 17.11 and 17.12. Washington D.C. U.S. Department of the Interior.
U.S. Fish and Wildlife Service. 1999. U.S. Listed Flowering Plant Species Index by Lead Region and Status as of November 30, 1999, 13 pp. Washington, D.C. U.S. Department of the Interior, U.S. Fish and Wildlife Service, Division of Endangered Species. Website: http://endangered.fws.gov.
U.S. Forest Products Laboratory. 1974. Wood Handbook: Wood as an Engineering Material. Agricultural Handbook No. 72. Washington, D.C. United States Department of Agriculture.
U.S. Geological Survey. 2000. Hurricane and Extreme Storm Impact Studies, Vol. 2000. U.S. Geological Survey, Center for Coastal Geology. Website: http://coastal.er.usgs.gov/hurricanes.
U.S. National Park Service. 1997. Big Cypress Official Map and Guide. Washington, D.C. U.S. Department of the Interior.
U.S. National Research Council. 1992. *Restoration of Aquatic Ecosystems: Science, Technology, and Public Policy*. Committee on Restoration of Aquatic Ecosystems. 552 pp. Washington, D.C. National Academy Press.
U.S. National Research Council. 1995. *Wetlands: Characteristics and Boundaries.* 307 pp. Committee on Characterization of Wetlands, Water Science and Technology Board, Board on Environmental Studies and Toxicology, Commission on Geosciences, Environment, and Resources. Washington, D.C. National Academy Press.
U.S. Natural Resources Conservation Service. 1991. Technical Requirements for Agricultural Wastewater Treatment. National Bulletin No. 210-1-17. Washington, D.C. Natural Resources Conservation Service, U.S. Department of Agriculture.
Uchino, A., Samejima, M., Ishii, R., and Ueno, O. 1995. Photosynthetic carbon metabolism in an amphibious sedge, *Eleocharis baldwinii* (Torr.) Chapman: modified expression of C_4 characteristics under submerged aquatic conditions. *Plant Cell Physiology* 36: 229–238.
Ueno, O. 1998. Induction of Kranz anatomy and C_4-like biochemical characteristics in a submerged amphibious plant by abscisic acid. *The Plant Cell* 10: 571–583.
Ueno, O., Samejima, M., Muto, S., and Miyachi, S. 1988. Photosynthetic characteristics of an amphibious plant, *Eleocharis vivipara*: expression of C_4 and C_3 modes in contrasting environments. *Proceedings of the National Academy of Science* 85: 6733–6737.
Urban, N.H., Davis, S.M., and Aumen, N.G. 1993. Fluctuations in sawgrass and cattail densities in Everglades Water Conservation Area 2A under varying nutrient, hydrologic and fire regimes. *Aquatic Botany* 46: 203–223.
Vaithiyanathan, P. and Richardson, C.J. 1999. Macrophyte species changes in the Everglades: examination along a eutrophication gradient. *Journal of Environmental Quality* 28:1347–1358.
Valiela, I. 1984. *Marine Ecological Processes*, 546 pp. New York. Springer-Verlag.
Valiela, I. and Teal, J.M. 1974. Nutrient limitation in salt marsh vegetation. In *Ecology of Halophytes.* R.J. Reimold and W.H. Queen, Eds. pp. 547–563. New York. Academic Press.

Valiela, I., Teal, J.M., and Sass, W.J. 1975. Production and dynamics of salt marsh vegetation and the effects of experimental treatment with sewage sludge. *Journal of Applied Ecology* 12: 973–982.

Valiela, I., Teal, J.M., and Persson, N.Y. 1976. Production and dynamics of experimentally enriched salt marsh vegetation: belowground biomass. *Limnology and Oceanography* 21: 245–252.

Valiela, I., Teal, J.M., and Deuser, W.G. 1978. The nature of growth forms in the salt marsh grass *Spartina alterniflora*. *The American Naturalist* 112: 461–470.

Valley, R.D. and Newman, R.M. 1998. Competitive interactions between Eurasian watermilfoil and northern watermilfoil in experimental tanks. *Journal of Aquatic Plant Management* 36: 121–126.

Van, T.K. and Vandiver, V. 1994. Response of hydrilla to various concentrations and exposures of bensulfuron methyl. *Journal of Aquatic Plant Management* 32: 7–11.

Van, T.K., Wheeler, G.S., and Center, T.D. 1999. Competition between *Hydrilla verticillata* and *Vallisneria americana* as influenced by soil fertility. *Aquatic Botany* 62: 255–233.

Vance, H.D. and Francko, D.A. 1997. Allelopathic potential of *Nelumbo lutea* (Willd) Pers to alter growth of *Myriophyllum spicatum* L. and *Potamogeton pectinatus* L. *Journal of Freshwater Ecology* 12: 405–409.

Van den Berg, M.S., Coops, H., Simons, J., and de Keizer, A. 1998. Competition between *Chara aspera* and *Potamogeton pectinatus* as a function of temperature and light. *Aquatic Botany* 60: 241–250.

van der Pijl, L. 1982. *Principles of Dispersal in Higher Plants*, 214 pp. Berlin. Springer-Verlag.

van der Putten, W.H. 1997. Die-back of *Phragmites australis* in European wetlands: an overview of the European Research Programme on Reed Die-back and Progression (1993–1994). *Aquatic Botany* 59: 263–275.

van der Valk, A.G. 1981. Succession in wetlands: a Gleasonian approach. *Ecology* 62: 688–696.

van der Valk, A.G. 1987. Vegetation dynamics of freshwater wetlands: a selective review of the literature. *Archivs für Hydrobiologie* 27: 27–39.

van der Valk, A.G. and Bliss, L.C. 1971. Hydrarch succession and net primary production of oxbow lakes in central Alberta. *Canadian Journal of Botany* 49: 1177–1199.

van der Valk, A.G. and Davis, C.B. 1978. The role of seed banks in the vegetation dynamics of prairie glacial marshes. *Ecology* 59: 322–335.

van der Valk, A.G. and Pederson, R.L. 1989. Seed banks and the management and restoration of natural vegetation. In *Ecology of Seed Banks*, pp. 329–346. New York. Academic Press.

van der Valk, A.G., Squires, L., and Welling, C.H. 1994. Assessing the impacts of an increase in water level on wetland vegetation. *Ecological Applications* 4: 525–534.

van der Valk, A.G., Bremholm, T.L., and Gordon, E. 1999. The restoration of sedge meadows: seed viability, seed germination requirements, and seedling growth of *Carex* species. *Wetlands* 19: 756–764.

Van Diggelen, J., Rozema, J., and Broekman, R. 1987. Growth and mineral relations of salt marsh species on nutrient solutions containing various sodium-sulfide concentration. In *Vegetation between Land and Sea*. A.H.L. Huiskes, C.W.P.M. Blom, and J. Rozema, Eds. pp. 259–266. Dordrecht. Dr. W. Junk Publishers.

Van Raalte, C.D., Valiela, I., and Teal, J.M. 1976. Production of epibenthic salt marsh algae: light and nutrient limitation. *Limnology and Oceanography* 21: 862–872.

Van Vierssen, W. 1982. Reproductive strategies of *Zannichellia* taxa in Western Europe. In *Studies on Aquatic Vascular Plants*. J.J. Symoens, S.S. Hooper, and P. Compere, Eds. pp. 144–149. Brussels. Royal Botanical Society of Belgium.

van Wijck, C., de Groot, C.J., and Grillas, P. 1992. The effect of anaerobic sediment on the growth of *Potamogeton pectinatus* L.: the role of organic matter, sulphide and ferrous iron. *Aquatic Botany* 44: 31–49.

van Wirdum, G. 1993. An ecosystems approach to base-rich freshwater wetlands, with special reference to fenlands. *Hydrobiologia* 265: 129–153.

Van Zon, J.C.J. 1977. Introduction to biological control of aquatic weeds. *Aquatic Botany* 3: 105–109.

Vartapetian, B.B. and Jackson, M.B. 1997. Plant adaptations to anaerobic stress. *Annals of Botany* 79: 3–20.

Vartapetian, B.B., Andreeva, I.N., and Kozlova, G.I. 1976. The resistance to anoxia and the mitochondrial fine structure of rice seedlings. *Protoplasma* 88: 215–244.

Verhoeven, J.T.A. 1982. Reproductive strategies of *Ruppia* taxa in Western Europe. In *Studies on Aquatic Vascular Plants.* J.J. Symoens, S.S. Hooper, and P. Compere, Eds. pp. 156–157. Brussels. Royal Society of Belgium.

Verhoeven, J.T.A., Koerselman, W., and Beltman, B. 1988. The vegetation of fens in relation to their hydrology and nutrient dynamics: a case study. In *Vegetation of Inland Waters.* J.J. Symoens, Ed. pp. 249–282. Dordrecht. Kluwer Academic Publishers.

Vesk, P.A. and Allaway, W.G. 1997. Spatial variation of copper and lead concentrations of water hyacinth plants in a wetland receiving urban run-off. *Aquatic Botany* 59: 33–44.

Vince, S.W. and Snow, A.A. 1984. Plant zonation in an Alaskan salt marsh. I. Distribution, abundance and environmental factors. *Journal of Ecology* 72: 651–667.

Vincent, W.F. and Downes, M.T. 1980. Variation in nutrient removal from a stream by watercress (*Nasturtium officinale*). *Aquatic Botany* 9: 221–236.

Vintéjoux, C. 1982. Particularités physiologiques et cytochimiques des hibernacles de quelques plantes aquatiques. In *Studies on Aquatic Vascular Plants.* J.J. Symoens, S.S. Hooper, and P. Compere, Eds. pp. 17–28. Brussels. Royal Botanical Society of Belgium.

Visser, J.M. and Sasser, C.E. 1995. Changes in tree species composition, structure and growth in a bald cypress-water tupelo swamp forest, 1980–1990. *Forest Ecology and Management* 72: 119–129.

Visser, E.J.W., Nabben, R.H.M., Blom, C.W.P.M., and Voesenek, L.A.C.J. 1997. Elongation by primary lateral roots and adventitious roots during conditions of hypoxia and high ethylene concentrations. *Plant, Cell and Environment* 20: 647–653.

Visser, J.M., Sasser, C.E., Chabreck, R.H., and Linscombe, R.G. 1999. Long-term vegetation change in Louisiana tidal marshes, 1968–1992. *Wetlands* 19: 168–175.

Vitousek, P.M. 1994. Beyond global warming: ecology and global change. *Ecology* 75: 1861–1876.

Vitousek, P.M., D'Antonio, C.M., Loope, L.L. and Westbrooks, R. 1996. Biological invasions as global environmental change. *American Scientist* 84: 468–478.

Vitt, D H. and Bayley, S. 1984. The vegetation and water chemistry of four oligotrophic basin mires in northwestern Ontario Canada. *Canadian Journal of Botany* 62:1485–1500.

Vitt, D.H. and Chee, W. 1990. The relationships of vegetation to surface water chemistry and peat chemistry in fens of Alberta, Canada. *Vegetatio* 89: 87–106.

Vivian-Smith, G. 1997. Microtopographic heterogeneity and floristic diversity in experimental wetland communities. *Journal of Ecology* 85: 71–82.

Voss, E. 1972. *Michigan Flora, Part I: Gymnosperms and Monocots,* 488 pp. Bloomfield Hills, MI. Cranbrook Institute of Science.

Voss, E. 1985. *Michigan Flora, Part II: Dicots (Sauraceae-Cornaceae),* 724 pp. Bloomfield Hills, MI. Cranbrook Institute of Science.

Voss, E. 1996. *Michigan Flora, Part III: Dicots (Pyrolaceae-Compositae),* 622 pp. Bloomfield Hills, MI. Cranbrook Institute of Science.

Vought, L.B.-M., Lacoursiere, J.O., and Voelz, N. 1991. Streams in the agricultural landscape. *Vatten* 47: 321–328.

Vought, L.B.-M., Dahl, J., Pedersen, C.L., and Lacoursiere, J.O. 1994. Nutrient retention in riparian ecotones. *Ambio* 23: 342–348.

Waisel, Y., Eshel, A., and Agami, M. 1986. Salt balance of leaves of the mangrove *Avicennia marina. Physiologia Plantarum* 67: 67–72.

Wakely, J.S. and Lichvar, R.W. 1997. Disagreements between plot-based prevalence indices and dominance ratios in evaluations of wetland vegetation. *Wetlands* 17: 301–309.

Walbridge, M.R. 1993. Functions and values of forested wetlands in the southern United States. *Journal of Forestry* 91: 15–19.

Walker, B.A., Pate, J.S., and Kuo, J. 1983. Nitrogen fixation by nodulated roots of *Viminaria juncea* (Schrad. and Wendl.) Hoffmans (Fabaceae) when submerged in water. *Australian Journal of Plant Physiology* 10: 409–421.

Walters, B.B. 1997. Human ecological questions for tropical restoration: experiences from planting native upland trees and mangroves in the Philippines. *Forestry Ecology and Management* 99: 275–290.

Walters, B.B. 2000a. Event Ecology in the Philippines: Explaining Mangrove Tree Cutting and Planting and Their Environmental Effects. Ph.D. dissertion. 371 pp. Graduate Program in Ecology and Evolution. New Brunswick, NJ. Rutgers, The State University of New Jersey.

Walters, B.B. 2000b. Local mangrove planting in the Philippines: are fisherfolk and fishpond owners effective restorationists? *Restoration Ecology* 8: 237–246.

Ward, L.G., Kemp, W.M., and Boynton, W.R. 1984. The influence of waves and seagrass communities on suspended particulates in an estuarine embayment. *Marine Geology* 59: 85–103.

Wardrop, D.H. 1997. The Occurrence and Impact of Sedimentation in Central Pennsylvania Wetlands. Ph.D. dissertation. University Park. Pennsylvania State University.

Wardrop, D.H. and Brooks, R.P. 1998. The occurrence and impact of sedimentation in central Pennsylvania wetlands. *Environmental Monitoring and Assessment* 51:119–130.

Warming, E. 1909. *Oecology of Plants. An Introduction to the Study of Plant Communities*. Oxford. Clarendon Press (updated English translation of 1886 text).

Wassen, M.J., Barendregt, A., Palczynski, A., De Smidt, J.T., and De Mars, H. 1990. The relationship between fen vegetation gradients, groundwater flow and flooding in an undrained valley mire at Biebrza, Poland. *Journal of Ecology* 78: 1106–1122.

Waters, I., Morrell, S., Greenway, H., and Colmer, T.D. 1991. Effects of anoxia on wheat seedlings. II. Influence of O_2 supply prior to anoxia on tolerance to anoxia, alcoholic fermentation and sugar levels. *Journal of Experimental Botany* 42: 1437–1447.

Weber, M. and Brandle, R. 1996. Some aspects of the extreme anoxia tolerance of the sweet flag, *Acorus calamus* L. *Folia Geobotanica Phytotaxonomica* 31: 37–46.

Weeden, C.R., Shelton, A.M., and Hoffman, M.P. 1996a. *Hylobius transversovittatus* (Coleoptera: Curculionidae), Biological Control: A Guide to Natural Enemies in North America. Ithaca, NY. Cornell University. Website: http//www.nyaes.cornell.edu/ent/biocontrol/index.html.

Weeden, C.R., Shelton, A.M., and Hoffman, M.P. 1996b. Weed-Feeders, Biological Control: A Guide to Natural Enemies in North America. Ithaca, NY. Cornell University. Website: http//www.nyaes.cornell.edu/ent/biocontrol/index.html.

Weeden, C.R., Shelton, A.M., and Hoffman, M.P. 1997. *Hydrilla verticillata*, Biological Control Insects for Aquatic and Wetland Weeds, Aquatic and Wetland Plant Information Retrieval System. Ithaca, NY. Cornell University. Website: http//www.nyaes.cornell.edu/ent/ biocontrol/index.html.

Weihe, P.E. and Neely, R.K. 1997. The effects of shading on competition between purple loosestrife and broad-leaved cattail. *Aquatic Botany* 59: 127–138.

Weiher, E. and Keddy, P.A. 1995. The assembly of experimental wetland plant communities. *Oikos* 73: 323–335.

Weiher, E., Wisheu, I.C., Keddy, P.A., and Moore, D.R.J. 1996. Establishment, persistence, and management implications of experimental wetland plant communities. *Wetlands* 16: 208–218.

Weisner, S.E.B. 1993. Long-term competitive displacement of *Typha latifolia* by *Typha angustifolia* in a eutrophic lake. *Oecologia* 94: 451–456.

Weisner, S.E.B. and Strand, J.A. 1996. Rhizome architecture in *Phragmites australis* in relation to water depth: implications for within-plant oxygen transport distances. *Folia Geobotanica Phytotaxonomica* 31: 91-97.

Weller, M.W. 1994. *Freshwater Marshes Ecology and Wildlife Management*, 154 pp. Minneapolis. University of Minnesota Press.

Welling, C.H., Pederson, R.L., and van der Valk, A. G. 1988a. Recruitment from the seed bank and the development of zonation of emergent vegetation during a drawdown in a prairie wetland. *Journal of Ecology* 76: 483–496.

Welling, C.H., Pederson, R.L., and van der Valk, A. G. 1988b. Temporal patterns in recruitment from the seed bank during drawdowns in a prairie wetland. *Journal of Applied Ecology* 25: 999–1007.

Wenerick, W.R., Stevens, S.E., Jr., Webster, H.J., Stark, L.R., and DeVeau, E. 1989. Tolerance of three wetland plant species to acid mine drainage: a greenhouse study. In *Constructed Wetlands for Wastewater Treatment*. D.A. Hammer, Ed. pp. 801–807. Chelsea, MI. Lewis Publishers.

Wentworth, T.R., Johnson, G.P., and Kologiski, R.L. 1988. Designation of wetlands by weighted averages of vegetation data: a preliminary evaluation. *Water Resources Bulletin* 24: 389–396.

Westerdahl, H.E. and Getsinger, K.D. 1988. Aquatic Plant Identification and Herbicide Use Guide, Vol. 2, Aquatic Plants and Susceptibility to Herbicides. 104 pp. Technical Report A-88-9. Vicksburg, MS. U.S. Army Corps of Engineers Waterways Experiment Station.

Westlake, D.F. 1967. Some effects of low-velocity currents on the metabolism of aquatic macrophytes. *Journal of Experimental Botany* 18: 187–205.

Westlake, D.F. 1975. Primary production of freshwater macrophytes. In *Photosynthesis and Productivity in Different Environments*. C.P. Cooper, Ed. pp. 189–206. Cambridge. Cambridge University Press.

Westlake, D.F. 1982. The primary production of water plants. In *Studies on Aquatic Vascular Plants*. J.J. Symoens, S.S. Hooper, and P. Compere, Ed. pp. 165–180. Brussels. Royal Botanical Society of Belgium.

Wetzel, P.R. and van der Valk, A.G. 1996. Vesicular-arbuscular mycorrhizae in prairie pothole wetland vegetation in Iowa and North Dakota. *Canadian Journal of Botany* 74: 883–890.

Wetzel, P.R. and van der Valk, A.G. 1998. Effects of nutrient and soil moisture on competition between *Carex stricta, Phalaris arundinacea*, and *Typha latifolia*. *Plant Ecology* 138: 179–190.

Wetzel, R.G. 1964. A comparative study of the primary productivity of higher aquatic plants, periphyton, and phytoplankton in a large, shallow lake. *Internationale Revue der gesamten Hydrobiologie* 49: 1–61.

Wetzel, R.G. 1966. Techniques and problems of primary productivity measurements in higher aquatic plants and periphyton. In *Primary Productivity in Aquatic Environments*. C.R. Goldman, Ed. pp. 249–267. Berkeley. University of California Press.

Wetzel, R.G. 1983a. *Limnology*, 767 pp. Philadelphia. Saunders College Publishing.

Wetzel, R.G. 1983b. Attached algal-substrata interactions: fact or myth, and when and how? In *Periphyton of Freshwater Ecosystems*. R.G. Wetzel, Ed. pp. 207–215. The Hague. Dr. W. Junk Publishers.

Wetzel, R.G. 1988. Water as an environment for plant life. In *Vegetation of Inland Waters*. J.J. Symoens, Ed. pp. 1–30. Dordrecht. Kluwer Academic Publishers.

Wetzel, R.G. and Hough, R.A. 1973. Productivity and role of aquatic macrophytes in lakes. An assessment. *Polskie Archiwum Hydrobiologii* 20: 9–19.

Wetzel, R.G. and Likens, G.E. 1990. *Limnological Analyses*, 391 pp. New York. Springer-Verlag.

Wetzel, R.G. and Pickard, D. 1996. Application of secondary production methods to estimates of net aboveground primary production of emergent aquatic macrophytes. *Aquatic Botany* 53: 109–120.

Wheeler, B.D. 1995. Introduction: restoration and wetlands. In *Restoration of Temperate Wetlands*. B.D. Wheeler, S.C. Shaw, W.J. Fojt, and R.A. Robertson, Eds. pp. 1–18. Chichester. John Wiley & Sons.

Whigham, D.F., McCormick, J., Good, R.E., and Simpson, R.L. 1978. Biomass and primary production in freshwater tidal wetlands of the middle Atlantic coast. In *Freshwater Wetlands Ecological Processes and Management Potential*. R.E. Good, D.F. Whigham, and R.L. Simpson, Eds. pp. 3–20. New York. Academic Press.

White, P.S. and Pickett, S.T.A. 1985. Natural disturbance and patch dynamics: an introduction. In *The Ecology of Natural Disturbance and Patch Dynamics*. S.T.A. Pickett and P.S. White, Eds. pp. 3–13. Orlando, FL. Academic Press.

Whitehead, A.J., Lo, K.V., and Bulley, N.R. 1987. The effect of hydraulic retention time and duckweed cropping rate on nutrient removal from dairy barn wastewater. In *Aquatic Plants for Water Treatment and Resource Recovery*. K.R. Reddy and W.H. Smith, Eds. pp. 697–702. Orlando, FL. Magnolia Publishing.

Whitehead, D.R. 1983. Wind pollination: some ecological and evolutionary perspectives. In *Pollination Biology*. L. Real, Ed. pp. 97–108. Orlando, FL. Academic Press.

Whittaker, R.H. 1953. A consideration of climax theory: the climax as a population and pattern. *Ecological Monographs* 23: 41–78.

Whittaker, R.H. 1962. Net production relations of shrubs in the Smoky Mountains. *Ecology* 43: 357–377.

Whittaker, R.H. 1966. Forest dimensions and production in the Great Smoky Mountains. *Ecology* 47: 103–121.

Whittaker, R.H. and Woodwell, G.M. 1968. Dimension and production relations of trees and shrubs in the Brookhaven Forest, New York. *Journal of Ecology* 56: 1–25.

Wieder, R.K. 1989. A survey of constructed wetlands for acid coal mine drainage treatment in the eastern United States. *Wetlands* 9: 299–315.

Wiegert, R.G. and Evans, F.C. 1964. Primary production and the disappearance of dead vegetation on an old field in southeastern Michigan. *Ecology* 45: 49–62.

Wiegert, R.G., Chalmers, A.G., and Randerson, P.F. 1983. Productivity gradients in salt marshes: the response of *Spartina alterniflora* to experimentally manipulated soil water movement. *Oikos* 41: 1–6.

Wiegleb, G. 1988. Analysis of flora and vegetation in rivers: concepts and applications. In *Vegetation of Inland Waters*. J.J. Symoens, Ed. pp. 311–341. Dordrecht. Kluwer Academic Publishers.

Wienhold, C.E. and van der Valk, A. G. 1989. The impact of duration of drainage on the seed banks of northern prairie wetlands. *Canadian Journal of Botany* 67: 1878–1884.

Wigand, C. and Stevenson, J.C. 1994. The presence and possible ecological significance of mycorrhizae of the submersed macrophyte, *Vallisneria americana*. *Estuaries* 17: 206–215.

Wigand, C., Stevenson, J.C., and Cornwell, J.C. 1997. Effects of different submersed macrophyes on sediment biogeochemistry. *Aquatic Botany* 56: 233–244.

Wijte, A.H.B.M. and Gallagher, J.L. 1996a. Effect of oxygen availability and salinity on early life history stages of salt marsh plants. I. Different germination strategies of *Spartina alterniflora* and *Phragmites australis* (Poaeceae). *Journal of Botany* 83: 1337–1342.

Wijte, A.H.B.M. and Gallagher, J.L. 1996b. Effect of oxygen availability and salinity on early life history stages of salt marsh plants. II. Early seedling development advantage of *Spartina alterniflora* over *Phragmites australis*. *American Journal of Botany* 83: 1343–1350.

Wilcove, D.S., Rothstein, D., Dubow, J., Phillips, A., and Losos, E. 1998. Quantifying threats to imperiled species in the United States. *BioScience* 48: 607–615.

Wilcox, D.A. 1995. Wetland and aquatic macrophytes as indicators of anthropogenic hydrologic disturbance. *Natural Areas Journal* 15: 240–248.

Wilcox, D.A. and Whillans, T.H. 1999. Techniques for restoration of disturbed coastal wetlands of the Great Lakes. *Wetlands* 19: 835–857.

Wilcox, D.A., Shedlock, R.J., and Hendrickson, W.H. 1986. Hydrology, water chemistry and ecological relations in the raised mound of Cowles Bog. *Journal of Ecology* 74: 1103–1117.

Wilen, B.O. and Tiner, R.W. 1993. Wetlands of the United States. In *Wetlands of the World I: Inventory, Ecology and Management*. D.F. Whigham, D. Dykyjova, and S. Hejny, Eds. pp. 515–636. Dordrecht. Kluwer Academic Publishers.

Wilhelm, G. and Ladd, D. 1988. Natural Area Assessment in the Chicago Region. Trans. 53rd North American Wildlife and Natural Resources Conference. pp. 361–375. Chicago, IL.

Williams, D.C. and Lyon, J.G. 1997. Historical aerial photographs and a geographic information system (GIS) to determine effects of long-term water level fluctuations on wetlands along the St. Marys River, Michigan, USA. *Aquatic Botany* 58: 363–378.

Willis, C.N. and Mitsch, W.J. 1995. Effects of hydrology and nutrients on seedling emergence and biomass of aquatic macrophytes from natural and artificial seed banks. *Ecological Engineering* 4: 65–76.

Wilson, E.O. 1992. The Diversity of Life, 424 pp. Cambridge, MA. Belknap Press.

Wilson, J.R. 1975. Comparative response to nitrogen deficiency of a tropical and temperate grass in the interrelation between photosynthesis, growth and accumulation of non structural carbohydrate. *Netherlands Journal of Agricultural Science* 23: 104–112.

Wilson, S.D. and Keddy, P.A. 1986. Measuring diffuse competition along an environmental gradient: results from a shoreline plant community. *The American Naturalist* 127: 862–869.

Wilson, S.D., Moore, D.R.J., and Keddy, P.A. 1993. Relationships of marsh seed banks to vegetation patterns along environmental gradients. *Freshwater Biology* 29: 361–370.

Winter, T.C. 1992. A physiographic and climatic framework for hydrologic studies of wetlands. In *Aquatic Ecosystems in Semi-Arid Regions: Implications for Resource Management*. R.D. Roberts and M.L. Bothwell, Eds. pp. 127–148. Saskatoon. Environment Canada.

Wisheu, I.C. and Keddy, P.A. 1994. The low competitive ability of Canada's Atlantic coastal plain shoreline flora: implications for conservation. *Biological Conservation* 68: 247–252.

Wójcik, W. and Wójcik, M. 2000. Data on the Biala River wetland and the results of the field experiments. In *Heavy Metals in the Environment: Using Wetlands for Their Removal*. H.T. Odum, W. Wójcik, J.L. Pritchard, S. Ton, J.J. Delfino, M. Wójcik, S. Leszczynski, J.D. Patel, S.J. Doherty, and J. Stasik, Eds. pp. 211–254. Boca Raton, FL. Lewis Publishers.

Wu, X. and Mitsch, W.J. 1998. Spatial and temporal patterns of algae in newly constructed freshwater wetlands. *Wetlands* 18: 9–20.

Xia, J.H. and Saglio, P.H. 1992. Lactic acid efflux as a mechanism of hypoxic acclimation of maize root tips to anoxia. *Plant Physiology* 100: 40–46.

Xia, J.H., Saglio, P.H., and Roberts, J.K.M. 1995. Nucleotide levels do not critically determine survival of maize root tips acclimated to a low-oxygen environment. *Plant Physiology* 108: 589–595.

Yamasaki, S. 1984. Role of plant aeration in zonation of *Zizania latifolia* and *Phragmites australis*. *Aquatic Botany* 18: 287–297.

Yan, J., Zhang, Y., and Wu, X. 1993. Advances of ecological engineering in China. *Ecological Engineering* 2: 193–216.

Yasumoto, E., Adachi, K., Kato, M., Sano, H., Sasamoto, H., Baba, S., and Ashihara, H. 1999. Uptake of inorganic ions and compatible solutes in cultured mangrove cells during salt stress. *In Vitro Cellular and Developmental Biology Plant* 35: 82–85.

Ye, Z.H., Baker, A.J.M., Wong, M.H., and Willis, A.J. 1997a. Zinc, lead, cadmium tolerance, uptake and accumulation by *Typha latifolia*. *New Phytologist* 136: 469–480.

Ye, Z.H., Baker, A.J.M., Wong, M.H., and Willis, A.J. 1997b. Zinc, lead, cadmium tolerance, uptake and accumulation by the common reed, *Phragmites australis* (Cav.) Trin. ex Steudel. *Annals of Botany* 80: 363–370.

Yeo, R.R., Falk, R.H., and Thurston, J.R. 1984. The morphology of Hydrilla (*Hydrilla verticillata* (L.f.) Royle). *Journal of Aquatic Plant Management* 22: 1–17.

Yoder, C.O. and Rankin, E.T. 1995. Biological response signatures and area of degradation value: new tools for interpreting multimetric data. In *Biological Assessment and Criteria: Tools for Water Resource Planning and Decision Making*. W.S. Davis and T.P. Simon, Eds. pp. 263–286. Boca Raton, FL. Lewis Publishers.

Yuan, C., Zhao, Q., and Zhen, J. 1993. Comparing crop-hog-fish agroecosystems with conventional fish culturing in China. *Ecological Engineering* 2: 231–242.

Zapata, O. and McMillan, C. 1979. Phenolic acids in seagrasses. *Aquatic Botany* 7: 307–317.

Zartman, R.E. and Fish, E.B. 1992. Spatial characteristics of playa lakes in Castro County, Texas. *Soil Science* 153: 62–68.

Zayed, A., Gowthaman, S., and Terry, N. 1998. Phytoaccumulation of trace elements by wetland plants. I. Duckweed. *Journal of Environmental Quality* 27: 715–721.

Zedler, J.B. 1977. Salt marsh community structure in the Tijuana Estuary, California. *Estuarine and Coastal Marine Science* 5: 39–53.

Zedler, J.B. 1980. Algal mat productivity: comparisons in a salt marsh. *Estuaries* 3: 122–131.

Zedler, J.B. 1993. Canopy architecture of natural and planted cordgrass marshes: selecting habitat evaluation criteria. *Ecological Applications* 31: 123–138.

Zedler, J.B. 2000a. Restoration of biodiversity to coastal and inland wetlands. In *Biodiversity in Wetlands: Assessment, Function and Conservation*, Vol. I, B. Gopal, W.J. Junk, and J.A. Davis, Eds. pp. 311–330. Leiden, The Netherlands. Backhuys Publishers.

Zedler, J.B., Ed. 2000b. *Handbook for Restoring Tidal Wetlands*, 400 pp. Boca Raton, FL. CRC Press.

Zedler, J.B. and Rea, N. 1998. Introduction, ecology and management of wetland plant invasions. *Wetlands Ecology and Management* 5: 161–163.

Zedler, J.B., Winfield, T., and Williams, P. 1980. Salt marsh productivity with natural and altered tidal circulation. *Oecologia* 44: 236–240.

Zhu, T. and Sikora, F.J. 1995. Ammonium and nitrate removal in vegetated and unvegetated gravel bed microcosm wetlands. *Water Science and Technology* 32: 219–228.

Zoltai, S.C. 1983. Wetlands in Canada: their classification, distribution, and use. In *Ecosystems of the World. Vol. 4A. Mires: Swamp, Bog, Fen, and Moor*, A.J.P. Gore, Ed. pp. 245–268. Amsterdam. Elsevier Science.

Zuberer, D.A. and Silver, W.S. 1978. Biological dinitrogen fixation (acetylene reduction) associated with Florida mangroves. *Environmental Microbiology* 35: 567–575.

Index

A

Acanthus ilicifolius, 112, 128
Accentria ephemerella, 303
Acer, 49
Aceraceae, 167, 169
Acer negundo, 50, 70,
Acer rubrum, 9, 50, 70, 73
  effects of flooding, 71, 72
  in delineation, 369
  nutrient content, 345, 347
  nutrient release, 347
Acer saccharinum, 50, 70
Acer saccharum, 168
Acetaldehyde, 108
Acid precipitation, 81
Acnida cannabina, 39
Acorus, 170
Acorus calamus, 39, 96
  seed banks, 252
  succession, 245
Acrostichum, 184
Adenosine triphosphate, see ATP
ADH, 106-107
Adventitious roots, 91-93
Aegialitis annulata, 112
Aegiceras corniculata, 112
Aerenchyma, 88-91
  definition, 17
  formation, 89-91
  function, 91
Aerobic respiration, 87-88
Aerobic sediments, 74
Aeschynomene, 116
Aeschynomene aspera, 169
Aesculus glabra, 50
African elodea, see *Lagarosiphon major*
Agasicles hygrophila, see South American alligatorweed flea beetle
Agelaius phoeniceus, see Red-winged blackbird
Agropyretum, 93
Agrostis stolonifera, 328
Alaska alkali grass, see *Puccinellia nutkaensis*
Albany pitcher plant, see *Cephalotus follicularis*
Alcohol dehydrogenase, see ADH
Alder, see *Alnus*, Betulaceae
Aldrovanda, 117
  range, 118
  seed dispersal, 171
Aldrovanda vesiculosa, 124, 126
Alisma, 19
  seed dispersal, 171, 172
  seeds, 173
  in treatment wetlands, 351
Alisma plantago-aquatica, 245
Alismataceae, 7, 40, 41
  asexual reproduction, 183
  fruit, 167, 169
Alismatales, 156
Alkaloids, 135, 265
Allelopathy, 240, 254, 265-266
Allen curve method, 205, 207, 208-209, 216-219
Alligators, 34
Alligatorweed stem borer, 297
Alligatorweed thrips, 297
Alnus, 9, 41, 116
Alnus glutinosa, 92
Alnus incarna, 115
Alternanthera philoxeroides
  asexual reproduction, 181, 183
  biological control, 297
  as invasive, 285, 288
  nutrient content, 344
  response to drawdown, 290
  response to herbicides, 292
Alternative end products hypothesis, 107
Althenia, 157
Ambrosia trifida, 39
  competition, 259
  as invasive, 284
American beech, see *Fagus grandifolia*
American coot, 350
American elm, see *Ulmus americana*
American water lotus, see *Nelumbo lutea*
Ammonification, 75-76
Amphibians, 34
Amphibolis, 157, 159, 162
Amphibolis anarctica, 159, 177
Amynothrips andersoni, see Alligatorweed thrips

Anacardiaceae, 135
Anaerobic metabolism, 87-88, 106
Anaerobic sediments, 75-79
Andromeda glaucophylla, 56, 58
Anemophily, see Pollination, wind
Angiosperm classification, 17-19
Anoxia, 75
- gas transport mechanisms, 97-102
- root adaptations, 91-95
  - adventitious roots, 91-93
  - pneumatophores, 93-95
  - prop and drop roots, 95
  - shallow rooting, 93
- stem adaptations, 95-96
  - hypertrophy, 96
  - rapid underwater shoot extension, 95-96
  - stem buoyancy, 96
- whole plant adaptations, 88-110
  - avoidance, 104
  - carbohydrate storage, 104
  - metabolic responses, 104-110, see also Metabolic processes under anoxia

Apiaceae, 41
- deterrents to herbivory, 135
- fruit, 167, 169

Apium prostratum, 328
Apomixis, 177
Aponogetonaceae, 167
Aponogeton, 171
Appertiella, 155, 156
Apple snail, 286
Aquatic acid metabolism, 130-131
Araceae, 7, 41, 167, 170
Arethusa bulbosa, 115
Arisaema triphyllum, 379
Armoracia aquatica, 179, 183
Arrow arum, see *Peltandra virginica*
Arrow grass, see *Triglochin maritimum*
Arrowhead, see *Sagittaria latifolia*
Arthrocnemum, 113
Arthrocnemum perenne, 239
Arthrocnemum subterminale, 264-265
Arum, see Araceae
Arundo donax, 135, 287
Asclepias incarnata, 135, 136, 386
Asclepias syriaca, 379
Asexual reproduction, 177-188
- evolutionary consequences, 19
- fragmentation, 178-180
- frequency, 186-188
- propagule types, 178-186
- turions, 180-182

Ash, see *Fraxinus*
Aster, 245
Asteraceae, 7, 167, 168
Aster novae-angliae, 9
Aster ontarionis, 259
Aster tripolii, 93
Aster tripolium, 340
Asters, see *Aster*, Asteraceae
Atlantic white cedar, see *Chamaecyparis thyoides*
ATP, 87-88
- factor in cell pH, 108-109
- production in aerobic respiration, 87
- production in anaerobic metabolism, 88, 104, 107

Atriplex, 113
Atriplex prostrata, 328
Atriplex triangularis, 174
Australian pine, see *Casuarina*
Auxin, 92
Avicennia, 44, 94, 95
Avicenniaceae, 12
Avicennia germinans, 44, 45
- gas transport, 101
- hurricanes, 270
- pneumatophores, 45, 48
- primary productivity, 230-232
- radial oxygen loss, 103, 114
- salinity tolerance, 48
- vivipary, 177
- zonation, 46, 47

Avicennia marina, 112, 116
Avicennia officinalis, 128
Azolla, 282
Azolla pinnata, 340

B

Bacopa monnieri, 340
Bald cypress, see *Taxodium distichum*
Baldellia ranunculoides, 168, 172, 183
Balsam poplar, see *Populus balsamifera*
Barnyard grass, see *Echinochloa crus-galli*
Basal area measurement, 222-223
Basin forests, 56
Baumea articulata, 342, 344, 351, 352
Bayeriola salicariae, 313
Beak rush, see *Rhynchospora*
Beavers, 34, 272
Betulaceae, 9
Berula erecta, 168
Betula glandulosa, 115, 137
Betula nigra, 50, 72, 175
Betula pumila, 115
Betula pumila var. *glandulifera*, 245
Bicarbonate
- availability underwater, 83
- polar model, 130
- uptake, 12, 83, 129-130

Bidens, 271
Bidens cernua, 249
Bidens laevis, 39
Biogeochemistry, 33
  effects of hydrology, 72-74
  processes in wetlands, 74-83
Biological controls, see Invasive species, controls, biological
Biological indicators, 4, 371-387
Biovularia, 118
Black crappie, 305
Black gum, see *Nyssa biflora*
Black mangrove, see *Avicennia germinans*
Black needlerush, see *Juncus roemerianus*
Black oak, see *Quercus velutina*
Black spruce, see *Picea mariana*
Black tern, 312
Black tupelo, see *Nyssa sylvatica*
Bladderworts, see Lentibulariaceae, *Utricularia*
Blueberries, see *Vaccinium*, *Vaccinium corymbosum*
Bluegill, 305
Blyxa, 164
Blyxa alterniflora, 166
Blyxa octandra, 151, 152
Boehmeria cylindrica, 379
Bog rosemary, see *Andromeda glaucophylla*
Bog turtle, 312
Bogs, 53-58, 80-81, 232, 235
Bolboschoenus, see Preface, 344
Bottomland hardwoods, 71-72, see also Southern bottomland hardwoods
Box elder, see *Acer negundo*
Bracharia mutica, 285, 292
Brachyelytrum erectum, 379
Branta bernicla, see Brent geese
Brasenia, 149, 171
Brasenia schreberi, 169, 170, 292
Brazilian pepper, see *Schinus terebinthifolius*
Brent geese, 271
Brome, see *Bromus inermis*
Bromus inermis, 361
Bruguiera, 44
  pneumatophores, 94, 95
  pollination, 150
  salt exclusion, 111
Bruguiera gymnorhiza, 128
Bryophytes, 7
Bulb-bearing water hemlock, see *Cicuta bulbifera*
Bulk flow, see Pressurized ventilation
Bulrushes, see *Scirpus*
Bur reeds, see Sparganiaceae, *Sparganium*
Burhead, see *Echinodorus tenellus*
Bushy pondweed, see *Najas*
Butomaceae, 41, 167
Butomus umbellatus, 131
Buttercup family, see Ranunculaceae
Butterwort, see *Pinguicula*
Buttonbush, see *Cephalanthus occidentalis*
Buttonwood, see *Conocarpus erectus*
Byblidaceae, range, 118
Byblis, 117, 118, 119

C

C_3 photosynthesis, 134, 265, 378
C_4 photosynthesis, 134, 265, 378
Cabomba
  aquarium plant, 282
  reproduction, 149
  seed dispersal, 171
Cabombaceae, 167
Cabomba caroliniana
  biological controls, 298
  pollination, 152
  response to drawdown, 290
  response to herbicides, 293
Cabomba furcata, 132
Calamagrostis, 74, 245
Calamagrostis canadensis, 386
Calamagrostis deschampsiodes, 271
Calamagrostis stricta, 386
Caldesia parnassifolia, 183
California pitcher plant, see *Darlingtonia californica*
California vernal pools, see vernal pools
Calla, 170, 171
Calla palustris, 170
Callitrichaceae, 13, 167, 169
Callitriche, 150
Callitriche cophocarpa, 68
Callitriche heterophylla, 166
Callitriche palustris, 131, 132
Callitriche platycarpa, 96
Callitriche verna, 166
Calopogon tuberosus, 115
Caltha natans, 169, 245
Caltha palustris, 90
Canna flaccida, 348, 351, 352
Caprifoliaceae, 9
Carbohydrate storage, 104
Carbon transformations, 76, 78, 83
Carbon dioxide
  availability underwater, 83
  submerged species' adaptations, 129-131, 262
Caretta caretta caretta, see Loggerhead turtles
Carex, 3, 27, 56, 74
  anoxia endurance, 104, 105
  competition, 259, 263
  Index of Vegetative Integrity, 385, 386, 387
  pollination, 153
  in prairie potholes, 362

primary productivity, 234
sedimentation tolerance, 379
seed viability, 333
succession, 245
in treatment wetlands, 352
Carex acuta, 342
Carex alternifolius, 105
Carex aquatilis, 137
Carex curta, 90
Carex flagellifera, 328
Carex gracilis, 101
Carex lacustris, 384
Carex lupilina, 168
Carex lyngbyei, 38, 264
Carex papyrus, 105
Carex prasina, 379
Carex ramenskii, 187, 264
Carex rostrata, 105
Carex stricta, 258, 259
Carex subspathacea, 271
Carex vulpinoidea, 379
Carnivorous plants, 117-126, 142-145
distribution, 54
facultative carnivory, 126
range, 117-118, 142
trap types, 118-126, 143, 144
use of prey, 126, 144-145
Carum verticillatum, 168
Carya aquatica, 70
Castor canadensis, see Beavers
Casuarina, 282, 286
Casuarina cunninghamiana, 282
Casuarina equisetifolia, 282
Casuarina glauca, 282
Catclaw mimosa, see *Mimosa pellita*
Cation exchange, 80-81
Cattails, see *Typha*, Typhaceae
Cellular pH, 108-109
Celtis laevigata, 70
Cephalanthus, 9, 41
Cephalanthus occidentalis, 10, 49, 92, 261
Cephalotaceae, 118
Cephalotus, 117, 118, 119
Cephalotus follicularis, 119, 120
Ceratophyllaceae, 13, 18, 41
fruit, 167
pollination, 157, 165
Ceratophyllales, 157
Ceratophyllum
evolution, 18, 165
leaf morphology, 127
pollination, 19, 157, 161, 162, 165
in prairie potholes, 272
in treatment wetlands, 353
Ceratophyllum demersum, 12
allelopathy, 266
asexual reproduction, 179, 180
biological controls, 298
competition, 262, 305
as cosmospolitan species, 16
growth after dredging, 289
as invasive, 283
and *Myriophyllum spicatum*, 303
nutrient content, 345
phytoremediation, 340
range, 280
response to herbicides, 293
succession, 245
in treatment wetlands, 351
Ceratopteris, 179
Cerbera, 172
Cercis canadensis, 50
Ceriops, 44, 94, 95
Chamaecyparis thyoides, 268
fire, 271
mycorrhizae, 115
succession, 245
Chamaedaphne, 9
Chamaedaphne calyculata, 56, 57
competition, 258
in New Jersey Pine Barrens, 268
primary productivity, 234, 235
succession, 245
Chara
biological control, 298
in Everglades, 277
growth after dredging, 289
Index of Vegetative Integrity, 384
in prairie potholes, 272
Chara aspera, 262
Chara vulgaris, 290
Charophyceae, 17
Chelonia mydas mydas, see Green sea turtles
Chen caerulescens caerulescens, see Lesser snow geese
Chenopodiaceae, 111
Cherrybark oak, see *Quercus falcata* var. *pagodifolia*
Chesapeake Bay, 39, 115, 200
decline of submerged plants, 82, 273
eutrophication, 273
invasives, 300, 302, 305
Chevroned water hyacinth weevil, 309
Chinese water chestnut, see *Eleocharis dulcis*
Chlidonias niger, see Black tern
Cicuta, 169, 179
Cicuta bulbifera, 135
Cicuta maculata, 135, 185
Cicuta virosa, 168
Cirsium arvense, 374, 379
Cladium, 171
Cladium jamaicense

adventitious roots, 92
fire, 270, 271
formation of aerenchyma, 90
hurricanes, 270
occurrence, 276, 277
primary productivity, 209
radial oxygen loss, 103
response to herbicides, 292
rhizomes, 184
Cladium mariscoides, 89
Clean Water Act, 371
Clearweed, see *Pilea pumila*
Clementsian succession, 238-239
Clemmys muhlenbergii, see Bog turtle
Clethra alnifolia, 261
Climate change, see Global warming
Clonal reproduction, see Asexual reproduction
Coastal Marshes, 36, see also Salt marshes, Tidal freshwater marshes
Colocasia esculenta, 285, 351
Combretaceae, 12
Common carp, 82, 174, 298
Common dodder, see *Cuscuta gronovii*
Common reed, see *Phragmites australis*
Compatible solutes, 110
Competition, 253-266, 276-277
allelopathy, 254, 265-266
definition, 253
interspecific, 255-265
intraspecific, 254-255
life history traits, 258-261
root-zone oxygenation, 257
self-thinning, 254-255
terminology, 253-254
Conium maculatum, 135
Conocarpus, 111
Conocarpus erectus, 46, 47
Convective throughflow, see Pressurized ventilation
Cordgrass, see *Spartina alterniflora*
Cornaceae, 9
Cornus, 41, 49
Cosmopolitan species, 16
Cotton rats, 286
Cottonwood, see *Populus*
Cotula coronopifolia, 328
Cotula dioica, 328
Cowardin classification system, 35, 48
Cowbane, see *Oxypolis rigidior*
Cranberries, see *Vaccinium*, *Vaccinium macrocarpon*
Crassulacean acid metabolism, 130
Creeping buttercup, see *Ranunculus repens*
Crinum americanum, 179, 185
Crocodiles, 34, 286
Crocodylus acutus, see Crocodiles
Cryptocoryne beckettii, 131
Cryptocoryne ciliata, 131
Cryptocoryne thwaitesii, 131
Cryptocoryne wendtii, 131
Cryptotaenia canadensis, 259
Ctenopharyngodon idella, see Grass carp, Triploid sterile grass carp
Curly pondweed, see *Potamogeton crispus*
Cuscuta gronovii, 39
Cyanobacteria, 116
Cymodocea, 157
Cymodoceaceae, 157, 167
Cynometra, 172
Cypella aquatica, 179, 181, 185
Cyperaceae, 7, 41
formation of aerenchyma, 90
fruit, 167, 168
Index of Vegetative Integrity, 386
mycorrhizae, 114, 116
nutrient translocation, 126
pollination, 152
Cyperus, 19
allelopathy, 265
competition, 263
nutrient uptake, 346
in prairie potholes, 271
in treatment wetlands, 351
Cyperus involucratus, 342, 344
Cyperus papyrus, 131
Cyperus rotundus, 179, 181, 184
Cyperus strigosus, 185
Cypress domes, 52
Cypress strands, 52
Cypress swamps, 51-52
Cypress, see *Taxodium*
Cyprinus carpio, see Common carp
Cypripedium, 115
Cytoplasmic acidosis, 88, 108-109

D

Darlingtonia, 118, 119
Darlingtonia californica, 117
Decodon verticillatus, 72
asexual reproduction, 179, 182, 183, 184, 187
succession, 241
Deepwater rice, see *Oryza sativa*
Delineation, 363-370
Delta Marsh, Manitoba, 71
Denitrification, 75-76, 336, 341, 369
Depressional marshes, 42-44
Des Plaines River Wetlands Demonstration Project, 68, 329, 330
Diameter at breast height, 221, 222, 223
Dicotyledons, see Eudicotyledons
Didiplis diandra, 131

Diffusion within plants, 97, 98
Dimension analysis, 221-225, 230-235
Dionaea, 118, 119
Dionaea muscipula, 117, 124
Dipsacus sylvestris, 379
Discolobium, 116
Distichlis spicata, 38, 317
  competition, 263, 264, 265
  primary productivity, 208, 228
  salt secretion, 111
Disturbances, 266-273
  animals, 271-272
  eutrophication, 276-277
  fire, 270-271
  floods, 269-270
  human-induced, 267-269, 272-273, 374
    and invasive species, 267-269, 276-277, 282, 283-284
  urbanization, 267-269
Diurnal dissolved oxygen method, 199-200
Diversity, effects of water level fluctuation, 70-72
Dogwoods, see *Cornus*, Cornaceae
Dominance ratio, 368-369
Dragon's mouth, see *Arethusa bulbosa*
Dreissena polymorpha, see Zebra mussels
Drop roots, 95
Drosera, 54
  range 117, 118
  trap type, 119, 123, 124
Drosera anglica, 117
Drosera binata var. *multifida*, 126
Drosera brevifolia, 117
Drosera capensis, 126
Drosera capillaris, 117
Drosera filiformis, 117
Drosera intermedia, 117, 126
Drosera linearis, 117
Drosera rotundifolia, 117, 124, 126, 268
Drosophyllum, 118, 119
Drosophyllum lusitanicum, 120, 122
Duckweeds, see *Lemna*, Lemnaceae
Dulichium arundinaceum, 386
Dune slacks, 27
Dwarf birch, see *Betula glandulosa*
Dwarf spike rush, see *Eleocharis parvula*

E

Eastern redbud, see *Cercis canadensis*
Echinochloa crus-galli, 43
  ethanol tolerance, 108
  germination, 253
  nutrient content, 347
  protein production, 109
  range, 280
  seeds, 173
Echinochloa phyllogogon, 109
Echinochloa stagnina, 168
Echinodorus, 16, 183
Echinodorus brevipedicellatus, 131
Echinodorus grisebachii, 131
Echinodorus paniculatus, 129
Echinodorus tenellus, 129, 131
Ecological integrity, 4, 371, 372
Ecological risk assessment, 382-383
Ecological succession, see Succession
Egeria, 152, 282
Egeria densa
  asexual reproduction, 186
  bicarbonate uptake, 129
  as invasive, 283, 286, 303
  response to herbicides, 293
Eichhornia, 172, 179
Eichhornia crassipes, 14, 306-309
  asexual reproduction, 177
  biological control, 297, 299
  biology, 306-307
  controls, 288, 309
  effects on humans, 287, 288, 308-309
  effects on native community, 285, 286, 308
  as invasive, 3, 21, 279, 280
  metal uptake, 340
  nutrient content, 344, 346
  radial oxygen loss, 348
  range, 307-308
  response to drawdown, 290
  response to herbicides, 292, 294, 309
  in treatment wetlands, 335, 340, 342, 351, 352, 353
Eichhornia paniculata, 187
Elaeagnus angustifolia, 50
Elatinaceae, fruit, 169
Elderberry, see *Sambucus canadensis*
Eleocharis
  allelopathy, 265
  asexual reproduction, 179
  in Everglades, 276, 277
  gas transport, 100
  seed banks, 252
  in treatment wetlands, 351
Eleocharis acicularis, 184, 260
Eleocharis baldwinii, 74, 134
Eleocharis cellulosa, 277
Eleocharis dulcis, 280, 351
Eleocharis elongata, 277
Eleocharis obtusa, 168
Eleocharis palustris, 105, 245
Eleocharis parvula, 312
Eleocharis rostellata, 179, 184
Eleocharis smallii, 260

Eleocharis vivipara, 134
Elodea
aquarium plant, 282
asexual reproduction, 178, 179
pollination, 156, 159, 160
Elodea canadensis, 12
asexual reproduction, 179, 180, 186, 187
bicarbonate uptake, 129, 130
biological control, 298
competition, 257, 258, 263
growth after dredging, 289
as invasive, 263, 283, 286, 302, 303
nutrient content, 345
response to drawdown, 290
response to herbicides, 293
in treatment wetlands, 351
Elodea nuttalli
carbon dioxide re-use, 131
competition, 257
as invasive, 302
pollination, 160
Elymus athericus, 271
Elymus pycnanthus, 328
Elymus virginicus, 259
Elytrigia pungens, 208
Emergent plants
competition, 263
definition, 7
primary productivity methods, 205-220
Endangered species, 23-27, 34
Endemic species, 16, 23
Enhalus, 155, 156, 162, 171
Enhalus acoroides, 170
Entomophily, see Pollination, insect
Environmental sieve model of succession, 248-250, 251
Ephemeral wetlands, 44
Epinasty, 88
Equisetum arvense, 379
Equisetum fluviatile, 245, 260
Ericaceae, 9, 57, 58
adpatations to water stress, 136-137
evergreen leaves, 127
mycorrhizae, 115
Eriocaulaceae, 169
Eriophorum angustifolium, 90
Eriophorum gracile, 386
Eriophorum vaginatum, 90
Erosion control, 33
Erythronium americanum, 379
Estuary seablite, see *Suaeda esteroa*
Ethanol
formation, 87
produced under anoxia, 105, 107
toxicity, 106, 107, 108
Ethylene
adventious root formation, 92
production, 90
rapid underwater shoot extension, 96
release from plants, 91
Eudicotyledons, 18
Euhrychiopsis lecontei, see North American weevil
Eupatorium maculatum, 379
Euphotic zone, 82
Eurasian watermilfoil, see *Myriophyllum spicatum*
European weevil, 313, 314
Euryale ferox, 100
Euthamia graminifolia, 379
Eutrophication, 267-269, 276-277
Evaporation, 64-67
Evapotranspiration, 62, 64-67
Everglades, 4, 39, 51, 70, 349, 350
communities, 71
competition, 261, 276-277
eutrophication, 276-277
Evolution, 17-20
Exotic species, see also Invasive species
terminology, 279
as threat to wetlands, 21-22

F

Fagus grandifolia, 50
Fanwort, see *Cabomba*, *Cabomba caroliniana*
Federal Noxious Weed Act of 1974, 288
Fens, 53-56, 73
Fern allies, see Pteridophytes
Ferns, see Pteridophytes
Festuca arundinacea, 328
Festuca rubra, 271
Filipendula ulmaria, 90, 105
Fimbristylis, 351
Fire, 270-271, 287
Fisheries, 34
Floating heart, see *Nymphoides*
Floating-leaved plants
adaptations to anoxia, 96
definition, 13
primary productivity, 220-221
Floating plants
definition, 14
primary productivity, 220-221
role of roots, 14
Flood control, 33
Floodplain wetlands, 48-52
Florida caerulea, see Little blue heron
Florida Everglades kite, 34, 286
Florida panthers, 34
Floristic Quality Assessment Index, 378-382
Flower structure, 147-148

Flowering rush family, see Butomaceae
Food chains, 4
Forested wetlands, 35, 44-52, see also Bottomland hardwoods, Cypress swamps, Inland forested wetlands, Mangroves, Northeastern floodplain wetlands, Southern bottomland hardwoods, Western riparian zones
Forget-me-not, see *Myosotis scorpioides*
FQAI, see Floristic Quality Assessment Index
Fragaria virginiana, 379
Frankenia grandifolia, 228
Fraxinus, 9
Fraxinus pennsylvanica, 50, 70, 71
Frogbit, see Hydrocharitaceae
Fulica americana, see American coot
Functions of wetland plants, see also Nutrients, Water Quality, Wetlands, functions
  food chains, 4
  habitat, 4
  hydrologic, 4
Fur trade, 34

G

Galerucella calmariensis, 313, 314
Galerucella pusilla, 313, 314
Garden pea, see *Pisum sativum*
Gas transport, 97-102
  diffusion, 97, 98
  pressurized ventilation, 97-100
  underwater gas exchange, 101
  Venturi-induced convection, 101-102
Genlisea, 118, 119
Giant cutgrass, see *Zizaniopsis milacea*
Giant foxtail, see *Setaria magna*
Giant sensitive plant, see *Mimosa pellita*
Glasswort, see *Salicornia*
Gleasonian succession, 238-239, 248
Glecoma hederacea, 379
Gleditsia triacanthos, 70
Global warming, 22-23, 284
Glyceria, 263
Glyceria aquatica, 254
Glyceria borealis, 386
Glyceria grandis, 245
Glyceria maxima, 90
  anoxia endurance, 104, 105
  primary productivity, 217
  succession, 240
  in treatment wetlands, 351, 352
Glyceria striata, 386
Glyceria stricta, 90
Glycolysis, 87
Glycophytes, 110
Gopher turtles, 286
Gopherus polyphemus, see Gopher turtles
Grass carp, 174, see also Triploid sterile grass carp
Grass of Parnassus, see *Parnassia glauca*
Grasses, see Poaceae
Great Lakes, 42
Great ragweed, see *Ambrosia trifida*
Greater duckweed, see *Spirodela*, *Spirodela polyrrhiza*
Green ash, see *Fraxinus pennsylvanica*
Green milfoil, see *Myriophyllum verticillatum*
Green sea turtles, 286
Groundwater
  chemical composition, 73
  discharge, 33
  fluctuations, 65
  recharge, 33
Growth forms, 7-15
Grus americana, see Whooping cranes
Gulf coast pitcher plant bogs, 117
Gulf coast spikerush, see *Eleocharis cellulosa*

H

Habenaria, 115
Halberd-leaved tearthumb, see *Polygonum arifolium*
Halimione portulacoides, 340
Halodule, 157, 159, 160, 163
Halodule pinifolia, 159, 160
Halophila, 156, 159, 162, 163, 173
Halophila ovalis, 111
Halophytes, 110
Haloragaceae, 13, 41, 299
Halosarica indica, 178
Heath family, see Ericaceae
Heliamphora, 118, 119
Helianthus tuberosus, 259
Herbicides, see Invasive species, controls, chemicals
Herbivory
  chemical defenses, 135, 266
  as disturbance, 271-272
  structural defenses, 135-136
Heritiera, 172
Heteranthera dubia, 266
Heteranthera reniformis, 280
Heterophylly, 13-14, 131-134
Heterozostera, 156
HGM, see Hydrogeomorphic setting
HGS, see Hydrogeologic setting
Highbush blueberry, see *Vaccinium corymbosum*
Hippopotamus amphibius, see Hippopotamus
Hippopotamus, 271
Hippuris vulgaris, 245
Holly, see *Ilex*

Honeysuckle, see Caprifoliaceae
Hordeum jubatum, 174
Horned pondweed, see *Zannichellia*, Zannichelliaceae
Hornwort, see Ceratophyllaceae, *Ceratophyllum demersum*
Hottonia, 171
Hottonia inflata, 149
Howellia aquatilis, 23
Humidity-induced pressurization, 100
Hunting, 34
Hurricanes, 45-46, 240, 270
Hydrarch succession, 241-245
Hydric soils, 30, 74-79
Hydrilla
  aquarium plant, 282
  pollination, 156, 159
Hydrilla verticillata, 29, 303-306
  allelopathy, 254
  aquatic acid metabolism, 131
  asexual reproduction, 182, 186, 187
  bicarbonate uptake, 129
  biological controls, 298, 299, 306
  biology, 303-304
  carbon dioxide re-use, 131
  in Chesapeake, 273
  competition, 262, 302
  controls, 305-306
  effects on native community, 305, 286
  as invasive, 3, 129
  pollination, 159, 149
  radial oxygen loss, 103
  range, 304-305
  reproduction, 149
  response to drawdown, 290
  response to growth regulators, 296
  response to herbicides, 293
  stem buoyancy, 96
  turions, 252, 304
Hydrocharis, 152, 171
Hydrocharitaceae, 13, 41
  canopy formation, 129
  controls, 290
  fruit, 167
  insect pollination, 152
  as invasives, 280, 283
  water pollination, 155, 156
Hydrochory, 19, 72, 171-172
Hydrocleys nymphoides, 100, 344
Hydrocotyle, 169, 292
Hydrocotyle umbellata
  nutrient content, 344, 346
  radial oxygen loss, 348
  in treatment wetlands, 351, 352
Hydrogeologic setting, 35
Hydrogeomorphic setting, 35
Hydrologic budget, 62-64
Hydrology, 61-74
  alterations, 21
  disturbances, 266-269, 275-276
  effects of plants, 4
  effects on biogeochemistry, 72-74
  effects on pH, 81
  effects on plant distribution, 69-70
  effects on primary productivity, 67-68, 228-235
  effects on species composition, 249
  functions of wetlands, 32-33
Hydropectis aquatica, 168
Hydroperiod, 21, 62
Hydrophily, see Pollination, water
Hygrorrhiza aristata, 340
Hylobius transversovittatus, see European weevil
Hymenocallis palmeri, 276
Hypertrophy, 88, 96, 97
Hypoxis, 179

I

IBI, see Index of Biotic Integrity
Ilex, 49
Impatiens, 259
Impatiens capensis, 39
  sedimentation tolerance, 379
  seed banks, 252
Index of Biotic Integrity, 371, 377-378
Index of Vegetative Integrity, 384-387
Indicator status, 364, 365
Indigo bunting, 312
Inland forested wetlands, 48-52
  primary productivity, 48
  terminology, 48
Inland marshes, 39, see also Lacustrine marshes, Playas, Prairie potholes, Vernal pools
Intermediate disturbance hypothesis, 72
Invasive species, 21-22, 279-321, see also Exotic species
  controls, 288-299
    biological, 296-299
    changes in hydrology, 289-290
    chemicals, 292-293, 294-296
    dredging, 289
    mechanical, 290-291, 294
    salt, 296
    shading sediments, 289
    shading water, 288-289
  effects of disturbance, 267-269, 276-277, 282, 283-284
  effects on ecosystem functions, 286-287
  effects on humans, 287
  effects on plant communities, 284-287

effects on seed banks, 286
hybrids, 285
on islands, 282-283
legislation, 288
terminology, 279
Ipomoea aquatica, 280, 342
Iridaceae, 41, 167
Iris
anoxia endurance, 104, 105
asexual reproduction, 179
Index of Vegetative Integrity, 386
seed dispersal, 171
Iris family, see Iridaceae
Iris pseudacorus, 105, 351
Iris versicolor, 351, 386
Iron, 76, 77, 103
Isoetes echinospora, 260
Isoetes lacustris, 89
carbon dioxide use, 131
radial oxygen loss, 102
Isolepis cernua, 328
Isotria verticillata, 115
Iva frutescens, 239

J

Juncaceae, 7, 41
fruit, 167, 170
Index of Vegetative Integrity, 386
mycorrhizae, 114
pollination, 152
Juncus, 3
anoxia endurance, 104, 105
formation of aerenchyma, 90
nutrient uptake, 346
seed banks, 252
seed dispersal, 172
in treatment wetlands, 351, 352
Juncus bufonius, 169
Juncus bulbosus, 131, 262
Juncus conglomeratus, 105
Juncus effusus, 64, 65
allelopathy, 265
anoxia endurance, 105
formation of aerenchyma, 90
Juncus gerardii, 38, 263, 265
Juncus maritimus, 239, 328
Juncus roemerianus, 10, 208
Justicia americana, 184, 292

K

Kandelia, 44

L

Labiatae, see Lamiaceae
Labrador tea, see *Ledum groenlandicum*
Lachnagrostis filiformis, 328
Lactic acid, 88, 108-109
Lacustrine marshes, 41-42
Ladies' tresses, see *Spiranthes*
Lady-slipper, see *Cypripedium*
Lagarosiphon
introduced species, 282
pollination, 155, 156, 158
Lagarosiphon major
asexual reproduction, 186
competition, 257
controls, 291
as invasive, 283, 303
stem buoyancy, 96
Laguncularia, 44
pneumatophores, 94, 95
salt exclusion, 111
seed dispersal, 172
Laguncularia racemosa, 44
cyanobacteria, 116
primary productivity, 230-232
succulence, 113
vivipary, 177
zonation, 46, 47, 48
Lamiaceae, 7, 41, 167, 169
Laportea canadensis, 259
Larger bur marigold, see *Bidens laevis*
Larix laricina, 6, 29, 56, 58
mycorrhizae, 115
primary productivity, 232, 233
succession, 244
Leatherleaf, see *Chamaedaphne calyculata*
Ledum groenlandicum, 234, 235
Leersia oryzoides
competition, 259
in prairie potholes, 362
response to herbicides, 292
Leguminosae
fruit, 167, 170
nitrogen fixation, 116
salt tolerance, 111
Lemna
nutrient content, 344, 346
response to herbicides, 292
seed dispersal, 172
in treatment wetlands, 351, 353
Lemnaceae, 14, 15, 41
asexual reproduction, 180
flowering frequency, 147
Lemna minor, 15
allelopathy, 265
as cosmopolitan species, 16

in treatment wetlands, 352, 353
phytoremediation, 340
Lentibulariaceae, 5, 13
fruit, 167
range, 118
Lenticels, 89
Lepilaena, 157, 159, 163
Lepilaena bilocularis, 161
Lepilaena cylindrocarpa, 159
Lepomis macrochirus, see Bluegill
Lepomis microlophus, see Redear
Leptochloa filiformis, 43
Lesser snow geese, 271
Light bottle/dark bottle method, 200-201, 204
Light
availability underwater, 81-82
competition, 262-263
submerged species' adaptations, 127-129
Light-footed clapper rail, 229, 359-360
Lilaeopsis, 282
Lilaeopsis novae-zelandiae, 328
Lilium michiganense, 258
Limnobium, 171, 179
Limnocharis, 171
Limnophyton, 172
Limonium, 111
Limonium carolinianum, 9
Liquidambar styraciflua, 49, 369
Liriodendron tulipfera, 49, 104, 347
Listera, 115
Little blue heron, 350
Littorella uniflora, 27, 89
aquatic acid metabolism, 131
carbon dioxide use, 131
heterophylly, 132
radial oxygen loss, 102
Lobelia, 171
Lobeliaceae, 149
Lobelia dortmanna, 131, 149, 177
Loggerhead turtles, 286
Lolium perenne, 328
Lomnicki et al. method, 205, 207, 208-209, 215-216
Long's bulrush, see *Scirpus longii*
Loosestrife family, see Lythraceae
Low birch, see *Betula pumila*
Ludwigia, 179, 292
Ludwigia arcuata, 131
Ludwigia natans, 129
Ludwigia palustris, 131
Ludwigia peploides, 344
Ludwigia repens, 131
Lumnitzera, 44, 111, 172
Luronium natans, 89, 102
Lycopersicon esculentum, 88, 104-105
Lycopus, 7
Lysigeny, 90
Lysimachia ciliata, 259
Lysimachia nummularia, 379
Lythraceae, 41, 167, 169
Lythrum flagellare, 186
Lythrum hyssopifolium, 280
Lythrum salicaria, 310-313, 314
biology, 310, 311
competition, 256, 260
controls, 313, 314
effects, 22, 312
heterostyly, 187
Index of Vegetative Integrity, 387
as invasive, 3, 22, 267, 269, 279
range, 310-312
in restored wetlands, 333
seed dispersal, 172
seed production, 280

M

Madder, see Rubiaceae
Magnolia virginiana, 345
Magnoliaceae, 115
Magnoliids, 17-18
Maidencane, see *Panicum hemitomon*
Maidenia, 155, 156
Maize, see *Zea mays*
Malate, 107
Manatees, see West Indian manatees
Manganese, 76-77
Mangroves, 44-48
definition, 10, 44
distribution, 44-46
hurricanes, 45-46, 240, 270
number of species, 10, 44
pneumatophores, 94-95
primary productivity, 221-225, 230-232
pulse stability, 240
restoration, 356-359
shoreline stabilization, 33
succession, 246, 248
timing of planting, 333
understory, 44
zonation, 46-48
Manna grass, see *Glyceria aquatica*
Marsh rabbits, 286
Marsilea mutica, 344
Mass flow, see Pressurized ventilation
Mayaca fluviatilis, 173
Mayaceae, 169
Meadow sweet, see *Spiraea*
Melaleuca, 353
Melaleuca quinquenervia
fire, 287

harvesting, 294
insect control, 297
as invasive, 280, 284, 285
root growth, 95
transpiration, 4, 282, 286
Melaleuca weevil, 297
Mentha, 7
anoxia endurance, 104, 105
seed dispersal, 171
Mentha aquatica, 90, 105
Mentha arvensis, 379
Menyanthaceae, 167, 169
Menyanthes, 171, 172
Mermaid weed, see *Proserpinaca palustris*
Mesquite, see *Prosopis*
Metabolic processes under anoxia, 104-110
anaerobic metabolism, 106
crop plants, 104-105, 108
cytoplasmic acidosis, 108-109
mitochondrial adaptations, 109-110
Pasteur effect, 106
Metal removal, 339-341
Milner and Hughes method, 205, 206, 208-209, 211-212
Mimosa pellita, 282
Mimosa pigra, see *Mimosa pellita*
Mimulens repens, 328
Mink, 34
Mints, see Lamiaceae
Mitochondrial adaptations to anoxia, 109-110
Monochoria, 169
Monocotyledons, 17-18
dominance in wetlands, 7
water pollination, 156
wind pollination, 152
Mosses, see *Sphagnum*, Bryophytes
Mottled water hyacinth weevil, 309
Muskgrass, see *Chara vulgaris*
Muskrats, 34, 271-272, 312
Mycorrhizae, 114-116
Mycteria americana, see Wood storks
Myosotis scorpioides, 171
Myrica gale
nitrogen fixation, 116
primary productivity, 225
root growth, 95
stomatal depression, 137
Myrica heterophylla, 179, 185
Myriophyllum, 19
allelopathy, 265
asexual reproduction, 178, 179, 180
deterrents to herbivory, 135
leaf morphology, 127
in prairie potholes, 272
response to drawdown, 290
seed dispersal, 172
self-pollination, 150, 166
Myriophyllum aquaticum
asexual reproduction, 186
biological controls, 298
carbon use, 129
response to herbicides, 293
Myriophyllum brasiliense, see *Myriophyllum aquaticum*
Myriophyllum exalbescens, see *Myriophyllum sibiricum*
Myriophyllum hippuroides, 129
Myriophyllum oliganthum, 12
Myriophyllum sibiricum
competition, 254, 262, 263
growth after dredging, 289
and *M. spicatum*, 303
Myriophyllum spicatum, 299-303
asexual reproduction, 180
bicarbonate uptake, 129
biological controls, 298, 302
biology, 299-300
competition, 258, 262, 263
controls, 290, 301-302
declines, 302-303
effects on native community, 22, 286, 301
as invasive, 22, 129
primary productivity, 217
range, 300-301
response to growth regulators, 296
response to herbicides, 293
stem buoyancy, 96
Myriophyllum verticillatum
asexual reproduction, 179, 181, 182
carbon use, 129
growth after flooding, 269

N

Naiads, see *Najas*, Najadaceae
Najadaceae, 6, 41
Najas
germination, 174
pollination, 161, 162
in prairie potholes, 272
reproduction, 147, 155
response to herbicides, 293
in treatment wetlands, 351, 353
Najas flexilis
growth after dredging, 289
life history type, 249
response to drawdown, 290
Najas marina, 174, 180
Nanophyes brevis, 313
Nanophyes marmoratus, 313
Narrow-leaved cattail, see *Typha angustifolia*
Narthecium ossifragum, 90

Nasturtium officinale, see *Rorippa nasturtium-aquaticum*
National Oceanic and Atmospheric Administration, 288
Nechamandra, 155, 156
Nelumbo, 19, 265
Nelumbo lutea, 14
- response to herbicides, 292
- seed dispersal, 171
- in treatment wetlands, 351

Nelombonaceae, 13
Nelumbo nucifera, 100
- competition, 254
- seeds, 173

Neobeckia aquatica, 178
Neochetina bruchi, see Chevroned water hyacinth weevil
Neochetina eichhorniae, see Mottled water hyacinth weevil
Neostapfia, 23
Nepenthaceae, 118
Nepenthes, 117, 119
Nepenthes alata, 121
Nepenthes mirabilis, 119, 121
Neptunia, 116
Neptunia oleracea, 169
New England aster, see *Aster novae-angliae*
Nitella, 291, 384
Nitrogen
- denitrification, 75-76, 336, 341, 369
- fixation, 116-117, 145
- limiting factor, 80
- nitrate ammonification, 75-76
- nitrification, 33, 336
- transformations, 33, 75-76, 336

Non-Indigenous Aquatic Nuisance Prevention and Control Act of 1990, 288
Non-indigenous species, see Exotic species
North American weevil, 302, 303
Northeastern bulrush, see *Scirpus ancistrochaetus*
Northeastern floodplain wetlands, 49-50
Northern pitcher plant, see *Sarracenia purpurea*
Northern white cedar, see *Thuja occidentalis*
Nuphar
- allelopathy, 265
- anoxia avoidance, 104
- leaves, 13
- pollination, 151
- response to drawdown, 290
- response to herbicides, 292, 294
- seed dispersal, 171

Nuphar advena, 39, 40
Nuphar lutea
- aerenchyma, 89
- allelopathy, 254, 265
- gas transport, 98-100
- radial oxygen loss, 102
- rhizomes, 184

Nuphar variegatum, 245
Nutrients
- adaptations to low levels, 114-127
  - carnivory, 117-126, 142-145
  - evergreen leaves, 127
  - mycorrhizae, 114-116
  - nitrogen fixation, 116-117
- availability, 78
- release from plants, 74
- tissue concentrations, 68, 343-348
- translocation, 126-127
- uptake, 4, 33, 81, 343-348

Nymphaea
- allelopathy, 265
- anoxia avoidance, 104
- fruit, 170
- gas transport, 100
- reproduction, 149
- seed dispersal, 171
- in treatment wetlands, 351

Nymphaeaceae, 5, 13, 40, 41
- allelopathy, 265
- fruit, 167, 170

Nymphaea alba, 13, 89
- radial oxygen loss, 102
- stem growth, 95

Nymphaea lotus, 182
Nymphaea odorata, 13, 41, 56
- asexual reproduction, 179
- in Everglades, 276
- primary productivity, 220
- response to herbicides, 292
- rhizomes, 184

Nymphoides
- pollination, 152
- reproduction, 149
- seed dispersal, 171
- in treatment wetlands, 351

Nymphoides aquatica, 179, 185
Nymphoides indica, 100
- asexual reproduction, 186
- nutrient content, 344

Nymphoides peltata, 13, 89
- gas transport, 100
- radial oxygen loss, 102
- seeds, 173
- stem growth, 95

Nypa, 172, 184
Nyssa aquatica, 8, 49, 51
- flood tolerance, 70
- fruit, 170
- seed dispersal, 172

seedlings, 175
Nyssa biflora, 250, 251
Nyssaceae, fruit, 167, 170
Nyssa sylvatica, 49
in delineation, 369
mycorrhizae, 115
nutrient content, 345, 347
Nyssa sylvatica var. *biflora*
ethanol production, 108
nutrient content, 345
nutrient translocation, 127

O

Oak, see *Quercus*
Oenanthe, 172
Ohio buckeye, see *Aesculus glabra*
Ondatra zibethica, see Muskrats
Onoclea sensibilis, 379
Orchidaceae, 115
Orchis, 115
Orconectes rusticus, see Rusty crayfish
Orcuttia, 23, 44
Orcuttia californica, 173
Oreochromis, see Tilapia
Organic acids, 80-81, 107
Orontium, 171, 172
Oryza rufipogon, 280
Oryza sativa, 34
in anoxia, 104-105
ethanol tolerance, 108
formation of aerenchyma, 90
gas transport, 101
germination, 253
lactate production, 109
mitochondria, 109
radial oxygen loss, 102, 103
seeds, 173
stem growth, 96
Osmo-regulation, 110
Osmunda, 245
Ottelia, 152, 166, 171
Ottelia alismoides, 166
Ottelia ovalifolia, 166
Oxidation-reduction potential, 75
Oxidized rhizosphere, 102-103, see also Radial oxygen loss
Oxyops vitiosa, see Melaleuca weevil
Oxypolis rigidior, 135

P

Palmae, 12
Panicum, 179, 276, 277
Panicum hemitomon, 74
fire, 270
response to herbicides, 292
in treatment wetlands, 351
Panicum repens, 292
Panicum virgatum, 361
Papyrus, 179
Paragrass, see *Brachiara mutica*
Parish's glasswort, see *Arthrocnemum subterminale*
Parnassia glauca, 54
Parrot feather, see *Myriophyllum aquaticum*
Parsley family, see Apiacae
Paspalum dilatatum, 292
Paspalum paniculatum, 292
Passerina cyanea, see Indigo bunting
Pasteur effect, 96, 106
Peak biomass
definition, 193
method, 205, 206, 208-209, 210-211
Peatlands, 52-58, 80-81
geographic distribution, 54
primary productivity, 232-235
succession, 241-245
Peliciera, 172
Peltandra, 170
Peltandra virginica, 39, 72
in Everglades, 277
seed banks, 252
Penman-Monteith equation, 66
Penman equation, 66
Periphyton
effects on macrophytes, 82, 273
primary productivity, 202-204
terminology, 202
Phalaris
allelopathy, 266
competition, 263
in treatment wetlands, 352
Phalaris arundinacea, 8, 20, 39
competition, 258, 259, 263
nutrient release, 347
in prairie potholes, 362
sedimentation tolerance, 379
in treatment wetlands, 334
Phenolic acids, 266
Phenotypic plasticity, 19, 20
Phleum pratense, 256
Phosphorus
mycorrhizae, 115, 116
in treatment wetlands, 337-338
Phragmites, 74
anoxia endurance, 104, 105
competition, 263
Index of Vegetative Integrity, 387

nutrient uptake, 346
in treatment wetlands, 352
Phragmites australis, 313, 315-320
allelopathy, 265
anoxia endurance, 105
asexual reproduction, 179, 184
belowground biomass, 220
biology, 315
competition, 257
controls, 317
declining populations, 317-320
distribution, 16
effects as invasive, 316
gas transport, 97, 100, 101-102
germination, 174
as invasive, 273, 280
life history type, 248
metal uptake, 340
name for reedswamps, 3
nutrient content, 344, 347
pollination, 153
radial oxygen loss, 348
range, 315-316
response to herbicides, 292
response to increased salinity, 296
in restored wetlands, 333
rhizomes, 179
seed banks, 252
shallow roots, 93
succession, 240
in treatment wetlands, 335, 342, 351, 352, 353
Phragmites communis, see *Phragmites australis*
Phyllanthus fluitans, 15
Phyllospadix, 157, 162
Phytoremediation, 340
Picea mariana, 56
evergreen leaves, 127
mycorrhizae, 115
primary productivity, 232, 233, 235
Picea marina, 244
Pickerelweed family, see Pontederiaceae
Pilea pumila, 39, 259
Pinguicula, 54
range, 118
trap type, 119, 122, 123
Pinguicula cearulea, 117
Pinguicula ionantha, 117
Pinguicula lutea, 117
Pinguicula planifolia, 117
Pinguicula primuliflora, 117
Pinguicula villosa, 117
Pinguicula vulgaris, 117
Pinus contorta, 70, 92
Pinus elliotti, 250
Pinus rigida, 268
Pistia stratiotes, 14, 15
effects on humans, 287
effects on native community, 285
as invasive, 288
in treatment wetlands, 351, 352, 353
nutrient content, 344, 346
radial oxygen loss, 348
range, 280
response to herbicides, 292
Pisum sativum, 108
Pitch pine, see *Pinus rigida*
Pitcher plant, see *Sarracenia*
Plaginathus divaricatus, 328
Plantago coronopus, 328
Plantago maritima, 271
Platanus occidentalis, 49, 50, 70
effects of flooding, 72
seedlings, 175
Platanus wrightii, 50
Playas, 43-44
Pneumatophores, 45, 48, 93-95
Poaceae, 7, 41
fruit, 167
Index of Vegetative Integrity, 386
mycorrhizae, 115
nutrient translocation, 126
pollination, 152
Poa laevis, 328
Podostemaceae, 169
Podestemum, 150, 166, 169
Pogogyne, 44
Pogogyne abramsii, 23
Pogonia ophioglossoides, 115
Poison hemlock, see *Conium maculatum*
Poison ivy, see *Toxicodendron radicans*
Poison sumac, see *Toxicodendron vernix*
Pollination, 150-166
evolution, 152, 164-165
insect, 150-152
pollen adaptations, 152-153, 162-163
self, 166
stigma adaptations, 163
water, 19, 20, 154-165
wind, 152-154
Polygonaceae, 7, 41, 167
Polygonum, 19, 43
allelopathy, 265
Index of Vegetative Integrity, 387
in prairie potholes, 271
reproduction, 149
response to herbicides, 292
Polygonum amphibium, 386
Polygonum arifolium, 39
Polygonum pennsylvanicum, 259
Polygonum punctatum, 39, 258, 259

Polygonum sagittatum, 386
Polygonum scandens, 386
Polypogon monspeliensis, 284, 296, 328
Polypompholyx, 118, 119, 123, 125
Polypompholyx tenella, 125
Pomacea paludosa, see Apple snail
Pomoxis nigromaculatus, see Black crappie
Pond cypress, 52, see also *Taxodium ascendens*
Pondweeds, see *Potamogeton*, Potamogetonaceae
Pontederia, 171, 292, 346
Pontederiaceae, 41
Pontederia cordata
  fire, 270
  in Everglades, 277
  in treatment wetlands, 351, 352
  pollination, 151
  radial oxygen loss, 348
Pontederia rotundifolia, 186
Pool moss, see *Mayaca fluviatilis*
Poplar, see *Populus*
Populus, 9, 50, 283
Populus balsamifera, 245
Populus deltoides, 50, 70, 347
Posidonia, 156, 166
Posidoniaceae, 156
Potamogeton, 13, 43
  asexual reproduction, 179, 187
  bicarbonate uptake, 129, 130
  biological controls, 298
  growth after dredging, 289
  leaf morphology, 127
  in prairie potholes, 272, 361
  reproduction, 149
  response to drawdown, 290
  response to growth regulators, 296
  response to herbicides, 293
  seed dispersal, 171
  self-pollination, 166
  in treatment wetlands, 351, 353
  water pollination, 156
Potamogetonaceae, 5, 13, 41
  fruit, 167, 169
  pollination, 156
Potamogeton americanus, 262
Potamogeton amplifolius, 303
Potamogeton crispus, 16
  allelopathy, 266
  fruit, 168
  nutrient content, 345
Potamogeton distinctus, 96
Potamogeton freisii, 303
Potamogeton illinoensis, 303
Potamogeton lucens, 130
Potamogeton natans, 245, 269
Potamogeton pectinatus, 16, 68
  allelopathy, 266
  competition, 254, 262
  lactate production, 109
  nutrient content, 345
  Pasteur effect, 106
  primary productivity, 217
  rapid stem growth, 96
  salt tolerance, 111
  succession, 240, 245
Potamogeton perfoliatus, 128, 204
Potamogeton polygonifolius, 128
Potamogeton praelongus, 303
Potamogeton pusillus, 165, 269
Potamogeton richardsonii, 245
Potamogeton zosteriformis, 245
Potentilla, 379
Potentilla fruticosa, 54
Potentilla palustris, 90
Prairie potholes
  cycles, 271-272
  distribution, 43
  habitat types, 43
  origin, 43
  restoration, 360-362
  seed banks, 252, 361-362
  species richness, 43
  succession, 248-249
Pressurized ventilation, 97-100
Prevalence index, 368-369
Primary productivity
  belowground biomass, 219-220
  effects of hydrology, 67-68
  errors, 197-198
  methods
    comparisons, 206-209
    emergents, 205-220
    floating and floating-leaved plants, 220-221
    forests, 225, 230-235
    moss, 226-227, 232-235
    periphyton, 202-204
    phytoplankton, 199-202
    shrubs, 225-226
    submerged plants, 204-205, 217-218
    trees, 221-225, 230-235
  terminology, 191-198
  of wetland types, 193
  of world ecosystems, 4, 192
Primulaceae, 167
Prop roots, 95
Proserpinaca palustris, 132, 133
Prosopis, 50
Proteins under anoxia, 109
Pseudalthenia, 157
Pteridophytes, 7
Puccinellia, 19

Puccinellia fasciculata, 328
Puccinellia maritima, 271
Puccinellia novae-zelandiae, 328
Puccinellia nutkaensis, 264
Puccinellia phryganodes, 38, 187, 271
Puccinellia stricta, 328
Pulse stability, 240
Puma concolor coryi, see Florida panthers
Purple loosestrife, see *Lythrum salicaria*

Q

Quaking bog, 241-243
Quercus, 9, 49
Quercus bicolor, 71
Quercus falcata var. *pagodifolia*, 175
Quercus lyrata, 70
Quercus macrocarpa, 70
Quercus nigra, 70
Quercus nuttalii, 70
Quercus palustris, 70
Quercus velutina, 73

R

Radial oxygen loss, 102-103, 141
  competition, 257
  definition, 17
  oxidation of sulfur, 113-114
  in treatment wetlands, 348
Rallus longirostris levipes, see Light-footed clapper rail
Ramenski sedge, see *Carex ramenskii*
Ramsar Convention, 31
Ranunculaceae, 41, 167
Ranunculus
  anoxia endurance, 104, 105
  asexual reproduction, 179
  competition, 262
  reproduction, 149
  response to herbicides, 293
  seed dispersal, 171, 172
  stem growth, 96
  succession, 245
Ranunculus flabellaris, 13
Ranunculus flammula, 90
  competition, 260
  heterophylly, 132, 133
Ranunculus lingua, 105
Ranunculus peltatus, 68
Ranunculus peltatus subsp. *baudotii*, 132
Ranunculus repens, 105, 171
Ranunculus sceleratus, 14
Rapid underwater shoot extension, 95-96
Red mangrove, see *Rhizophora mangle*
Red maple, see *Acer rubrum*
Red sprangletop, see *Leptochloa filiformis*
Red-winged blackbird, 312
Redbelly tilapia, 298
Redear, 305
Redox, see oxidation-reduction potential
Reduced sediments, 75-79
Reed canary grass, see *Phalaris arundinacea*
Reedswamps, 3, see also *Phragmites australis*
Regnellidium diphyllum, 96
Rein orchid, see *Habenaria*
Remote sensing, 370
Reproduction, see Sexual Reproduction, Asexual Reproduction
Reptiles, 34
Respiration, 87-88, 194
Restoring wetlands, 325-333, 356-362
  planting, 329-331, 332-333
  seed banks, 331-332
Rhizophora, 19, 44
  roots, 95
  salt exclusion, 111
Rhizophoraceae, 12, 176
Rhizophora mangle, 10, 11, 44, 45
  cyanobacteria, 116
  as invasive, 284
  primary productivity, 230-232
  prop roots, 46-47
  radial oxygen loss, 103
  succulence, 113
  vivipary, 176
  zonation, 46
Rhizophora mucronata, 357, 358
Rhizophora stylosa, 357, 358
Rhus radicans, see *Toxicodendron radicans*
Rhus vernix, see *Toxicodendron vernix*
Rhynchospora, 276
Riccia fluitans, 386
Rice cutgrass, see *Leersia oryzoides*
Rice, see *Oryza sativa*
Riparian buffer strips, 353-354
Riparian wetlands, 72, see also Western riparian zones, Southern bottomland hardwoods, Northeastern floodplain wetlands, Bottomland hardwoods
River birch, see *Betula nigra*
River bulrush, see *Scirpus fluviatilis*
Riverine marshes, 42
Rivers and Harbors Act of 1899, 288
Roadgrass, see *Eleocharis baldwinii*
Root-to-shoot ratios, 220
Rootless submerged species, 12, 124
Rorippa nasturtium-aquaticum, 92
Rorippa sylvestris, 179, 185

Rosa, 41
Rosaceae, 9
Rose pogonia, see *Pogonia ophioglossoides*
Roses, see *Rosa*, Rosaceae
Rostrhamus sociabilis, see Florida Everglades kite, Snail kite
Rotala indica, 131
Rotala rotundifolia, 131
Rubiaceae, 9
Rubus occidentalis, 386
Rudbeckia laciniata, 259
Rumex
  adventitious roots, 92
  ethanol production, 107
  in prairie potholes, 271
Rumex maritimus, 90
Ruppia
  seed dispersal, 171
  self-pollination, 150, 166
  water pollination, 156, 159, 163
Ruppia cirrhosa, 240
Ruppia marina, 163
Ruppia maritima
  biological controls, 298
  germination, 174
  response to herbicides, 293
  salt tolerance, 111
Rushes, see *Juncus*, Juncaceae
Russian olive, see *Elaeagnus angustifolia*
Rusty crayfish, 299

S

Sagittaria, 14, 16
  fire, 270
  heterophylly, 131
  nutrient uptake, 346
  in prairie potholes, 271, 361
  seed dispersal, 172
  in treatment wetlands, 351
Sagittaria cuneata, 132
Sagittaria falcata, 208
Sagittaria lancifolia, 276
Sagittaria latifolia, 39, 40
  asexual reproduction, 179
  radial oxygen loss, 348
  stem growth, 95
  in treatment wetlands, 352
Sagittaria montevidensis, 16
Sagittaria pygmaea, 95
Sagittaria sanfordii, 16
Sago pondweed, see *Potamogeton pectinatus*
Salicornia, 19, 112, 113
Salicornia bigelovii, 74
Salicornia dolichostachya, 39
Salicornia europaea, 38, 39, 174
Salicornia subterminalis, 296
Salicornia virginica, 38, 74
  competition, 264-265
  formation of aerenchyma, 90
  primary productivity, 228, 229
Salinity
  adaptations, 110-113
  effects on plants, 79-80
  exclusion, 111
  levels in wetlands, 79
  osmo-regulation, 110
  secretion, 111-113
  shedding, 113
  succulence, 112, 113
Salix, 9, 41, 49, 50
  adventitious roots, 92
  ethanol production, 108
  in western riparian zones, 283
Salix bebbiana, 245
Salix exigua, 239
Salix interior, 70
Salix lutea, 245
Salix nigra, 347
Salix planifolia, 137
Salix reticulata, 137
Salt marsh hay, see *Spartina patens*
Salt marsh rush, see *Juncus gerardii*
Salt marsh water hemp, see *Acnida cannabina*
Salt marshes
  competition, 263-265
  definition, 36
  distribution, 36-39
  disturbance, 267
  dominant plants, 37-39
  restoration, 359-360
  species zonation, 38
  succession, 246-247
Saltcedar, see *Tamarix*, *Tamarix chinensis*, *Tamarix ramosissima*
Salvinia
  effects on native community, 285
  introduced species, 282
  nutrient uptake, 346
  response to herbicides, 292
  in treatment wetlands, 351, 353
Salvinia auriculata, 287
Salvinia minima, 280
Salvinia molesta, 280, 286, 344
Salvinia natans, 340
Sambucus canadensis, 49
Sameodes albiguttalis, see Water hyacinth borer
Samolus repens, 328
Sarcocornia quinqueflora, 328
Sarracenia, 54

range, 117, 118, 177
trap type, 118, 119
Sarraceniaceae, 118
Sarracenia purpurea, 142-145
occurrence, 268
range, 117
Scapania undulata, 340
Scheuchzeria, 171
Schinus terebinthifolius, 282, 284
Schizogeny, 90
Schoenoplectus, see Preface, 100
Schoenoplectus pungens, 328
Schoenus nigricans, 27, 90
Schoenus nitens, 328
Scirpus, 43, 56
anoxia endurance, 104, 105
Index of Vegetative Integrity, 387
name changes within genus, see Preface
nutrient uptake, 346
and *Phragmites australis*, 316
in prairie potholes, 271, 361
response to herbicides, 292
seed dispersal, 171
in treatment wetlands, 351, 352
Scirpus acutus, 260, 384
Scirpus americanus, 105, 260
Scirpus ancistrochaetus, 23
Scirpus californicus, 254, 352
Scirpus cyperinus, 8
Scirpus fluviatilis, 105, 184
Scirpus grossus, 179
Scirpus lacustris
anoxia endurance, 105
belowground biomass, 220
carbon dioxide use, 131
stem growth, 96
Scirpus longii, 312
Scirpus maritimus, 96
Scirpus microcarpus, 245
Scirpus pungens, 348
Scirpus subterminalis, 131
Scirpus tabernaemontani
anoxia endurance, 105
Index of Vegetative Integrity, 386
nutrient content, 344
radial oxygen loss, 348
in treatment wetlands, 342, 352
Scirpus validus, see *Scirpus tabernaemontani*
Scrophylaria aquatica, 172
Scutellaria, 171
Scutellaria galericulata, 386
Scyphiphora, 172
Sea lavender, see *Limonium carolinianum*
Sea level rise, 22
Seagrasses, 154, 177, 265
Secondary metabolites, 135
Sedge meadow, 43
Sedge family, see Cyperaceae
Seed banks
in prairie potholes, 361-362
in restored wetlands, 331-332
succession, 252-253
Seedlings
dispersal, 175-176
establishment, 175-176
vivipary, 176-177
Seeds
dispersal, 171-172, 174
dormancy, 173-174
germination, 104, 173-174, 354
types, 167-170
Selliera radicans, 328
Senecio aquaticus, 91
Senecio aureus, 379
Sesbania, 116
Setaria magna, 292
Sexual reproduction, 147-177, see also Pollination, Seeds, Seedlings
obstacles, 148-150
timing, 19, 20
Shallow rooting, 93
Shellfish, 34
Shoreline stabilization, 33
Shrubby cinquefoil, see *Potentilla fruticosa*
Shrubs
common wetland species, 9
of bottomland forests, 49
of depressional wetlands, 41
of peatlands, 56-58
primary productivity, 225-226
Sigmodon hispidus, see Cotton rats
Silver maple, see *Acer saccharinum*
Sium, 169, 172
Sium suave, 39
Slender waterweed, see *Elodea nuttalli*
Slim spikerush, see *Eleocharis elongata*
Smalley method, 205, 206, 208-209, 213, 228
Smartweeds, see *Polygonum*, Polygonaceae
Smilacina trifolia, 232
Snail kite, 34, 286
Soft rush, see *Juncus effusus*
Soft-stemmed bulrush, see *Scirpus tabernaemontani*
Solidago, 379
Solidago patula, 379
Solidago uliginosa, 379
Sonchus uliginosus, 245
Sonneratia, 44
pneumatophores, 94, 95
pollination, 150
salt exclusion, 111

Sonneratiaceae, 12
South American alligatorweed flea beetles, 297
Southern bottomland hardwoods, 48-49
Southern cattail, see *Typha domingensis*
Sparganiaceae, 7, 41, 170
Sparganium
  Index of Vegetative Integrity, 387
  in prairie potholes, 271, 361
  seed dispersal, 171
Sparganium erectum, 173
Sparganium eurycarpum, 14, 260
Spartina, 19
  anoxia endurance, 104, 105
  in treatment wetlands, 352
Spartina alterniflora
  asexual reproduction, 184
  belowground biomass, 220
  competition, 258, 265
  decline, 273, 296, 316, 317
  distribution, 37-38
  dominant of salt marshes, 37
  ethanol diffusion, 108
  germination, 174
  hybridization, 265, 285
  nitrogen limitation, 140
  planting instructions, 333
  primary productivity, 139, 140, 208-209, 210, 211, 212, 213, 228
  radial oxygen loss, 141
  salt exclusion, 111
  salt marsh inhabitant, 10
  salt secretion, 111
  seedlings, 175
  shoreline stabilization, 33
  short and tall forms, 38, 139-142
  succession, 246
  turnover, 195
Spartina anglica, 39
  anoxia endurance, 105
  asexual reproduction, 187
  competition, 265
  hybridization, 285
  primary productivity, 208
  radial oxygen loss, 340
  salt tolerance, 328
Spartina cynosuroides, 208
Spartina foliosa, 38, 359-360
  primary productivity, 228-229
  salt secretion, 111
Spartina maritima, 39
  hybridization, 265, 285
  succession, 239
Spartina patens, 38, 90, 317
  ADH activity, 107
  competition, 263, 264, 265
  primary productivity, 208
  salt secretion, 111
  succession, 246
Spartina pectinata, 168
Spartina townsendii, 111, 265, 285
Spatterdock, see *Nuphar*
Species richness, 72
Speckled alder, see *Alnus incarna*
Spergularia media, 328
Sphagnum, 52, 54, 80
  primary productivity, 226-227, 232-235
  succession, 241, 244, 245
Sphagnum papillosum, 53
Spike grass, see *Distichlis spicata*
Spiraea, 41
Spiraea alba, 386
Spiranthes, 115
Spirodela, 14, 351
Spirodela polyrrhiza, 15
  asexual reproduction, 181
  phytoremediation, 340
  response to herbicides, 293
Spotted cowbane, see *Cicuta maculata*
Spotted touch-me-not, see *Impatiens capensis*
Standing crop, 191-192
Stem buoyancy, 96
Stem elongation, 95-96
Sticky tofieldia, see *Tofieldia glutinosa*
Stomata, 17
  in submerged plants, 128
  on floating leaves, 13
Stratiotes, 152
Stratiotes aloides, 186
Straw-colored cyperus, see *Cyperus strigosus*
Stream channelization, 21
Suaeda, 113
Suaeda esteroa, 74
Suaeda novae-zelandiae, 328, 329
Submerged plants
  adaptations, 127-134
  competition, 262-263
  definition, 12-13
  leaf morphology, 127-129
  nutrient uptake, 12
  primary productivity methods, 204-205, 217-218
Subularia aquatilis, 166
Succession, 237-253, 275-276
  coastal wetlands, 246-248
  history, 238-240
  hydrarch, 241-245
  models, 241-250, 251
  oxbows, 245
  peatlands, 241-245
  seed banks, 250, 252-253
  terminology, 237-238

Sulfur
adaptations, 113-114
in coastal wetlands, 77
oxidation, 103, 113-114
toxicity, 141
transformations, 76, 77-78, 81
Summed shoot maximum method, 205, 207, 208-209, 219
Sundew, see *Drosera*
Swamp black gum, see *Nyssa sylvatica*
Swamp loosestrife, see *Decodon verticillatus*
Swamp milkweed, see *Asclepias incarnata*
Swamp pine, see *Pinus elliotti*
Swamp thistle, see *Cirsium arvense*
Sweet gum, see *Liquidambar styraciflua*
Sweetflag, see *Acorus calamus*
Switchgrass, see *Panicum virgatum*
Sycamore, see *Platanus occidentalis*, *Platanus wrightii*
Sylvilagus palustris, see Marsh rabbits
Symplocarpus foetidus, 379
Syringodium, 157, 161

T

Tamarack, see *Larix laricina*
Tamarix, 50, 283
Tamarix chinensis, 51, 286
Tamarix ramosissima, 51
effects on native community, 285
transpiration, 286-287
Taraxacum officinale, 168
Taro, see *Colocasia esculenta*
Taxodium ascendens, 51
nutrient content, 345
nutrient translocation, 127
succession, 250, 251
Taxodium distichum, 3, 7, 8, 11, 49, 51-52
adaptations to water stress, 136
cypress knees, 93-94
effects of invasives, 284, 285
fire, 271
flood-tolerance, 70
formation of aerenchyma, 90
hypertrophy, 96, 97
nutrient content, 345, 346
radial oxygen loss, 102
seed dispersal, 172, 173
seedlings, 175
shallow roots, 93
in treatment wetlands, 335, 353
Taxodium distichum var. *nutans*, 51, see also *Taxodium ascendens*
Terminalia, 172
Terpenoids, 265
Thalassia, 156, 161, 162
Thalassia testudinum, 161, 169
Thalassodendron, 157, 163
Thelypteris noveboracensis, 379
Thermal transpiration, 99
Thornthwaite equation, 66
Threatened species, 23-27
Three-leaved Solomon's seal, see *Smilacina trifolia*
Thuja, 244
Thuja occidentalis, 74, 245
Thuja picata, 92
Tidal freshwater marshes, 39
Tilapia, 174
Tilapia zillii, see Redbelly tilapia
Timothy, see *Phleum pratense*
Tofieldia glutinosa, 54, 55
Tomatoes, see *Lycopersicon esculentum*
Torpedograss, see *Panicum repens*
Toxicodendron radicans, 135, 137
Toxicodendron vernix, 135, 137
Toxins, 79
Transpiration, 64, 65-67
Trapa, 171, 172
Trapa bispinosa, 34, 254
Trapa natans
as invasive, 280
lactate production, 109
response to herbicides, 292
Treatment wetlands, 34, 68, 333-355
agricultural wastewater, 334, 335
harvesting, 353
metal removal, 339-341
nitrogen removal, 336-337
pathogen removal, 338-339
phosphorus retention, 337-338
plant recommendations, 350-354
planted vs. unplanted, 342
planting, 354-355
role of plants, 341-349
stormwater, 335
wastewater, 334
wildlife habitat, 349
Trichechus manatus, see West Indian manatees
Trichorphorum, see Preface
Trichophorum cespitosum, 90
Triglochin maritimum, 54, 55, 264
Triglochin striatum, 328
Triploid sterile grass carp, 298
Triticum aestivum
salt tolerance, 328
in anoxia, 104-105
Tsuga heterophylla, 92
Tuctoria, 23
Tuctoria greenia, 173
Turtle grass, see *Thalassia*, *Thalassia testudinum*
Tussock sedge, see *Carex stricta*

Twayblade, see *Listera*
Twig rush, see *Cladium mariscoides*
Typha, 3, 19, 39, 74
  anoxia avoidance, 104
  asexual reproduction, 179
  competition, 261, 263
  distribution, 42, 43
  Index of Vegetative Integrity, 387
  as invasive, 279
  and *Lythrum salicaria*, 312
  nutrient content, 344
  nutrient uptake, 346
  pollination, 153, 154
  and *Phragmites australis*, 316
  in prairie potholes, 271
  response to herbicides, 292
  in restored wetlands, 333
  seed banks, 253
  in treatment wetlands, 351, 352
Typhaceae, 7, 41
Typha angustifolia, 29
  anoxia endurance, 105
  asexual reproduction, 184
  competition, 256
  Index of Vegetative Integrity, 384
  as invasive, 267
  pollination, 154
  in restored wetlands, 330, 333
Typha domingensis, 66, 90
  adventitious roots, 92
  in Everglades, 277
  radial oxygen loss, 103
Typha glauca, 248, 337
Typha latifolia
  allelopathy, 265
  anoxia endurance, 105
  asexual reproduction, 181, 184
  belowground biomass, 220
  competition, 256, 258, 259, 260
  in Everglades, 277
  gas storage capacity, 91
  Index of Vegetative Integrity, 384
  metal uptake, 340
  nutrient release, 347
  primary productivity, 208, 209, 210, 218
  radial oxygen loss, 239, 348
  rapid stem growth, 96
  in restored wetlands, 330
  sedimentation tolerance, 379
  succession, 239, 245
  in treatment wetlands, 342, 352
  turnover, 195
Typha orientalis, 342, 347

U

U.S. Army Corps of Engineers, 30, 364, 366
U.S. Fish and Wildlife Service, 31, 34, 35, 48, 288, 366
Ulmaceae, 167, 169
Ulmus americana, 50
  adventitious roots, 92
  effects of flooding, 72
  seedlings, 175
Ulmus crassifolia, 168
Umbelliferaceae, see Apiaceae
Underwater gas exchange, 101
Urbanization, 267-269
Urtica dioica, 259
Urtica major, 245
Utricularia, 54
  asexual reproduction, 178, 179, 180
  biological controls, 298
  in Everglades, 277
  range 117, 118
  response to herbicides, 293
  seed dispersal, 171
  self-pollination, 150, 166
  trap type, 119, 123, 125
Utricularia macrorhiza, 56, 125
  Index of Vegetative Integrity, 386
  prey use, 126
  response to drawdown, 290
Utricularia purpurea, 277
Utricularia radiata, 149, 150
Utricularia tenella, 125

V

Vaccinium, 9, 34, 245
Vaccinium corymbosum, 56, 57, 248
Vaccinium macrocarpon, 56
Valiela et al. method, 205, 206, 208-209, 212
Vallisneria, 19
  allelopathy, 265
  aquarium plant, 282
  pollination, 155, 156, 158, 163
  seed dispersal, 171
Vallisneria americana, 12
  allelopathy, 266
  asexual reproduction, 181-182
  competition, 262, 305
  germination, 174
  mycorrhizae, 115
  and *Myriophyllum spicatum*, 303
  pollination, 155, 156
  primary productivity, 217
  radial oxygen loss, 103
  reproduction, 149
  response to herbicides, 293

seedlings, 175
in treatment wetlands, 351
turions, 252
Vegetative reproduction, see Asexual reproduction
Venturi-induced convection, 101-102
Venus' flytrap, see *Dionaea muscipula*
Vernal pools, 44
germination, 173, 280
global warming, 284
threatened species, 23
Viburnum, 9
Vibernum americanum, 285
Viburnum opulus, 285
Victoria, 171
Victoria amazonica, 96, 100
Viminaria juncea, 95, 116
Viola, 379
Vivipary, 176-177
Vogtia malloi, see Alligatorweed stem borer

W

Wastewater treatment wetlands, 34, 68, 333-355
Water budget, 62-64
Water buttercup, see *Ranunculus*
Water celery, see *Vallisneria americana*
Water chestnut, see *Trapa bispinosa*
Water cress, see *Rorippa nasturtium-aquaticum*
Water dispersal, see Hydrochory
Water fern, see *Salvinia minima*
Water hemlock, see *Cicuta maculata*
Water hyacinth borer, 309
Water hyacinth, see *Eichhornia crassipes*
Water lettuce, see *Pistia stratiotes*
Water level fluctuations, 70-72, 131-134
Water lily, see *Nymphaea*, Nymphaeaceae
Water lobelia, see *Lobelia dortmanna*
Water lotus, see *Nelumbo lutea*, Nelumbonaceae
Water meal, see *Wolffia*
Water milfoil, see *Myriophyllum*, Haloragaceae
Water parsnip, see *Sium suave*
Water paspalum, see *Paspalum paniculatum*
Water pennywort, see *Hydrocotyle*
Water plantains, see Alismataceae
Water potential, 110
Water primrose, see *Ludwigia*
Water Quality, see also Treatment wetlands, Wastewater treatment wetlands
improvement with plants, 4
functions of wetlands, 33-34
Water Resources Development Act of 1986, 288
Water shield, see *Brasenia schreberi*
Water smartweed, see *Polygonum punctatum*
Water spinach, see *Ipomoea aquatica*
Water starwort, see Callitrichaceae
Water stress, 136-138
Water tupelo, see *Nyssa aquatica*
Water velocity, 68
Water velvet, see *Azolla pinnata*
Water weed, see *Elodea canadensis*
Water willow, see *Justicia americana*
Waterfowl, 16, 34
Watergrass, see *Paspalum dilatatum*
Waterwheel plant, see *Aldrovanda vesiculosa*
West Indian manatees, 286
Western riparian zones, 50-51, 283
Wet meadows, 70
Wet prairies, 43, 209
Wetland delineation, see Delineation
Wetland restoration, see Restoring wetlands
Wetlands
definition, 29-32
classification, 31, 34-36
diminishing area, 20, 23
distribution, 16, 20
functions, 32-34, 369-370
values, 32
wildlife habitat, 34, 349
Wheat, see *Triticum aestivum*
White mangrove, see *Laguncularia racemosa*
White water lily, see *Nymphaea odorata*
Whooping cranes, 34
Whorled pogonia, see *Isotria verticillata*
Wiegert and Evans method, 205, 206, 208-209, 213-215
Wigeon grass, see *Ruppia cirrhosa*, *Ruppia maritima*
Wild red rice, see *Oryza rufipogon*
Wild rice, see *Zizania aquatica*, *Zizania latifolia*
Willow, see *Salix*
Wolffia, 14, 15
response to herbicides, 293
seed dispersal, 172
Wolffiella, 14
Wood storks, 34
Wooly bulrush, see *Scirpus cyperinus*

X

Xanthocephalus xanthocephalus, see Yellow-headed blackbird
Xylocarpus, 19, 94, 95
Xyridaceae, fruit, 169

Y

Yellow poplar, see *Liriodendron tulipfera*
Yellow-headed blackbird, 8

Z

Zanichellia
  fruit, 170
  pollination, 157, 161, 163, 166
  response to herbicides, 293
Zannichelliaceae, 41, 157
Zea mays, 88, 104-105
Zebra mussels, 288
Zizania aquatica, 39, 173
Zizania latifolia, 257, 351, 352
Zizania palustris, 384
Zizaniopsis milacea, 292, 351, 352
Zonation
  effects of hydrology, 69-70
  effects of salinity, 37-38, 48, 74
  effects of soil chemistry, 73-74
Zostera, 157, 162, 173
Zostera capensis, 111
Zosteraceae, 156
Zostera japonica, 111
Zostera marina, 90
  allelopathy, 265
  pollination, 162
  salt tolerance, 111